ECOLOGY AND BIODIVERSITY OF INDIAN MANGROVES

PART — II
(MORPHO-ANATOMY OF MANGROVES)

Kumudranjan Naskar, Ph.D.
&
Rathindranath Mandal, Ph.D.

1999

DAYA PUBLISHING HOUSE
DELHI – 110035

ISBN 81 - 7035 – 190 – 1 (Set)
ISBN 81 - 7035 – 191 – X (Vol 1)

Published by : **DAYA PUBLISHING HOUSE**
1123/74, Deva Ram Park
Tri Nagar, Delhi-110035
Phone : 7103999
Fax: (011) 7198902

Laser Composed by : **DELITE PRINTERS**
Delhi-110035, Ph: 7182133, 7182135

Printed at : **DP's Impressive Impressions**
New Delhi-110059

PRINTED IN INDIA

ACKNOWLEDGEMENT

The authors wish to express their deep sence of gratitude to Dr. R.S. Paroda, Director General, ICAR & Secretary, DARE, New Delhi for very kindly writing the Foreword Notes and the authors also are grateful to the Director, Central Inland Capture Fisheries Research Institute, Barrackpore-743101 for permitting the authors to work in the Laboratory of the Institute.

Authors are also grateful to the following undermentioned distinguished Scientists and other fellow colleagues for their constant encouragements, suggestions, valuable guidelines and necessary help for carrying out this research work both in the fields and the laboratories and to the Publisher for publishing this research work in the form a book. These respected, distinguished, beloved scientists and friends are Dr. D. N. Guha Bakshi, Ex Deputy Director, Botanical Survey of India, Dr. G. G. Maity, Senior Lecturer, Kalyani University, Dr. M. Sinha, Director, Central Inland Capture Fisheries Research Institute, Dr. Ajoy Kumar Ghosh, Principal Scientist & Head, Estuarine Division, CICFRI, Mr. P. Sanyal, Chief Environment Officer, Government of West Bengal, Dr. P. K. Pandit, Senior Scientist, CICFRI, Dr. S. Santra, Professor, Kalyani University, Dr. P. R. Sen, Ex-Principal Scientist, CIFA, Dr. A. K. Mahapatra, Senior Scientist, CRIJAF, Mrs. Sandhya Mahapatra, Deputy Director, Botanical Survey of India, Dr. G. N. Chattopadhyay, Senior Soil Scientist, Sriniketan, Viswa Bharati, Mr. Anil Mittal, Publisher, Daya Publishing House, Delhi and many others.

We are very much grateful specially to Smt. Dolan Champa Sarkar, Senior Research Fellow, National Fellow Scheme, CICFRI for analysing the soil and water samples and to Miss Neera Sen, Junior Research Fellow, N.F. Scheme, CICFRI, Sri Amal Sarkar, Research Associate, N.F. Scheme, CICFRI, Sri Anup Biswas for their constant help and analysing the algal samples and also helping to draw the maps, sketches, typing and photography works of the manuscript for the books. Thanks are also due to Sri P. K. Chakraborty, Senior Scientist, CICFRI, Sri R. K. Das, Senior Scientist, CICFRI, Sri Amoy Barui, Sri Debasis Saha, Sri Tarun Kabasi, Sri Joydev Patra, Sri A. M. Patra, Sri H. K. Pramanik, Sri Ratan Das and Sri Bindu Debnath of CICFRI and to Sri Girindranath Mridha, James Pur, Gosaba, Sundarbans for helping in the field and laboratory.

We are grateful to Mrs. Sipra Naskar, Mrs. Biva Naskar, Mrs. Gouri Mandal, Mrs. Usha Rani Sarkar, Mrs. Purnima Mandal, Sri Manindranath Mandal, Sri Dinataran Sarkar, Sri Kalipada Seth, Head Master and Tusar Bhakta, Teacher, Mayna Purnananda Vidaypith, Midnapore for helping the authors by several ways and encouraging them for carrying out the research investigation.

Kumudranjan Naskar
&
Rathindranath Mandal

PREFACE

MANGROVES—the threatened coastal intertidal halophytic flora play very dominant and important roles in the estuarine mouths, sea-land interphase areas or deltaic ecosystems of both the tropical and sub-tropical zones, especially in the highly populated South-East Asian countries, several Pacific Islands and Australian coasts. Inspite of their important roles and immense ecological impacts, these mangroves and the mangrove ecosystems have faced both biotic and abiotic threat and these highly productive mangrove zones have been cleared or the mangrove ecosystems have also been degraded very rapidly during the last two centuries. Only during the last three decades, these mangroves and mangrove ecosystems have attracted the interest of the Scientists, Ecologists and Planners. The indiscriminate exploitation and degradation of these mangals have been made after clearing forest lands for the development of agricultural fields, brackishwater fisheries, prawn farms, salt pans and several other developmental purposes; all these have been noted very much alarming and dangerous for the sustainable productivity of these sea-land interphase areas or estuarine ecosystems.

In fact, for the last three decades good numbers of workers have worked and highlighted several important features of these mangals through several valuable publications and media. But till date, the correct definition of these mangroves or mangals and their impacts or importances have not been well circulated and accepted by the scientists from all the corners of the Globe. Their halophytic adaptations in these physiological drier zones and the harsh ecological nitches have not studied or highlighted much, particularly the mangals of the Indian Sundarbans.

Besides these, the present trends of reclamation of several hectares of these Sundarbans mangals and excavation of brackishwater fish and prawn farms in these coastal areas have been increasing day by day for the high demand of export varieties tiger prawn *(Penaeus monodon)* and for dollar earning. These resulted rapid degradation of these mangals of the Sundarbans and other biotic components of these estuarine environment. Along with these, wood cutting and other means of deforestation activities may not be tackled by mere plantation of few quick growing mangrove plant species on the newly silted up river flats or river beds. These afforested river flats or river bed mangroves must not fulfil all these true ecological balances of these composite mangrove ecosystems, comprising of several important and interacted mangrove forest flora and mangrove inhabitant fauna.

A good number of less known Indian mangrove flora and several of such mangrove habitant fauna are still existed in these major Indian mangrove zones of the Sundarbans, West Bengal. All these threatened and endangered flora, fauna need urgent protection and conservation measures. Accordingly several important National and International Projects have now been undertaken for the conservation of these mangrove ecosystems of the Sundarbans.

In these perspectives, the mangroves and other halophytic plant samples of these Indian Sundarbans have been collected, identified, studied and described these flora in relations to their morphometric features and anatomical peculiarities. Their halophytic adaptations and some of their ecological importances have also been studied and consulted the literatures on mangroves to know their distributions and impact in the global perspectives and also investigated their degradations trends. Several important photographs, maps, tables and sketches have also been prepared based on these collections and studies and all these features are documented or focussed to highlight the importances and conservation needs, on an urgent basis.

23.01.1999

Authors

CONTENTS

PART-1
Global Status

PART - II

PART — II
Morpho-Anatomy of Mangroves

20

MORPHOMETRIC FEATURES OF THE MANGROVES & ASSOCIATED FLORA OF THE SUNDARBANS

1. KEY TO THE FAMILIES

1A Plants generally trunkless or inconspicuous trunk; underground roots mostly fibrous; seed with single cotyledon; leaves with parallel venation 25 (MONOCOTYLEDON)

1B Plants with trunk or conspicuous shoot; underground roots—tap root system; seed with two cotyledons; leaves with reticulate venation 2 (DICOTYLEDON)

2A (1B) Plants generally grow in the intertidal zone; saline habitat, on the deltaic area of estuarine mouth; generally aerial roots and viviparous or crypto-viviparous or incipient viviparous germination are conspicuous 3 (MAJOR MANGROVES)

2B (1B) Plants generally grow in the above tidal zone; prefer non-saline habitat and above the normal tidal reaches, generally aerial roots absent and viviparous, crypto-viviparous, incipient viviparous germination are inconspicuous 11 (MOSTLY MINOR MANGROVES)

3A (2A) Plants normally tree or shrub, profusely or greogariously grow in the intertidal zone, purely on saline habitat and have the aerial or above ground conspicuous root systems and characteristic types of seed germination, *viz.*, viviparous, crypto-viviparous or incipient viviparous 4

3B (2A) Plants normally tree or shrub, grow in the frequent tidal zone or above the normal tidal zone and occasionally have aerial roots and viviparous, crypto-viviparous or incipient viviparous germination, but these characters are generally infrequent or absent 7

4A(3A) Plants normally grow beyond the frequent tidal zone, with aerial roots, *viz.*, stilt root, proproot, knee root, broom like stilt root or root buttress or conical roots fused at trunk bases and with conspicuous viviparous germination; hypocotyles ranges from 8.0 cm to 75.0 cm and remain as hanging hypocotyle seedling in the mother plants for considerably longer period RHIZOPHORACEAE (Mangrove Rhizophoraceae)

4B (3A) Plants normally grow in the intertidal river flats or river slopes, regularly inundated with saline tidal water; negatively geotropic roots, pneumatophores from the horizontal cable roots. Seed germination normally non-viviparous but occasionally, crypto-viviparous, incipient viviparous or seed germination take place within very short period after the fruit maturation 5

5A (4B) Pneumatophores only facultitatively develop looping, one seeded drupe like fruit, without any endosperm; generally grow above the frequent tidal zone, without any viviparous, crypto-viviparous or incipient viviparous germination seed germination performed immediately from the mature fruit, directly from the mother plant.

COMBRETACEAE (*Lumnitzera*)

5B (4B) Pneumatophores very much conspicuously develop from the horizontal spreading roots, fruits one seeded or several seeded; plants generally grow in newly silted up soil, inundated with every tidal inundation 6

6A (5B) Plants with numerous erect pencil like, upto 30 cm high pneumatophores, without any secondary growth, occasionally with few fibrous aerial roots or infrequent stilt roots form from the bending trunk bases; grow in the newly silted up river slopes or flats, fruit one seeded and seed germination incipient viviparous

AVICENNIACEAE

6B (5B) Plants with numerous conspicuous, tall, upto 3 m high conical pneumatophores, with secondary growth but no stilt root present; normally grows on the river slopes and river flats, after secondary deposition of sediments and mostly inundated with high tidal water' fruits globose with more than 4 seeds, germination not incipient viviparous nor viviparous

SONNERATIACEAE

7A (3B) Plants shrub or small tree, aerial roots normally absent but stem base thick, swollen, slightly conical, grow in the intertidal zone, occasionally vivi-parous germination present, flower in cluster, white, small, fragrant 9

7B (3B) Plant medium to large size tree, grow in the ridge forest, normally in the upper strata of the mature soil and inundated during high tides or spring tides; leaves dark green, partially deciduous 8

8A (7B) Plants grow in the mature soil, infrequent tidal zone, may also grow away from the tidal influences; partially deciduous, unisexual, ovary tri-lobed with milky latex and without any aerial roots or buttress, but tidal water very often washed out the base soil and spreading cable roots get exposed; ball like several galls very often develop from the periphery of the trunk bases of aged plant

EUPHORBIACEAE *(Excoecaria)*

8B (7b) Plants grow in the mature soil, infrequent tidal zone, may not grow normally away from the tidal influences (except-*Xylocarpus moluccensis*); partially or fully deciduous, bisexual, ovary not tri-lobed, fruit size upto 20 cm diameter, with hard woody fruit coat and 10-15 pyramidal seeds compactly arranged within the woody fruit coat; viviparous germination absent but seed remain viable for a longer period floating in tidal water, trunk base with prominent buttress, thick buffalo horn like erect upto 70 cm long woody pneumatophores and knee roots conspicuous

MELIACEAE *(Xylocarpus)*

9A (7A) Plants shrub, grow in the intertidal river flats with dark green foliages, leaves ovate-oblong and flower small white cluster in cyme, fruit small dumble shaped upto 1.5 cm long, without any aerial roots or viviparous germination

RUBIACEAE *(Scyphiphora)*

Note : Other herbaceous sand dune species with succulent stem, leaves and pink coloured flower collected from sandy beach of the coast, this family Rubiaceae and the species identified as

Hydrophylax maritima L.

9B (7A) Plants shrub or small tree grow occasionally in the ridge forest, with dark green foliages, without aerial roots, stem bases swollen, thick; conical fruit long, upto 5 cm, cylindrical or quadrangular and crypto-viviparous gemination 10

10A(9B) Plants shrub, upto 7 m height; leaves isobilateral, rotundiform lamina, upto 10 cm diametre, long petiole, with sheathing petiolar base, enclose the stem, flower thickly clustered raceme on the apical branches. Fruit has no pointed end, quadrangular upto 5 cm long, cryptovivi-parous germination

AEGIALITIDACEAE *(nom. alt. Plumbaginaceae)*

10B (9B) Plants shrub to small tree, upto 4 m height, grow in the river bank, on the frequently intertidal zone, 3 cm long, tapering leaf base, dark green, thick, flower small, white, fragrant with uniform arranged umbell, fruit curved, upto 4 cm long with tapering pointed end and with thin fruit coat

MYRSINACEAE *(Aegiceras)*

11A (2B) Plants medium to tall tree, prefer freshwater, though occasionally grow in the infrequent tidal zones of brackishwater, aerial root occasionally present but generally absent the horizontal surface root, root may get exposed due to soil erosion 12

11B (2B) Plants erect or twining shrub or herb, mostly grow in less saline brackishwater zone or freshwater inundated areas, frequently present in the river slopes, river banks or river flats or sandy soil, in xerophytic habitat 14

12A(11A) Tall tree with thick elongate to lanceolate foliages, white tomentum beneath, prefer freshwater zone, plant can grow in the saline habitat and form thick erect flattened pneumatophores or knee like bending pneumatophores, stem buttresses conspicuous; fruit winged, woody and can float in tidal water, without any viviparous germination STERCULIACEAE (*Heritiera*)

12B (11A) Plants tall shrub, small tree or tree, grow just above the frequently tidal inundated zone, prefer non-saline brackishwater zone, without any viviparous or crypto-viviparous germination; lamina and tender stem with tomentum or pubescent and without any aerial roots 13

13A (12B) Tall shrubs or small trees, branches slightly twining habit, with long elongated, lanceolated leaves with tapering ends, acute, leaves pubescent, rusty or pale green; flowers small, cluster on the axis of the branches; fruits globose, small without viviparous germination TILIACEAE (*Brownlowia*)

13B (12B) Medium trees, broadly ovate leaves with tapering ends, acuminate, prefer freshwater zone, away from the frequent inundated areas, flowers mostly large, showy, long petiolate, colour of flower change depending on time and maturation, fruits globose, woody, without viviparous or crypto-viviparous germination MALVACEAE

14A(11B) Plants herb or shrub, normally grow on the sandy soil, away from the tidal reaches, though inundate very rarely during spring tides. 15

14B (11B) Plants herb or shrub grow in the frequently tidal inundated silty clay soil or as epiphytes, frequent on the river banks and ridge forests and river side thickets 18

15A(14A) Plants herbaceous creeping or twining on the coastal sand, grow completely in the dry xerophytic condition, leaves thick, bilobed, parallel vein conspicuous, flower large, bell shaped, pinkish, 4-5 cm radius CONVOLVULACEAE (*Ipomoea*)

15B (14A) Plants twining or tall shrub, grow on the dry habitat, on the above tidal zone, completely xerophytic nature, flower medium to large, non-viviparous or not crypto-viviparous 16

16A (15B) Plants tall shrub or occasionally small tee, grow on the sandy river bank, away from the normal tidal reaches, profusely branched and with much dissected minute leaves; inflorescence catkin, pendulous flowers not showy, pale, fruit small upto 1.5 cm TAMARICACEAE (*Tamarix*)

16B (15B) Plants twining or erect shrub, grow in the dry soil or sand, spiny or prickle, flowers medium to large , showy, fruit medium to large, non-viviparous or not crypto-viviparous 17

17A (16B) Plant twining shrub with bent prickle on leaves and stem, climbing on river side thickets or hedges on dry habitat, leaves rusty, tomentose, pubescent, generally trilobed; flower violet, upto 1.5 cm diameter, fruit globose, berry, 0.5 cm diameter SOLANACEAE (*S. trilobatum*)

17B (16B) Erect shrub on sandy, dry soil on the coastal area; stem converted to flat fleshy phylloclades, with milky latex, flat, oval, jointed, leaves converted to straight long spines on the phylloclades; flowers yellow, large upto 7 cm radius, showy, fruit upto 4 cm long, green, with prominently ribbed OPUNTIACEAE (*nom. alt. Cactaceae*)

18A (14B) Plants epiphytic on the mangrove trees, leaves and nodes jointed, flower red or greenish, variable, not having any typical characters of mangroves though these plants grow along with mangroves plants LORANTHACEAE

18B (14B) Plants shrub or herb, grow within the direct tidal influences in the tidal saline soil of river bank or river slopes 19.

19A (18B) Plants with erect habits with armed on the tidal inundated silty clay soil having saline resistance or salt tolerant capacity 20

19B (18B) Plants twining or climbing habit, shrubs or herbs, grow generally on the river flats or river side thickets 21

20A(19A) Plants shrub, either erect or climbing on the mangrove plants of the river slope, stem woody, slender, twining habit, having stilt roots with pointed armed at stem and leaves; leaves broad, hard, dark green, wavy margin with sharp spines; flowers violet or whitish to white, cluster raceme, fruits capsule, upto 2.5 cm long ACANTHACEAE

20B (19A) Plants erect shrub with bifurcated straight spines, leaves coriaceous, glossy upto 3 cm long, lemon scented flowers white, fruit triangular, fragrant, grow on the intertidal river bank and moderately saline zone. RUTACEAE

21A (19B) Plants twining, climbing armed without milky latex, occasionally shrubs or trees or unarmed; flower papillonaceous with characteristic pods FABACEAE

21B (19B) Plants shrubs or herbs, grow on intertidal river flats or slopes, with or without milky latex, leaves broad or minute succulent 22

22A (21B) Plants shrubs, twining habitat; leaves upto 8 cm long and 5 cm broad with milky latex, fruits folicule, upto 8 cm long ASCLEPIADACEAE

22B (21B) Plants purely herbaceous or undershrub, erect or twining, fully turgid; grow on frequently inundated saline soil; inflorescence axillary, flowers minute, 1 cm diameter 23

23A (22B) Plants herbaceous undershrub with minute flowers in the axile of the leaves, yellow or greenish colour CHENOPODIACEAE

23B (22B) Plants herbaceous, fleshy, grow in the intertidal salt crusted saline soil, leaves fleshy, fully succulent, reddish, dark green 24

24A (23B) Plants grow in intertidal zone reddish, succulent, turgid leaves, fully herbaceous with creeping branches, flower solitary with pink colour, upto 1 cm AIZOACEAE

24B (23B) Plants grow on the river dyke, in the salt crusted soil, purely herbaceous, dark green fleshy, turgid, dark green, apical inflorescence helicoid, with minute flowers BORAGINACEAE

25A (1A) Plants with stem or culm, woody, grow within the intertidal zone, leaves pinnately compound upto 3 m ARECACEAE

25B (1A) Plants stemless with culm, grow in the intertidal silted up soils, regularly inundated with tidal water, leaves long 26

26A (25B) Plants with fibrous roots, grow in the intertidal zone as pioneer species, leaf margin with sharp toothed, herb and spikelet inflorescences on apical portion of the branches POACEAE

26B (25B) Plants with rhizomatous stem, leaves long, slender, dark, green, parallel veins, grow in the intertidal river flat 27

27A (26B) Plants with white flower on long peduncle; grow on the intertidal zone AMARYLLIDACEAE

27B (26B) Plants spathe inflorescence in the axile of leaves, pollination cleistogamy CRYPTOGAMS, ARACEAE

28A Plants rhizomatous, without flowers, pinnately compound leaves, grow on the tidal zone, sorus on ventral surface PTERIDACEAE

28B Plants filamentous or branched thallus attached on the pneumatophores, stilt roots, trunk bases, grow as periphytic form The periphytic algae studied

2. TAXONOMIC ENUMERATION OF THE TRUE MANGROVE FAMILIES

FAMILY— I. *RHIZOPHORACEAE*

(Mangrove Rhizophoraceae)

Rhizophoraceae, the family name had derived after the **Type Genus—*Rhizophora*** L. Sp. Pl. 443.1753. This is a pantropical family, consisting of about 16 genera and 120 shrub or tree species. Most of these species under the family — Mangrove Rhizophoraceae are distributed in the Old World Tropics (Tomlinson, 1986). Previously, this family—Rhizophoraceae was placed under the Natural Order-Myrtales, but Dahlgren (1980) has separated the family— Rhizophoraceae and ranked it as a distinct family under a separate Order — **Rhizophorales.**

Rhizophoreae Blume is one of the tribes of the family— Rhizophoraceae, which includes 4 true mangrove genera, *viz.*, ***Rhizophora*** **L.**, ***Bruguiera*** **Lamark,** ***Ceriops*** **Arnold and** ***Kandelia*** **Wight & Arnold.** All these four true mangrove genera of the Rhizophoreae (Mangrove—Rhizophoraceae) Tribe have the typical characteristics, like aerial roots and viviparous germination. Most species under these four genera are distributed or grow in the typical mangrove habitats of the Old World Tropics and a few species are distributed in the New World Tropics, *i.e.*, in both the coasts of Atlantic Ocean and East Pacific Ocean regions. Naskar & Guha Bakshi (1987) have described nine species under the four true mangrove genera of Tribe-Rhizophoreae (Mangrove- Rhizophoraceae) from the intertidal zones of the tidal estuaries and deltas at the Sundarbans mangrove forests of the lower Ganga delta, *viz.*, *Rhizophora apiculata* Blume, *R. mucronata* Lamk., *Bruguiera gymnorrhiza* (L.) Lamark, *B. sexangula* (Lour.) Poir, *B. parviflora* W. & A. ex Griffith, *B. cylindrica* (L.) Blume, *Ceriops tagal* (Perr.) Robinson. *C. decandra* (Griff.) Ding Hou and *Kandelia candel* (L.) Druce.

Mukerjee (1966) had also reported and described *Bruguiera hainesii* Rogers after collecting this from the Sundarbans; but Mitra & Banerjee (1979) have reported that the collection of Mukerjee (Mukerjee-6244, CAL) from the Swan Islands of Sundarbans is nothing but *B. cylindrica* (L.) Blume.

These four genera of the Mangrove - Rhizophoraceae (Tribe- Rhizophoreae) can be distinctly identified with the critical studies of several key characters and habitats, as well.

Members of the *Tribe-Rhizophoreae* (Mangrove- Rhizophoraceae) are the true mangroves and Tomlinson (1986) has ranked these as *"Major Elements of Mangroves"*.

These 'major elements of mangroves' or the 'true mangrove species' perfectly give the real shape of this unique and interesting ecosystem and these mangrove species make most productive and typical characteristic mangrove ecosystem of World renowned.

TRIBE—*RHIZOPHOREAE*

The Tribe—Rhizophoreae (*Mangrove - Rhizophoraceae*) consists of four mangrove genera, *viz.*, *Rhizophora* L., *Bruguiera* Lamk., *Ceriops* Arnold and *Kandelia* W. & A. All these true mangroves under these genera mostly grow in the intertidal silty clay and loam soil, frequently inundated with high saline tidal sea water or brackishwater; their salt resistance efficiencies are very high for having several morphological and anatomical features of these true mangrove species.

Medium to tall trees, occasionally shrubs or small tree; much spreading, diffuse woody branches, bark grayish to whitish gray and mostly contain high percentage of red tannin. Aerial roots-stilt roots, bow roots, prop roots, knee roots and plank roots or flanges; these roots mostly mechanically support the trunks in the clayey silted up sediments within the intertidal zones of the mangrove habitats.

Their shoots have differentiated into nodes and congested internodes; internodes 0.1 cm.- 1.0 cm long; apical buds 8.0 cm - 12.0 cm long, initially covered by brown thin papery covers; apical shoot green, in the shoot apex the green leaves are clustered; rough brown layers encircling the nodes and continued, below it two leaf scars present on both sides.

Phyllotaxis bijugate and leaves are mostly arranged densely on the apical part of the shoots and distinct leaf scars are present on the basal part of the branches. Leaves mostly broad, thick, leathery, highly cuticularised, waxy, dark green, long petioled.

Leaves opposite decussate, simple, exstipulate, petiolate; petiole 2 cm - 5 cm long, slightly flattened at the base; 0.3 cm - 0.45 cm broad, glabrous, green, pulvinous, flexible; lamina broadly ovate elliptic, entire, acute or mucronate; tapering both apically and basally; 12.0 cm - 20.5 cm long x 5.0 cm - 8.0 cm broad; dorsiventral, dorsal side waxy, shiny, coriaceous, dark green; mid vein distinct in ventral side; unicosted reticulate venation inconspicuous.

Flower solitary cyme, aggregate cyme, mostly 0.5 cm to 4.0 cm long with persistent calyx teeth; fruit capsule. Germination viviparous and hypocotyles ranges 5.0 cm — 75.0 cm long, pendulous from the mother plants. Pollination by wind, insects and birds.

Inflorescence cymes, opposite decussate, ebracteate, peduncle 1.8 cm long-5.5cm. long, 2-flowers or 7-flowers on each peduncle; peduncle erect, slightly bent. Flowers ebracteate, pedicellate or sessile, pedicel upto 2.0 cm long, complete, bisexual, regular, erect, odourless, upto 4 cm long, globose; sepals - 4, polysepalous, superior, ovate, entire, acute, fleshy, valvate, persistent in the fruit, upto 1.4 cm long x 0.7 cm broad; petals- 4 polypetalous, superior, lanceolate, entire, acute, deciduous, alternate to sepal, 1.0 cm long x 0.25 cm broad, shorter than sepals; stamens 8-32, free, filaments short or sessile, anther bi-lobed, inserted, introrse, sagittate, basifixed, upper surface same levelled, yellowish, superior, longitudinally dehiscent; carpels - 2–3, syncarpy; ovary globose, inferior, 2-4 chambers, 2-6 ovules in each chamber, axile placentation; style- 1, terminal, short, stigma- 2, bifid, short.

Fruits capsule, 2.0 cm - 7.0 cm long, 7.0 - 9.5 cm diam., oval, broad in the middle and tapering towards both ends, brownish in dry condition, with four persistent sepals; seed one, covered by thick indehiscent pericarp; after maturation the seed bifurcate transversely within the fruit; upper portion with tapering appendages, logged into furrow bascate like cavity. Seed 1.8 cm long x 2.2 cm diam.; hypocotyle cylindrical; germination viviparous, epigeal.

Key Characters of the Tribe— Rhizophoreae (Mangrove Rhizophoraceae)

1a Tree with aerial stilt roots or looping pendulous roots or flanges on the trunk bases; shrubs or small to medium trees (2)

1b Tree with knee like aerial roots or pneumatophores, develop from looping horizontal roots or without any pneumatophores, trunk base conical by fused broom like still roots (3)

2a(1a) Medium trees; stilt roots long upto 3 m long or more; leaves broadly ovate, acute or mucronate apex; flowers tetramerous, not lobed, petals with marginal hairy or glabrous; hypocotyle not ridged, upto 75 cm long; ovary 2- celled, stamens 8-12 **1. RHIZOPHORA** L.

2b(1a) Shrubs or small trees; broom like stilt roots or flanges densely encircling the trunk bases, attaining 50 cm long, leaves ovate-lanceolate, obtuse or rounded apex; flowers pentamerous, petals with apical three apendages or apical ciliated, stamen 10-12, ovary 3-celled; hypocotyle ridged, 5 cm—25 cm long. **3. CERIOPS** Arnold

3a(1b) Medium to tall trees, trees with knee like pneumatophores develop from the looping horizontal roots; leaves largely ovate, lanceolate, acute, reddish, yellowish or dark green; calyx lobes varies from 8-16, reddish, yellowish or green, petal number equal to sepals; stamens 16-32; ovary 2—4 celled; hypocotyle not more than 25 cm long, ridged or smooth **2. BRUGUIERA** Lamk.

3b (1b) Medium tree, without any aerial roots pneumatophores absent; leaves broadly lanceolate, acute; flowers pentamerous, petals bilobed and each lobe apically multi-fid; stamens indefinite with unequal filament length; ovary 1- celled; hypocotyle 45 cm long with sharp end and glabrous surface. **4. KANDELIA** W. & A.

GENUS - 1. *RHIZOPHORA* L. Sp. Pl. 443.1753

Linnaeus (1753) first designated all these mangrove plants under the genus ***Rhizophora*** *L.*, though Rumphius (1743) used the term ***"Mangium"*** for the mangroves; Linnaeus (1753) has listed seven mangrove species under the Genus-*Rhizophora L.* and latter on Linnaeus (1762) had also included other two species under *Rhizophora* L. Out of these, six mangrove species which were mentioned by Linnaeus (1753) initially has now considered under the genera *Bruguiera* Lamk., *Ceriops* Arnold, *Kandelia* W. &. A., *Aegiceras* Gaertner and *Sonneratia* Linn. f.; but only one originally described species of Linnaeus, (1753) is now treated under *Rhizophora* L., *i.e., R. mangle* L. 1753 (Tomlinson, 1986).

Rhizophora L. is a pantropical genus and have nine species (including three putative hybrids), which are very uniform and distinct in their vegetative features (Tomlinson, 1986).

Species of this genus are medium sized trees, upto 15 m high in Sundarbans; sometime more tall trees in other parts of the world mangroves; perennial evergreen tree with dark foliages; woody, erect, branched, deliquiscent, bark grey to whitish grey, longitudinally fissured generally, but not smooth.

Stilt roots looping from lower branches or trunk bases, prof roots, slightly corky, woody, cylindrical, upto 3.0 m long, 5.0 cm- 8.0 cm diam., forming bow roots, mechanically support the trunks and also serve the respiratory functions through lenticels.

Differentiated into nodes and congested internodes; internodes 0.1 cm- 1.0 cm long; apical buds 8.0 cm- 12.0 cm long, initially covered by brown thin papery covers, apical shoot green in the shoot apex, the green leaves are clustered; rough brown layers encircling the nodes and continued, below it two leaf scars present on both sides.

Leaves opposite decussate, simple, exstipulate, petiolate; petiole 2 cm - 4 cm long, slightly flattened at the base; 0.3 cm - 0.45 cm broad, glabrous, green, pulvinous, flexible ; lamina broadly ovate, elliptic, entire, acute or mucronate; tapering both apically and basally ; 12.0 cm - 18.0 cm long X 5.0 cm, - 8.0 cm broad; dorsiventral, dorsal side waxy, shiny, coriaceous, dark green; mid-vein distinct in ventral side; unicosted reticulate venation inconspicuously.

Inflorescence cymes, opposite decussate, ebracteate, peduncle 1.8 cm - 5.2 cm long, 2 flowers or 4 flowers on each peduncle; peduncle erect, slightly bent. Flowers ebracteate, pedicellate or sessile, pedicel upto 0.5 cm long, complete, bisexual, regular, erect, odourless, tetramerous, upto 1.7 cm long, globose; sepals- polysepalous, superior, ovate, entire, acute, fleshy, valvate, persistent in the fruit, upto 1.4 cm long × 0.7 cm broad; petals - 4, polypetalous, superior, lanceolate, entire without appendages, acute, deciduous, alternate to sepal, 1.0 cm long × 0.25 cm broad, shorter than sepals; stamens 8-12, free, filaments short or sessile, anther bi-lobed, inserted, introrse, sagittate, basifixed, upper surface same levelled, yellowish, superior, longitudinally dehiscent; carpels - 2, syncarpy; ovary globose, inferior, 2-chambers, 2- ovules in each chamber, axile placentation; style-1, terminal, short, stigma -2, bifid, short,

Fruit capsule, 2.0 cm - 7.0 cm long, 7.5 - 9.5 cm diam., oval shaped, broad in the middle and tapering towards the both ends, brownish in dry condition with four persistent sepals; seed one, covered by thick indehiscent pericarp ; after maturation the seed is bifurcated transversely within the fruit; upper portion with the tapering appendages, locked into furrow basket like cavity. Seed 1.8 cm long × 2.2 cm diam., cylindrical hypocotyle, germination viviparous, epigeal.

Wind Pollinated Flowers

Tomlinson (1986) reported total 9 species (including three hybrids) under the Genus - *Rhizophora* L.; of these, four species and two hybrids are reported from the Eastern zone of the World *i.e.*, Old World tropics and 2 species and one hybrid are reported from the Western zone of the World, *i.e.*,New World tropics.

Key to the Species of Genus - *Rhizophora* L.

1a. Leaves with acute apex; flowers - 2 on each peduncle; mature flowers blossom below the leafy cluster; peduncle upto 2 cm long, straight, flowers small to medium, 1.2 cm long, sessile, petals glabrous margin, stamens 11-12, sessile 2. ***R. apiculata*** Blume

1b. Leaves with mucronate apex ; flowers 4 on each peduncle ; mature flowers blossom within the leafy cluster, peduncle long upto 5.0 cm, slightly bent, flowers medium upto 1.7 cm long, pedicellate; petals hairy in the margin, stamens 8 ; filaments short 1. ***R. mucronata*** Lamark

1.1. *Rhizophora mucronata* Lamark Encycl. 6 : 189. 1804.

R. macrorrhiza Griff. ; *R. longissima* Blanco & *R. mucronata* Lamk. var. *typica* Schimp. Chromosome No. $2n = 36$ (Patil, 1958 & Sidhu, 1968)

Common Name : Long fruited stilted-mangrove.

Local Name (Bengali) : Garjan or Khamo, Bhora (D. Prain, 1903a)

(Plate— 81 and Fig. 1.)

Medium to tall tree, upto 15 m high, deliquiscent branched; branches spreading horizontally; trunk not conspicuous and supported by profuse prop roots and stilt roots; bark brownish to whitish grey; longitudinally fissured.

Stilt roots profuse, looping and supporting the lower trunk regions and the lower branches; corky, woody, slender, cylindrical upto 3 m or more long, 2.5. cm - 8.0 cm diam.

Stem differentiated into nodes and inter-nodes, much branched, deliquiscent, terete ; green leaves clustered in the shoot apex; thick rough brown scaly layer encircling the nodes; leaf- scars prominent, elliptic-shape, 0.7 cm broad and 0.4 cm long, internodes, 0.1 cm - 0.5 cm in length, clustered, closely arranged, apical shoot green, solid, glabrous, apical bud 9.0 cm long, old leaf-scars brown and adjacent internode grey in colour and rough surface.

Leaves opposite decussate, simple, exstipulate, petiolate, 14 cm - 16 cm long shoot apex, covered with leaves, petiolate, petiole 4.0 cm long, slightly flattened, 0.3 cm broad, glabrous, green pulvinous, flexible, mature leaves turn yellow; lamina - ovate-elliptic, entire, mucronate, rolled upto 0.6 cm long, tapering, cup-shaped lamina base; 16.0 cm long × 7.5 cm broad; dark green, dorsiventral, coriaceous; shining dorsal side; pale green, slightly shining in the ventral side, reticulate venation inconspicuous.

Inflorescence upto 5.2 cm long, cyme, opposite decussate, ebracteate, main peduncle having generally 4 flowers, peduncle 2.8 cm long, dichotomously branched, one branch of the peduncle having 2 flowers, peduncle green, glabrous, flattened slightly, 0.25 cm broad, pulvinous, apex of the peduncle cup-shaped, with two branches flattened, 0.8 cm diam., grooved; each peduncle develop from leaf axis, each shoot apex having 2-4 inflorescence stalk; peduncle erect or bent slightly.

Flowers ebracteate, pedicellate, pedicel 0.5 cm long, flattened, green, glabrous; complete, regular, hermaphrodite, regular, erect, odourless, 1.7 cm long, globose. *Calyx* persistent in the fruits; sepals - 4, polysepalous, sepal ovate, entire, acute, fleshy, thick, green, glabrous, valvate, 1.2 cm long, 0.7 cm broad, persistent, superior, equal length, upper surface same levelled. *Corolla* white; petals- 4, polypetalous, superior, petal lanceolate, entire, acute, irregular pubescent, curved margin, white, thick, fleshy, valvate, 1.0. cm long, 0.25 cm broad, deciduous,

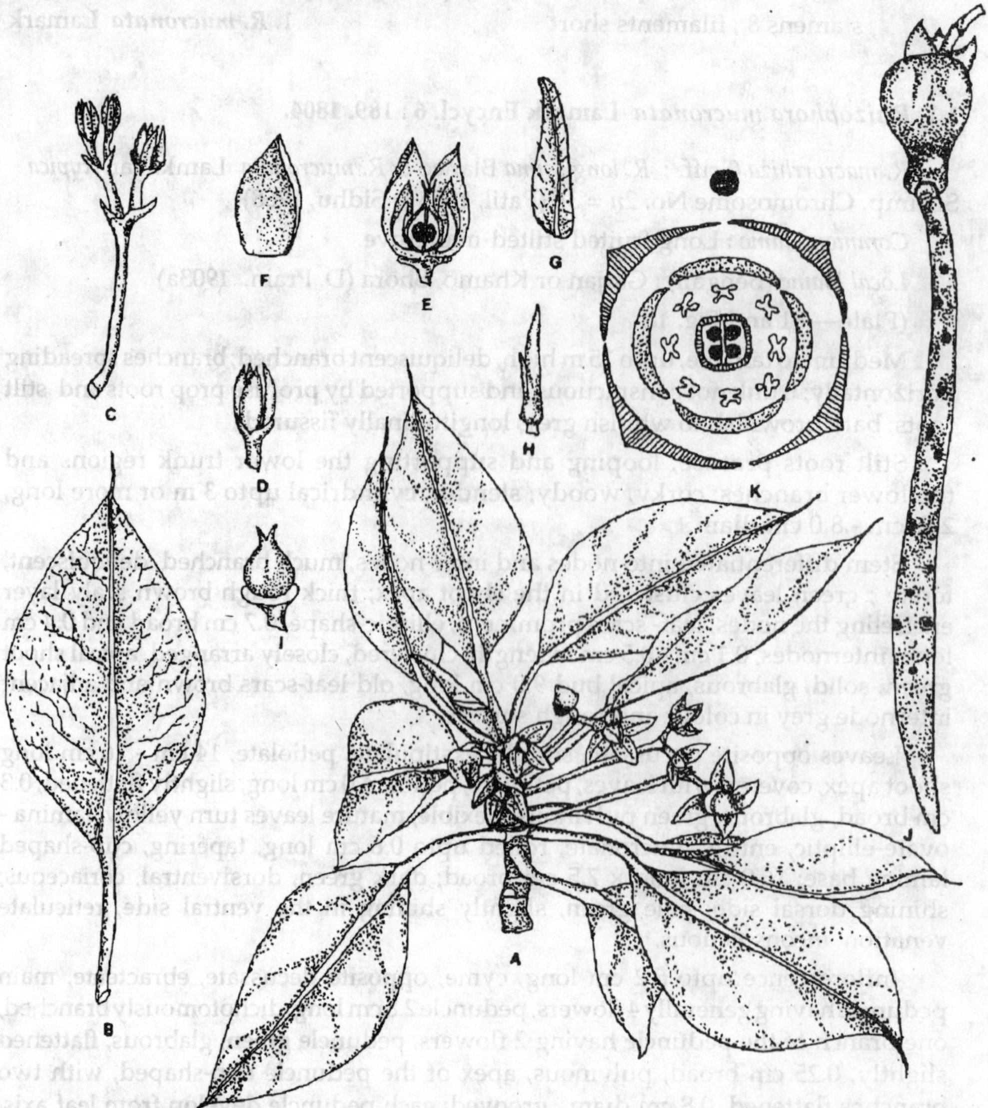

Fig. 1 : ***Rhizophora mucronata*** **Lamk.**

A. Portion of flowering twig x 1/3; B. Leaf : petiole - 4.0 cm long, lamina - 14.5 cm long x 7.5 cm width ; C. Inflorescence - 5.02 cm long; D. Flower 1.7 cm long ; E. L. S. of flower; F. Sepal - 1.2 cm long x 0.7 cm width; G. Petal - 1.0 cm long x 0.25 cm width; H. Stamen-0.7 cm long; I. Gynoecium - 0.4 cm long ; J. Fruit with vivipary 75 cm long; K. Floral diagram.

odourless, alternate to sepal, shorter than sepal. *Androecium* : stamens - 8, free, 4-stamens opposite to petal and 4-stamens opposite to sepal, superior, filaments short, 0.1 cm long, round, white, basifixed, anther bilobed, introrse, inserted, 0.7 cm long, sagittate, longitudinally dehiscent after maturation; anther dissociate from outer thin cover, equal length, upper surface same levelled, yellowish colour, superior. *Gynoecium* : carpels - 2, syncarpous, ovary globose, inferior, 0.3 cm long, 2 - chambered, 2 ovules in each chamber, ovary wall greenish, glabrous, axile placentation; style, one, terminal short, less than 1.5 cm, stigma sessile, two, bifurcate, 0.1 cm long.

Wind Pollinated Flowers and viviparous germinated seeds

Capsule 5.0 cm - 7.0 cm long and 9.5 cm periphery with four persistent sepals; after maturation the fruit turn to brownish with rough surface, seed one, 1.8 cm long and 2.2 cm broad; covered by hard fleshy pericarp; transversely separated, upper part having enlarged structure, which is locked into lower furrow basket like cavity; collar present on mature stage at the base of hypocotyle, spreading upto 2.5 cm, encircling the hypocotyle base, corky in nature.

Flowering and fruiting are normally observed during February to October; but viviparous germinated hypocotyle hang in the mother plants, almost throughout the year. Hypocolyle length upto 75 cm. Shoot initiation conspicuous during the monsoon months, when the soil and water salinity falls down with monsoon precipitation.

This typical and best known mangroves are common throughout. Very common on the river side slopes and tidal canal sides; but occasionally it grow on the flat river sides and ridge forests within the intertidal limits, as this species require frequent tidal inundation. Salinity tolerance is much more, as such this species grow on the mid- and outer estuaries, facing the sea also.

Distributed throughout in the Indian mangals, both in the east and west coasts and Andaman & Nicobar Islands; in Bangladesh, Burma, Sri Lanka, S. E. Asia, E. Africa and Madagascar to Micronesia, South Pacific and North Australia ; an Old World Species, introduced in Hawaii.

Seedling development same as *Rhizophora apiculata* Blume, discussed latter on.

Act as soil binder, add biomass, supply fuel wood and timber and tanin; act as the nesting place of several avi-fauna, estuarine fauna—like shell fish, fin fish species.

Stem bark contain high percentage of tannin (20%) which is useful in leather industries; wood used for construction purposes charcoal making. Bark is also astringent and said to be medicinally used in hemorrhage, hematuria and angina.

1.2. *Rhizophora apiculata* Blume En. Pl. Jav. 1 : 91. 1827.

R. conjugata Arn. Ann. Mag. Nat. Hist. 1 : 363. 1838 (non Linn.) and *R. candelaria* DC.

Common Name : Tall-Stilted Mangrove

Chromosome No. *2n = 36* (Subramaniam, 1980)

Local Name (Bengali) : Garjan or Khama, Bhora (D. Prain, 1903a)

(Plate- 82 and Fig. 2.)

Medium tree, upto 12 m high, deliquiscent branched; branches spreading profusely; trunk not conspicuous and supported by profuse stilt roots on prop roots; bark brownish to whitish brown or grey, fissured.

Stilt roots looping from lower branches and trunk bases, slightly corky, woody, cylindrical, upto 2.5 m long. 5.0 - 7.5 cm diam. forming bow like bending roots and mechanically support the tree in the loosely concentrated tidal soil.

Stem differentiated into nodes and internodes, branched deliquiscent, terete; in shoot apex green leaves clustered; thick rough brown scaly layer encircling the nodes; leaf-scars prominent, elliptic shape, 0.6 cm broad and 0.3 cm long, brown layer continued below, 2 leaf-scars in opposite side; internode 0.1 - 0.5 cm in length, clustered, closely arranged, apical shoot green, solid, glabrous, apical bud 8.0 cm long, old leaf-scars brown and adjacent to internode, grey in colour, rough surface.

Leaf opposite decussate, simple, exstipulate, petiolate, 12 cm shoot apex covered by the leaves, petiole 4.0 cm long, slightly flattened, 0.35 cm broad, glabrous, green, pulvinous, flexible, lamina ovate-elliptic, entire, acute, tapering, 12 cm - 15 cm long and 7 cm broad, green, dorsiventral, coriaceous, dorsal side deep, dark green, slightly shiny, compared to ventral side, unicostate reticulate venation, inconspicuous venation.

Inflorescence cyme, opposite decussate, ebracteate, main peduncle 1.8 cm- 2.0 cm long, peduncle unbranched, 2 flowers on each peduncle, peduncle brownish-green with terminal cup shaped bracteole, glabrous, terete, 0.5 cm diam.; inflorescence 2.5 cm long, each peduncle develop from mature branches, each shoot apex having 6-8 inflorescence stalk, peduncle erect.

Flowers ebracteate, sessile, glabrous, complete, hermaphrodite, regular, erect odourless, 0.8 - 1.2 cm long, globose. *Calyx* inconspicuously persistent in the fruits; sepals - 4, polysepalous, superior, sepal ovate, entire, acute, fleshy, thick, green glabrous, valvate, 0.6 - 0.8 cm long, 0.35 - 0.40 cm broad, persistent, superior, equal length, upper surface same levelled. *Corolla*-white; petals- 4, polypetalous, petals lanceolate, entire, acute, glabrous, white, herbaceous, fleshy, 0.5 cm-0.7 cm long, 0.1 cm - 0.25 cm broad, deciduous, odourless, alternate to sepal, shorter than sepal. *Androecium* : stamens; 11 - 12, free, superior, sessile, anther bilobed, introrse, inserted, 0.5 cm- 0.6 cm long, sagittate, longitudinal dehiscent, equal length, upper surface same levelled, whitish, white yellowish colour, superior. *Gynoecium* - carpels - 2, syncarpy, ovary globose, inferior, 0.20- 0.25 cm in long., 2-chambered, 2-ovules in each chamber, ovarian surface white-greenish, glabrous, axile placentation, style - 1, terminal, short, stigma- 2, bifurcate. Mostly wind pollinated.

Flowering and fruiting are common during March to September, but frequently found the hanging viviparous germinated hypocotyle in the mother plants throughout the year. Vegetative growth is conspicuous during the monsoon months, through the apical buds. Frequently found in the core area of the intertidal mangrove forests of the Sundarbans; in the river slopes and side of the tidal creeks or canals.

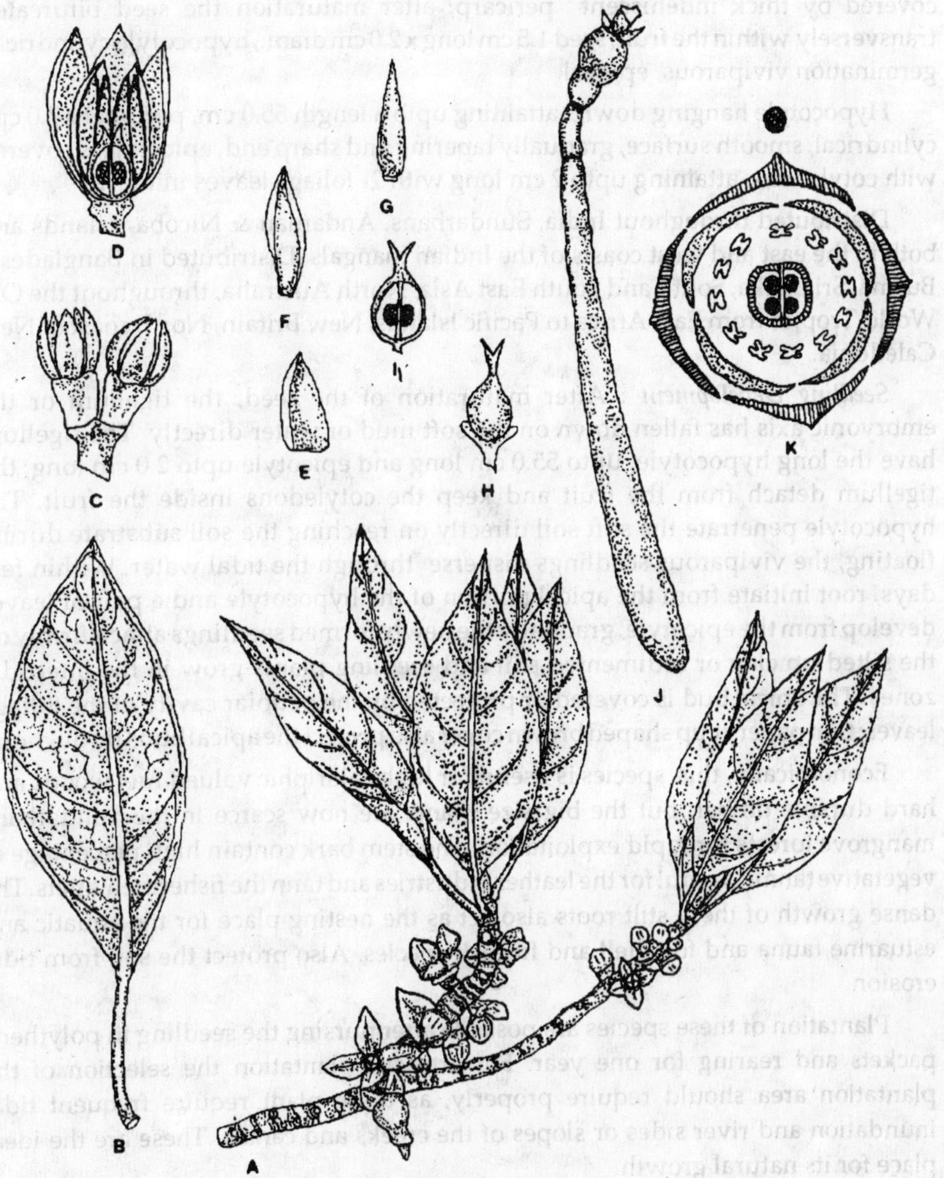

Fig. 2 : *Rhizophora apiculata* Blume

A. Portion of flowering twig x 1/3; B. Leaf : petiole 4.0 cm long, lamina- 12.5 cm long x 6.5 width; C. Inflorescence ; D. L/ S. of flower; E. Sepal- 0.7 cm long x 0.35 cm width; F. Petal 0.5 cm long x 0.2 cm width; G. Stamen - 0.5 cm long; H. Gynoecium- 0.2 cm long; I. L. S. Gynoecium; J. Fruit with vivipary-45 cm long; K. Floral diagram.

Fruit capsule, 2.0 cm - 4.0 cm long, 5.0 cm- 6.5 cm periphery, oval, tapering towards apex, brownish in dry condition, with four persistent sepals; seed one, covered by thick indehiscent pericarp; after maturation the seed bifurcated transversely within the fruit; seed 1.5 cm long x 2.0 cm diam., hypocotyle cylindrical; germination viviparous, epigeal.

Hypocotyle hanging down, attaining upto a length 55.0 cm, periphery 7.0 cm, cylindrical, smooth surface, gradually tapering and sharp end, epicotyle is covered with cotyledons, attaining upto 2 cm long with 2- foliage leaves initially.

Distributed throughout India, Sundarbans, Andaman & Nicobar Islands and both in the east and west coasts of the Indian mangals. Distributed in Bangladesh, Burma, Sri Lanka, South and South East Asia, North Australia, throughout the Old World Tropics from East Africa to Pacific Islands, New Britain, North coast of New Caledonia.

Seedling Development : After maturation of the seed, the tigellum or the embryonic axis has fallen down on the soft mud or water directly. The tigellum have the long hypocotyle, upto 55.0 cm long and epicotyle upto 2.0 cm long; the tigellum detach from the fruit and keep the cotyledons inside the fruit. The hypocotyle penetrate the soft soil directly on reaching the soil substrate during floating, the viviparous seedlings disperse through the tidal water. Within few days, root initiate from the apical portion of the hypocotyle and a pair of leaves develop from the epicotyle; gradually the newly formed seedlings attach firmly on the silted up clay or sedimented soil and seedling plants grow in the intertidal zones. The apical bud is covered or protected by the petiolar cavity of the foliage leaves; thin papery cup shaped brown cover also protect the apical buds throughout.

Economically, this species is useful for high caloriphic valued fuel woods and hard durable timber; but the big size plants are now scarce in the Sundarbans mangrove forests for rapid exploitation. The stem bark contain high percentage of vegetative tannin, useful for the leather industries and tann the fishermen's nets. The dense growth of these stilt roots also act as the nesting place for the aquatic and estuarine fauna and for shell and fin-fish species. Also protect the soil from tidal erosion.

Plantation of these species are possible after nursing the seedling in polythene packets and rearing for one year. For artificial plantation the selection of the plantation area should require properly, as these plant require frequent tidal inundation and river sides or slopes of the creeks and canals. These are the ideal place for its natural growth.

Experiment : Viviparous germinated seedlings were collected from the sandy beach of Sundarbans and transplanted in the non-saline soil in the experimental garden. Non-saline tubewell water was applied and this seedling survived and grow in this experimental garden for more than 3 years with feeble growth of apical buds during monsoon months.

Table 56 : Comparative Morphology of the Two Species of *Rhizophora* L. of Indian Sundarbans

Characters	*R. apiculata* Blume	*R. mucronata* Lamk.
Habit	Medium tree, upto 12 m high.	Medium- tall tree, upto 20 m high.
Root (Aerial)	Stilt root upto 2.5 m long, diam. 5.0 cm - 7.5 cm.	stilt root upto 3.0 m or more diam. 6.0 cm - 8.0 cm.
Stem bark	Gray brownish	Grey to whitish gray
Leaf	Petiole-4.0 cm long; lamina with acute apex, lamina - 13 cm long x 6 cm broad.	Petiole - 4.0 cm long; lamina with mucronate apex; lamina- 18 cm long x 7.5 cm broad.
Inflorescence	2-flower, in each peduncle; Peduncle straight, flowers below the leafy cluster.	4-flowers in each peduncle; peduncle bent slightly, flowers within the leafy cluster.
Flowers	Sessile, small to medium, 0.8 - 1.2 cm long. Sepals 0.6 - 0.8 cm long; Petals - 0.7 cm long, glabrous margin. Stamens - 11-12, 0.6 cm long Filament - sessile. Ovary- 0.25 cm long.	Short pedicellate, medium, 1.7 cm long. Sepals - 1.2 cm long; petals - 1.0 cm long; hairy margin. Stamens - 8, 0.7 cm long. Filament - short, ovary - 0.30 cm long.
Fruit	2.0 cm - 4.0 cm long x 5.0 cm - 6.0 cm periphery	5.0 cm - 7.0 cm long x 9.5 cm periphery
Seed	1.5 cm long x 2.0 cm diam.	1.8 cm long x 2.2 cm diam.
Embryonic	Epicotyle 2.0 cm long; hypocotyle 55.0 cm long; 7.0 cm periphery, smooth surface, gradually tapering but sharp end	Epicotyle 2.5 cm long; hypocotyle 75 cm long; 9.0 cm periphery, warted surface, gradually tapering, blunt end.
Distribution	Less in Sundarbans	Common in the Sundarbans.

Besides these two species of *Rhizophara* L., *R. stylosa* Griff. and *Rhizophora Xlamarckii* Montr. were also reported from the Gahirimata estuary of Orissa coast and Andaman & Nicobar Islands by Banerjee, *et al.* (1989) and Dagar *et al.* (1991), respectively. Tomlinson (1986) has stated the distribution of *R. stylosa* Griff. from S. E. Asia and Australia only,while Tomlinson & Womersley (1976) have also mentioned that *R. Xlamarckii* Montr. is a sterile F_1 hybrid between *R. apiculata* and also considered this hybrid as endemic to N. W. coast of New Caledonia.

GENUS-2. *BRUGUIERA* Lamark Encycl. Meth. Bot. 4 : 696. 1797-8.

Till date, six species are described under the genus -*Bruguiera* Lamk. (Tomlinson, 1986). These six species are distributed in the Old World Tropics, *i.e.*, East Coast of Africa to Pacific Islands, at South East Asia and North Australia. The members of the Genus- *Bruguiera* Lamark can be distinctly grouped under two categories, *i.e.*, *Group-1*, flowers large, generally with solitary flower (exception, *B. exaristata* Ding Hou - having large but two flowers in each inflorescence), and *Group-2*, flowers small, generally with many flowers in each inflorescence (exception, *B. hainesii* Rogers- having small but solitary flower).

Bruguiera gymnorrhiza Lamk., *B. sexangula* (Lour.) Poir. have single but large flower, ranging 2 cm - 4 cm long; while *B. parviflora* W. & A. and *B. cylindrica* (L.) Blume have many small flowers in each inflorescence, ranging 0.05 cm - 2.0 cm long.

In the Indian Sundarbans, only four species, *viz.*, *B. gymnorrhiza* Lamk., *B.*

sexangula (Lour.) Point., *B. cylindrica* (L.) Blume and *B. parviflora* W. & A. are distributed; though Mukerjee (1966) has reported *B. hainesii* Rogers from the Indian Sundarbans. Generally, the plants under the Genus- *Bruguiera Lamk.* attain the height upto 40 m in the other S. E. Asian mangals; but in the Indian Sundarbans the height of *Bruguiera* spp. attain upto 15 m due to large scale exploitation of the large sized trees, as well as having, higher saline condition of the present day's mangals of the Indian Sundarbans. These higher salinity ranges in soil hamper their vigorous growth.

The members of the *Bruguiera* Lamk. are distinct from other members of the Rhizophoreae (Mangrove- Rhizophoraceae) in respect of its distinct and conspicuous persistent calyx lobes in its fruits and hypocotyle, as well; besides these, the viviparous seedlings of these *Bruguiera* spp. also detach from the mother plant along with the germinated fruit, viviparous hypocotyles.

Members of this genus are medium to tall trees, evergreen, erect, deliquiscent branching or pyramid shaped branching, woody, dark green leaves, yellowish or reddish green, very showy, bark deep black grey, fissured.

Horizontal roots much spreading, aerial roots are knee like pneumatophores, root-buttress or occasionally plank like roots fused on the trunk bases; as such, trunk bases much thick for fusing the basal roots.

Branches—differentiated into nodes and internodes, apical shoots green, yellowish or reddish; internodes more or less equal length upto 1 cm long, leaf-scars present but not distinct like *Rhizophora* spp. Leaves; simple, opposite decussate, cauline, exstipulate, petiolate, petiole 0.8 cm- 5.0 cm long, terete, glabrous, green, yellowish or reddish black, pulvinous, base slightly flattened; lamina broadly ovate lanceolate, 5.0 cm - 20.5 cm long × 2.0 cm - 6.6 cm broad, entire, acute, tapering base, dorsiventral, coriaceous, dorsal surface shining, waxy, odourless, unicosted reticulate venation inconspicuously, mid-vein distinct in the ventral side.

Inflorescence solitary cymes or upto 7 flowers in groups on each peduncle; flowers small to medium or upto 4 cm long, ebracteate, pedicellate, pedicel 1.0 cm - 2.0 cm long; complete, bisexual, glabrous; sepals 8-16, polysepalous, 8-16 lobes above, beneath long slender tube, persistent lobes acicular, terminally pointed, valvate; petals 8-16, polypetalous, each folded, margin hairy or glabrous, terminally 3-4 apical cilia present or not, coriaceous, deciduous; stamens 16-32, 2 stamens in each petal, *i.e.*, stamen numbers are twice than the numbers of petals, 2 stamens with unequal filaments in each pair; carpels - 3, syncarpy, ovary inferior, cup shaped, 2-4 locules, axile placentation; style - 1, terminal, stigma - 3 fids.

Pollination - by birds and butterflies in the day time.

Fruit-capsule, pendulous; viviparous seedling 8-27 cm long, 0.4 - 3.0 cm diam. hypocotyle stought., ridged or smooth, generally blunt end and tapering on both sides, cylindrical, with persistent calyx tube even after hypocotyle dispersed.

Flowering and fruiting times for the species of the *Bruguiera* Lamk. are mostly during the months of March to October, but the pendulous viviparous seedlings in the mother plants are visible almost throughout the year. During monsoon months when the water and soil salinity come down, the shoot initiation take place.

The seedling hypocotyles can float in the tidal water and dispersed far away with the help of tidal water current; viability of these viviparous germinated hypocotyle seedlings remain for long period, even for few months.

Key to the Species of the *Bruguiera* Lamk.

1a. Inflorescence solitary; flower large single, upto 4.0 cm long, reddish or yellowish, persistent calyx-tube, 12-16 calyx lobes (2)

1b. Inflorescence 3-7 flowers in group on each peduncle; flowers small to medium, upto 2.0 cm long, yellowish or dark green with persistent calyx tube, calyx lobes - 8 (3)

2a (1a) Medium tree, 12 m - 15 m, reddish distinct mid-vein in the leaves; sepal lobes 13-16, reddish, petals terminated into 4 - cilia; stamens 26-32; hypocotyle upto 18 cm long, with reddish persistent calyx tube

1. *B. gymnorrhiza* (L.) Lamark

2b (1a) Small tree, upto 10.0 m height, yellowish distinct mid-vein; sepal lobes 10-12, yellowish, petals terminally not ciliated; stamens 20-24, hypocotyle upto 12.0 cm long, with yellowish persistent calyx tube

2. *B. sexangula* (Lour.) Poir

3a (1b) Large tree, upto 20 m height, dark green, 3-flowers in each group; flowers upto 1.3 cm long; petals white ; hypocotyle upto 15 cm long with dark green persistent calyx tube, calyx lobes folding inversely

3. *B. cylindrica* (L.) Blume

3b(1b) Medium tree, upto 12 m height, slightly yellowish, 4-7 flowers in groups, flowers upto 0.8 cm long; petals yellowish, hypocotyle upto 25 cm long with yellowish persistent long calyx tube; calyx lobes straight.

4. *B. parviflora* W. & A.

2.1. *Bruguiera gymnorrhiza (L.)* Lamark Encycl. Meth. Bot. 4 : 696.1797-1798.

Rhizophora gymnorrhiza L. Sp. Pl. 443.1753, *Bruguiera rheedii* Blume En. Pl. Jav. 92.1827. *B. conjugata* Merr. Philipp. J. Sci. Bot. 9 : 118. 1914. Chromosome No. *2n=36* (Sidhu, 1968 & Gill, 1978)

Common Name : Large-Leaved Orange Mangrove

Local Names (Bengali) : Kankra

(Plate-83 and Fig. 3)

Tree medium, 12 m - 15 m high, evergreen, erect, much diffused spreading branches; reddish- dark green; bark blackish grey, fissured.

Root—under ground roots not deep sunken, profusely spreading from the surface layer; aerial roots form knee like pneumatophores developing from bending horizontal underground roots; root buttress thickened the trunk bases.

Branches are differentiated into distinct nodes and internodes; glabrous, blackish green, terete, solid, jointed at the nodes, leaf-scars present inconspicuously.

Leaves opposite decussate, simple, cauline, exstipulate, petiolate, petiole upto 5.0 cm long, round, glabrous, pulvinous, reddish, flattened at base slightly, lamina broadly ovate-lanceolate, 13.0 cm - 20.5 cm long × 5.0 cm- 6.5 cm broad, entire, acute angled, dorsiventral, coriaceous, mature lamina reddish green and young lamina dark green, shiny, glabrous, odourless, unicostate venation inconspicuous; reddish mid-vein distinct in the ventral surface.

Inflorescence—solitary, single flower in each peduncle; flowers ebracteate, pedicellate, pedicel 2.0 cm long, round, glabrous; complete, bisexual, regular, spiral, pendulous, elliptical, 3.5 cm long; *sepals* 13-16, polysepalous, 13-16 sepal lobes above, beneath long tube, reddish, glabrous, slightly ridged, valvate, cup-shaped, lobes acicular, terminally pointed, 1.5 cm long, reddish, persistent, glabrous; *petals* 13-16 , polypetalous, stiff, each petal folded into two parts, each part 1.0 cm long, 0.2 cm broad, white, terminally 3-4 apical cilia, cilia white, 0.15 - 0.40 cm long, basal and margin silky hairy, oblong, coriaceous, upper surface slippery; *stamens* 26-32, free, two stamens pair within one petal, pair stamens with unequal filament lengths, but anthers same in length; filament upto 0.25 cm long, round, glabrous, basifixed, superior, anther bilobed, longitudinally dehiscent, introrse, inserted, 0.25 cm long ; *carpels* - 3, syncarpy, inferior, cup-shaped, 2-4 locules, 2- ovules in each chamber, style 1, terminal, terete, glabrous, 0.75 cm long, stigmas 3, 3-lobed, oval, entire, glabrous.

Pollination caused by birds during day time.

Fruits capsule, pendulous with persistent calyx ; seed one, hypocotyle 18.0 cm long, 2.0 - 3.0 cm diam., blackish, dark- green; vertical narrow ridges, showy, with distinct persistent reddish calyx. Germination viviparous, very conspicuous in the Indian Sundarbans.

Flowering and fruiting times are during March to October, but very often flowers and hypocotyles are seen throughout the year. The vegetative growth mainly restricted during monsoon and postmonsoon months.

Seed and seedlings dispersed during post-monsoon months and directly penetrate on the silty clay soil; sometimes float on the water and dispersed with the tidal current; the hypocotyle initiates the root on soil, penetrated apical portion and two foliage leaves develop within very short period from the apical portion, *i.e.*, plumule. The hypocotyle seedlings are viable for longer period in the floating condition on tidal saline water and foliage leaves develop shortly on fixing the seedling in the silty clay soil, during monsoon months.

This species is common in the intertidal river banks, river slopes and ridge forests of the mangrove forests in the Sundarbans; it also grow in the mid-and outer-estuaries, facing the sea or shelter coast. This species grow along with *Rhizophora* spp., *Ceriops* spp., *Aegiceras* sp., *Excoecaria* sp., *Xylocarpus* spp. and alongwith the other true mangrove species.

Bruguiera gymnorrhiza (L.) Lamark is the type species of the Genus - *Bruguiera* Lamark, Encycl. Meth. Bot. 4 : 696. 1797- 8. It is common in the Indian parts of the Sundarbans and throughout the east and west coast of India and Andaman & Nicobar Islands. Also distributed in Bangladesh, Sri Lanka, Burma and other parts of S. E. Asian mangals and in North Australia.

Bark tannin percentage is comparatively low.

Bark is said to be useful in diarrhoea. Leaves and hypocotyle cooked occasionally as green vegetable.

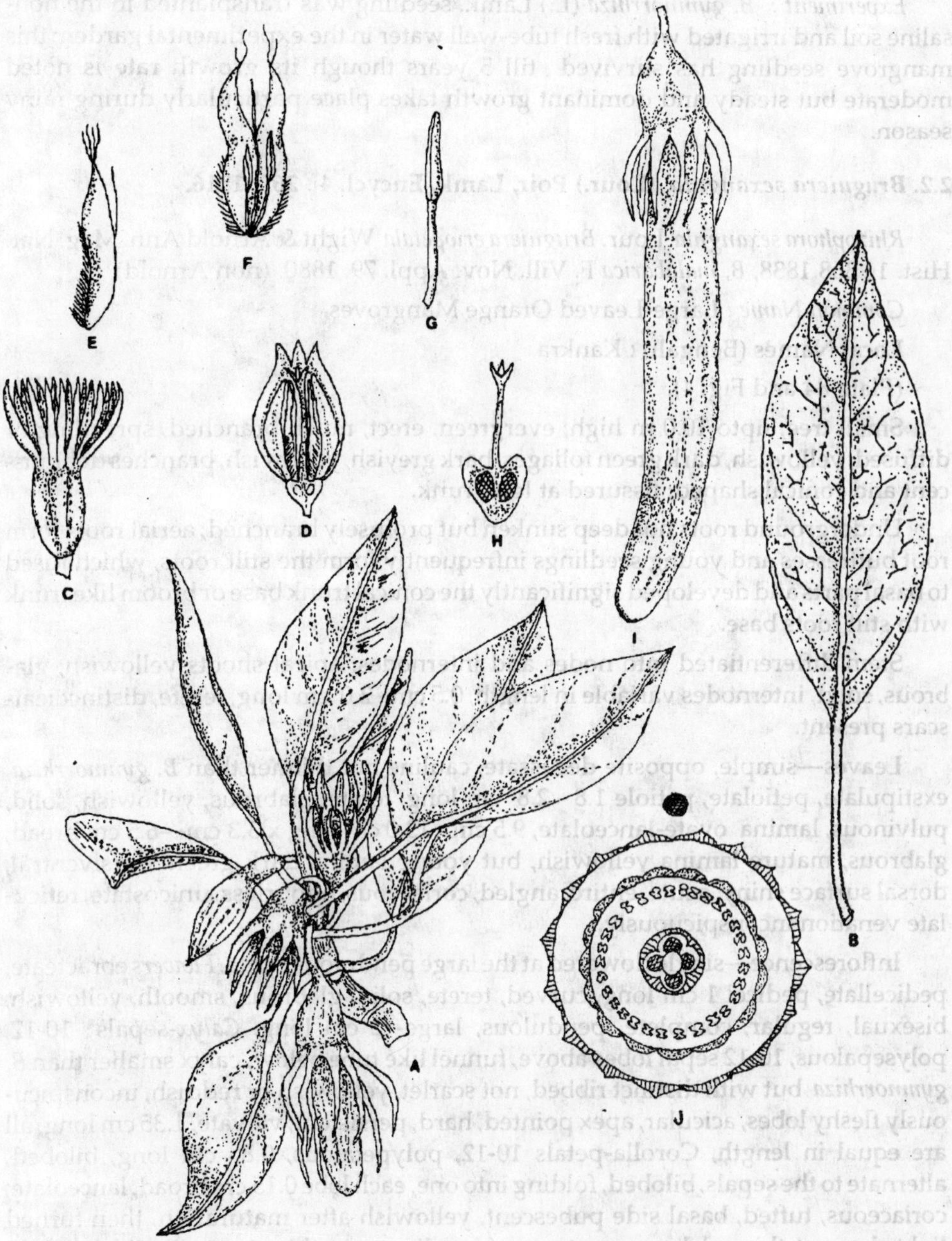

Fig. 3 : ***Bruguiera gymnorrhiza*** **(L.) Lamk.**

A. Portion of flowering twig x 1/3; B. Leaf : petiole- 5.0 cm long, lamina- 20.0 cm long x 6.5 cm width; C. Flower-2.5 cm long; D. L. S. of flower; E. Petal - 0.9 cm long; F. Petal showing stamens; G. Stamen; filament- 0.25 cm long; anther- 0.25 cm long; H. L.S. of Gynoecium : ovary - 0.5 cm long, style - 1.0 cm long; I. Fruit with vivipary 15 cm long; J. Floral diagram.

Experiment : *B. gymnorrhiza* (L.) Lamk. seedling was transplanted in the non-saline soil and irrigated with fresh tube-well water in the experimental garden; this mangrove seedling has survived till 5 years though its growth rate is noted moderate but steady and dominant growth takes place particularly during rainy season.

2.2. *Bruguiera sexangula* (Lour.) Poir, Lamk. Encycl. 4 : 262. 1816.

Rhizophora sexangula Lour. *Bruguiera eriopetala* Wight & Arnold Ann. Mag. Nat. Hist. 1 : 368.1838, *B. malabarica* F. Vill. Nov. Appl. 79. 1880. (non Arnold).

Common Name : Large-Leaved Orange Mangroves.

Local Names (Bengali) : Kankra

(Plate-84 and Fig. 4).

Small tree, upto 10.0 m high; evergreen, erect, much branched, spreading or diffused; yellowish, dark green foliages; bark greyish, yellowish, branches deliquiscent and conical shaped; fissured at low trunk.

Underground roots not deep sunken but profusely branched; aerial roots form root buttresses and young seedlings infrequently form the stilt roots, which fused to basal parts and developed significantly the conical trunk base or broom like trunk with still roots base.

Stem differentiated into nodes and internodes, apical shoots yellowish, glabrous, solid, internodes variable in length, 0.5 cm - 2.0 cm long, terete, distinct leaf-scars present.

Leaves—simple, opposite decussate, cauline but thinner than *B. gymnorrhiza*, exstipulate, petiolate, petiole 1.8 - 2.8 cm long, terete, glabrous, yellowish, solid, pulvinous, lamina ovate-lanceolate, 9.5 cm - 13.0 cm long x 3.3 cm—6.5 cm broad, glabrous, mature lamina yellowish, but young lamina dark green. dorsiventral, dorsal surface shiny, acute, entire, angled, coriaceous, odourless, unicostate, reticulate venation inconspicuously.

Inflorescence—single flowered at the large pendulous cyme. *Flowers* ebracteate, pedicellate, pedicel 1 cm long, curved, terete, solid, glabrous, smooth, yellowish, bisexual, regular, complete, pendulous, large- 2 cm long. *Calyx*-sepals 10-12, polysepalous, 10-12 sepal lobes above, funnel like tube below, calyx smaller than *B. gymnorrhiza* but with distinct ribbed, not scarlet, yellowish or reddish, inconspicuously fleshy lobes, acicular, apex pointed, hard, persistent, valvate, 1.35 cm long, all are equal in length, Corolla-petals 10-12, polypetalous, 0.8 cm long, bilobed, alternate to the sepals, bilobed, folding into one, each lobe 0.15 cm broad, lanceolate, coriaceous, tufted, basal side pubescent, yellowish after maturation, then turned light brown, thin, glabrous, outer surface slippery, white inner surface slightly roughed, each petal contain two stamens of unequal filament length. *Androecium*—stamens 20-24, free, but 10-12 groups, 2 stamens into each folding petal, unequal filament, upto 0.25 cm long and anther 0.2 cm long but anthers same in length, inserted introse, basifixed, smooth, the base of filament is becoming hairy, white, round, soft, anther bilobed, longitudinal dehisced, yellowish. *Gynoecium* - carpels 3, syncarpy, ovary 0.5 cm long, superior, cup-shaped, 2-4 celled, two ovules in each

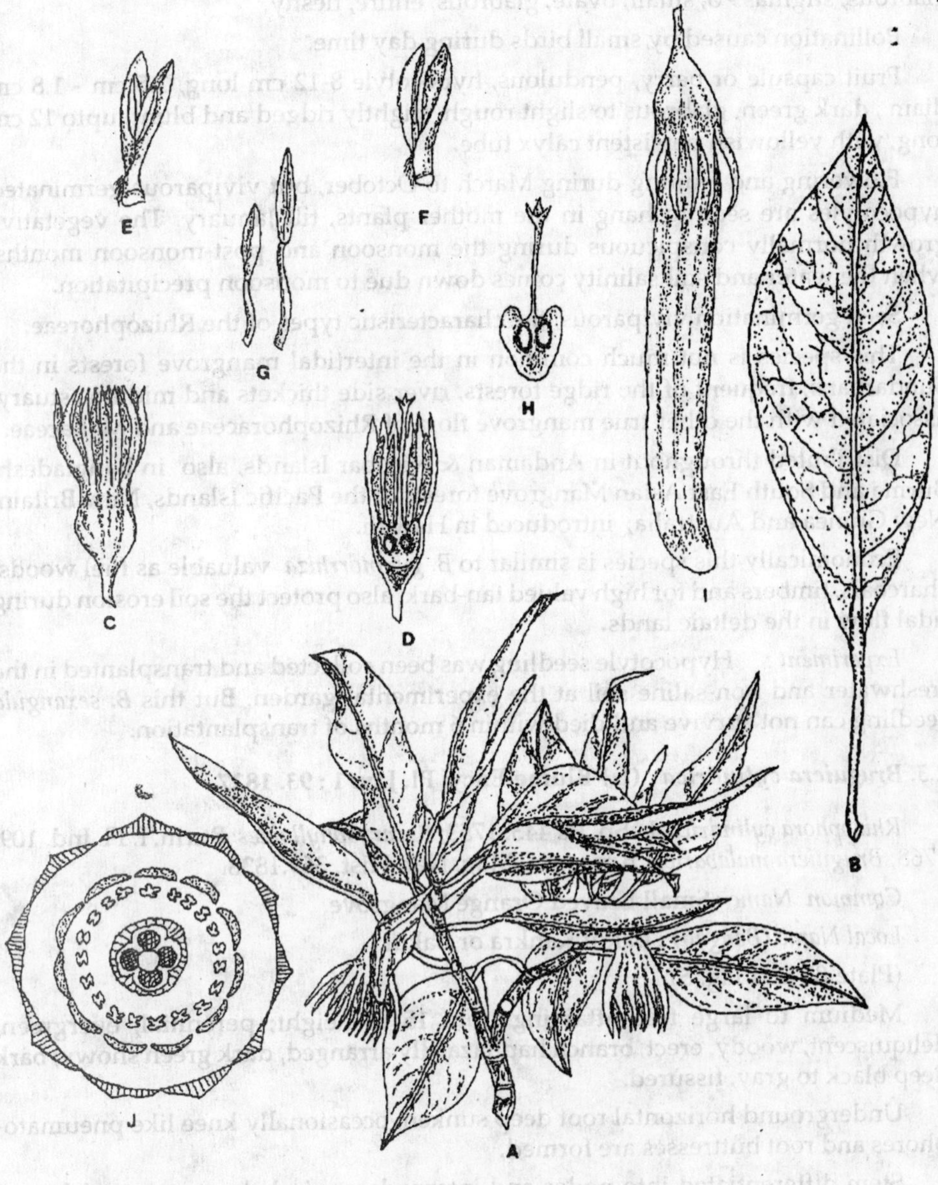

Fig. 4 : ***Bruguiera sexangula*** **(Lour.) Poir**

A. Portion of flowering twig x 1/3; B. Leaf : petiole - 2.8 cm long, lamina - 10.2 cm long x 6.5 cm width; C. Flower- 2.0 cm long; D. L.S. of flower; E. Defolded petal showing stamens. F. Petal - 0.8 cm long; G. Stamens: filament - 0.25 cm long, anther- 0.2 cm long; H. L.S. of Gynoecium : ovary- 0.5 cm long, style- 1.0 cm long; I. Fruit with vivipary 10 cm long ; J. Floral diagram.

chamber, axile placentation, style - 1, terminal, 1.0 cm long, round, soft, white, glabrous, stigmas - 3, small, ovate, glabrous, entire, fleshy.

Pollination caused by small birds during day time.

Fruit capsule or berry, pendulous, hypocotyle 8-12 cm long, 1.5 cm - 1.8 cm diam., dark green, glabrous to slight rough, slightly ridged and blunt, upto 12 cm long, with yellowish persistent calyx tube.

Flowering and fruiting during March to October, but viviparous germinated hypocotyles are seen to hang in the mother plants, till January. The vegetative growth normally conspicuous during the monsoon and post-monsoon months, when the water and soil salinity comes down due to monsoon precipitation.

Seed germination viviparous, the characteristic types of the Rhizophoreae.

This species is not much common in the intertidal mangrove forests in the Sundarbans; frequent of the ridge forests, river side thickets and middle estuary, associated with the other true mangrove flora of Rhizophoraceae and Meliaceae.

Distributed throughout in Andaman & Nicobar Islands, also in Bangladesh, Burma and South East Asian Mangrove forests of the Pacific Islands, New Britain, New Guinea and Australia; introduced in Hawaii.

Economically this species is similar to *B. gymnorrhiza* valuable as fuel woods, charcoals, timbers and for high valued tan-bark, also protect the soil erosion during tidal flow in the deltaic lands.

Experiment : Hypocotyle seedling was been collected and transplanted in the freshwater and non-saline soil at the experimental garden. But this *B. sexangula* seedling can not survive and died within 6 months of transplantation.

2.3. *Bruguiera cylindrica* (L.) Blume Ency. Pl. Jav. 1 : 93. 1827.

Rhizophora cylindrica L. Sp. Pl. 443. 1753; *R. caryophylloides* Burm. f. Fl. Ind. 109. 1768; *Bruguiera malabarica* Arn. Ann. Mag. Nat. Hist. 369.1838.

Common Name : Small Leaved Orange Mangrove.

Local Name (Bengali) : Bakul Kankra or Bakul.

(Plate-85 and Fig. 5).

Medium to large tree attaining upto 15 m height; perennial, evergreen, deliquiscent, woody, erect, branch haphazardly arranged, dark green showy; bark deep black to gray, fissured.

Underground horizontal root deep sunken, occasionally knee like pneumatophores and root buttresses are formed.

Stem differentiated into nodes and internodes, apical shoot green, glabrous, solid, internodes, more or less equal in length, 1.0 cm leaf scar present but not very much conspicuous.

Leaves simple, opposite decussate, cauline, exstipulate, petiolate; petiole 2-4.5cm long, terete but base slightly flattened, reddish-dark green, solid, pulvinous, lamina ovate-lanceolate, elliptic, 10-12 cm long x 4-5 cm broad, entire, acutely or bluntly pointed, angled, dorsiventral, coriaceous, odourless unicostate reticulate venation inconspicuous.

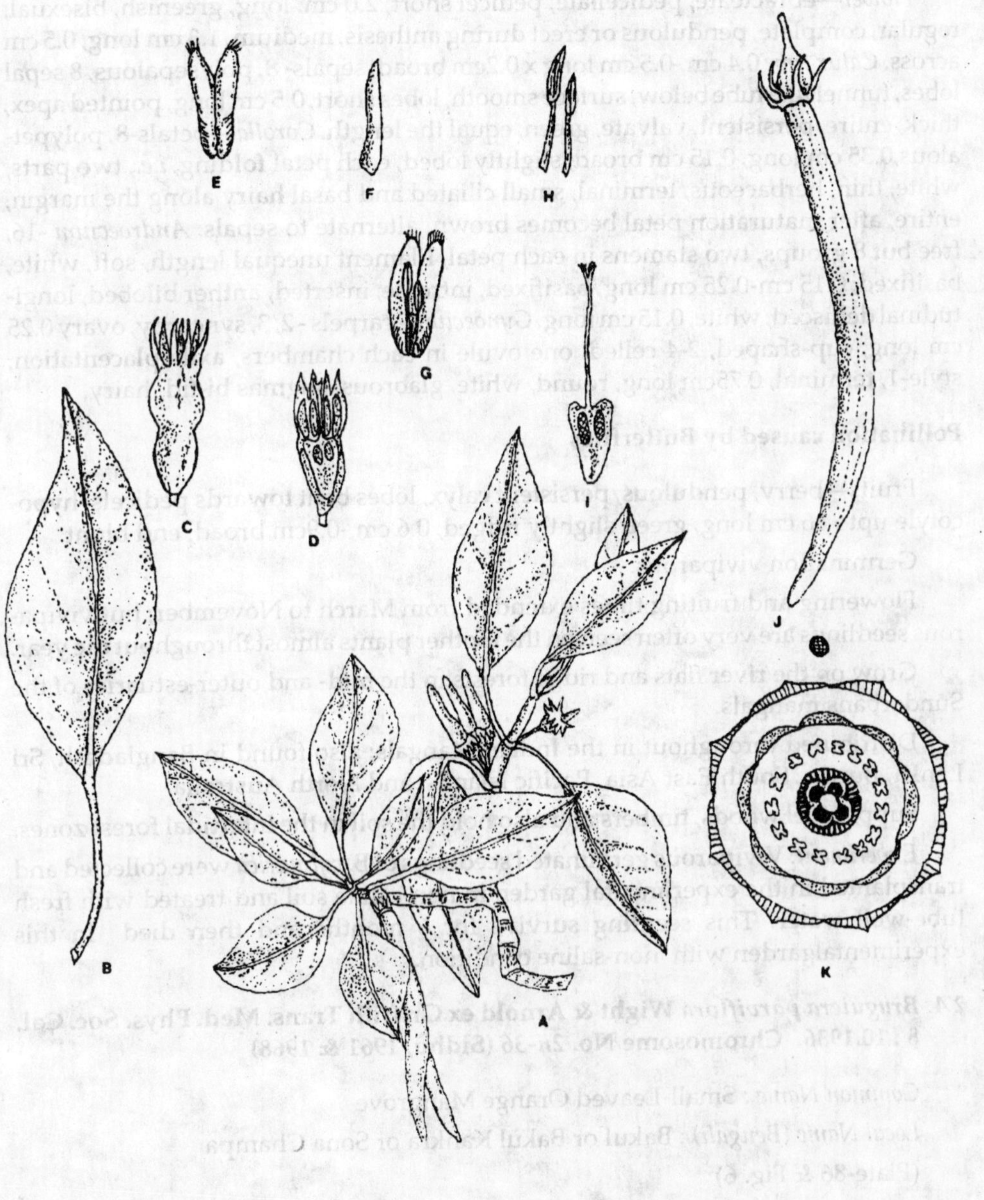

Fig. 5 : ***Bruguiera cylindrica* (L.) Blume**

A. Portion of twig with flower and viviparous fruit x 1/3; B. Leaf : petiole- 4.5 cm long, lamina- 10.5 cm long x 5.0 cm width : C. flower - 1.3 cm long; D. L. S. of flower ; E. Petal- 0.35 cm long; F. Sepal- 0.5 cm long; G. Defolded petal showing stamens; H. Stamens: filament- 0.25 cm long, anther - 0.1 cm long; I. L. S. of Gynoecium : ovary- 0.25 cm long, style - 0.75 cm long ; J. Fruit with vivipary 15.3 cm long ; K. Floral diagram.

Inflorescence cyme, 3-floweres in each peduncle, peduncle long, terete, green, glabrous.

Flower—ebracteate, pedicellate, pedicel short, 2.0 cm. long, greenish, bisexual, regular, complete, pendulous or erect during anthesis, medium, 1.3 cm long, 0.5 cm across. *Calyx tube* 0.4 cm -0.5 cm long x 0.2cm broad; sepals- 8, polysepalous, 8 sepal lobes, funnel like tube below, surface smooth, lobes short, 0.5 cm long, pointed apex, thick, entire, persistent, valvate, green, equal the length. *Corolla*—petals-8, polypetalous,0.35 cm long, 0.15 cm broad, slightly lobed, each petal folding, *i.e.*, two parts, white, thin, herbaceous, terminal, small ciliated and basal hairy along the margin, entire, after maturation petal becomes brown, alternate to sepals. *Androecium* -16, free but 8 groups, two stamens in each petal, filament unequal length, soft, white, basifixed, 0.15 cm-0.25 cm long, basifixed, introrse, inserted, anther bilobed, longitudinal dehisced, white, 0.15 cm long. *Gynoecium* - carpels - 2, 3, syncarpy, ovary 0.25 cm long, cup-shaped, 2-4 celled, one ovule in each chambers, axile placentation, style-1, terminal, 0.75cm long, round, white, glabrous, stigmas bi-fid, hairy.

Pollination caused by Butterflies

Fruits—berry, pendulous, persistent calyx, lobes bent towards pedicels, hypocotyle upto 16 cm long, green slightly ridged, 0.6 cm -0.9cm broad, end blunt.

Germination viviparous.

Flowering and fruiting times extended from March to November; but viviparous seedlings are very often seen on the mother plants almost throughout the year.

Grow on the river flats and ridge forests in the mid- and outer-estuaries of the Sundarbans mangals.

Distributed throughout in the Indian Mangals; also found in Bangladesh, Sri Lanka, Burma, South East Asia, Pacific Islands and North Australia.

Supply fuel woods, timbers and also hold the soil in the intertidal forest zones.

Experiment: Viviparous germinated seedlings of B. *cylindrica* were collected and transplanted in the experimental garden in non-saline soil and treated with fresh tube-well water. This seedling survive for 6 months and then died in this experimentalgarden with non-saline condition.

2.4. *Bruguiera parviflora* Wight & Arnold ex Griffith Trans. Med. Phys. Soc. Cal. 8 : 10.1936. Chromosome No. *2n-36* (Sidhu, 1961 & 1968)

Common Name : Small-Leaved Orange Mangrove

Local Name (Bengali) : Bakul or Bakul Kankra or Sona Champa

(Plate-86 & Fig. 6)

Medium tree, attain a height upto 15 m; perennial, evergreen, deliquiscent, woody, erect, branching, slightly pyramidal shape, yellowish, showy, bark blackish grey, but not fissured.

Underground roots not deep sunken; inconspicuous root buttress and root knee.

Differentiated into nodes and internodes, in the erect branches, apical shoot looks like other species of the genus, internodes variable, 0.25 cm -1.0 cm, terete, leaf scars present in the branches.

Leaves simple, opposite decussate, cauline, exstipulate, petiolate, petiole 0.8 cm- 4.5 cm long, glabrous, terete, yellowish, solid, pulvinous; lamina elliptic-oblong, entire, acute, angled, 5.0 cm - 10 cm long x 2.0 cm -5.0 cm broad, glabrous, dorsiventral, dorsal side shiny, coriaceous, odourless, unicostate reticulate venation.

Inflorescence-cyme, 4-7 flowers on each peduncle, 2.0 cm- 4.0 cm long, glabrous, terete, yellowish. *Flowers*-ebracteate, pedicellate, pedicel upto 0.8 cm long, terete, glabrous, smooth, yellowish, solid, bisexual, regular, complete, pendulous, small, 0.85 cm long. *Sepals*-8, sepalous, 8-sepal lobes above, below cylindrical tube, acicular, pointed apex, persistent, 0.25 cm long, equal length. *Corolla*-8, polype-talous, 0.1cm -0.2 cm long x 0.05 cm broad, yellowish white, at first whitish and gradually turning to yellowish through maturation; petals folded into two parts, at the terminal part 3 long cilia, petal margin slightly hairy, thin, herbaceous, glabrous, each petal contain two stamens of unequal filament length. *Androecium* -stamens-16, free, but 8 groups, 2 stamens in each folding petal, unequal filament length of two stamens, 0.10 cm - 0.15 cm long, glabrous, soft, terete, white, basifixed, anther bilobed, longitudinal dehisced, introrse, inserted, yellowish white. *Gynoecium*-carpels 3, syncarpy, ovary upto 0.15 cm long, but after maturation it beocmes 1.5 cm long, attached with fruit, cylinder shaped, 2-celled, axile placentation, style 1, 0.15 cm -0.25 cm long, white, soft, glabrous, stigma 2, bifid, short.

Pollination caused by butterflies in day times.

Fruits—berry, pendulous, upto 2.0 cm long, 0.4 cm-0.6 cm diam., hypocotyle light yellow-green, sepal lobes short, persistent but long calyx tube, end slightly pointed; hypocotyle length ranges upto 25 cm.

Flowering and fruiting generally during April to October, but viviparous germinated seedlings are very often seen in the mother plant, almost throughout the year.

Frequently grow inside the intertidal ridge forest in the Sundarbans mangals and very occasionally grow on the river banks, almost as pioneer species.

Distribution : Mostly grow in Andaman & Nicobar Island, Mahanadi Delta and the Sundarbans, also reported from Tropical Asia to Australia and Melanesia; introduced in Hawaii.

Experiment : Viviparous germinated seedlings of *Bruguiera parviflora* were collected and transplanted in the experimental garden of non-saline soil and has been treated or watered with freshwater of tube-well. But this seedling could not survive beyond 6 months.

GENUS-3. *CERIOPS* Arnold Ann. Nat. Hist. 1 : 363.1838

Species under the genus- *Ceriops* Arn. are true mangroves belongs to the inner mangrove communities on the intertidal ridge forests, which mostly form the pure strands. Two distinct species, *viz., C. tagal* (Perr.) Robinson and *C. decandra* (Griff.)

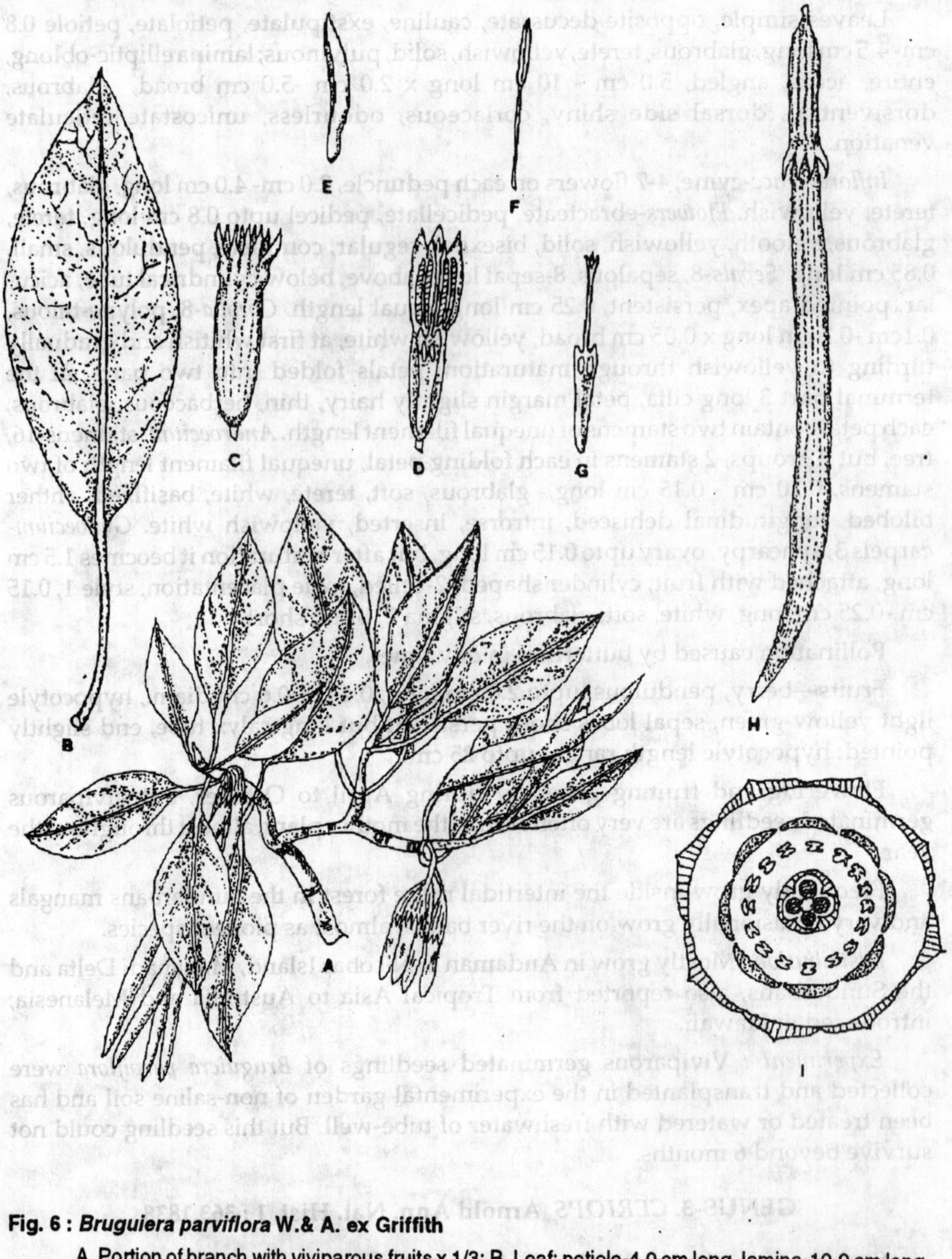

Fig. 6 : ***Bruguiera parviflora*** **W.& A. ex Griffith**

A. Portion of branch with viviparous fruits x 1/3; B. Leaf: petiole-4.0 cm long, lamina-10.0 cm long x 4.5 cm width; C. Flower 0.85 cm. long; D.L.S. of flower; E. Petal -0.2 cm long x 0.05 cm width; F. Stamen-0.15 cm long; G.L.S. of Gynoecium; H. Fruit with vivipary-25 cm long; I. Floral diagram.

Table 57 : Comparative Morphology of the Different Species of *Bruguiera* Lamk.

Habits & Habitat	*B. gymnorrhiza*	*B. sexangula*	*B. cylindrica*	*B. parviflora*
Habit	Medium tree 12 m - 15 m high	Small tree, 10 m high	Medium tree, 15 m high	Medum tree, 15 m high
Colour of tree	Reddish green	Yellowish green	Dark green	Yellowish green
Leaf petiole	5.0 cm long	1.8 cm - 2.8 cm long	2.0 cm- 4.5 cm long	0.8 cm - 4.5 cm long
Lamina	Upto 20.5 cm long and upto 6.5 cm broad; reddish mid-vein	Upto 13.0 cm long and upto 6.5 cm broad; yellowish mid-vein	Upto 12.0 cm long and Upto 5.0 cm broad; dark-green mid-vein	Upto 10.0 cm long and 5.0 cm broad; yellowish mid-vein
Inflorescence	Solitary, reddish flower	Solitary yellowish flower	3-flowers in group, green	4-7 flowers in group, yellowish
Flower	3.5 cm long, pedicel 2.0 cm long, sepal lobes 12-16, reddish, 1.5 cm long; petals-12-16, 1.0 cm long, each lobe terminated into 4 cilia; stamens-24 - 32	2.0 cm long, pedicel 1.0 cm long. sepal 10-12, lobes, yellowish, 1.35 cm long; petals-10-12, yellowish, terminated without cilia, apex blunt; stamen - 20-24	1.3 cm long, pedicel 0.2 cm long, sepal lobes 8, greenish, 0.5 cm long ; petals--8, white, pubescent margin; stamens-16	0.85 cm long, pedicel 0.8 cm long, sepal lobes-8, yellowish, sepals cylindrical tube, upto 0.25 cm long petals yellow, marginal hairy - 8, stamens - 16
Fruit	3.8 cm long	3.2 cm long	1.8 cm long	1.0 cm long
Hypocotyle	15.0 cm long with reddish persistent calyx tube	10.0 cm long, with yellowish persistent calyx tube	16 cm long with green invarsely folded peristent calyx tube	25 cm long with yellowish straight persistent long calyx tube.

Ding Hou are reported under the Genus- *Ceriops* Arnold; these species are wide spread in the Old World Tropics, *i.e.* Indo-west Pacific Regions. White (1926) has reported two varieties of *C. tagal* , *i.e.*, (i) *C. tagal* Robinson var. *tagal* and (ii) *C. tagal* Robinson var. *australis* from the Queensland, with the characteristic difference between the two varieties like—(i) distinctly fluted seedling hypocotyle for the first and (ii) terete, not fluted, seedling hypocotyles for the 2nd, respectively.

In the Indian mangrove ecosystems, both these two species are present, particularly in the Sundarbans, Mahanadi delta and Andaman & Nicobar Islands.

Plants evergreen shrub to small tree, upto 5 m tall in the Sundarbans; though these species are reported upto 10 m high in other parts of the Old World Tropics. Perennial, deliquiscent branched, bushy or pyramidal appearance.

Stilt roots developed from the base of the stem, upto 50 cm upward and form broom encircling the stem, very often fused to buttresses.

Trunk distinct, stem differentiated into swollen nodes and internodes; more frequently branched, branching upwardly arranged, internode upto 8.0 cm long, woody, glabrous, yellowish green; bark insignificantly fissured, grey to whitish.

Leaves opposite decussate, simple, exstipulate, cauline, petiolate, petiole upto 2.7cm long, green, solid, glabrous, pulvinous, basal flattened, lamina elliptic, obovate, ovate-ellipltic, oblong, entire, obtuse or round apex, tapering towards base, upto 9.5 cm long x upto 5.5 cm broad, glabrous, coriaceous, dorsiventral, dorsal surface shining, unicostate reticulate venation, vein inconspicuously found, mid-vein prominent in ventral side.

Inflorescence cyme, opposite decussate, axillary within the leafy cluster or below the leafy cluster; axis short to moderate, 2-10 peduncles on each shoot apex. Flower covered with thick bracts and bracteoles, sessile or pedicellate, in compact bifurcated cyme; bisexual, pentamerous, complete, regular, 0.75cm long x 0.60 cm across, odourless, sepals -5, polysepalous, superior, imbricate, elliptic, or elliptic-lanceolate, entire, acute, upto 0.45 cm long x upto 0.25 cm broad, fleshy, thick; after maturation sepals become brown, persistent, valvate or imbricate; petals- 5, polypetalous, imbricate or valvate, thin, keeled, herbaceous, white, 0.3 cm long, 0.08 cm broad, terminally 3 clavate appendages or ciliated; stamens 10, free, anther bi-lobed, equal to filament length or shorter than filament length, 0.25 cm long; carpels-2, syncarpy; ovary inferior, 0.25 cm long, ovary l-2-1oculed, 1-2 ovules in each locule, axile placentation; style-1, terminal, stigmas-3, trifid.

Pollinations explosive or non-explosive, during night time by moth or short tongued insects.

Fruits conical, brown and rough on maturity, with persistent worty calyx tube, upto 3 cm long; germination viviparous, epigeal; hypocotyles attain the length upto 25 cm and 4.0 cm - 4.5 cm diam. ; terete or ridged.

Flowering and fruiting is conspicuous during summer to monsoon months,*i.e.*, April to September; but flowers, fruits and viviparous seedling hypocotyles occasionally found almost throughout the year.

Germination, Seedling Morphology and Seeds Dispersal

Germination is viviparous, epigeal; tigellum or embryonic axis hanged down mostly from terminal branches; hypocotyles attain length upto 25.0 cm, longitudinally ridged, blunt or sharp end, turning green even after maturation; tigellum detached from the fruit and or cotyledons. The tigellum fall down on sticky mud or clay soil or water, which can also float several days; after floatings and carrying by water current it attach in substratum on which it develop new seedlings and mature plants. Rooting start immediately after reaching the substratum; the hypocotyle becomes straight and short epicotyle enlarge in size; two foliage leaves including one apical bud gradually increase in size and attain upto 3.0 cm length x 2.5 cm broad as mature shoot having apical leaves.

The two species of *Ceriops* Arnold can be easily separated or identified with the following key characters:

Key to the Species of *Ceriops* Arnold

1a. Small to moderate tree, branches pyramidal appearance; leaves elliptic, peduncle 2-4, flowers within leafy cluster; petals terminally 3-clavate appendages; anther shorter than filament length; hypocotyle upto 25 cm long with sharp end

2. C. *tagal* (Perr.) Robinson

1b. Shrub, deliquiscent branching; leaves ovate, elliptic, peduncles 6-10; flowers within and below the leafy cluster; petals terminally ciliated; anther equal to the filament length; hypocotyle upto 12 cm long, apex blunt

1. C. *decandra* (Griff.) Ding Hou

3.1. *Ceriops decandra* (Griff.) Ding Hou, Fl. Males. Ser. I. 5 (4) : 471.1958.

Bruguiera decandra Griff. Trans. Med. Phys. Soc. Calc. 8 : 10.1836. *Ceriops roxburghiana* Arn. Ann. Mus. Nat. Hist. 1: 364.1838: *Rhizophora decandra* Roxb. (Hort. Beng. 36: 1814, nomen ex Griff. Natul 4 : 663.1854.

Common Name : Yellow Mangrove.

Chromosome No. $2n = 36$ (Sidhu, 1968)

Local Name (Bengali) : Jale-Goran or Jhamti-Goran

(Plate-87 and Fig.7).

Evergreen shrub or small tree, attain 3.0 m -5.0 m height, deliquiscent, bushy, more diffused branched.

Stilt roots developed from the base of the stem, upto 30 cm upward, and form broom encircling the stem; aerial roots fused in the stem base.

Differentiated into swollen nodes and inter-nodes; deliquiscent, more diffused branches, inter-nodes 3.0 cm - 8.0 cm length, solid, wood hard, glabrous, terete,

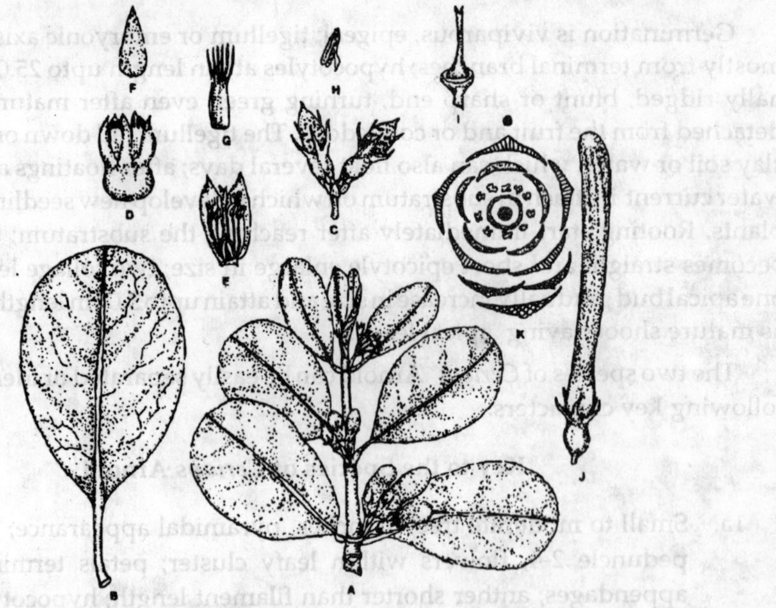

Fig. 7 : ***Ceriops decandra*** **(Griff.) Ding Hou**

A. Part of flowering twig with fruits × 1/3; B. Leaf: petiole-1.9 cm long, lamina- 9.4 cm long x 5.3 cm width; C. Inflorescence; D. Flower-0.6 cm long x 0.5 cm across; E. L.S. of Flower; F. Sepal-0.4 cm long x 0.15 cm width; G. Petal-0.25 cm long x 0.1 cm width; H. Stamen-0.25 cm long; I. Gynoecium- 0.15 cm long; J. Fruit with vivipary- 12 cm long; K. Floral diagram.

yellowish-green, generally each node having several branches . Bark pale yellow or dark, smooth.

Leaves opposite decussate, exstipulate, cauline, simple, petiolate, petiole 2.5 cm long, round, yellowish-green, glabrous, pulvinous, lamina ovate, ovate-elliptic, oblong, entire, obtuse or round apex, upto 9.5 cm long x 5.5 cm broad, green, mid-vein yellowish, dorsiventral, glabrous, slightly shining dorsal side, pale beneath, coriaceous, unicosted reticulate venation; vein inconspicuously present.

Inflorescence—1.0 cm long, having single stalk on both sides of the nodes, umbel like, one peduncle having 16-flower buds but 4-6 of them are matured; 6-10 peduncles on each shoot apex, within or below the leafy cluster; 16 flower buds aggregate on each head, bracteate, 2 bracts adjoining with 4-buds; each of these 2- buds have another bracteoles, buds arranged in series accordingly, *i.e.,* 1 +1 = 2, 2 + 2 = 4, 4 + 4 = 8, 8 + 8 =16. *Flower*—short, sessile, bisexual, regular, pentamerous complete, glabrous, green, 0.75 cm long × 0.60 cm across, rosaceous; bracts 0.2 cm long and 0.2cm broad. *Sepals*-5, polysepalous, 0.4 cm long, 0.15 cm broad, persistent, whitish yellow, fleshy, thick, elliptic, entire, acute, mature sepals turn brown, rough, terminal pointed, valvate. *Petals*-5, polypetalous, oblong, entire, 0.28 cm long, 0.08 cm broad, white; terminally ciliated, numerous, white, soft-hyphae like structure on terminal portion, not showy, glabrous, herbaceous, shorter than calyx, imbricate, deciduous, alternate to sepal. *Stamens*-10, free, filament length equal to anther, stamens 0.25 cm long, while filament 0.14 cm long, whitish, soft, round,glabrous, anther dorsifixed, 0.11 cm long, bilobed, longitudinally dehisced, introrse, inserted, each stamen in equal length, shorter than petal. *Carpels* -3, syncarpy, ovary 0.2 cm long and 0.2 cm diam. inferior, single chambered, 4-ovules in each chamber, 1-survive, anther 3-degenerate, axile placentation, style-1, terminal, greenish-white, 0.1 cm -0.5 cm long, round, glabrous, brown stigmas-3, shortly trifid.

Pollination non-explosive, the pollination agents are short tongued insect, Pollination is performed during night time.

Fruit—capsule, 3 cm long with 5 persistent calyx, brown with rough surface; blunt apex. Seed germination epigeal, viviparous.

Flowering and fruiting are during February-September. But the main vegetative growth is restricted during monsoon and post-monsoon months. The viviparous germinated seedlings almost are visible throughout the year. Viviparous hypocotyles ranging the length between 8 cm and 12 cm.

Distributed in the ridge mangrove forests, throughout in the Sundarbans; associated with the other mangroves also distributed in Andaman & Nicobar Islands and other mangrove forests/ecosystems in India. A Old World Tropical Species of S.E. Asia, Kalimantan, Philippines, New Guinea.

Economically this species is similar to *Ceriops tagal* (Perr.) Robinson for its hard and high calorific valued fuel wood. Tannin of stem bark is largely used.

Experiments: Viviparous germinated seedling have collected from the tidal mangrove forest and transplanted in the experimental garden on non-saline soil and has been watered with fresh tube well water. This seedling cannot survive beyond 8 months.

3.2. *Ceriops tagal* (Perr.) Robinson Philipp. J. Sci. Bot. 3 : 306. 1908

Rhizophora tagal Perr. Mem Soc. Linn. Paris 3 : 138. 1824. *Ceriops candolleana* Arn. Ann. Mag. Nat. Hist. 1: 364.1838. *C. timoriensis* Domin Bibl. Bot. 89: 444. 1928. *C. boiviniana* Tulasne Ann. Sci. Nat. 6: 112. 1856.

Common Names : Yellow Mangrove

Local Name (Bengali) : Mat-Goran or Jat-Goran

(Plate-88 and Fig. 8)

Small to moderate tree, much branched, pyramidal, upto 8 m high; evergreen, perennial.

Aerial roots broom like and form buttresses on trunk-base.

Differentiated into swollen nodes and inter-nodes; nodes slightly ridged; shoot apex, inter-nodes short, 0.1 cm-0.3 cm long, greyish encircling rough layer at the nodes with prominent leaf-scars below the encircling layer; leaf-scars 0.25 cm diam., rough, inter-nodes terete, green, solid, 4.5 cm long shoot apex, flattened covered with clustered leaves.

Leaves opposite decussate, simple, cauline, exstipulate, petiolate, petiole 2.7cm

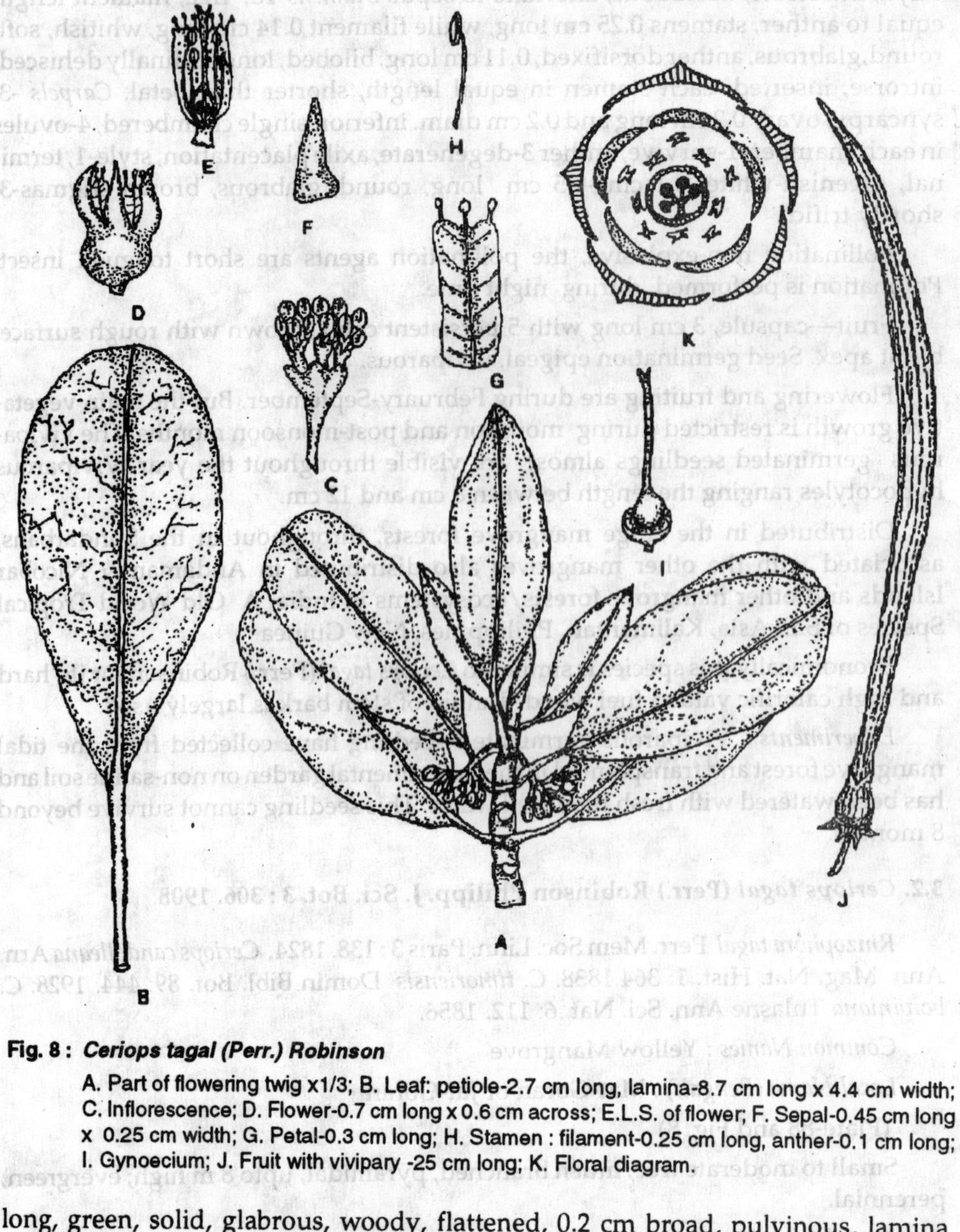

Fig. 8 : *Ceriops tagal (Perr.) Robinson*

A. Part of flowering twig x1/3; B. Leaf: petiole-2.7 cm long, lamina-8.7 cm long x 4.4 cm width; C. Inflorescence; D. Flower-0.7 cm long x 0.6 cm across; E.L.S. of flower; F. Sepal-0.45 cm long x 0.25 cm width; G. Petal-0.3 cm long; H. Stamen : filament-0.25 cm long, anther-0.1 cm long; I. Gynoecium; J. Fruit with vivipary -25 cm long; K. Floral diagram.

long, green, solid, glabrous, woody, flattened, 0.2 cm broad, pulvinous, lamina elliptic, obtuse, entire, tapering, 9.4 cm long x 4.4 cm broad, coriaceous, dorsiventral, dorsal side green, shiny in comparison to ventral side, glabrous, unicostate reticulate venation, vein inconspicuously found.

Inflorescence—2.0 cm long within leafy cluster, cyme, opposite decussate, ebracteate, peduncle develop from the leaf base, peduncle erect but bent to downwards, 1.0 cm long, glabrous, green, flattened, 2-4 peduncles in one shoot apex, peduncle terminate in 4-6 short branches, each branch having 2-3 flowers. *Flowers*—pedicellate; pedicel short, 0.1 cm long, terete, glabrous, green, bisexual, complete, regular, 0.7 cm long, 0.6 cm diam. rosaceous, pentamerous, minute, 10-12 flowers on single peduncle, *Sepals* - 5, polysepalous, superior, elliptic-lanceolate, entire, thick, green, acute, 0.45 cm long x 0.25 cm broad at base, persistent, after maturation it become like spine, tapering end, woody, equal length, upper surface same levelled, upper 0.6 cm in diam. *Petals*-5, polypetalous, valvate, petals white, thin, herbaceous, inconspicuous veinlets present, thick clavate appendages are present on apex, appendage short with globose head, oblong, entire, each petal slightly attached at mid-side to one another; alternate to sepal, equal length, upper surface same level shorter than calyx, superior, petal 0.3 cm long; *stamens*-10, free, superior, filament 0.25 cm long; terete, soft, white, dorsifixed, anther bilobed, 0.1 cm long, sagittate, introrse, inserted, filaments unequal length, 5 shorter and 5 tall, upper surface not to be same level, anther longitudinally dehisced, light yellowish; *carpels* 3, syncarpous, ovary inferior, elliptic globose, 0.15 cm long, 2-chambered, axile placentation, 2-ovules in each chamber, but 1 survive and 3 degenerate, 0.3 cm diam.; style-1, terminal, 0.3 cm long, terete, short, greenish white, glabrous, stigma-3, shortly tri-fid.

Pollination by moths, visitors during night time and mechanism explosive.

Fruit—capsule, 2.5 cm long, with 5-persistent calyx, after maturation it turns brown and rough surface with sharp apex; single seeded. Germination epigeal, viviparous.

Flowering and fruiting times are during April to September; but the viviparous germinated seedlings are also visible in the mother plant almost throughout the year. Main vegetative growth through the apical buds is significant during the monsoon months and continuing till the post-monsoons. Hypocotyle length is attaining upto 25 cm, with sharp apex, prominently ridged.

Common species are in the ridge tidal forests, both in inner and mid-estuaries in the Sundarbans. This species prefer the well drained silted up clayey loam and sandy soil; the other associated flora of *C. tagal* are *Excoecaria* sp., *Phoenix* sp., *Aegiceras* sp., *Aegialitis* sp.

Produces high calorific valued fire wood, fishing stalks and poles and cottage timbers; also used for fencing the river dykes and gardens. The wood is very hard and durable, not damaged by white ants, worms, wood-borers or not spoiled in water or direct sunlight for a long period. Bark produce high percentage of tannin for fishermen nets and leather industries. Very conspicuous species are in the mangrove habitats and protect the intertidal soil from erosion and add profuse biomass to the nature. Medicinally this species are important. Stem bark decoction used to stop haemorrhages; bark is said to be useful for malignant ulcers; also used for cottage and rural house construction.

Distributed widely in Sundarbans, Andaman & Nicobar Islands and other parts of the Indian mangove in East and West Coasts. Also distributed in the Old World Tropics, in the S. E. Asian countries, East Africa, Madagascar, Scychelles, Australia and Micronesia.

Table 58 : Comparative Morphology of the two species of *Ceriops* Arn.

Habits & Habitats	*C. tagal* Robinson	C. decandra Ding Hou
Habits	Small to moderate tree pyramidal, upto 8.0 m tall.	Shrub or small tree, diffused branched, 3 m- 5 m tall.
Leaf petiole	2.7 cm long.	2.5 cm long.
Lamina	Elliptic, 9.4 cm long, 4.4 cm broad.	Ovate elliptic, 9.5 cm ong, 5.5 cm broad.
Inflorescence	Within leafy cluster, pendulous, 2-4 peduncles on each shoot.	Within and below the leafy cluster, straight, 6-10 peduncles on each shoot.
Flower	0.7 cm long, 0.6 cm across, sepals 0.45 cm long, 0.25 cm broad, petals with three clavate appendages terminally, 0.3 cm long, filament 0.25 cm long, 5 short and 5 long, upper surface not same level, anther 0.1 cm long, ovary 0.15 cm long and 0.3 cm diam.	0.75 cm long, 0.60 cm across, sepals, 0.40 cm long, 0.15 cm broad, petals multiciliated terminally, 0.28 cm long, 0.08 cm broad, filament 0.14 cm long, anther 0.11 cm long, ovary 0.2 cm long and 0.2 cm diam.
Fruit	Capsule 2.5 cm long.	Capsule 3.0 cm long.
Hypocotyle	Upto 25.0 cm long, distinctly ridged with sharp end, green.	Upto 12.0 cm long, ridged with blunt apex, greenish yellow.

GENUS-4. KANDELIA Wight & Arnold Prodromus 310.1834.

Kandelia W. & A. a monotypic genus of the Tropics, in Old World Mangals.

Kandelia candel (L.) Druce is the only true mangrove species of the genus *Kandelia* W. & A. of Mangrove Rhizophoraceae, distributed in the South- East Asian mangrove communities. This species normally has no aerial roots, like other members of the Rhizophoraceae. But, the viviparous germinated seedling is much slender upto 45 cm long and tapering sharply towards the end.

4.1. *Kandelia candel* (L.) Druce Rep. Bot. Exch. Club Brit. Isles 1913 (3) : 420 : 1914.

Rhizophora candel L. Sp. Pl. 443. 1753. *Kandelia rheedii* Wright & Arnold, Prodromus 311. 1834.

Local Name (Bengali) : Goria

(Plate-89 and Fig. 9)

Medium evergreen tree, attain a height upto 7.5 m, diffused or deliquiscent branched. Bark greyish or reddish brown, smooth.

Aerial roots normally absent; but the underground roots are much branched and holding the soil firmly. Trunk bases occasionally conical with fused broom like roots.

Differentiated into nodes and inter-nodes, terete branches, nodes slightly ridged and flattened, brown, thin layer encircling the node, below it two leaf-scars and upper two peduncle scar present. Internodes-1.0 - 4.0 cm long alternately, solid

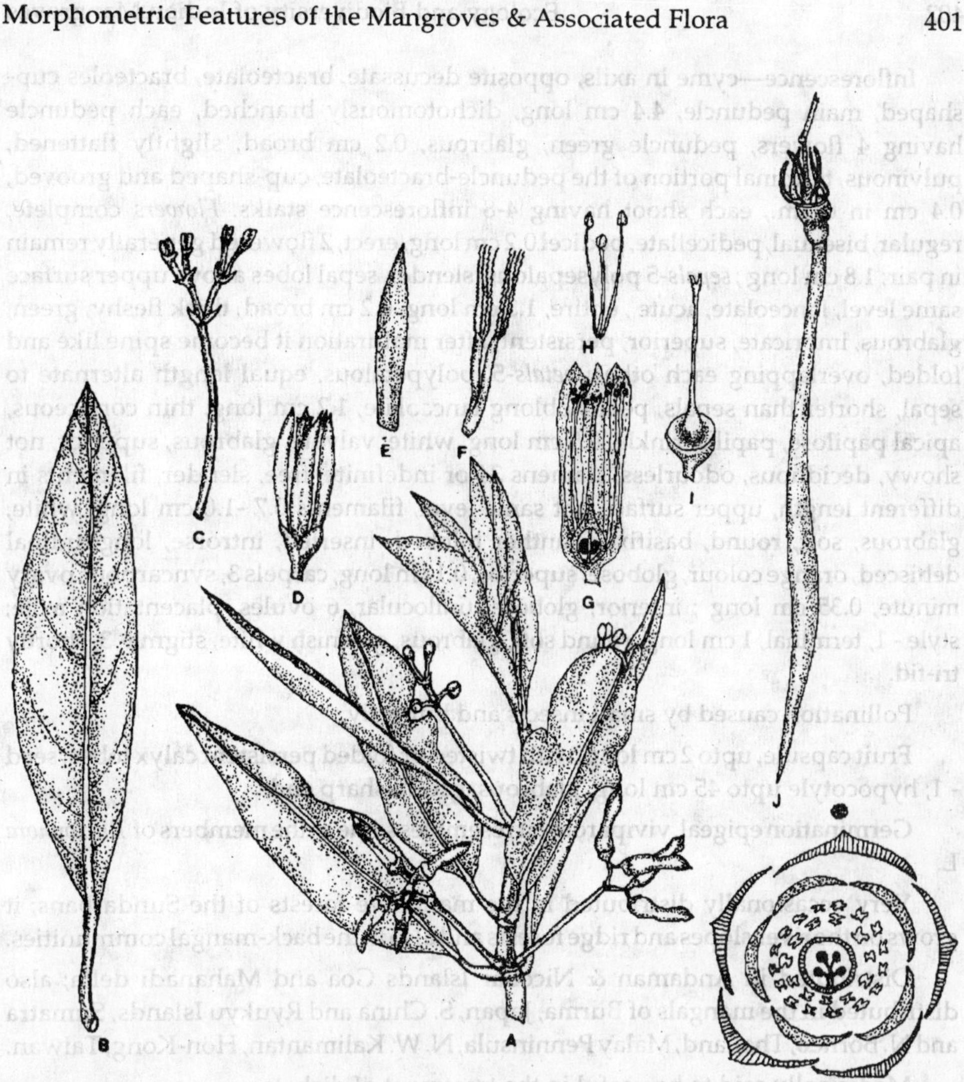

Fig. 9 : ***Kandelia candel*** (L.) Druce

A. Portion of flowering twig x 1/3. B. Leaf : petiole-2.0 cm long, lamina- 12.0 cm long x 4.0 cm width; C. Inflorescence - 6.0 cm long; D. Flower - 1.7 cm long; E. Sepal- 1.3 cm long x 0.2 cm width ; F. Petal- 1.2 cm long; G. L.S. of flower; H. Stamen : filament-1.0 cm long, anther- 0.1 cm long; I. Gynoecium; : ovary- 0.35 cm long, style- 1.0 cm long; J. Fruit with vivipary-40 cm long; K. Floral diagram.

glabrous, leaf-scars 0.4 cm diam.; upto 25.0 cm long shoot apex cover with leafy cluster.

Leaves opposite decussate, simple, stipulate, petiolate, petiole 2.0 cm long, terete, green, glabrous, pulvinous. Lamina 10.0 cm - 14.0. cm long x 4.0 cm - 5.0 cm broad; oblong-lanceolate, acute, entire, tapering, glabrous, coriaceous, slightly fleshy, dorsiventral, dorsal side dark green and slightly shining, unicosted reticulate venation inconspicuously.

Inflorescence—cyme in axils, opposite decussate, bracteolate, bracteoles cup-shaped, main peduncle, 4.4 cm long, dichotomously branched, each peduncle having 4 flowers, peduncle green, glabrous, 0.2 cm broad, slightly flattened, pulvinous, terminal portion of the peduncle-bracteolate, cup-shaped and grooved, 0.4 cm in diam., each shoot having 4-8 inflorescence stalks. *Flowers* complete, regular, bisexual, pedicellate, pedicel 0.2 cm long, erect, 2 flowered generally remain in pair; 1.8 cm long ; *sepals*-5 polysepalous, slender sepal lobes above upper surface same level, lanceolate, acute , entire, 1.3 cm long, 0.2 cm broad, thick fleshy, green, glabrous, imbricate, superior, persistent, after maturation it become spine like and folded, overlapping each other; *petals*-5, polypetalous, equal length alternate to sepal, shorter than sepals, petals oblong lanceolate, 1.2 cm long, thin coriaceous, apical papilose, papilae rinkle, 0.5 cm long, white, valvate, glabrous, superior, not showy, deciduous, odourless; stamens 34 or indefinite, free, slender, filaments in different length, upper surface not same level, filaments 0.7 -1.0 cm long, white, glabrous, soft, round, basifixed, anther bilobed, inserted, introrse, longitudinal dehisced, orange colour, globose, superior, 0.1 cm long; carpels 3, syncarpous ovary minute, 0.35 cm long ; inferior, globose, unilocular, 6 ovules, placentation axile; style - 1, terminal, 1 cm long, round soft, glabrous, greenish white, stigma- 3, shortly tri-fid.

Pollination caused by small insects and butterfly.

Fruit capsule, upto 2 cm long, with twisted or folded persistent calyx lobes, seed - 1; hypocotyle upto 45 cm long, glabrous, green, sharp end.

Germination epigeal, viviparous; it resembles same as the members of *Rhizophora* L.

Very occasionally distributed in the mangrove forests of the Sundarbans; it grows on the river slopes and ridge forests and also in the back-mangal communities.

Distributed in Andaman & Nicobar Islands Goa and Mahanadi delta; also distributed in the mangals of Burma, Japan, S. China and Ryukyu Islands, Sumatra and N. Borneo, Thailand, Malay Penninsula, N. W. Kalimantan, Hon-Kong, Taiwan.

Medicinally said to be useful in the treatment of diabetes.

FAMILY- II. *AVICENNIACEAE*
(Grey Mangrove Family)

Avicennia L. Sp. Pl. 1 : 110-1753 is the only and Type Genus of the Family-Avicenniaceae, which was treated initially under the family-Verbenaceae. Avicenniaceae was segregated as distinct family by van Tieghem (1898), based on the characteristics wood- and leaf-anatomy, aerial roots, axile placentation, crypto -or incipient viviparous germination and seedling morphology. The Genus - *Avicennia* L. was originally described (*l.c.*), based on the type species, *Avicennia officinalis* L. Sp. Pl. 1 : 110. 1753 and the generic epithet was coined after the name of the famous Arabian philosopher, mathematician, physician and naturalist, 'Iban Sina' (Tan & Keng, 1977).

GENUS - 5. AVICENNIA L. Sp. Pl. 1 : 110. 1753.

Tomlinson (1986) has described 10 species, *viz.*, *Avicennia officinalis* L., *A. marina* (Forsk.) Vierh., *A. alba* Blume, *A. lanata* Ridley, *A, eucalyptifolia* (Zipp. ex Miq.) Moldenke, *A. balanophora* Stapf & Mold. ex Mold., *A. germinans* (L.) Stearn, *A. africana* Palisot de Beauv., *A. bicolar* Standley and *A. schaueriana* Stapf. & Leechman ex Moldenke and has also seven taxonomical putative varieties or ecotypes. Out of these 10 distinct taxa, 6 species and few varieties belong to the Indo-West Pacific Islands, *i.e.*, mangals of East Africa and South East Asian mangrove ecosystems, the rest 4 species and few putative varieties or ecotypes also belong to New World, *i.e.*, Atlantic and East Pacific.

In the Indian coasts (both east the west) and Andaman & Nicobar Islands, only three species of *Avicennia* L. *viz.*, *A. officinalis* L., *A. marina* (Forsk.) Vierh. and *A. alba* Blume and few putative varieties are quite significant mangrove flora, commonly known as the grey or 'white mangals' (Naskar & Guha Bakshi, 1987 and Banerjee, *et. al.,* 1989).

All these species of the Genus —*Avicennia* L. are the true mangroves and mostly small to medium tree with profuse branching and dense canopy, occasionally shrubs. Tomlinson (1986) has categorised these *Avicennia* L. speices as 'Major Elements of Mangal' or 'Major Components of Mangals'.

Underground much spreading roots and aerial respiratory breathing pneumatophores or stilt roots are the characteristic features of the genus. Underground roots support mechanically the trunk/tree in the newly silted up, less-consolidated soil of the inter-tidal zone by developing cable roots and several lateral branches of the horizontal roots. Occasionally, stilt roots also support these trees mechanically. Aerial pencil like pneumatophores and several much branched, fibrous absorbing roots or nutritious roots are also developed from the spreading horizontal roots. These roots provide or support the root zone respiratory function and nutrient intake and other physiological function in these hyperspaline inter-tidal zones.

Both these apical and terminal meristematic tissues are significant for the shoot growth, which function all the seasons of the year, except less functional during the winter and drier seasons. During late winter their reproductive growth initiate. The resting shoot buds are well protected in the petiolar grooves. Lateral and terminal twigs are jointed and the base of the young twigs are with swollen, almost 4-angled, terete nodes; inter-nodes varies considerable depending upon the nutrient status of the soil phase, salinity of soil and water phases and nature of the shade of other plants in the strands. Their branches are developed acropetally and growth of the individual species is governed completely by the salinity and environmental factors, *i.e.*, rain fall, temperature, wind flow, humidity, tidal impact, shade of the other tree and biotic inter actions.

Leaves dense and decussately arranged, petiole length varies between 0.8 cm - 3.5 cm, grooved with marginal hairs; leaf-blades shining, leathery, somewhat fleshy; mid-rib prominent in the ventral side, glabrous and light to dark green; 2.5 cm - 12.5 cm long and 1.5 cm - 7.5 cm broad, ovate, narrowly elliptic to lanceolate, apex round, acute or slightly emerginate, attenuate, entire and slightly thickened.

Inflorescence long spicate or globose, panicles in the distal terminal and lateral branches, clustered terminally and subtended by the foliage leaves.

Flowers perfect, generally tetramerous, usually with 4-12 pairs flowers, decussately arranged, dense spicate to capitate units. Each flower base is subtended with one triangular bract and two bracteoles, imbricate and enclosed the base. *Sepals* -5, free, ovate, imbricate, covered by bracteoles; both calyx and bracteoles are persistent in fruit. *Corolla* tubular in the base and lobed in the apex, lobes erect or reflexed. *Stamens*-4, anther 0.1 cm-0.5 cm long, alternating with corolla lobes, dehiscing by longitudinal slits and adhere the dehisced pollen, pollination by insects. *Stamens* attached basally on corolla, slender, unequal filaments 0.2 cm - 0.3 cm long; anthers sessile or with short filament. Ovary completely hairy on the middle, superior, unilocular globose with style, stigma bilobed. Placentation axile, 4 ovules, pendulous from central stalk.

Fruit 1.0 cm- 4.0 cm long, globose capsule, 1- seeded, leathery surface, ellipsoid to flattened, ovoid on maturity, with a prominent or inconspicuous stylar beak; pericarp thin, green or grayish green. Seed solitary, testa thin, leathery; endosperm absent, two fleshy folded cotyledons in the ripe fruits; radicle blunt and exposed with tuft hairs; cotyledons purple occasionally.

Germination epigeal; incipient or crypto viviparous, *i.e.*, germination started in the mother plant itself but not like the members of mangrove - Rhizophoraceae.

The flowers of *Avicennia* spp. are comparatively smaller in size, as such small insects are the effective pollination agent. Broad corolla lobes act as the platform for the insects, especially the honey bee, *Apis dorsata*. These insects transmit the pollen from flower to flower. Sweet nectar is originated in the small glands of the corolla lobes and accumulated or stored in the corolla tubes. Honey bee, during its visit (for the sweet nectar collection) transmit the pollen from flower to flower and accelerate the process of fertilization. Inter specific hybridization, *viz.*, *A. alba* x *A. marina* has resulted the mixed characters and prominent in leaf morphology.

Seed dispersal is also a very important phenomenon and *Avicennia* spp. seeds or incipient viviparous germinated seedlings are dispersed through the tidal water flow. These seeds or seedlings can float in the water and their leathery testa protect them from water absorption, during floating. These floating seeds and seedlings can anchor and soon germinate in the silted up soil, where they settle; these seedlings grow very gregariously in the protected or sheltered bays.

Avicennia spp. are the common mangrove flora in diverse ecological conditions and different saline zones; these are common both in the Old World and New World Tropics, sub-Tropics and occasionally in Temperate Zone. For their varied distributional features and maximum salt tolerance ability, these *Avicennia* L. are cosmopolitan in their distributions. In India, particularly in the Sundarbans mangrove ecosystem only three species of *Avicennia* L. are present; all these three species are collected and studied in detail, in relation to their morphology, anatomical peculiarities, successions, adaptations and economic importances.

Key Characters for *Avicennia* L.

1a. Medium to tall tree; pneumatophores longer than other members of *Avicennia* L. Stem bark whitish, smooth, leaves obovate-oblong, round apically, not much silvery beneath; flowers 1.0 cm - 1.5 cm in across; fruit 2.5 cm - 3.0 cm. long with a short narrow beak almond shaped 1. *Avicennia officinalis* L.

1b Shrub or small to medium tree; pneumatophores shorter; stem bark dark or peeling; leaves lanceolate or linear with pointed apex; silvery or whitish beneath; flowers less than 0.5 cm in across; fruit with or without narrow beak (2)

2a (1b) Leaves lanceolate or linear, upto 12.5 cm long; ventral surface bright silvery; bark turning black with age; flowers dull yellow, less than 0.4 cm in across; fruit upto 4.0 cm long, conical, tapering towards the apical portion

2. *Avicennia alba* Blume

2b (1b) Leaves ovate-lanceolate; upto 11.0 cm long ; not much silvery or white like *A. alba* ; flowers orange yellow, 0.4 cm - 0.5 cm in across ; mature stem bark peeling like guava ; most common plant throughout the mangals

3. *Avicennia marina* (Forsk.) Vierh.

5.1. *Avicennia officinalis* L. Sp. Pl. 1 : 110.1753.

A. tomentosa Willd. Sp. Pl. 3 (1) : 395.1800.
Common Name : White Mangrove
Local Name (Bengali) : Jat ban, Jat bani, Jat byne
(Plate - 90 and Fig. 10)

Evergreen perennial tree, generally attain a height between 15.0 m and 20.0 m; bark smooth, whitish; deliquiscent, irregularly spreading branches, having dense crowned globose appearance on the intertidal river flats and ridge forests; trunk base 80 cm - 100 cm diam. in the normal full grown tree. Stem bark not scaly but occasionally papery peeling, grey to light coloured, lenticellate, not fissured, wood somewhat hard, brownish grey and strong.

Aerial roots well developed, pneumatophores numerous, spongy, pencil like erect, developed from the spreading horizontal (underground) roots; lenticels frequent, conspicuous, appeared on the outer corky surface; pneumatophores very often with blunt apex, little or without any secondary growth; reaching upto a height of 30 cm above the flat ground; number of the pneumatophores varies between 500-

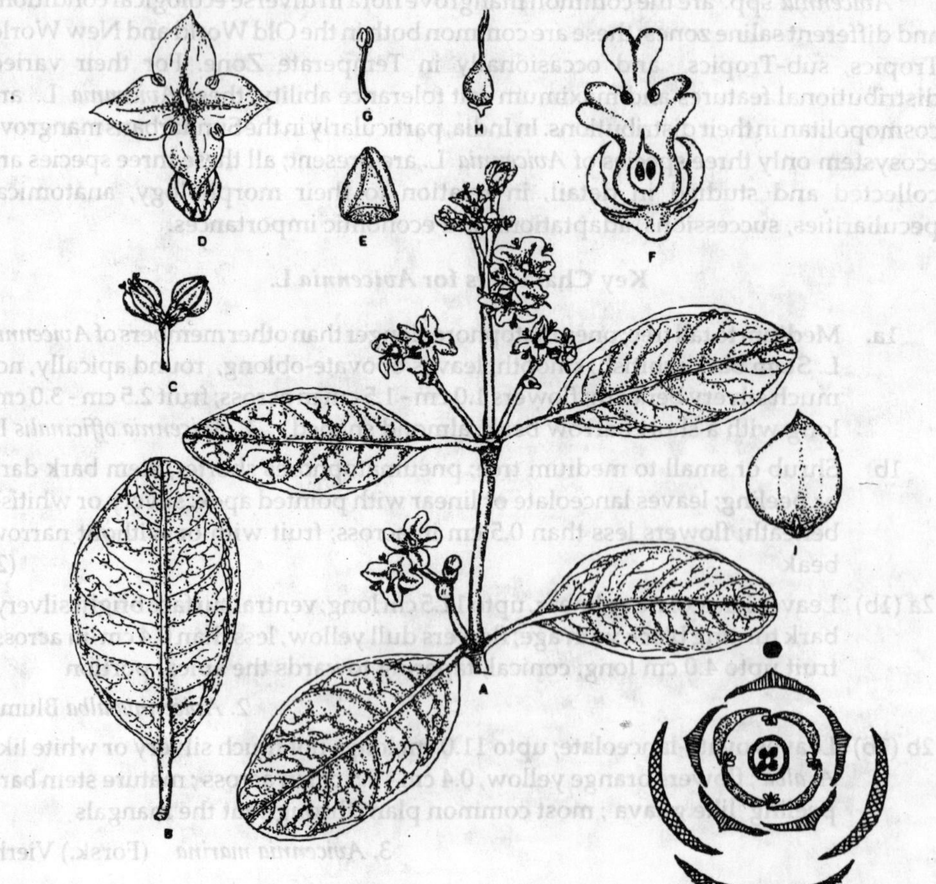

Fig. 10 : *Avicennia officinalis* L.

A. Portion of flowering twig x 1/3; B. Leaf: petiole-1.5 cm long, lamina-12.0 cm long x 5.5 cm width; C. Part of inflorescence; D. Flower- 1.0 cm long x 1.5 cm across; E. Sepal - 0.6 cm long x 0.3 cm width; F. L.S. of flower; G. Stamen : filament - 0.2 cm long, anther-0.15 cm long; H. Gynoecium : ovary-0.7 cm long, style - 0.7 cm long; I. Fruit with persistent calyx - 3.0 cm long x 2.8 cm broad; J. Floral diagram.

1200 or more average in an adult normal grown up tree; diameter 20.0 mm - 25.0 mm; these aerial breathing roots densely grow around the tree trunk, prop roots haphazardly grow in stressed condition and/or in the eroded river banks or river flats; the other type of aerial root are stilt roots, appear on the trunk base; older tree occasionally formed the insignificant trunk buttresses in the mature well developed soil.

Nodes of the shoots slightly ridged, internodes 7.0 cm-8.0 cm length in the apical branches, terete, glabrous; branching opposite decussate, solid.

Leaves opposite decussate, exstipulate, simple, cauline, petiolate, petiole 1.3 cm - 1.5 cm length, slightly flattened, base hairy, green, leathery, glabrous, pulvinous, lamina elliptic obovate, broadly ovate or obovate-oblong, entire, obtuse or occasionally rounded apex; curved, coriaceous, odourless, glabrous smooth or shining in dorsal side and brownish pubescent on ventral side, never glaucous white; lamina length 8.0 cm - 12.0 cm and 4.5 cm - 5.5 cm broad, unicostate, 7-10 reticulately venation conspicuously ; leaves turning black on drying.

Inflorescence compound spikes, peduncle upto 25.0 cm long, trichotomously panicles, upto 7.5 cm long, flower buds glabrous round brown ; flowers somewhat zygomorphic, sessile, dull yellow, 0.8 cm - 1.5 cm across, slightly foetid with 10-12 flowers in each branch, capitate globose, bracteate; bracts small with fimbriate margin, bracts 2 ciliate, concave with age, turn sessile and blackish.

Flowers complete, bisexual, regular, cyclic, hypogynous, rosaceous, erect, 0.5 cm - 1.0 cm long and 1.0 cm - 1.5 cm across in open flower; foetid smell and with two bracteoles. *Sepals* 5, polysepalous, imbricate, inferior, elliptical, 0.5 cm - 0.6 cm in length and 0.25 cm - 0.30 cm broad, entire, acute, coriaceous, persistent, pubescent. *Petals* 4, gamopetalous, inferior, imbricate, orange yellow, petals 0.6 cm - 0.8 cm in length and 0.5 cm - 0.6 cm broad, elliptical, entire, acute, curved in the tips, herbaceous, rosaceous, regular, deciduous, 4 lobes above, short tube from beneath, turning blackish on maturity. *Stamens* 4, epipetalous, 0.35 cm long, fertile, filament attached with the petal base, but anthers remain free, filament 0.2 cm long, round, white, glabrous, inferior, basifixed, extrorse, exserted, exceeding the anther; anther bilobed longitudinally dehiscent, 0.15cm long, yellow. *Ovary* superior, 0.7 cm long, elliptical or sligtly oblique, pubescent or with shortly hairy throughout, one loculed; style one, below the adaxial corolla or terminal, 0.15 cm - 0.75 cm long, round, glabrous, greenish white stigmas shortly or 3-fid, lanceolate, entire, glabrous.

Capsule fleshy, ovoid, opaque, green, densely hairy, 3.0 cm long and 2.8 cm broad, flattened, wrinkled at the base and with a short apical beak. Seed single, dark-green, completely occupy the fruit cavity, shining, pubescent, cotyledons large fleshy, green or purple, partially germinate in the mother plant.

Germination epigeal, crypto-viviparous, leaves of the seedling plants are acute in the initial stages of development; grow throughout the year.

Flowering and fruiting times in the Sundarbans are March to early October; seedling gregariously grow during late monsoon, *i.e.*, October to December.

Average size of *A. officinalis* L. is the largest among three *Avicennia* spp. commonly grow in the Sundarbans mangrove forests. This species very often grow in the intertidal, well developed silted up flat lands or in the ridge forests and very occasionally form pure strand vegetation.

Distributed throughout India, both in the East and West coasts and Andaman & Nicobar Islands; also in Bangladesh, Sri Lanka, Burma, Malaya, Java, China, Phillipines, Australia and other S.E. Asian mangrove forests.

A. officinalis L. is comparatively less frequent among these three species of *Avicennia* L., which normally grow in the mid-estuary and its salinity tolerance is very high. Seeds and seedlings are dispersed with the help of tidal flow and the new seedling plants grow very gregariously in the silted up clay soil in the shelter places. The rate of seed production of these *Avicennia* L. is profuse, which float on the tidal water.

Experimental Plantation : Crypto-viviparous germinated seeds/seedlings were collected and transplanted in the non-saline soil and flooded with fresh tube-well water, regularly. These transplanted seedlings of *A.officinalis* grow and survive more than 2 years in the freshwater and non-saline soil and the air-breathing pneumatophores were also developed from the spreading underground roots.

Medium quality timber and largely used for domestic purposes, fuelwood and for the paper pulp; kernals bitter taste but edible; green fruits medicinally used as poultice for boils.

5.2. *Avicennia alba* Blume, Bijdr. Fl. Ned. Ind. 14 : 821. 1826.

A. marina (Forsk.) Vierh. var. *alba* (Blume) Bakh. Bull Jard. Bot. Buitenz, Series 3(3) : 205, 207-219. 1921.

Chromosome *No.* $2n = 66$ (Raghaban & Arora, 1958)

Common Name : White Mangrove

Local Name (Bengali) : Kalo Ban or Kal Ban or Kala Byne or Kali Bain (Plate-91 and Fig. 11)

Evergreen medium tree, perennial, generally 10-12 m high (in Sundarbans), but occasionally attain the height upto 20 m, deliquiscent, much branched, irregularly spreading, pale or blackish, woody trunk, in the base the diam. ranges between 15 cm and 30 cm in an average full grown tree. Bark greyish black or somewhat brown, turning dark with age, frequently lenticellate and slightly roughed but not fissured, often sooty mould.

Underground roots much branched, divergent, spreading horizontally from the depth of 0.3 m - 2.5 m in the silted up soil and goes upto 15 m away encircling the trunk base. Several pencil like aerial breathing roots, *viz.*, pneumatophores developed from the superficial horizontal spreading roots; generally pneumatophores are unbranched with pointed, narrow apex, but occasionally branched on injury on the apical zone. Cortical layer spongy, attain the height upto 30.0 cm and diam. upto 1.5 cm, with little or without any secondary growth. Several much branched fibrous, nutritious, anchoring roots are also developed from the horizontal roots and average 800-1500 pencil like aerial breathing pneumatophores are arranged linearly along with the horizontal roots in a full grown tree.

Leaves opposite, decussate, cauline, simple exstipulate, petiolate, petiole 1.5 cm - 2.0 cm long, pulvinous, glabrous, ventral or abaxial portion of the base frequently pubescent, lamina lanceolate, 10.5 cm - 12.5 cm in length and 1.5 cm -4.5 cm broad, entire, acute, cuneate base, whitish green glabrous or pale green in the dorsal side and bright silvery white or silvery grey with papillose in the ventral side, coriaceous, odourless, unicostate, reticulate venation conspicuously, leaves dense in branches and terminal branches often pendulous.

Inflorescence spicate raceme, long spikes 2.5 cm -5.0 cm long, bracteate, bracteole 0.25 cm long, branched, branching opposite decussate, acropetally arranged in clustered 8-32 flowers, dull yellow, flowers sessile.

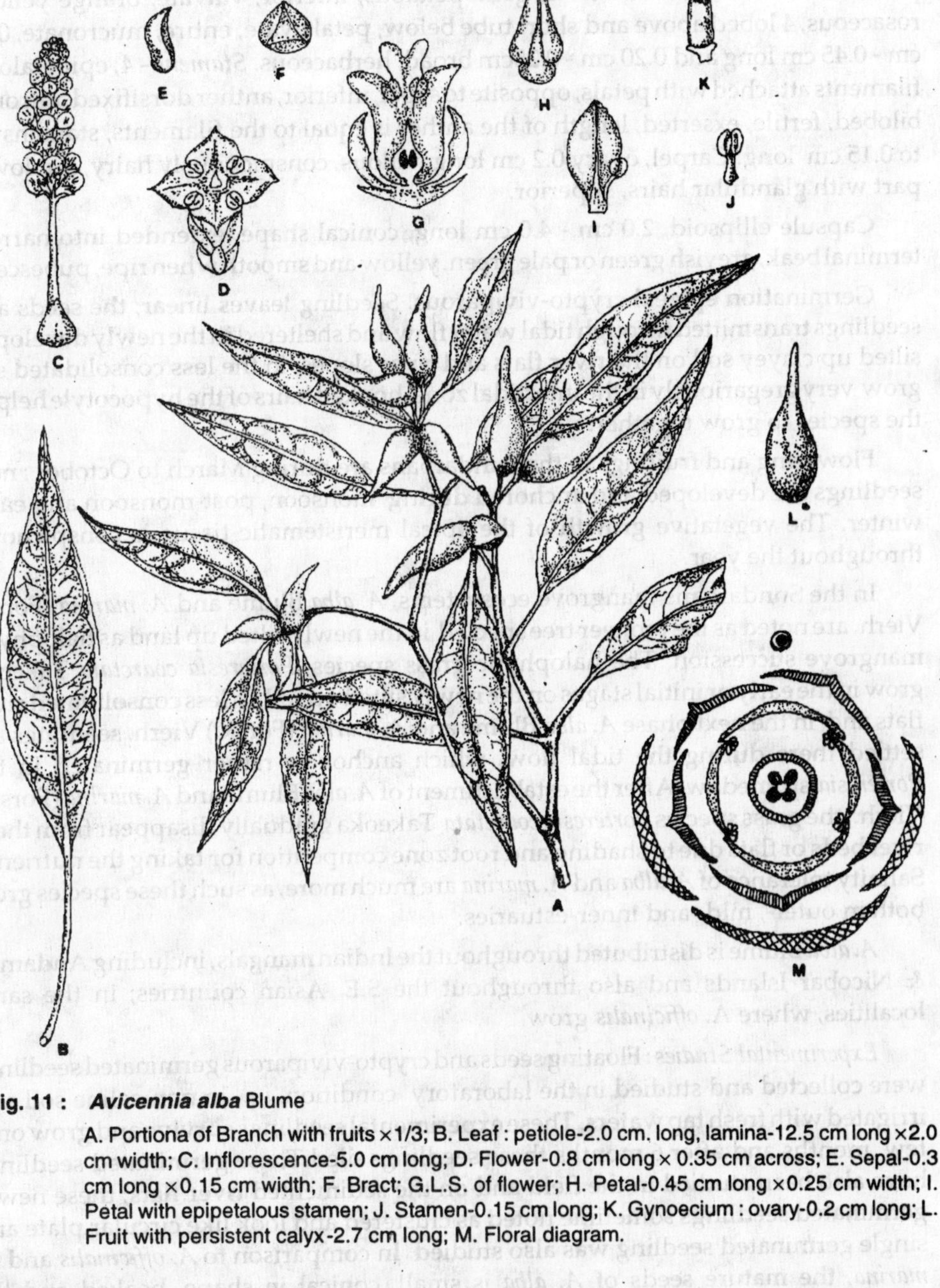

Fig. 11 : ***Avicennia alba*** **Blume**

A. Portiona of Branch with fruits ×1/3; B. Leaf : petiole-2.0 cm. long, lamina-12.5 cm long ×2.0 cm width; C. Inflorescence-5.0 cm long; D. Flower-0.8 cm long × 0.35 cm across; E. Sepal-0.3 cm long ×0.15 cm width; F. Bract; G.L.S. of flower; H. Petal-0.45 cm long ×0.25 cm width; I. Petal with epipetalous stamen; J. Stamen-0.15 cm long; K. Gynoecium : ovary-0.2 cm long; L. Fruit with persistent calyx-2.7 cm long; M. Floral diagram.

Bracteate, sessile, bisexual, regular, flower 0.5 cm - 0.8 cm long and 0.35 cm across, mild fragrant, cyclic, hypogynous, erect, small, one bract and two bracteoles.

Sepals 5, polysepalous, ovate, entire, mucronate, imbricate, coriaceous, pubescent, inferior, 0.20 cm - 0.30 cm long and 0.1 cm - 0.15 cm broad, green, rosaceous, persistent in the fruit. *Petals* - 4, gamopetalous, inferior, valvate, orange yellow, rosaceous, 4 lobed above and short tube below, petal ovate, entire, mucronate, 0.40 cm - 0.45 cm long and 0.20 cm - 0.25cm broad, herbaceous. *Stamens* - 4, epipetalous, filaments attached with petals, opposite to sepal, inferior, anther dorsifixed, introrse, bilobed, fertile, exserted, length of the anther is equal to the filaments, stamens up to 0.15 cm long. Carpel, ovary 0.2 cm long, villous, conspicuously hairy, but lower part with glandular hairs, superior.

Capsule ellipsoid, 2.0 cm - 4.0 cm long, conical shape, extended into narrow terminal beak, greyish green or pale green, yellow and smooth when ripe, pubescent.

Germination epigeal crypto-viviparous. Seedling leaves linear; the seeds and seedlings transmitted through tidal water flow and sheltered in the newly developed silted up clayey soil on the river flats and river slopes, in the less consolidated soil grow very gregariously in the intertidal zone, hooked hairs of the hypocotyle helped the species to grow together.

Flowering and fruitings in the Sundarbans are during March to October ; new seedlings are developed and anchored during monsoon, post-monsoon and early winter. The vegetative growth of the apical meristematic tissue is conspicuous, throughout the year.

In the Sundarbans mangrove ecosystems, *A. alba* Blume and *A. marina* (Forsk) Vierh. are noted as the pioneer tree specied in the newly silted up land as halophytic mangrove succession. The halophytic grass species, *Porteresia coarctata* Takeoka grow in the early or initial stages on the newly silted up soil, in less consolidated river flats and in the next phase *A. alba* Blume and *A. marina* (Forsk.) Vierh. seedlings are settled there during the tidal flow, which anchor or rather germinated in the *Porteresia* sp. medow. After the establishment of *A. alba* Blume and *A. marina* (Forsk.) Vierh., the grass species *Porteresia coarctata* Takeoka gradually disappear from these river beds or flats due to shading and root zone competition for taking the nutrients. Salinity tolerance of *A. alba* and *A. marina* are much more, as such these species grow both in outer-, mid- and inner-estuaries.

A. alba Blume is distributed throughout the Indian mangals, including Andaman & Nicobar Islands and also throughout the S.E. Asian countries; in the same localities, where A. *officinalis* grow.

Experimental Studies: Floating seeds and crypto-viviparous germinated seedlings were collected and studied in the laboratory condition, *i.e.*, in non-saline soil and irrigated with fresh tap waters. These experimental seedlings survive and grow only few months and after 6 months these seedlings died. These germinated seedlings have also been studied in the field and on the sedimented river flats; these newly germinated seedlings sometime noted as clustered and look like circular plate and single germinated seedling was also studied. In comparison to *A. officinalis* and *A. marina*, the mature seeds of *A. alba* is small, conical in shape, beaked and the cotyledonary leaves are also pointed or narrow in the apex.

Economically used as *A. officinalis* for its wood and timber.

5.3. *Avicennia marina* (Forsk.) Vierh., Denkschr. Akad. Wiss. Wien Math. Nat. 71 : 435. 1907.

A. intermedia Griff. Trans. Linn. Soc. Lond. Bot. 20 : 1. 1846.

Chromosome No. 2n = 36 (Subramanian, 1980)

Common Name : Grey Mangrove

Local Name (Bengali): Peyara Ban or Peyara Bani or Peyara Byne

(Plate - 92 and Fig.12)

Shrub to small tree, bushy, evergreen, perennial, generally 3.0 m - 8.0 m high (in the Sundarbans), occasionally attain the height upto 10m or more, deliquiscent, much branched, irregularly spreading; woody trunk, the base 15.0 cm - 20.0 cm diam. in an average full grown tree. Bark smooth, flaky, yellowish brown or light yellowish, occasionally lenticellate, but not fissured; peeling in patches like guava (*Psidium guajava* L.) in mature bark ; as such the local name for this plant is Peyara Bain (Local name of *P. guajava* L.).

Underground roots much branched, spreading horizontally, runs upto 15 m away from the trunk base. In the river banks and river flats, sometime the underground roots are exposed due to soil erosion. Pencil like pneumatophores arising upward from the horizontal spreading roots. The unbranched aerial pnumatophores arising upward from the horizontal spreading roots. The pneumatophores are slightly spongy, pointed narrow apex, attain the height upto 50 cm and the diameter is 0.80 cm. without any secondary growth. From the horizontal roots several anchoring and nutritive roots are developed. A full grown tree may form 1000-2000 such pneumatophores, which are arranged linearly along with the horizontal roots.

Leaves opposite decussate, exstipulate, simple, cauline, petiolate, petiole 1 cm long, slightly flattened, glabrous, pulvinous, lamina elliptic-oblong or elliptic-ovate, 8- cm -11 cm in length and 2.5 cm - 3.5 cm broad, glabrous, entire, acute, cuneate or tapering towards the base, coriaceous, odourless, smooth, shining in dorsal side and whitish pubescent in ventral side, unicostate, reticulate vein conspicuously.

Inflorescence globose, compound spike, peduncle long upto 10 cm infrequently branched, glabrous, round, condensed terminal cyme, flower heads capitate.

Flower—4 lobed, actinomorphic, condensed in terminal or axillary cyme, sessile, 0.5 cm - 0.9 cm long and 0.5 cm across in open condition, pale yellow or orange yellow, complete, bisexual, regular, cyclic, hypogynous, rosaceous, sweet scented. Bract and bracteoles are persistent in the fruits. *Sepals* - 5, polysepalous, imbricate, inferior, elliptical, 0.25 cm - 0.35 cm in length and 0.15 cm - 0.2 cm broad, entire, blunt, coriaceous, persistent, outer surface hairy, tomentose. *Petals* - 4, gamopetalous, inferior, valvate, rosaceous, orange yellow, lobes ovate, acute, hairy on the outer and glabrous within, 0.40 cm - 0.45 cm in length, regular and 0.25 cm - 0.30 cm width. *Stamens* - 4, epipetalous, 0.2 cm long, fertile, filament attached with the petal at its mouth, but anthers remain free, filament round, white, glabrous, basifixed, extrorse, exserted, anther bilobed, longitudinally dehisced, yellow. *Carpel* - 3, ovary superior, 0.3 cm long, oblong, style short, terminal, stigmas - 2, glabrous below but hairy above.

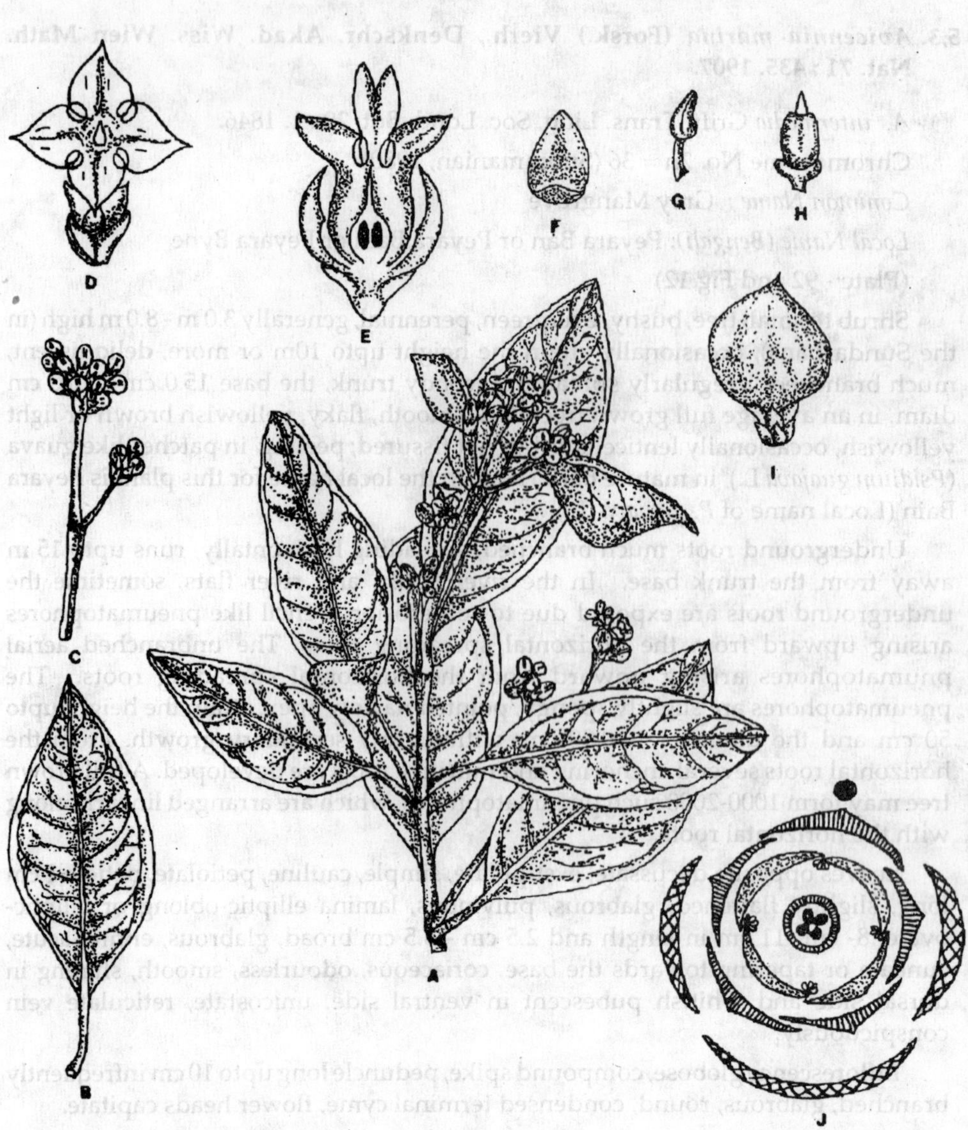

Fig. 12 : *Avicennia marina* (Forsk.) Vierh.

A. Portion of flowering twing × 1/3; B. Leaf : petiole-1.0 cm long, lamina-11.0 cm long × 3.5 cm width; C. Inflorescence; D. Flower-0.9 cm long × 0.5 cm across; E. L.S. of flower; F. Petal-0.45 cm long × 0.3 cm width; G. Stamen-0.2 cm long; H. Gynoecium-0.3 cm long; I. Fruit with persistent calyx-2.5 cm long; J. Floral diagram.

Capsule spherical, ovoid, 1.2 cm - 2.5 cm long, apiculate, greyish tomentose, compressed, shortly beaked, silvery grey green. Each capsule contain 1 seed, seed 0.2 cm - 0.3 cm in length, long and flattened with thin testa. Incipient or crypto-viviparous germination.

Germination epigeal, crypto-viviparous, Seedling hypocotyle hairy only at its base, in the region of short radicle.

Flowering and fruiting in the Sundarbans during March to October. Active growth period is early monsoon to late monsoon.

Wood is used as domestic timber, fuelwood and timber; bitter aromatic juice is said to be useful as an abortifacient.

These species, *A. marina* Vierh. grows almost throughout the Sundarbans mangrove forests, along with the other species of *Avicennia* L. It is the pioneer species in the halophytic adaptation; grow in the intertidal zones and on the mangrove ecosystems, almost everywhere. Distributed throughout in India and Andaman & Nicobar Islands.Grow on the extreme South Australia and Japan. These species can stand in wide salinity range and diverse temperature regime.

Morphometric Differences Among the Three Species of *Avicennia* L.

These three species of *Avicennia* L. are common in the mangrove forests of the Sundarbans and are very much distinct in regard to their shape, size and adaptations in particular. *A. officinalis* L. is the largest tree, among the three species and its leaf, length of trunk and diameter, pneumatophores, flowers, fruits and seeds are largest in comparison to the other two species. *A. officianlis* L. normally grow on the more mature soil and on the upper strata or stretches in the ridge forests. But, *A. marina* Vierh. is the smallest plant among these three species, though its flowers, fruits have shown the intermediate in shape, size and structure. *Avicennia* spp. have wide salinity tolerance and *A. marina* has the highest salinity tolerance among these three species. Both these species prefer newly silted up river flats or river slopes. But, *A. alba* Blume is intermediate in size among these three species; flowers and fruits are smallest in comparison to the other two species. This species also grow in the similar habitat like *A. marina* Vierh.

For these intermediate flower and fruit shape and size or structures, sometimes *A. marina* was considered as the hybrid between these two species *viz.*, *A. alba* Blume and *A. officinalis* L. (Ibrahim & Ahmad Shah, 1989). But based on the thorough morpho-anatomical studies, it is now clearly noted that these three species are distinct and this view also supported by Tan & Keng (1969) after thorough studies among these three above mentioned species of Malayasian region. (Table-59).

Fruit and Seed Morphology Among These Three Species of *Avicennia* L.

Among these three species, the fruits of *A. officinalis* L. are largest, flattened, terminated into a short beak, much hairy, pale green or greenish, leathery texture; while the fruits of *A. alba* Blume are conical, terminated into long beak, smallest in size, not flattened and blackish green. The fruits of *A. marina* Vierh. are intermediate in respect of size and shape among these three species; ovoid to spherical, slightly flattened, slightly or short beaked.

Fruits are one chambered, single seeded, seed dicotyledonous, showing peculiar morphological cotyledonary appearances. Two cotyledons are not to be placed side by side, but the cotyledons become folded; one folded cotyledon cover another completely and remain open in one longitudinal suture. Within the base of the two cotyledons embryo is placed; plumule remains inside, while radicle generally situated outside the cotyledons, but within fruit coat in case of mature seed. Radicle blunt, much hairy, shows incipient vivipary; after falling on the ground the fruit coat rupture and the cotyledons become loosened. Seed at fast attached with the substratum, the newly silted up sedimented soil on which it fall or reach after floating. Radicle soon develops roots and simultaneously anchor the soil; since it permanently settle there.

Nomenclatural Peculiarities Among these Three Species of *Avicennia* L.

Carolus Linnaeus (1753) described the Genus — *Avicennia* L. based on the Asian (Type) species, *A. officinalis* L. This specific epithet was derived after the word *'officinal'*, means medicinal use of these trees and for *A. marina* Vierh. the specific epithet is also derived from the *'marine habitat'* flora. The specific epithet for *A. alba* Blume is derived from the white and silvery ventral surface of the leaf.

The local (Bengali) name 'Jat Bani' is made for the bigger sized true and common Bani tree of the Sundarbans mangrove forests *A. officinalis* L. ; the Bengali name 'Peyara Bani' comes from the Peyara (Guava) tree like bark / peeling trunk and lastly; the 'Kale Bani' (Black Bani) is also comes from the black mature trunk bark of the species *A. alba* Blume.

Besides all these three true mangroves, *A. marina* var. *acuitissima* Stapf & Moldenke ex Moldenke (1940) has also reported by Tomlinson (1986) from Bombay and other Indian workers have also reported this species from the other mangrove ecosystems of India, including the Sundarbans or Ganga delta (Banerjee, *et. al.*, 1989).

These leaf structures among these three species of *Avicennia* L. are largely variable, *viz.*, *A. marina* with acute lanceolate leaves, *A.officinalis* with obtuse leaves, oval in shape and *A.alba* leaves are long, linear and elongated acute tip. But during field investigations a great variation have observed among these leaves structure and size. These variations are, *A. marina* with elongated but acute leaf, *A. officinalis* with ovoid but acute leaves. All these variations may lead to confusion among these species and it confer the possibility for inter specific hybridization among these species.

Afforestation of *Avicennia* spp. in the Sundarbans

During the recent time, large scale afforestation programmes of the mangrove plants in the tidal river-flats, river-slopes have been undertaken by the Forest Departments, Sundarbans Development Board, different NGOs and other organisations. During the last twenty five years, these Sundarbans mangrove forests have received much attention and an alround consciousness have been developed. As such, several large scale afforestation programmes have been undertaken in these tidal river-flats, river-banks and river-beds, as well. In most of these cases, the common mangrove species, *Avicennia alba, A. marina* and sometimes *A. officinalis* seedlings have been transplanted for having their wide salt tolerance, variable tidal

Table 59 : Comparative Morphology of the three Species of *Avicennia* L., commony grow in Sundarbans

Characters	*A. alba Blume*	*A. marina Vierh.*	*A. officinalis L.*
Habit	Medium tree, reaching upto 15 m height.	Medium tree, 5-8 m high, but very bushy appearance.	Medium to tall tree, 15-20 m high, globose.
Root (Aerial)	Spongy, 20-30 cm high and girth 9-10 mm thick and pencil like pneumatophores with pointed apex.	Spongy, same as A. alba but light yellowish, 30-50 cm high and girth 6-8 mm thick pneumatophores.	Corky, blunt apex, 25-30 cm high and girth 1.0 m - 1.5 cm, distinctly lenticellate pneumatophores; occasionally stilt/proproots in stresses.
Stem Bark	Greyish black or somewhat brownish.	Yellowish brown or light yellowish with guava like peeling.	Grey to light coloured, whitish.
Leaf	Lanceolate, lamina 10.0 cm- 12.5 cm long x 1.5 cm - 4.5 cm width, petiole 1.5 cm - 2.0 cm long, acute apex or acuminate.	Ellipitc-oblong or elliptic ovate, petiole 0.8 cm - 1.0 cm long, lamina 8 cm - 11 cm long x 2.5 cm- 3.5 cm width, acute apex.	Ovate-oblong, petiole 1.3 cm- 1.5 cm long, lamina 8 cm - 12 cm long x 4.5 cm- 5.5 cm width obtuse apex.
Inflorescence	Raceme, compact spike : peduncle upto 5 cm long.	Compound spike, termianl capitate : peduncle upto 10.0 cm long.	Compound and much branched spike : peduncle upto 25 cm long.
Flowers	Small, 0.5 cm - 0.8 cm long, 0.25 cm - 0.35 cm across., Sepal 0.2 cm - 0.3 cm long x 0.1 cm - 0.15 cm width petal orange yellow, 0.4 cm - 0.45 cm long x 0.20 cm- 0.25 cm width, stamen upto 0.15 cm long, ovary dumble shaped, 0.20 cm long.	Medium 0.5 cm - 0.9 cm long, 0.4 cm - 0.5 cm across, sepal 0.25 cm - 0.35 cm long x 0.15 cm- 0.20 cm width petal light orange-yellow, 0.4 cm - 0.45 cm long x 0.20 cm - 0.30 cm width stamen upto 0.20 cm long, ovary dumble shaped but slightly flattened, 0.30 cm long.	Comparatively large, 0.5 cm - 1.0 cm long, 1.0 cm - 1.5 cm across, sepal 0.5 cm - 0.6 cm long x 0.25 cm- 0.30 cm width, petal yellow, 0.6 cm- 0.8 cm long x 0.5 cm - 0.6 width cm stame upto 0.35 cm long, ovary conical shaped, 0.70 cm long.
Fruit	Conical, extended with terminal beak upto 2.7 cm long and upto 1.0 cm width.	Spherical to ovoid, with short apical beak, flattened upto 2.5 cm long and upto 2.0 cm width.	Flattened with short apical beak upto 3.0 cm long and upto 2.8 cm width.
Seed	2.5 cm long and 0.8 cn width.	2.2 cm long and 1.7 cm width.	2.75 cm long and 2.5 cm width.

inundation preferences, easy seeds and seedlings availability and comparatively rapid growth rates. For these reasons, the *Avicennia* spp. plantations have turned or formed very dense and artificial green forest belts in and around the Sundarbans rivers and creek sides. But, for insignificant growth rate, the other mangrove species are very scarce in these artificial forest zones. These artificial afforested forest zones may serve some important functions, but the question is the question is can these monotypic forest fulfil the requirements of these natural multi-species mangrove forests.

FAMILY—III. *SONNERATIACEAE*

Sonneratia Linn. f. (Suppl. Sp. Pl. 38. 1781, *nom cons.*) is the true mangrove Genus under the family-Sonneratiaceae. This genus was treated previously under the family-Lythraceae, for having their similar seedling structures, 4-angled branches, opposite leaves, sepals connate in calyx with free, petals few or rarely absent, stamens indefinite, ovary many celled with many seeded fruits and exalbuminous like *Lythrum* L. Other non-mangrove genus of the family—Sonneratiaceae is *Duabanga* Ham. Tomlinson (1986) has described only 5 species under the genus-*Sonneratia* Linn. f., *viz.*, *S. apetala* Buch. Ham., *S. griffithii* Kurz, *S. caseolaris* (L.) Engler., *S. alba* J. Smith and *S. ovata* Baker. The members of this genus are distributed in the mangrove habitats of the Old World Tropics, particularly in the Indo-West Pacific region, including North Australia. In the Indian sub-continent, the first 4 species, *i.e.*, except *S. ovata* Baker are very often distributed throughout the mangrove ecosystems (Backer & van Steenis, 1951). According to Banerjee *et. al.* (1989) *S . alba* J. Smith is only distributed in the Andaman & Nicobar Islands and West Coast of India. However, the rest three species were collected and studied from the Sundarbans (Ganga delta) mangrove ecosystem of India.

GENUS-6. SONNERATIA Linn. f. Sp. Pl. 38. 1781, *nom cons.*

All these 5 species, described by Tomlinson (1986) under the genus- *Sonneratia* Linn. f. are distributed from East Coast of Africa to South East Asia, Pacific Islands, North Tropical Australia, particularly in Micronesia, Melanesia, Malaysia, within the Tropical ranges. Out of these 5 species, only 3 species were collected from the Sundarbans during the present studies, *i.e.*, (1) *S. caseolaris* (L.) Engler— an inner estuarine Mangrove species; (2) *S. apetala* Buch. Ham. distributed throughout the Sundarbans and (3) *S. griffithii* Kurz is the mid-and outer extuarine species of the Sundarbans.

All these three true mangrove species are having conspicuously developed pneumatophores, upto 3 m high (Tomlinson, 1986) and salt tolerance characteristics. These species are lacking the viviparous or incipient viviparous germination mechanisms or abilities, like other true mangroves.

These intertidal mangrove species have dark, but dense leaves, which attract the fire fly during night; as such, it is commonly known as the *'fire fly mangroves'*.

Members of this genus are studied from the Sundarbans mangrove ecosystems are small to medium trees, attain the height upto 15 m with deliquiscent branching and dark, dense evergreen leaves. Branches are woody, occasionally pendulous and bushy; trunk base of the average mature full grown trees varies between 20 cm and 30 cm in girth. Bark is thin or thick on maturity, brownish. Underground roots are much spreading, horizontally cable roots and erect, slender conspicuous; aerial breathing roots pneumatophores also come out from the horizontal spreading roots; pneumatophores are tapering towards the end, normally 1.0 m-1.5 m high (occasionally upto 3.0 m high) and its girth attain upto 8.0 cm diameter. For having the secondary growth, the girth of the pneumatophores occasionally attain upto 10 cm or more, though initially it is greenish and pale-brown; on maturity-spongy, outer layer with thin peeling in the outer surface. Leaves simple, opposite decussate, alternate, exstipulate, cauline, short petiolate, petiole 0.5 cm - 1.0 cm long, flattened, glabrous, lamina ovate, elliptic, oblong, obovate, 5.0 cm - 12.0 cm long x 1.5 cm - 5.5 cm width, blunt or round, entire, fleshy, glabrous, dorsiventral, shining in both surface, distinct mid-rib, unicostate reticulate venation inconspicuously. Inflorescence terminal in apical or lateral branches, solitary cyme. Flowers ebracteate, pedicellate, pedicel 6.0 cm long, jointed, showing differentiation into nodes and internodes, bisexual, complete or incomplete, actinomorphic, 1.2 cm - 6.0 cm long and 1.5 cm - 7.0 cm across, perigynous; sepals 4-8 lobed, valvate, persistent in fruit on oblong, elliptic, 1.5 cm - 4.0 cm long x 0.8 cm - 1.5 cm width, flattened, green; petals 6 or occasionally absent (*S. apetala* and *S. griffithii)* or vestigeal in nature, thin deciduous, herbaceous, not showy; stamens many, exserted, introrse, filament long, glabrous, terete, upto 4.0 cm long; anther bilobed; ovary multilocular with many ovules, axile placentation, 0.6 cm - 3 cm long, upto 6 cm long, calyx persistent in the fruit, stigma capitate. Flowers are mostly ephemeral and expand in Sunset.

Pollination caused by the small birds, visitors for collect of nectars. Bats, *Macroglossus minimus* and Hawk moths, Sphingidae are the common visitors (Tomlinson, 1986).

Fruits berry, green, stalked, flattened or globose with persistent style, seeds 4-8 per fruit or numerous, generally tri-angular, light brown on maturity, exalbuminous.

Bat, *Macroglossus* sp. very often eat the sour fruits and help to disperse the seeds in distant places and also through the tidal current fruits are dispersed place to place. It has also reported that the carnivorous fish, *Pungasius pungasius* ate the fruits of *A. apetala* Buch. Ham. Though its consequence was not studied too.

Germination epigeal, not viviparous or incipient viviparous.

The main flowering and fruiting times are February to October, though *S. caseolaris* (L.) Engler is very often seen almost throughout the year with either flowers or fruits; these cover all the seasons of the year. Main vegetative growth observed during monsoon to post monsoon months, when the soil and water salinity come down. These species are mostly grown in the intertidal mangrove flats or ridge forests, just behind or next strata of the habitats of *Avicennia* spp. These species occasionally grow as the pioneer trees in the newly silted up river-flats.

These three species of the genus —*Sonneratia* Linn. f. can be distinctly separated with their distinct characters of fruits, leaf morphology and habitat peculiarities or on their distributions record.

Key Characters of *Sonneratia* Linn. f.

1a Small to medium tree in the mid-and outer-estuary; leaves narrow, tapering towards the apex, shining green, with thick waxy cuticle; apical and lateral branches pendulous; flowers on the end of the branches, calyx - 4 ; apetalous, fruits clustered in branch, like *Areca* nut clustered in the pendulous branches; green 1.5 cm - 2.0 cm diam.

1. *S. apetala* Buch. Ham.

1b Small to medium tree, grow either on the inner or on the mid-estuary; leaves broad; elliptic-oblong, ovate or obovate, round apex, shining green, waxy, apical and lateral branches not pendulous like the above and size of the flower much bigger than the above; mature fruits globose with long pointed persistent stigma, greenish, leathery with the conspicuous pointed calyx lobes (2)

2a(1b) Leaves elliptic-oblong or oval, obovate, petioles conspicuous but short, midrib red at the base., vein inconspicuous but not prominent, persistent calyx 6 lobed, petals linear lanceolate, dark red to reddish, inner side greenish or yellow, stamens numerous, white.

3. *S. caseolaris* (L.) Engler

2b (1b) Leaves obovate, petiole insignificant with greenish mid-rib, veins conspicuous prominent in dorsal surface; calyx lobes 4-8, smooth throughout, not ribbed, flowers apetalous, stamens numerous, white, species of mid- or outer-estuary. 2. *S. griffithii* Kurz.

5.1. *Sonneratia apetala* Buch. Ham. Syems, Embassy Ava 3 : 477. 1800.

Kambala apetala Rafin. Sylv. Tell. 19-1838. *Blatti apetala* O.K. Rev. Gen. 1 : 238. 1891. Chromosome No. *2n - 18* (Raghaban, 1959) and *2n* = *24* (Sidhu, 1961)

Common Name : Fire fly Mangrove

Local Name (Bengali) : Keora or Tak-Keora

(Plate - 93 and Fig. 13).

Small to medium tree, upto 15 m high, crown large, deliquiscent branching. Branches or twigs pendulous or drooping, from distance look very much similar to *Eucalyptus* sp. ; trunk base in an average full grown tree ranges between 20 cm and 50 cm diam.; branches not dense but loosely arranged downward. Bark thin, light brown, irregularly fissured.

Pneumatophores conspicuous, upto 1.5 m high with secondary thickening and upto 8.0 cm diam, arising from long spreading horizontal roots, spreading around—upto 15 m from the base of the trunk; aerial breathing pneumatophores are hard, corky and scaly outer surface; generally 200-500—pneumatophores have developed per mature tree.

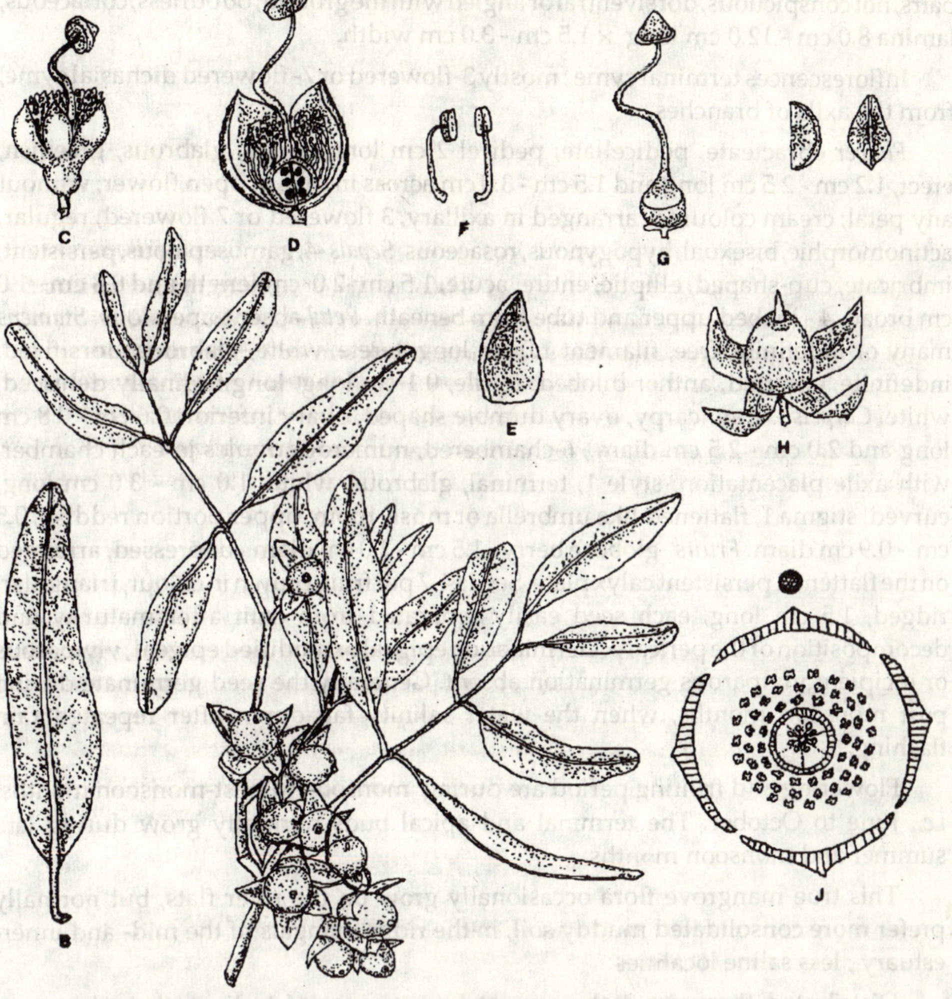

Fig. 13 : ***Sonneratia apetala*** Buch. Ham..

A. Portion of branch with fruits x 1/3; B. Leaf : petiole - 0.5 cm long, lamina - 12.0 cm long x 3.0 cm width; C. Flower- 2.5 cm long; D. LS. of flower; E. Sepal - 1.5 cm long x 0.8 cm width; F. Stamens : filament- 1.0 cm long., anther - 0.1 cm long; G. Gynoecium : ovary - 0.6 cm long, style - 1.4 cm long, stigma - 0.5 cm long; H. Fruit with persistent calyx; I. Seeds - 1.3 cm long; J. Floral diagram.

Internodes 3 cm - 4 cm length, jointed, easily separated from each other; branches smooth, not furrowed.

Leaves opposite decussate, exstipulate, simple, cauline, petiolate, petiole 0.5 cm long, pulvinous, glabrous, lamina narrow gradually, tapering towards the apex or oblong-lanceolate, glabrous, shining, entire, obtuse, cuneate, main nerve 10-12 pairs, not conspicuous, dorsiventral or angled with the ground; odourless, coriaceous; lamina 8.0 cm - 12.0 cm long × 1.5 cm - 3.0 cm width.

Inflorescences terminal cyme ; mostly 3-flowered or 7- flowered dichasial cyme, from the axils of branches.

Flower ebracteate, pedicellate; pedicel 2 cm long, terete, glabrous, greenish, erect, 1.2 cm - 2.5 cm long and 1.5 cm - 3.0 cm across in case of open flower; without any petal; cream coloured, arranged in axillary, 3 flowered or 7 flowered; regular, actinomorphic, bisexual, hypogynous, rosaceous. *Sepals* -4, gamosepalous, persistent, imbricate, cup-shaped, elliptic, entire, acute, 1.5 cm -2.0 cm length and 0.8 cm - 1.0 cm broad, 4 - lobbed upper and tube from beneath. *Petal* absent-apetalous. *Stamens* many or indefinite, free, filament 1.0 cm long, terete, white, glabrous, dorsifixed, indefinite, exserted, anther bilobed, fertile, 0.1 cm long, longitudinally dehisced, white. *Carpels* - 4, syncarpy, ovary dumble shaped, ovary inferior, 0.6 cm - 0.8 cm long and 2.0 cm - 2.5 cm diam., 6-chambered, numerous ovules in each chamber, with axile placentation; style 1, terminal, glabrous, white, 1.0 cm - 3.0 cm long, curved, stigma 1, flattened like umbrella or mush-room, upper portion reddish, 0.5 cm - 0.9 cm diam. *Fruits* globose berry , 1.5 cm - 2.0 cm diam., depressed, arranged on the flattened persistent calyx plate. Seeds 4-7 per fruit, brown in colour, triangular, ridged, 1.5 cm long, each seed easily separated from fruit after maturity and decomposition of the pericarp. Germination epigeal or modified epigeal ; viviparous or incipient viviparous germination absent. Generally the seed germinate during post monsoon months, when the water salinity fall down after repeated rain flashing.

Flowering and fruiting period are during monsoon to post-monsoon months, *i.e.,* June to October. The terminal and apical buds normally grow during late summer and monsoon months.

This true mangrove flora occasionally grow on the river flats, but normally prefer more consolidated muddy soil, in the ridge mangals of the mid- and inner-estuary ; less saline localities.

Distributed throughout the east and west coast of India and Andaman & Nicobar Islands; in Bangladesh, Sri Lanka, Burma and other mangrove areas of the S.E. Asian Countries, an Old World Species.

Wood medium quality timber and fruits cooked as condiment.

6.3. *Sonneratia griffithii* Kurz., J. Asiat. Soc. Bengal 40 (2) : 56. 1871.

Sonneratia acida var. *griffithii* King. J. Asiat. Soc. Bengal 67 (2) : 11.1998.

Common Name : Mangrove Apple

Local Name (Bengali) : Ora

(Plate - 94, Plate - 95 and Fig. 14)

Medium tree, upto 15 m high, deliquiscent spreading branching and occasionally drooping branched, evergreen, woody, erect, perennial. Bark deep brownish or pale brown, young tree hairy smooth bark but after maturity it becomes fissured.

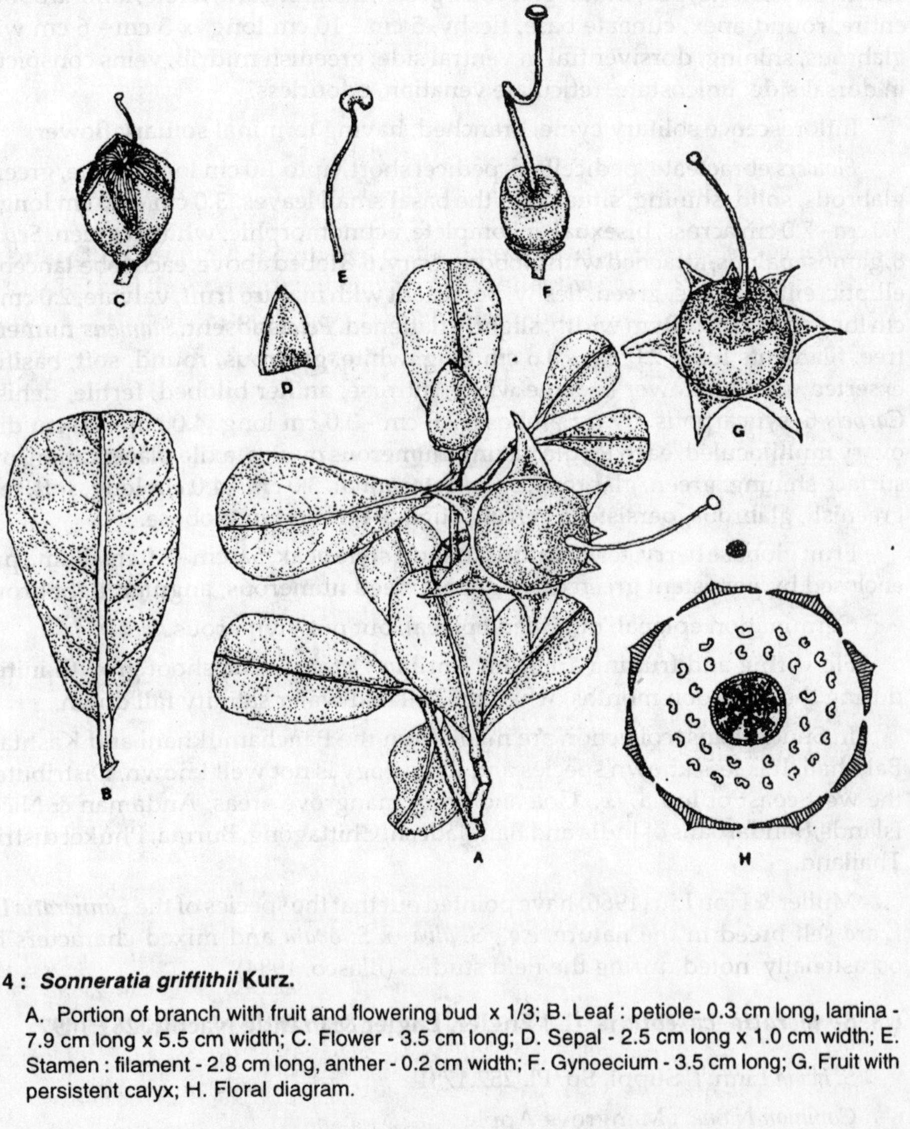

Fig. 14 : ***Sonneratia griffithii*** **Kurz.**

A. Portion of branch with fruit and flowering bud x 1/3; B. Leaf : petiole- 0.3 cm long, lamina - 7.9 cm long x 5.5 cm width; C. Flower - 3.5 cm long; D. Sepal - 2.5 cm long x 1.0 cm width; E. Stamen : filament - 2.8 cm long, anther - 0.2 cm width; F. Gynoecium - 3.5 cm long; G. Fruit with persistent calyx; H. Floral diagram.

Underground much spreading, repeatedly branched horizontal roots; erect breathing roots-pneumatophores with tapering ends, arising from the horizontal roots in rows, upto a height of 0.5 m and 1.0-1.5 cm diam., with secondary growth and corky appearance, 150-200 pneumatophores developed in a full grown mature tree.

Internodes 2.0 cm - 6.0 cm long, nodes ridged and grooved, jointed, terete, glabrous, shining, solid, green at apical part.

Leaves simple, exstipulate, opposite decussate, cauline, short petiolate, petiole

0.1cm - 0.3 cm long, pulvinous, glabrous, green, insignificant, terete, lamina obovate, entire, round apex, cuneate base, fleshy, 5 cm - 10 cm long x 3 cm - 6 cm width. glabrous, shining, dorsiventral in ventral side; greenish midrib, veins conspicuous in dorsal side; unicostate, reticulate venation, odourless.

Inflorescence solitary cyme, branched, having terminal solitary flower.

Flowers ebracteate, pedicellate, pedicel short, upto 1.0 cm long, terete, greenish, glabrous, solid, shining, situated at the basal small leaves, 3.0 cm - 5.0 cm long and 5.0 cm - 7.0 cm across, bisexual, incomplete, actinomorphic, whitish green. *Sepals* 6-8, gamosepalous, attached with globose ovary, 6-8 lobed above, each lobe lanceolate-elliptic, entire, acute, green, fleshy, persistent with mature fruit, valvate, 2.0 cm - 4.0 cm long x 0.5 cm - 1.0 cm width, slightly flattened. *Petals* absent. *Stamens* numerous, free, filaments long, 2.0 cm - 3.6 cm long, white, glabrous, round, soft, basifixed, exserted, showy, flower become white, introrse, anther bilobed, fertile, dehisced, *Carpels* 6, syncarpous, ovary globose, 2.0 cm - 3.0 cm long, 4.0 cm –5.0 cm diam., ovary multiloculed, each locule having numerous ovules, axile placentation, ovary surface shining, green, glabrous, style-1, terminal, 3.0 cm - 4.0 cm long, soft, terete, greenish, glabrous, persistent in fruit, stigma-1, round in globose.

Fruit globose berry, terminally into short style beak, 5.0 cm - 7.0 cm diam., partly enclosed by persistent green calyx at base; seed numerous, angular, light brown.

Germination epigeal, modified epigeal, but not viviparous.

Flowering and fruiting are from April to October. The shoot growth initiated during the monsoon months, when the soil and water salinity fall down.

In Sundarbans, collection are made from the Panchamukhani and Kashtala of Bakkhali. It is less known species and its ecology is not well known. Distributed in the west coast of India, *i.e.*, Goa and other mangrove areas; Andaman & Nicobar Islands, Sundarbans of India and Bangladesh, Chittagong, Burma, Phuket district in Thailand.

Muller & Hou Liu (1966) have pointed out that the species of the *Sonneratia* Linn. f. are self breed in the nature, *e.g.*, *S. alba* x *S. ovata* and mixed characters have occasionally noted during the field studies (Blasco, 1984).

6.3. *Sonneratia caseolaris* (L.) Engler, Engler & Prantle Nachtr. 261.1897.

S. acida Linn. f. Suppl. Sp. Pl. 252.1781.

Common Name : Mangrove Apple

Local Name (Bengali) : Chak-Keora

(Plate - 96 & Fig. 15)

Small evergreen tree, upto 8.0 m high; deliquiscent branching, woody, erect, branches horizontal, twig slender, slightly pendulous but not drooping like *A. apetala* Buch. Ham., slightly bushy, lateral branches slightly pendulous. Pale brown bark, smooth or furrows at maturity.

Underground, much spreading branched horizontal roots develop erect slender aerial breathing roots - pneumatophores, tapering towards the ends and the height upto 1.0 m and 1.5 cm - 2.0 cm diam., with spongy outer surface and secondary

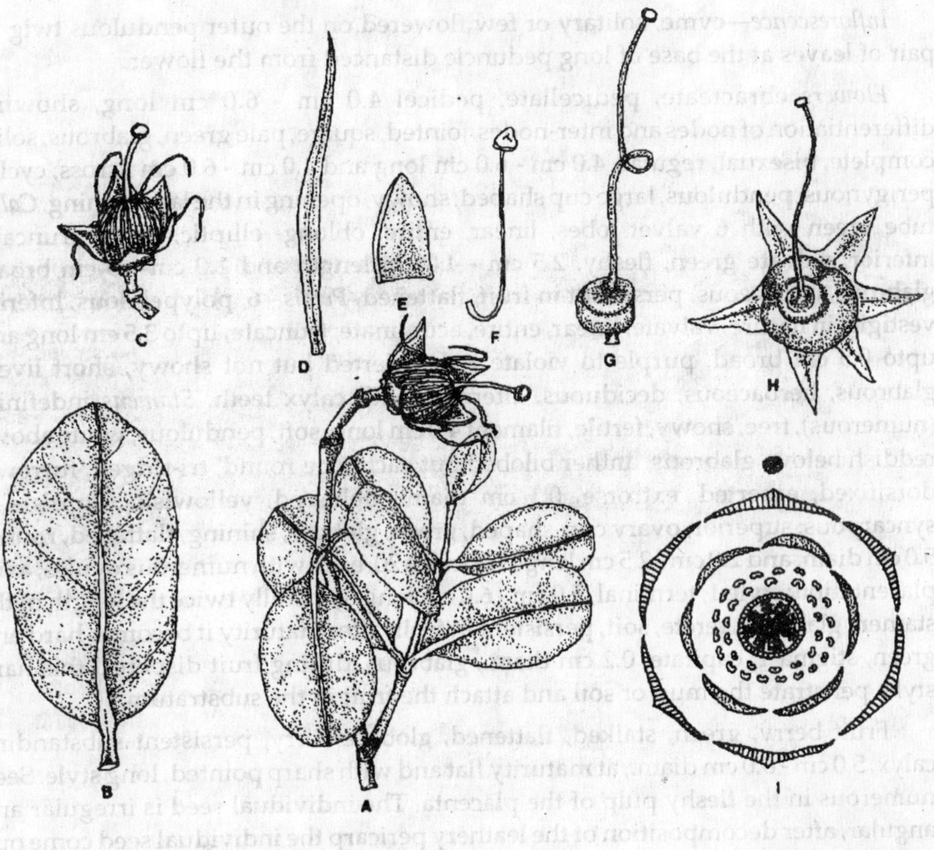

Fig. 15 : ***Sonneratia caseolaris*** **(L.) Engler**

A. Portion of flowering twig x 1/3; B. Leaf : petiole - 0.5 cm long, lamina - 8.3 cm long x 4.6 cm width; C. Flower - 6.0 cm long; D. Petal- 3.2 cm long x 0.4 cm width; E. Sepal - 2.5 cm long x 1.2 cm width; F. Stamen : filament - 4.0 cm long., anther - 0.1 cm long ; G. Gynoecium 6.0 cm long; H. Fruit with persistent calyx; I. Floral diagram.

growth; green initially but pale brown at maturity, several periphytic algal flora are also attached on the aerial pneumatophores; generally 250-400 pneumatophores are common on a full grown, mature tree.

Internodes 3.0 cm long, nodes slightly ridged in appearance, square at branches, glabrous, pale green, solid in the initial stages.

Leaves opposite decussate, exstipulate, simple, cauline, short petiolate, petiole 0.5 cm - 1.0 cm length, flat, glabrous, pale-green or reddish, pulvinous; lamina 8.0 cm - 12.0 cm in length and 4.0 cm - 5.0 cm broad, ovate-elliptic, entire; prominent reddish mid-rib, slight angled, green glabrous, coriaceous, fleshy, dorsiventral, odourless; unicostate reticulate venation; apex become acute at young stage but occasionally degenerate at maturity.

Inflorescence—cyme, solitary or few flowered on the outer pendulous twig ; a pair of leaves at the base of long peduncle distanced from the flower.

Flowers ebracteate, pedicellate, pedicel 4.0 cm - 6.0 cm long, showing differentiation of nodes and inter-nodes, jointed, square, pale green, glabrous, solid; complete, bisexual, regular, 4.0 cm - 6.0 cm long and 4.0 cm - 6.0 cm across; cyclic, perigynous, pendulous, large cup shaped; showy, opening in the late evening. *Calyx* tube green with 6 valvet lobes, linear, entire, oblong- elliptic, acute, truncate, inferior, valvate green, fleshy, 2.5 cm - 4.0 cm length and 1.0 cm-1.5 cm broad, glabrous, coriaceous, persistent in fruit, flattened. *Petals* - 6, polypetalous, inferior vestigial in nature, valvate, linear, entire, acuminate, truncate, upto 3.5 cm long and upto 0.5 cm broad, purple to violate red, exserted but not showy, short lived, glabrous, herbaceous, deciduous, alternate with calyx teeth. *Stamens* indefinite (numerous), free, showy, fertile, filament 4.0 cm long, soft, pendulous, white above, reddish below, glabrous, anther bilobed but each lobe round, tri-ridged, furrows, dorsifixed, exserted, extrorse, 0.1 cm diam., dehisced, yellowish. *Carpels* - 6, syncarpous, superior, ovary cup-shaped, green, globose, shining, flattened, round, 5.0 cm diam. and 2.0 cm - 2.5 cm length ; locules 20, each with numerous ovules; axile placentation; style 1, terminal, 5.0 cm - 6.0 cm long, generally twice the length of the stamen, glabrous, terete, soft, persistent in fruit, after maturity it becomes hard and green, stigma 1, capitate, 0.2 cm diam., glabrous, during fruit dispersal this hard style penetrate the mud or soil and attach the fruit in the substratum.

Fruit berry, green, stalked, flattened, globose-berry, persistent substanding calyx, 5.0 cm - 8.0 cm diam., at maturity flat and with sharp pointed, long style. Seed numerous in the fleshy pulp of the placenta. The individual seed is irregular and angular, after decomposition of the leathery pericarp the individual seed come out, float and disperse with the help of tidal flow.

Germination epigeal. Not having any viviparous or incipient viviparous germination.

Flowering and fruitings are common during March to October; but occasionally noted its flowers and fruits throughout the year.

Collected from the inner estuary, at Kulty-Minakhan areas of the North Sundarbans; where tidal water mixed with the sewage effluent; frequent on muddy soil along slow moving river sides; at the high tide zones. This species, *S. caseolaris* is not reported from the Sundarbans by many of the earlier workers; but Naskar & Guha Bakshi (1987) and Naskar (1993) have reported this species, this is threatened in this part of Sundarbans, which need for conservation.

Distributed in both west and east coasts of the Indian mangals. Also found in the Bangladesh part of the Sundarbans, Sri Lanka, Burma and throughout the South East Asian mangals, Northern Australia, Solomon Islands, New Hebrides.

Fruits edible, cooked as condiment or for flavouring curries; also used for the preparation of vinegar; pericarp said to be useful as vermifuge.

During field collections and thorough investigation of the fruits of *S. caseolaris* from the Kulti area of Indian Sundarbans and Cumbarjua canal side of Goa, it was noted that the fruits collected from Sundarbans are flattened-globose while the fruits collected from Goa are completely globose and the colour of the flowers are bright red in the specimens of Goa than that of the specimens of the Sundarbans.

Note : The species *S. alba* Backer is also collected from Goa, while during this field investigation *S. alba* was not collected from the Sundarbans mangals. (Plate-97)

Table 60 : Comparative Morphology of the three species of *Sonneratia* Linn. f. commonly found in Sundarbans.

Characters	*S. apetala* Buch. Ham.	*S.caseolaris* (L.) Engl.	*S. griffithii* Kurz.
Habit	Small to medium tree, upto 15 m high.	Small tree, upto 8.0 m high.	Medium tree, upto 15 m high.
Root (Pneumatophore)	Pneumatophores hard, corky and scaly outer surface, 0.5 m-1.5 m high and 2.0 cm - 3.0 cm diam.	Pneumaiophores hard, corky and spongy outer surface, tapering, 0.5 m - 1.0 m high and 1.5 cm -2.0 cm diam.	Pneumatophores hard, corky with tapering ends, upto 0.5 m high and 1.0 cm - 1.5 cm diam.
Stem bark	Thin, light brown, irregularly fissured.	Pale brown, smooth or furrows at maturity.	Deep brown or pale brown; smooth in young and fissured at maturity.
Leaf	Obtuse, narrow gradually, lamina 8 cm - 12 cm long x 1.5 cm -3.0 cm width; obtuse, cuneate.	Ovate-elliptic, broad, not narrow gradually, lamina 8.0 cm - 12.0 cm long x 4.0 cm - 5.0 cm width.	Obovate, round apex and cuneate base, 5 cm - 10 cm long x 3.0 cm - 6.0 cm width glabrous.
Inflorescence	Terminal cyme from branch axis.	Solitary cyme or few flowers on outer pendulous twig.	Solitary cyme on terminal and lateral branches.
Flowers	Medium, 1.2 cm - 2.0 cm long, 1.5 cm - 2.0 cm across; sepals -4, 1.5 cm - 2.0 cm long and 0.8 cm - 1.0 cm broad; petal-0; stamens indefinite 1.0 cm - 1.0 cm long; carpels - 4, ovary dumble shaped, 0.6 cm - 0.8 cm long and 2.0 cm - 2.5 diam,; style upto 3.0 cm.	Large, 4.0 cm - 6.0 cm long, 4.0 cm- 6.0 cm across; sepals - 6 , 2.5 cm - 4.0 cm long and 1.0 cm, - 1.5 cm broad; petals - 6 polypetalous, upto 3.5 cm long; stamens numerous, 4.0- cm long; carpel-6; ovary cup-shaped, 2.0 cm - 2.5 cm long and 5 cm- diam.; Style upto 6.0 cm long.	Large, 3.0 cm - 5.0 cm long, 5.0 cm - 7.0 cm across; sepals 6-8, 2 cm - 4 cm long x 0.5 cm - 1.0 cm width; petals - 0; stamens numerous, 2.0 cm- 3.6 cm long; carpels - 6, ovary multi loculed; style 1 , terminal 3.0 cm - 4.0 cm long, persistent.
Fruits	Capsule, 1.5 cm - 2.0 cm diam.; depressed globose berry with persistent 4 calyx plate.	Berry, 5.0 cm- 8.0 cm diam.; flattened globose with long style and peristent 6 calyx plate.	Globose berry, 5.0 cm- 7.0 cm diam. with persistent calyx.
Seed	4-7 in each fruit, brown, triangular, ridged, 1.5 cm long.	Numerous in the fleshy pulp of the placenta, angular and irregular.	Seed numerous, angluar, light brown.
Germinaton	Epigeal or modified epigeal.	Epigeal or modified epigeal.	Epigeal germination.
Distribution In Sundarbans	Inner and mid-estuaries.	Inner estuary	Mid- and outer-estuaries.

Morphometric differences among the three species of *Sonneratia* Linn. f.

These three species under the Genus - Sonneratiaceae in the Indian Sundarbans were studied by Naskar (1983) and Naskar & Guha Bakshi (1987); among these species the fruits of *Sonneratia apetala* Buch. Ham. are smallest and globose, very much similar to date palm fruit, but the fruits of *Sonneratia caseolaris* (L.) Engl. and

S. griffithii Kurz. are larger in size, globose (*S. griffithii*) and depressed-globose or flat-oval in shape (*S. caseolaris*). These green fruits are often eaten by the local people and cooked as condiment for having there sour taste. Their wood are minor timber and occasionally used as fuel-wood. All these three species are less abundant, particularly during field visit *S. caseolaris* was found from the inner estuary in the Sakha Bidyadhari river flats at Kulti and Minakhan and *S. griffithii* was collected from the river flats at mid- and outer-estuaries only. But *S. apetala* was also noted to distribute almost throughout the mid-estuary and ridge forest zones of the Sundarbans. The distinct characters of these species are discussed in the keys and the comparative tables.

As their seeds are embedded in pulp within the indehiscent berry, the seeds are very occasionally seen and which are released on decomposition of the pericarp. Comparatively, the rates of seed germination are less and very scarce, seedling development have been noted in these natural tidal forests.

FAMILY - IV. *COMBRETACEAE*

The family - Combretaceae has 20 genera and 500 species and distributed throughout the tropics. Out of these 20 genera, only 2 genera belong to the mangrove habitat; these are *Lumnitzera* Willd. and *Laguncularia* Gaertn. f. Furthermore, the genus-*Lumnitzera* has two true mangrove species, while *Laguncularia* Gaertn. f. is the monotypic genus (Tomlinson, 1986). *Lumnitzera* spp. are really the back mangal species, distributed in the S.E. Asian mangal zones; but the *Laguncularia racemosa* (L.) Gaertn. f. is dominantly distributed in the West Africa and Caribbean Islands mangal communities.

The family ***Combretaceae*** is distinct for its characters, like inferior unilocular ovary with pendulous ovules, one seeded drupe-like fruit, without any endosperm.

The other two genera, *i.e.*, *Conocarpus* L. and *Terminalia* L. were also reported by Tomlinson (1986) as mangrove associates; though *T. catappa* and *T. arjuna* are the terrestrial plant in the Lower Ganga Delta, but not the back mangals.

GENUS - 7. LUMNITZERA Willd. Neue Schr. Ges. Naturf. Fr. Berl. 4: 186. 1803.

The genus *Lumnitzera* Willd. has also two species and one putative hybrid, *viz.*, *L. littorea* (Jack.) Voigt., *L. racemosa* Willd. and *L. rosea* (Gaud.) Presl. The last one *L. rosea* (Gaud.) Presl. is reported as hybrid of *L. littorea* (Jack.) Voigt × *L. racemosa* Willd. (Tomlinson, 1986). *L. littorea* (Jack.) Voigt. has the spike inflorescence, attain upto length of 3.0 cm and flower is reddish, slightly zygomorphic; while *L. racemosa* Willd. has axillary raceme, attain a length of 8.0 cm and flower is white and actinomorphic.

Both *Lumnitzera littorea* (Jack.) Voigt. and *L. racemosa* Willd. are ecologically similar, as the mangal associate and are distributed from East Africa to Western Pacific Islands, Tropical Australia and Indo-China. *L. rosea* (Gaud.) Presl. is having the intermediate characters, between its putative parents and was reported from Hinchinbrook Island and Queensland (Tomlinson, 1986).

In the Indian mangrove ecosystems, the species *L. racemosa* Willd. is common throughout the mangals and *L. littorea* (Jack.) Voigt, was also reported from the mangals of Andaman & Nicobar Islands (Banerjee, *et al.*, 1989).

In the Sundarbans mangrove ecosystem, only *L. racemosa* Willd. is distributed and grow gregariously in the tidal and intertidal zones of the Western fringes of the mangals, facing thereby as true mangrove species. This species is also found in and around the degraded mangal zones of Bakkhali and Jammu dweep.

Lumnitzera racemosa Willd. is the type species, described by Willd. in Neue Schr. Ges. Naturf. Fr. Berl. 4 : 187. 1803 to ascertain the genus description.

7.1. *Lumnitzera racemosa* Willd. Neue Schr. Ges. Naturf. Fr. Berl. 4 : 187. 1803.

Bruguiera madagascariensis DC. Prodr. 3 : 23. 1828.

Petalema alternifolia Roxb. Fl. Ind. 2 : 371. 1832.

Common Name : Black Mangrove.

Local Name (*Bengali*) : Kripa or Kripal

(Plate-98 and Fig. 16).

Evergreen tree, height upto 10 m, deliquiscent and diffused, erect and much branched, dark green. This species is reported to attain an height upto 30 m in other S. E. Asian mangal communities; but in Sundarbans it is a small tree.

Aerial roots are pneumatophores, looping from lateral roots and conspicuously developed; above the tidal zone it has no aerial roots.

Differentiated into nodes and inter-nodes, much branched, inter-nodes short, 0.2 cm - 0.4 cm in length, fissured, brown, shoot apex green, terete, solid, woody, cauline, in short apex, slight leaf- scars also present.

Leaves cyclic, simple, exstipulate, petiolate, petiole 0.5 cm in length, round, flattened, green, glabrous, pulvinous, lamina subulate, entire or slightly wavy, obtuse or slightly notched in apex; lamina 7.0 cm - 8.5 cm long × 3.0 cm - 8.0 cm broad, dorsiventral, glabrous, coriaceous, fleshy, deep green, dorsiventral, unicostate, reticulate venation, vein inconspicuously present.

Inflorescence axillary raceme, peduncle 6.0 cm - 7.0 cm long; peduncle green, round, soft. *Flowers* pedicellate, pedicel short, bracteate; bracts 0.1 cm long and 0.1 cm broad, ovate, entire, acute, green, coriaceous, hermaphrodite, actinomorphic, complete, erect, odourless, pentamerous, epigynous. *Calyx*- sepals 5, gamosepalous, forming a tube like structure, from base to half portion of calyx tube, flattened and at that portion two short bracteoles like structure conspicuous, slightly upward downward level, other portion form tube, upper 5 sepal teeth, calyx tube 0.8 cm in

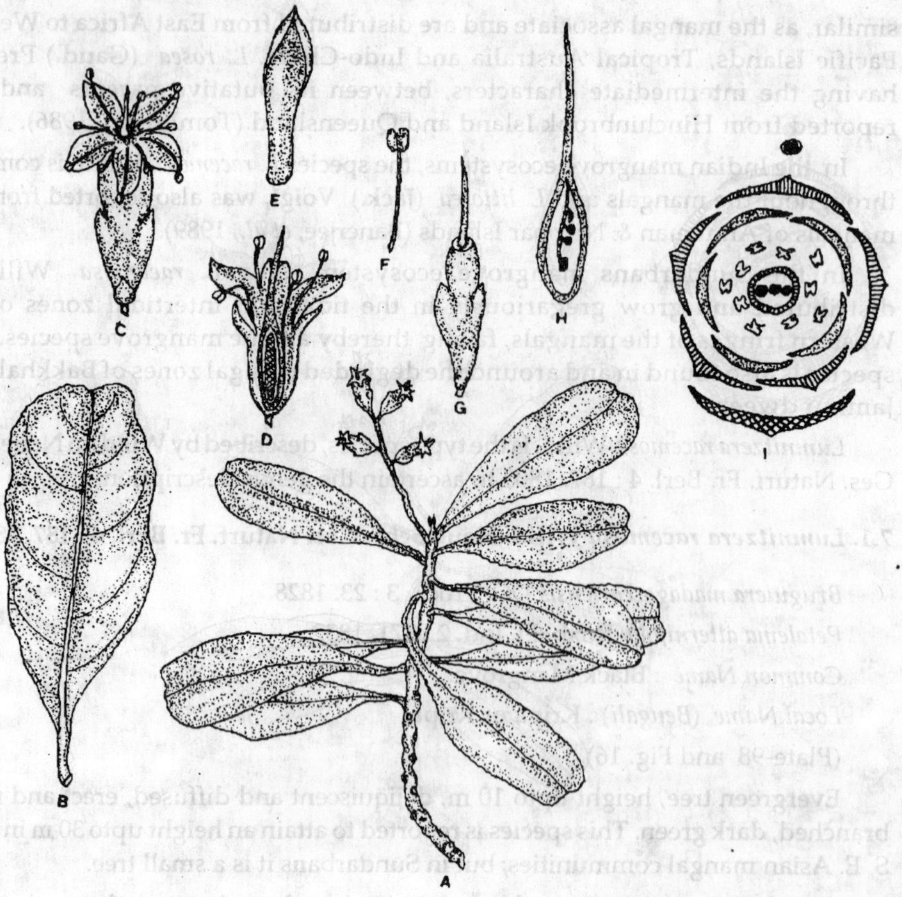

Fig. 16 : ***Lumnitzera racemosa*** **Willd.**

A. Portion of flowering twig x 1/3; B. Leaf : petiole - 0.5 cm long, lamina 8.5 cm long x 3.8 cm width; C. Flower - 1.0 cm long; D. L. S. of flower; E. Petal - 0.6 cm long x 0.2 cm width; F. Stamen : filament - 0.5 cm long, anther- 0.05 cm long;l G. Gynoecium : ovary - 0.8 cm long, style - 0.7 cm long; H. L. S. of carpel; I. floral diagram.

length, deep green glabrous, superior, fleshy, imbriacate, persistent. *Corolla* : petals-5, polypetalous, oblong-lanceolate, entire, acute, 0.6 cm in length and 0.2 cm in breadth, white, coriaceous, deciduous, longer than sepals, equal length, superior, imbricate, alternate to sepal. *Androecium*- stamens 10, free, arranged two whorls, 5 stamens developed at the base of petal, other 5 stamens develop at the base of sepal teeth : 5 - stamens at the petal base, low in height than others in upper surface level, though each having equal length : filament 0.5 cm in length, basifixed, round, soft, white, anther bilobed, extrorse, exserted, longitudinally dehisced, superior. *Gynoecium* - carpels 1, ovary inferior, 1.0 cm long, one chambered, 3 ovules, hanging from upper angle of the ovarian chamber, style 1, terminal, white, gradually tapering 0.7 cm in length, terete, glabrous, soft, stigma absent.

Pollination mechanism of this species is performed by several visitors like - bees, butterflies and moths, during day time. The actinomorphic flowers and shallow calyx cup are favour for that insect pollinators.

Flowering season is started during May and continuing upto November ; fruits are also common during the monsoon and post-monsoon months. Active vegetative growth is observed during the monsoon months.

Fruits are flattened, 1-seeded drupe : with persistent sepals and style, green, glabrous; at the mid-side of the fruit two distinct notches or oval appendages are present. Seed germination hypogeal and not viviparous; seeds dispersed through the water current and germinate on the sandy loose soil, after decaying the outer seed coat. The seeds lost its viability after floating for long days; as such the direct and immediate seed germination is more effective for rapid regeneration; salinity resistance for this species is very high and grow abundantly in the coastal sandy soil. Seed germination have not any special or extra mechanism like that of the other true mangrove species.

This species is well adapted in the outer estuaries; as it can tolerate saline condition ; it grows along with *Aegialitis rotundifolia, Avicennia* spp., *Ceriops* spp., *Phoenix paludosa* and *Sarcolobus* spp., facing the coast in the western part of the Sundarbans.

In India, it is distributed throughout the East and West Coasts and Andaman & Nicobar Islands, also found from E. Africa to N. Australia, throughout the Old World Mangroves in Indo-West Pacific Island.

Economically it is the good timber producing plant : but for indiscriminate exploitation the big size tree is now very scarce in Sundarbans and as such, used as fire wood in the western fringes of the Sundarbans.

Medicinally the fluid of the stem is said to be applied to itches and herpes.

FAMILY -V. *ARECACEAE* (nom. alt. *PALMAE*)

The family - Arecaceae (Palmae) has about 217 genera, including total about 2600 species are mostly distributed in the tropical and temperate regions; though most of these genera are localised, some genera are cosmopolitan. This family is characterised by berry or drupe and in case of drupe the endosperm usually united to one seed.

The common mangrove habitat palm of the mangrove back mangal zones are *Nypa* Steck, *Phoenix* L., *Calamus* L. and *Oncosperma* Bl.; but in the mangrove habitats of the Sundarbans, only the mangrove Palm - *Nypa* Steck and halophytic or salt tolerant palm *Phoenix* L. are common.

Key Characters to the Mangrove Habitat Palm of Sundarbans

Members of Arecaceae in the Sundarbans

1a Unbranched, caudex palm, looks like *Phoenix sylvestris* palm, single spathe, male inflorescence not catkin; carpels -3, united, fruit small, upto 1.5 cm long 2. *Phoenix* L.

1b Unbranched, rhizomatous palm, looks like sunken coconut palm, several spathe, male inflorescence catkin-like, carpels 3, aggregate, fruits medium upto 12 cm long 1. *Nypa* Steck.

GENUS - 8. *NYPA* Steck De Sagu 15.1757.

The **genus - *Nypa* Steck** is monotypic and well known as mangrove-palm.

8.1. *Nypa fruticans* (Thunb.) Wurmb., Verh. Batav. Gensotsch 156. 1781.

Common Name : Mangrove Palm or Water Coconut

Local Name (Bengali) : Gol-pata

(Plate - 99 and Fig.17)

Dichotomously branched, rhizomatous stem, underground, fibrous roots ; no aerial roots but the leaf-rachis or leaf base are spongy and act the function as the aerial roots for gaseous exchange. Looks like sunken coconut plant.

Leaves pinnately compound, spirally arranged on the top of the rhizomatous stem ; leaves upto 5 m long, rachis stout, glabrous, sheathing base upto 60 cm diam.; terete, shiny, green, glabrous, isobilateral, distinct mid-vein, unicostate parallel venation.

Inflorescence, spadix on a long peduncle, upto 50.0 cm long, cylinder shaped, several spathe enclosed flowers, peduncle erect, stout, glabrous, green. *Male* (♂) flowers gregariously on each peduncle, covered by spathe ; tepals-6, polytepalous in two whorls, upto 0.8 cm long and 0.5 cm broad, ovate elliptic, entire, acute, tufted, glabrous, shiny ; stamens-3, united to form central column, anthers bilobed, upto 0.4 cm long, short, *female flowers* (♀) enclosed by spathe, tepals 6 as male tepals, carpels 3, free, angular but funnel shaped, upto 1.0 cm high, unilocular ; single ovule, basal placentation, style and stigma 0.

Pollination caused by wind.

Fruits drupe, glabrous surface, dark brown epicarp, upto 12.0 cm long and 25.0 cm periphery, ridged and angular, mesocarp fibrous, endocarp thick.

Seeds grooved adaxially.

Germination incipient viviparous.

Flowering and fruiting occur during November to July, but fruits are very often seen throughout the year.

Commonly grow in Sundarbans in the eastern part of the mangrove forests, on the sides of tidal creeks, rivers and very often grow on the intertidal flat lands, river flats and marshy places; prefer less saline soil and it is now considered as threatened species in the Indian Sundarbans.

Distributed in the Sundarbans, Andaman & Nicobar Islands and also common in the South East Asian countries, particularly Burma, Malay Peninsula, Indonesia to New Guinea, Solomon Islands, Philippines, Ryukyu Islands, Queensland.

Fossil fruits of *Nypa* sp. and pollens were also reported from Paleocene in Brazil and Eocene in Europe, Africa and India (Tomlinson, 1986) ; by these fossil evidence it is assessed that its old distribution were different from the present.

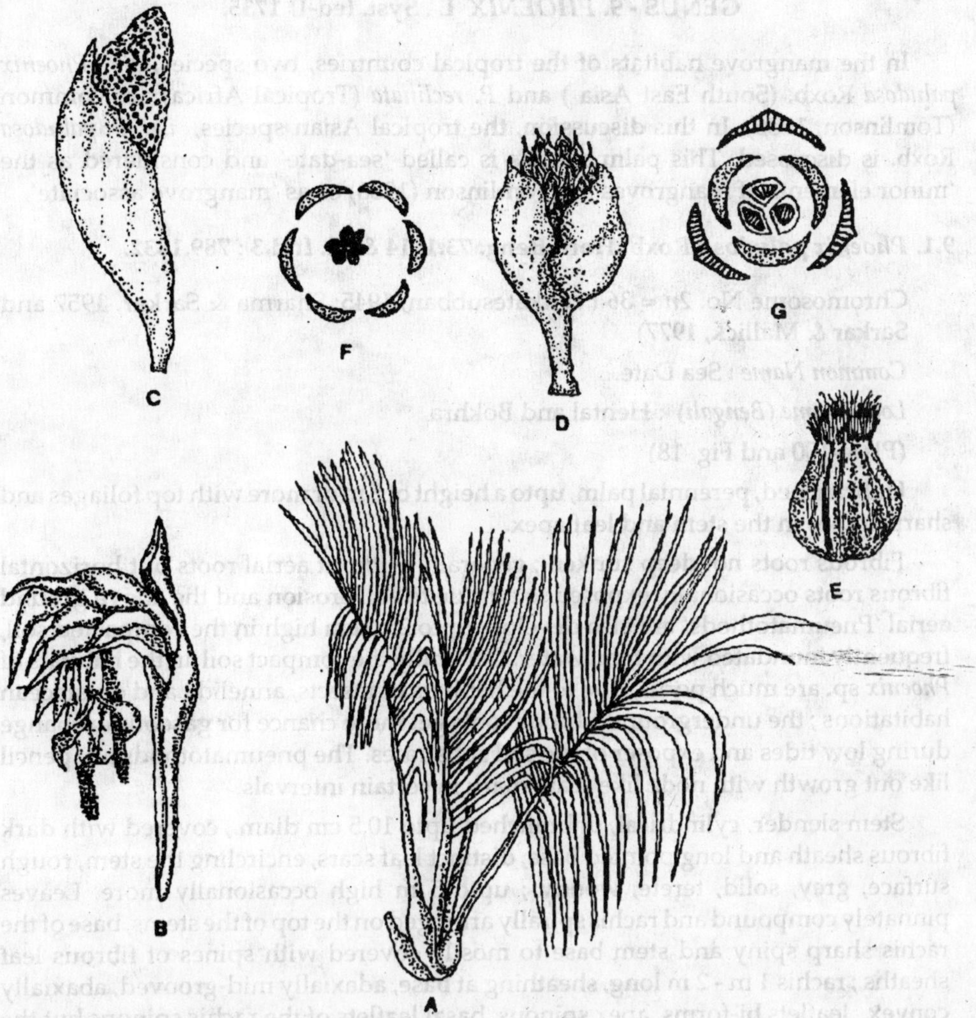

Fig. 17 : ***Nypa fruticans*** **(Thunb.) Wurmb.**

A. Entire plant showing habit (10 ft.); B. Inflorescence; C.Part of Inflorescence (Male); D. Part of Inflorescence (Female); E. Fruit; F. Floral diagram (Male); G. Floral diagram (Female).

Isothermic leaves (attap) are much used for thatching cottages and alcoholic drinks or mollases are also prepared from the fluid of young inflorescences in Philippines and Malayasia. Tender leaves are also used for Cigarette wrappers in Malayasia.

Experiment : Seedling plants were collected from the Sundarbans mangrove forest and transplanted in the non-saline soil and flooded with fresh tap-water regularly for three years. Growth of this experimental seedling is noted very slow, but new leaf appeared andor developed in an interval of 3-4 months and this seedling plant survived for more than 5 years in this experimental garden.

GENUS - 9. *PHOENIX* L . Syst. (ed-I) 1735.

In the mangrove habitats of the tropical countries, two species, *viz., Phoenix paludosa* Roxb. (South East Asia) and *P. reclinata* (Tropical Africa) are common (Tomlinson, 1986). In this discussion, the tropical Asian species, *i.e., P. paludosa* Roxb. is discussed. This palm species is called 'sea-date' and considered as the 'minor elements of mangroves' by Tomlinson (1986) or as 'mangrove associate'.

9.1. *Phoenix paludosa* Roxb. Hort. Beng. 73.1814 & Fl. Ind.3 : 789.1832.

Chromosome No. $2n = 36$ (Venkatesubban, 1945; Sharma & Sarkar, 1957 and Sarkar & Mallick, 1977)

Common Name : Sea Date.

Local Name (Bengali) : Hental and Bokhra

(Plate-100 and Fig. 18)

Unbranched, perennial palm, upto a height of 5 m or more with top foliages and sharp spines in the stem and leaf apex.

Fibrous roots not deep sunken ; generally without aerial roots but horizontal fibrous roots occasionally exposed to air due to soil erosion and the above ground aerial 'Pneumatothods' are also develop, upto 25.0 cm high in the compacted soil, frequently inundated with tidal water. Generally the compact soil in the habitats of *Phoenix* sp. are much porus, due to the habitats of insects, annelida and crustacean habitations ; the underground fibrous roots get more chance for gaseous exchange during low tides and exposer of the soil substrates. The pneumatothods are pencil like out growth with node like scars mark in certain intervals.

Stem slender, cylindrical, unbranched upto 10.5 cm diam., covered with dark fibrous sheath and long pointed base, distinct leaf scars, encircling the stem, rough surface, grey, solid, terete, woody ; upto 5 m high occasionally more. Leaves pinnately compound and rachis spirally arranged on the top of the stems, base of the rachis sharp spiny and stem base to mostly covered with spines of fibrous leaf sheaths ; rachis 1 m - 2 m long, sheathing at base, adaxially mid-grooved, abaxially convex, leaflets bi-forms, apex spinous, basal leaflets of the rachis spinous but the terminal leaflets gradually looks like leafy, upto 20.0 cm long and 2.0 cm broad, slightly folded, isobilateral, green, glabrous, shiny, distinct mid-vein, unicostate, parallel venation.

Spadix upto 45.0 cm long , peduncle branched, spathe upto 45 cm long, boat shaped, stalk 20 cm long, yellow, spongy, glabrous, deciduous, covered the inflorescence at young stages; gradually it ruptures into 2-3 halves ; flowers (♂) sessile, erect, 0.65 cm long and 0.45 cm across, not showy, odourless; *sepals*-3, gamosepalous, cup-shaped, 0.15 cm long and 0.3 cm broad, glabrous, valvate, stout; *petals*-3, polypetalos, ovate-lanceolate, entire, acute, curved, thick, glabrous, imbricate, 0.6 cm long, and 0.2 cm broad, yellow , stout; *stamens*-6, free, sessile, each two of 6 stamens on common ridged plate, anther bilobed, 0.3 cm long, yellow, longitudinally dehisced, twining, extrorse, exserted; flowers (♀) - sessile, erect 0.5 cm long, 0.35 cm

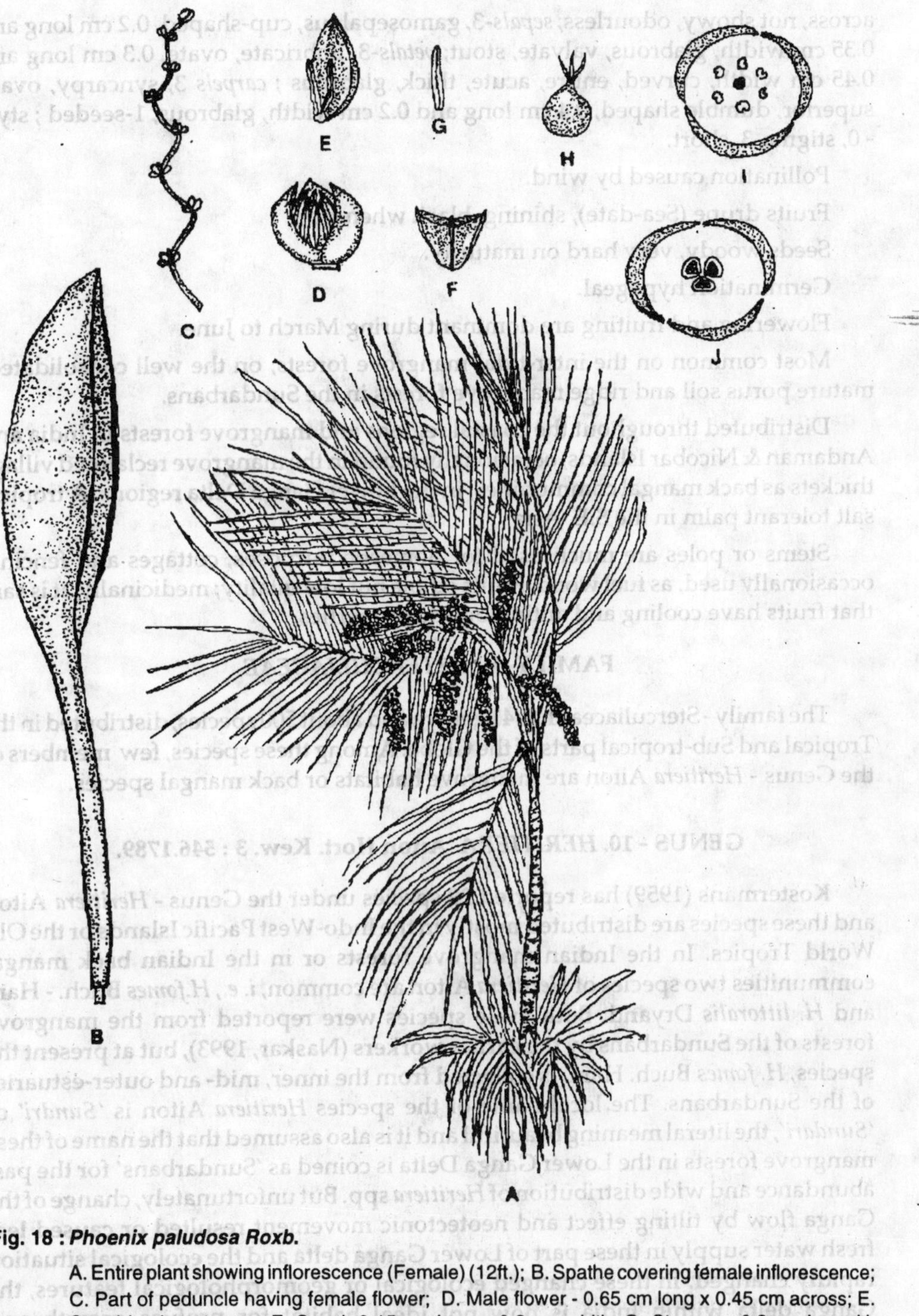

Fig. 18 : *Phoenix paludosa Roxb.*

A. Entire plant showing inflorescence (Female) (12ft.); B. Spathe covering female inflorescence; C. Part of peduncle having female flower; D. Male flower - 0.65 cm long x 0.45 cm across; E. Sepal having stamen; F. Sepal.-0.15 cm long x 0.3 cm width; G. Stamen - 0.3 cm long; H. Gynoecium - 0.4 cm long; I. Floral diagram (Male); J. Floral diagram (Female).

across, not showy, odourless; *sepals*-3, gamosepalous, cup-shaped, 0.2 cm long and 0.35 cm width, glabrous, valvate, stout; *petals*-3, imbricate, ovate, 0.3 cm long and 0.45 cm width, curved, entire, acute, thick, glabrous ; *carpels* 3, syncarpy, ovary superior, dumble shaped, 0.4 cm long and 0.2 cm width, glabrous, 1-seeded ; style - 0, stigma-3, short.

Pollination caused by wind.

Fruits drupe (Sea-date), shining, black when ripe.

Seeds woody, very hard on maturity.

Germination hypogeal.

Flowering and fruiting are dominant during March to June.

Most common on the inter-tidal mangrove forests, on the well consolidated, mature porus soil and ridge mangrove forests in the Sundarbans.

Distributed throughout the coastal regions and mangrove forests in India and Andaman & Nicobar Islands, occasionally found in the mangrove reclaimed village thickets as back mangal communities, in the Lower Ganga Delta regions. A tropical salt tolerant palm in the S.E. Asia.

Stems or poles are much used for domestic purposes, cottages and fencing; occasionally used, as fuel wood. Fruits edible but low quality; medicinally, it is said that fruits have cooling and antiphlogistic properties.

FAMILY- *VI . STERCULIACEAE*

The family - Sterculiaceae has 41 genera and about 500 species, distributed in the Tropical and Sub-tropical parts of the Globe. Among these species, few members of the Genus - *Heritiera* Aiton are mangrove habitats or back mangal species.

GENUS - 10. *HERITIERA* Aiton Hort. Kew. 3 : 546.1789.

Kostermans (1959) has reported 29 species under the Genus - *Heritiera* Aiton and these species are distributed mostly in the Indo-West Pacific Islands or the Old World Tropics. In the Indian mangrove forests or in the Indian back mangal communities two species of *Heritiera* Aiton are common, *i. e., H.fomes* Buch. - Ham and *H. littoralis* Dryand. Both these species were reported from the mangrove forests of the Sundarbans by the earlier workers (Naskar, 1993), but at present the species, *H. fomes* Buch. Ham. is collected from the inner, mid- and outer-estuaries of the Sundarbans. The local name of the species *Heritiera* Aiton is *'Sundri'* or *'Sundari'*, the literal meaning beautiful and it is also assumed that the name of these mangrove forests in the Lower Ganga Delta is coined as 'Sundarbans' for the past abundance and wide distribution of *Heritiera* spp. But unfortunately, change of the Ganga flow by tilting effect and neotectonic movement resulted or caused less fresh water supply in these part of Lower Ganga delta and the ecological situation rapidly changed. In these changed ecological or geomorphological features, the Ganga delta within India is now not ideal habitat for profuse growth and regeneration of these *Heritiera* spp. and *Heritiera* spp. are now scarce.

The species *H. fomes* is only distributed presently in the Ganga delta, the Sundarbans and Mahanadi delta of the Orissa State ; but *H. littoralis* is also present only from the Andaman & Nicobar Islands and also from Mahanadi delta (Banerjee, *et al.*, 1989). Majumdar & Banerjee (1985) have reported a new species, *i. e.*, *H. kanikensis* from Bhitarkanika, Orissa; it is distinct from two other species with globose fruits devoid of ridge.

The genus bear distinct leaf characters, like ventral surface of the leaves with scaly indumentum. The fruits are samara and the wings reduced to form vestigial keels for getting help during seed dispersal.

The less dominate species, *H. fomes* Buch. Ham. is collected and studied from the Sundarbans mangrove forests, which is described here. The other species, *H. littoralis* is also collected from the Wandoor Beach of the Port Blair, which are also shown (photographs of seed and buttresses) for comparison among these two species (Plate - 102).

10.1. *Heritiera fomes* Buch. Ham. Symes Kingdom Ava (ed.2)3 : 319-320.1800.

Chromosome No. $2n = 38$ (Sidhu, 1968)

Heritiera minor Roxb. (Hort. Beng. 50.1814) & Fl. Indica 3 : 142. 1832.

Common Name : Looking-glass Mangrove

Local Name (Bengali) : Sundri or Sundari

(Plate - 101, Plate - 102 and Fig. 19)

Evergreen tall to medium tree, attain upto 25 m height; but in Sundarbans, the present ecological condition is not congenial for the luxuriant growth of this species. Moreover, the timber of *Heritiera* spp. is much valued and as such, it is exploited indiscriminately. For this reason the big sized tree is not visible, at least from the outside or river banks. Here, in the Indian part of the Sundarbans mainly the plant attain a height upto 10 m. Bark-inner brown and turning reddish after peeling the bark, outer surface grey brown.

Root much spreading, blunt end peg-like pneumatophores present and exposed plank like horizontal roots and with stem buttresses in mature tree ; the aerial roots-pneumatophores are almost similar to *Xylocarpus* spp.; with blind root-suckers.

Stem differentiated into nodes and internodoes, branches deliquiscent, not much dense and pale colour; internodes 0.8 cm - 1.2 cm long. solid, grey brown, longitudinal narrow inconspicuous furrow or groved, axile and apical vegetative buds brown; inconspicuous leaf-scars present on apical shoots.

Leaf-alternate, simple, stipulate, stipules in pair in each node, petiolate; petiole 1.0 cm long, terete, grey, slightly woody, inconspicuous encircling depression found in petiole, pulvinous slightly; lamina elliptic—lanceolate, 8 cm-15 cm long × 3 cm - 5 cm width, entire, acute, tapering, coriaceous, slightly hard, dorsiventral, dorsal surface pale green glabrous, ventral surface whitish grey, rough surface with scaly indumentum; reddish or orange yellow, midvein conspicuous in the dorsal side, unicostate reticulate venation.

Inflorescence mixed, ebracteate, peduncle bent and pendulous, much branched;

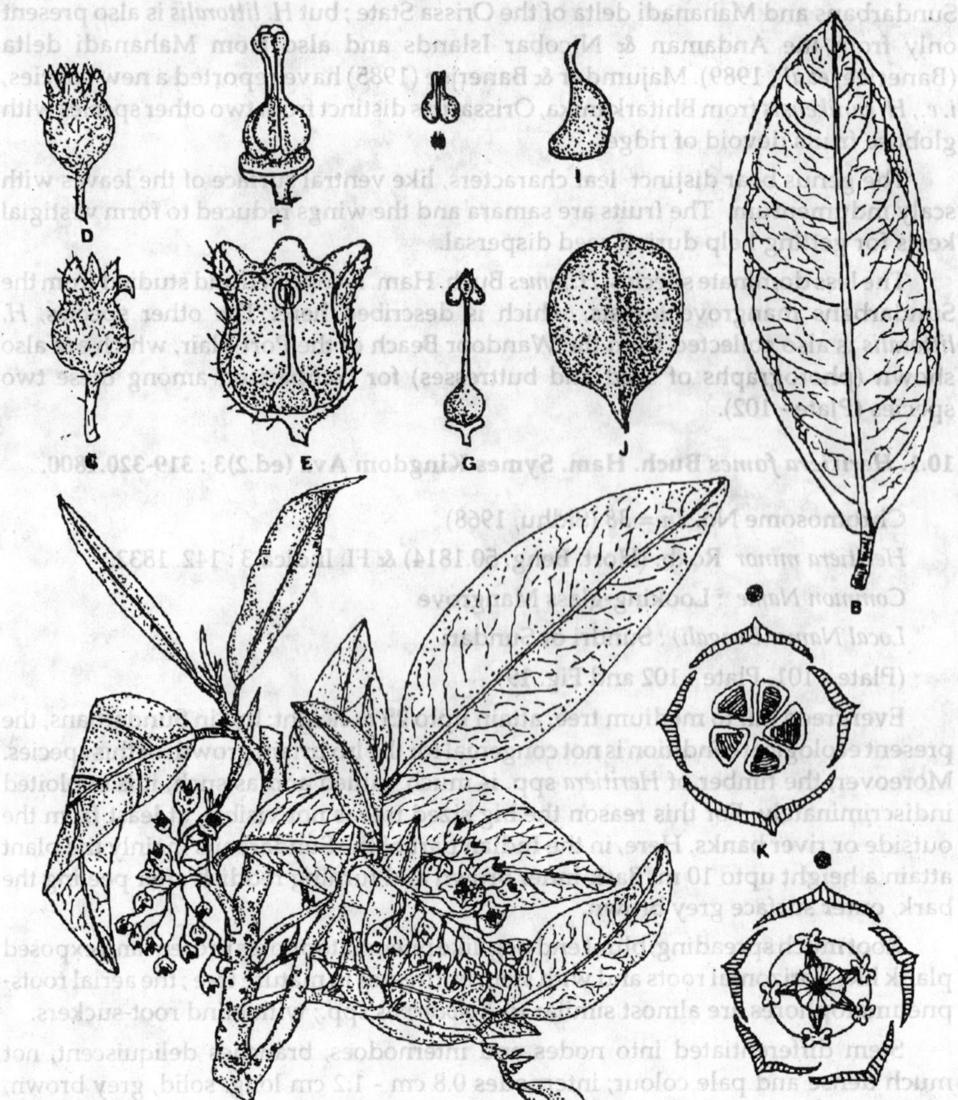

Fig.19 : *Heritiera fomes* Buch. Ham.

A. Portion of flowering twig x 1/3; B. Leaf : petiole - 1.0 cm long, lamina - 9.3 cm long x 3.5 cm width; C. Flower (Male) - 0.4 cm long; D. Flower (Female) - 0.4 cm long x 0.3 cm across; E. L.S. of male flower; F. Gynoecium - 0.25 cm long; G. Androecium; H. Stamen; I. Carpel - 0.25 cm long; J. Fruit; K. Floral diagram (Female); L. Floral diagram (Male).

peduncle upto 5.9 cm long, having both mature and young flowers. *Flowers* ebracteate, regular, unisexual, achlamydous, odourless, erect and pendulous, small, pedicellate, pedicel 0.65 cm long, terete, pubiscent, soft, easily dissociate from the thalamus, 1.0 cm long × 0.5 cm across. *Sepals* four-five, gamosepalous, globose, cup-shaped, tube beneath, upper 4-5 apical lobes; orange yellow outer surface but reddish-orange towards base in inner surface; lobes ovate, entire acute, pubescent, both outer and inner surface, leathery, 0.4 cm long, inferior, imbricate, upper surface same level. Petals absent. *Male* (♂) - flower : stamens 5, fused to form cylinder dumble known as pistillode, anther encircling the apical portion of the pistillode and to form globose structure, *anther* - distinctly bifurcated with apical white beak, yellow, longitudinally dehiscent, minute, introrse, inserted. *Female* (♀) flower: carpels 4-5, apocarpus, loosely attached, 0.25cm long, each ovary flattened, superior, closely aggregated to form globose structure; single chambered with 1-ovule, basal placentation, white, pubescent, style terminal 0.1 cm long, white, but after maturation it becomes brown, in young stages stigma remain united to form globose structure but after maturation it is separated to each other ; carpel shorter than calyx tube.

Pollination caused by *Apis dorsata* (rock-bee) and other insect fauna.

Fruit-capsule, sub-globose, 2.0 cm long, perifery upto 5.0 cm diam., pericarp brown, hard, woody divided in two halves by distinct ridged, adjoining into longitudinal circular keeled, 1-seeded.

Flowering and fruiting are during summer months and monsoon months, *i. e.*, from May to September. But mature seeds and fruits are very often seen on the coastal areas at Bakkhali, which germinate in the sands on the coastal areas frequently. These seeds reach this coastal area by floating, through the river and sea water. Vegetative growth are manifested during monsoon months. Germination is hypogeal.

Seedling Development Studies

Pericarp along with the cotyledons enlarge or swell up on imbibing sea or river water and the dormant seed respond to germination without taking prolong period of dormancy. In this stage, the size of the seeds, *i. e.*, cotyledons enlarge and change its shape almost round ; simultaneously the circular suture of the seed rupture in one side. Through this ruptured suture an inconspicuous pore between the two cotyledons becomes prominent. Within 2-3 days, the radicle comes out through the cotyledonary pore and gradually penetrate into the loose soil and anchor firmly. During the next 5-6 days, the radicle length increase considerably, attaining upto 20.0 cm long and two cylindrical portions (about 0.05 cm diam) of these two cotyledonary leaves come out through the well ruptured suture, along with the joined plumule base. The tip of the plumule is protected by these cotyledonary leaves intially and gradually elongate perpendicularly with the ground. Within 30 days of germination the tender, orange yellow plumules come up and attain upto 50 cm in height. During this stage 5-8 alternate leaves initiate in the apical portion of the plumule. In the axile of each alternate leaves a pair of needle shaped slender stipules develop initially, which abscissed soon after the maturation of the leaves. At the junction of the embryonic axis, the two cotyledonary connections remain attached till the plumule well matured with the development of greenish leaves and act the function of photosynthesis. The radicle becomes thick with several branched spreading roots. On well establishment of the seedling plant, the cotyledonary leaves along with the pericarp are separated or detached from the seedling. For the entire seedling development, it take about 35-50 days time, which is also variable for differential nature of substratum, soil phases, abundance of freshwater and presence of sunlight to these new seedling plant. Terminal buds gradually develop and a series of short bud-scars remain conspicuous.

H. fomes prefer the well drained less saline soil in the core area on river banks and ridge forests of the Sundarbans, where freshwater supply is abundant.

But, unfortunately, the Sundarbans mangrove forests are now lacking enough fresh water supply from the upstream due to neotectonic movement and change of flow of the Ganga towards the east ; as such, the ecological condition of the present day's is not much potential for the natural growth and regeneration of *H. fomes.* Besides all these changed ecological factors, the exploitation rate of *H. fomes* also causes its declining from the Sundarbans of West Bengal.

This species is only reported from the Ganga-Brahmaputra delta, the Sundarbans and from the Irrawaddy delta of Burma (Tomlinson, 1986).

Economically, the strong, durable timber is much favoured by the local people for boat-building and cabinate works ; as such, the exploitation of this species hampered its natural growth and distribution. The timber is said to be much durable and stronger than that of teak, *Tectona grandis.*

Experimental Studies : Water floating seeds were collected and these germinated seedlings are transplanted in the non-saline soil and treated with tube-well water. All these 10 seedlings survived and these transplanted seedling plant grow upto 6.5 m within 3 years in the experimental garden at non-saline condition these encouraging growth has been observed (Plate—101). Certain caterpiller and other insects have also been collected, which eat the tender leaves (Plate—102).

Notes : The earlier wokers (Prain, 1903a, 1903b, 1903c, and 1905) and Naskar & Guha Bakshi (1987) have also reported the existence of *Heritiera littoralis* Dryand from the Sundarbans of West Bengal; though it is now very scarce in this changed and highly populated areas of Sundarbans, West Bengal. It is now abundant in the Andaman & Nicobar Island in the back -mangal communities. This *H. littoralis* is much tall, with prominent and elaborate buttresses and the size of the seeds is bigger than *H. fomes,* 4 cm - 5 cm diam. Their comparative structures are represented through photographs (Plate - 102) .

FAMILY - VII. *MELIACEAE*

Meliaceae is a tropical family, which have about 50 genera and more than 100 species; these are distributed throughout the globe. Members of the Tribe—Xylocarpeae and Tribe—Aglaieae of the family Meliaceae generally grow as mangrove associated species in the intertidal regions of the Indo-West Pacific Islands. In the Sundarbans mangrove forest of India, these common mangrove tree species under-Meliaceae are *Xylocarpus granatum* Koenig and *Xylocarpus mekongensis* Pierre; while the other members like, *Xylocarpus molluccensis* (Lamk.) Roem. and *Aglaia cucullata* (Roxb.) Pellegrin are also reported from the Sundarbans mangrove forest by earlier workers (Naskar& Guha Bakshi, 1987); but these species are very rare in their distribution in Sundarbans. These above mentioned mangroves or back mangals are medium to large sized trees and timber of these species are fine grained, hard and durable; these timbers are used for valuable house hold articles, furnitures and boat building purposes. As such, exploitation of these species are enormous in the Sundarbans and elsewhere and the big sized trees are now very rare in the Sundarbans mangrove forests.

The two genera of the family Meliaceae (Mangroves) can be distinctly separated with the following morphometric key characters :

Key to the Genera of Meliaceae

1a. Tall tree, with large globose or round fruits, stem with buttresses and aerial roots, most common 1. *Xylocarus* Koen.

1b. Small to medium tree and the fruits are smaller in size, stem without any buttresses and not with aerial roots, uncommon 2. *Aglaia* Lour.

GENUS-11. *XYLOCARPUS* Koen. Natur. 20 : 2. 1784.

Three species are reported under the genus *Xylocarpus* Koen., *i.e. X. granatum* Koen., *X. mekongensis* Pierre and *X. molluccensis* (Lamk.) Roem.; the first two species are reported as true mangal, while the third species ,*X.molluccensis* is considered as non-mangrove or back mangal and were reported from the back mangal communities of Andaman & Nicobar Islands and Mahanadi delta in the sandy or rocky beach, associated with *Heritiera littoralis* Dryand Banerjee *et al.* (1989), Tomlinson (1986) has recognised these Genus—*Xylocarpus* Koen. as the 'minor elements or components of mangroves', though these species grow in the intertidal zones, along with other 'major mangrove elements'.

In the intertidal mangrove forests of the Sundarbans in India, X. *granatum* Koen. and X. *mekongensis* Pierre mostly prefer silty clayey soil and these are very common; on the other hand, *X. molluccensis* (Lamk.) Roem. prefer the sandy and rocky areas away from the frequent tidal inundation. As such, these first two species are considered by Narkar (1943) as the 'true mangals' or 'minor elements of managals' and the third species is considered as 'back mangal'. Moreover, *X. molluccensis* (Lamk.) Roem. is now very rare in the Sundarbans area but common in the Andaman & Nicobar Islands and Mahanadi delta of Orissa (Banerjee, *et al.*, 1989).

Regarding the nomenclature and abundance or presence of these three species of *Xylocarpus* Koen. there are several confusing reports. Prain (1903a) has reported in the Bengal Plants, p. 222, *Carapa obovata* Blume (Syn. *C. molluccensis)* and coined the Bengali name as 'Dhundul', as a small tree, with a large spherical fruit and latter on Prain (1903a) has also reported in Bengal Plants, pp. 962-963 (Rep. ed. 1963) *Carapa moluccensis* Lamk. var. *gangetica* with the Bengali name as 'Passur', a small tree. Naskar & Guha Bakshi (1987) have described both these two species, 'Dhundul' as X. *granatum* Koen. and 'Passur' as X. *molluccensis* (Lamk.) Roem. Banerjee *et al.* (1989) have reported that the species X. *molluccensis* (Lamk.) Roem. is absent from the Sundarbans mangals and have also remarked that both X. *granatum* Roem. and X *mekongensis* Pierre are pressent in the Sundarbans. But in their description, Banerjee, *et al.* (1989) have mentioned the fruit size of X. *granatum* Roem. is 30 cm-40 cm across, which is not reported in any other publications on mangroves of India and abroad. Moreover, during these present field studies no such big sized fruits are observed from the mangals of the Sundarbans.

These above mentioned three species of *Xylocarpus* Koen. can be identified in the field by studying the following under mentioned morphometric key characters :

Key to the Species of *Xylocarpus* Koen.

1a. Medium tree, grow in the intertidal silty clay soil, as minor components of mangroves, forming well developed buttresses surrounding the trunk base, but aerial pneumatophores and root suckers absent; bark yellowish or whitish with thin flakes; mature fruits hang on the mother plants—almost throughout the year; diam. of the fruit up to 20.0 cm or slightly more, yellowish brown fruit coat, completely round with long woody stalk

1. X. *granatum* Koen.

1b. Medium tree, grow either in the silty clay soil or on the sand or rocky bay as minor elements of mangals or back mangals; buttresses absent but well developed aerial blunt end pneumatophores or root suckers are present; bark reddish with thick flakes; fruits hang on the mother plant almost throughout the year, diam. of the fruit generally not exceeding 15cm, greenish, not completely round as 2

2a(1b). Grow generally on the intertidal silty but consolidated clay as true mangroves; buttresses occasionally formed but very inconspicuous, well developed root suckers or aerial pneumatophores present; bark peeling in narrow strips; inflorescence exceeding 8 cm

2. X. *mekongensis* Pierre

2b(1b). Grow generally on the sandy or rocky bay, away from the frequent tidal inundation, buttresses absent and with no aerial rootsuckers or pneumatophores; bark longitudinally fissured; inflorescence less than 8.0 cm long

3. X. *molluccensis* (Lamk.) Roem.

11.1. *Xylocarpus granatum* Koenig, Naturf. 20: 2.1784.

Carapa obovata Blume Bijdr. 179. 1825. *Xylocarpus obovatus* (Blume) Juss. Mem. Mus. Paris 19:244.1830. *X. benadirensis* Mattei Boll. Ort. Bot. Palermo 7:99.1908.

Chromosome No. $2n = 42$ (Gill, 1978) and $2n = 52$ (Styles & Vosa, 1971).

Common Name : Cannon ball Mangrove

Local Name (Bengali): Dhundul

(Plate- 103 and Fig. 20)

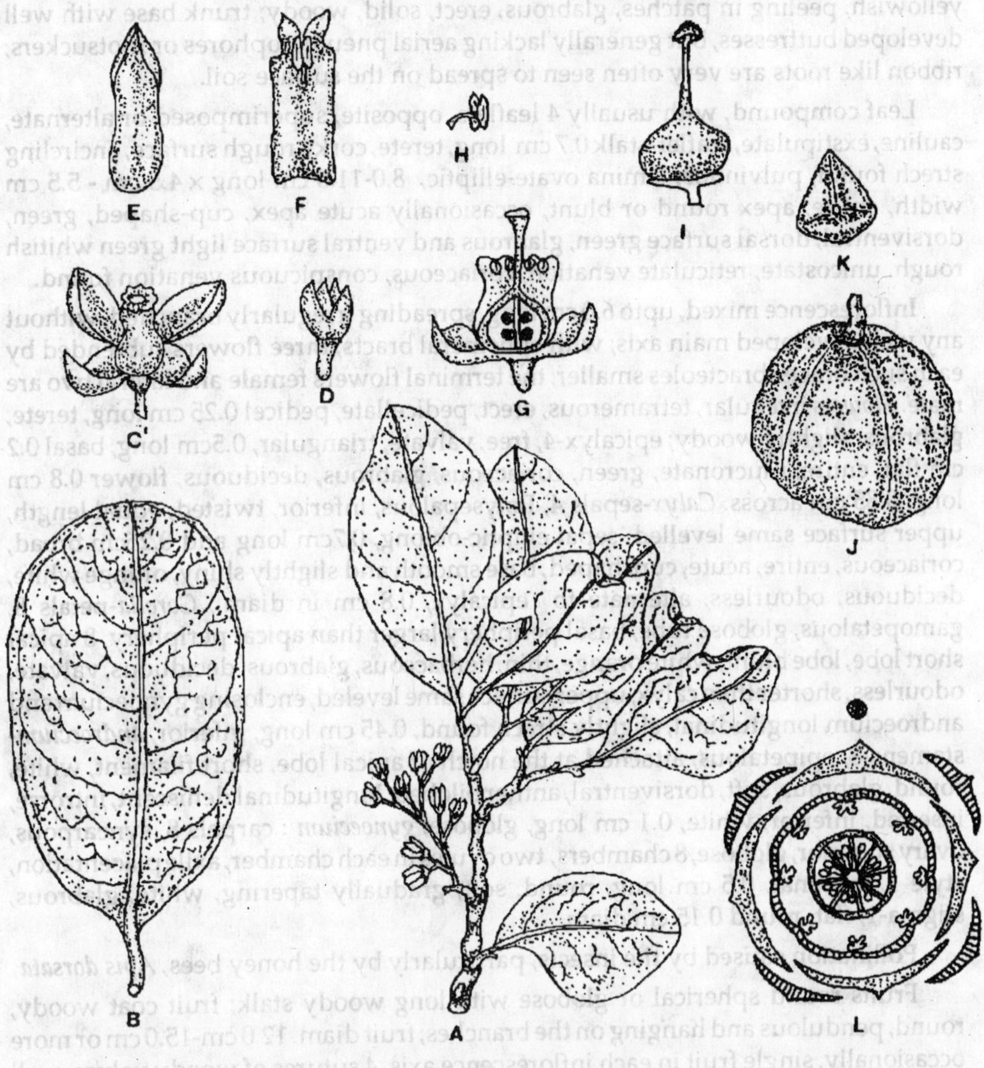

Fig. 20 : *Xylocarpus granatum* Koen.

A. Portion of flowering twig x1/4; B. Leaf.: Petiole-0.7 cm long, lamina-9.8cm long x 5.1cm width; C. Flower; D. Epicalyx -0.5cm long; E. Sepal-0.7 cm long x 0..25cm width; F. Petal with epipetalous stamen; G. L.S.of flower; H. Stamen : anther-0.1 cm long; I. Gynoecium-0.5cm long; J. Fruit- 15.0 cm across; K. Seed; L. Floral diagram.

Much branched, dark green, deciduous tree on the intertidal ridge forests and river banks; a mangrove species and the Type species of the genus—Xylocarpus Koenig.; attain the height upto 20 m or occasionally more, with smooth bark and dark foliages.

Branches differentiated with nodes and internodes, deliquiscent, nodes not ridges, internode long, 8.0 cm-12.0 cm long, terete, smooth surface, brownish or yellowish, peeling in patches, glabrous, erect, solid, woody; trunk base with well developed buttresses, but generally lacking aerial pneumatophores or rootsuckers; ribbon like roots are very often seen to spread on the surface soil.

Leaf compound, with usually 4 leaflets, opposite, superimposed or alternate, cauline, exstipulate, leaflet-stalk 0.7 cm long, terete, corky rough surface, encircling strech found, pulvinous, lamina ovate-elliptic, 8.0-11.0 cm long × 4.5 cm - 5.5 cm width, entire, apex round or blunt, occasionally acute apex, cup-shaped, green, dorsiventral, dorsal surface green, glabrous and ventral surface light green whitish rough. unicostate, reticulate venation, coriaceous, conspicuous venation found.

Inflorescence mixed, upto 6.0 cm long, spreading irregularly branched, without any well developed main axis; with ephemeral bracts, three flowers subtended by each such bracts; bracteoles smaller; the terminal flowers female and lateral two are male. *Flowers* irregular, tetramerous, erect, pedicellate, pedicel 0.25 cm long, terete, glabrous, slightly woody; epicalyx-4, free, valvate, triangular, 0.5cm long, basal 0.2 cm flat, entire, mucronate, green, coriaceous, glabrous, deciduous, flower 0.8 cm long, 0.85 cm across. *Calyx*-sepals 4, Polysepalous, inferior, twisted, equal length, upper surface same levelled, sepal elliptic-oblong, 0.7cm long and 0.25cm broad, coriaceous, entire, acute, cupshaped, base smooth and slightly shiny, orange white, deciduous, odourless, alternate to epicalyx, 0.8 cm in diam.; *Corolla*-petals 8, gamopetalous, globose tube, basal periphery larger than apical periphery, 8 apical short lobe, lobe blunt, white orange, thin, herbaceous, glabrous, deciduous, valvate, odourless, shorter than calyx, upper surface same leveled, enclosing gynoecium and androecium longitudinal, slightly strech found, 0.45 cm long, inferior, *androecium*-stamens-8, epipetalous, attached at the notch of apical lobe, short filament, white, round, glabrous, soft, dorsiventral, anther bilobed, longitudinal dehiscent, introrse, inserted, inferior, white, 0.1 cm long, globose; *gynoecium* : carpels 8, syncarpous, ovary superior, globose, 8 chambers, two ovules in each chamber, axile placentation, style 1, terminal, 0.5 cm long, round, soft, gradually tapering, white, glabrous, stigma-1, flat, round 0.15 cm diam.

Pollination caused by the insects, particularly by the honey bees, *Apis dorsata*

Fruits round spherical or globose with long woody stalk; fruit coat woody, round, pendulous and hanging on the branches; fruit diam. 12.0 cm-15.0 cm or more occasionally, single fruit in each inflorescence axis, 4 sutures of woody valves, well filled with 8-20 seed. Seeds angular but outer surface which is attached with the inner fruit coat or wall is convex, often look pyramidal with corky testa, the mark of plumule and radicle point are prominent on the convex side of the seed, as small boss. Seeds come out from the fruit on separation or after rupturing the fruit wall or valves, as and when the mature fruit falls down on the muddy substratum or water; seeds can float for long period in the water and germination start on the floating condition, floating seeds can survive 4-6 months or more.

Flowering and fruiting time almost throughout the year; but the normal and profuse flowering time is April to August and the fruits are very often seen to hang throughout the year in the mother plant. Active growth of the plant was noticed during the monsoon and post-monsoon months.

Commonly grow on the intertidal river slopes, ridge forests and core area of the

Sundarbans mangrove forests in association with *Excoecaria agallocha, Avicennia* spp., *Bruguiera* spp., *Ceriops* spp. etc.

Experiments: Seed germination starts during floating stages of the woody seed and as soon the seeds reach the substratum, the radicle extruded from the convex side of the seed *i.e.* from the small boss mark; seed germination hypogeal, tap roots developed from the radicle mark; the seedling axis initially develop and extend upto 30 cm or long within few weeks, with several spirally arranged scale leaves and latter on foliage leaves develop on the terminal portion of the seedling axis; the cotyledons remain closed within the corky seed coat, which remain attached with the seedling more than 16 months and during this stages much branched tap roots developed.

Germinated seeds are transplanted in non-saline soil and treated with freshwater in these experimental plantations the seedling plant survive more than three years and active growth observed during summer and monsoon months.

11.2 .*Xylocarpus mekongensis* Pierre, Fl. For. Cochinch. Plate 359B, 1897.

Carapa moluccensis Watson. Malay. For. Rec. 6: 70.1928.

Xylocarpus gangeticus I (Prain) Parkinson, Ind. For. 60: 140.1934.

X. *australasicus* Ridley, Kew Bull. 1928: 29.1938.

Common Name : Cedar mangrove

Local Name (Bengali) : Passur

(Plate-104 and Fig. 21)

Much branched, deciduous tree attain upto 20 m high, a mangrove species; buttresses very insignificant or absent, but well developed aerial root-suckers or pneumatophores with blunt end are present.

Stem with well differentiated nodes and inter-nodes, deliquiscent, not ridged, internodes terete, dark brown, glabrous, rough, erect, solid, woody, fissured with narrow strip of peeling.

Leaf compound with usually 4 leaflets, paripinnate, opposite, superimposed or alternate, cauline, exstipulate, leaf let stalk-0.7 cm long, terete, dark brown, glabrous, rough surface, pulvinous, lamina ovate-elliptic or elliptic, 10 cm - 14 cm long × 4.5 cm - 6 cm width, entire, slightly mucronate, cup-shaped, dark green, dorsiventral, dorsal dark green, glabrous and ventral whitish, rough, coriaceoous, unicostate reticulate venation conspicuous.

Inflorescence united cymes, about 8 cm long, longer with distinct main axis. *Flowers* irregular, tetramerous, erect,pedicellate, pedicel 0.4 cm long, terete, glabrous, woody in maturation, brown in colour, green in young, 0.7 cm long × 0.3 cm across., epicalyx- 4, free, valvate, ovate, 0.15 cm long and 0.2 cm broad, green, blunt, entire, coriaceous, glabrous, deciduous; *calyx*-sepal 4, polysepalous, inferior, twisted, equal length, upper surface same level, sepal ovate, 0.3 cm long and 0.25 cm broad, herbaceous, thin, white, deciduous, odourless, alternate to epicalyx, inconspicuous veins found, slight curved, close the other parts of flower; *corolla* - petals 8, gamopetalous, globose tube, basal periphery larger than the apical periphery, 8 apical short lobe, lobe ovate, petal white, thin, herbaceous, glabrous, deciduous,

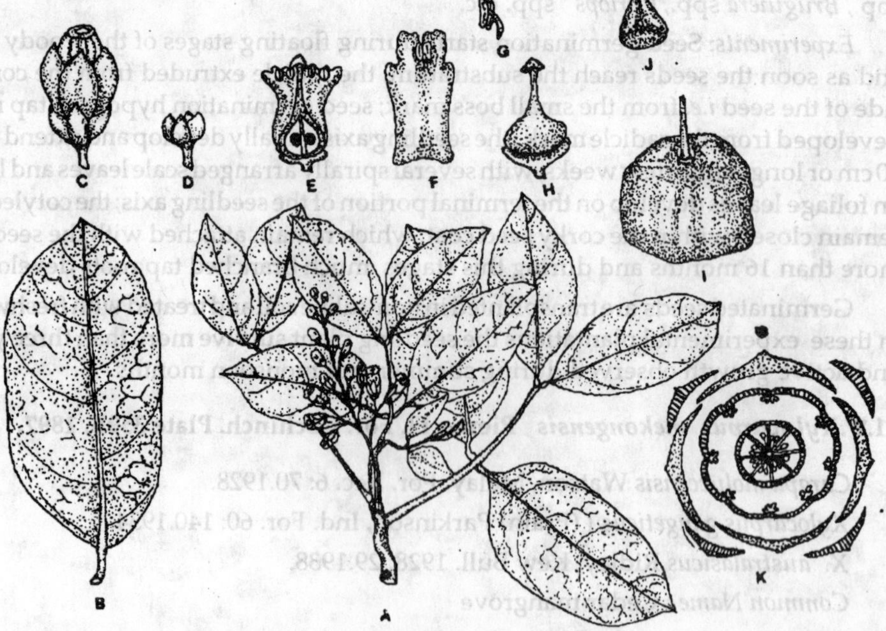

Fig. 21 : ***Xylocarpus mekongensis*** **Pierre**

A. Portion of flowering twig x 1/4; Leaf: leaf let-stalk-0.7 cm long, lamina-12.0 cm long x 5.0 cm width; C.Flower 0.7 cm long x 0.3 cm across; D. Epicalyx; E.L.S. of flower; F. Petal with epipetalous stamen; G. Stamen- 0.07 cm long; H. Gynoecium: ovary- 0.2 cm diam., style- 0.1 cm long; J. Fruit 8.0. cm across; J. Seed; K. Floral diagram.

shorter than calyx, upper surface same level, enclosing gynoecium and androecium, 0.15 cm long, inferior; *androecium*- stamens-8, epipetalous, attached at the notch of apical lobe, short filament, white, round, glabrous, soft, dorsifixed,, anther bilobed, longitudinal dehisced, introrse, globose, inserted, inferior, white, 0.07 cm long; *gynoecium*-carpels 8, syncarpous, ovary superior, globose, 0.2 cm diam., 8-chambers, two ovules in each chamber, axile placentation, style -1, terminal 0.1 cm long, round, soft, gradually tapering, white, glabrous, stigma-1, flat, 0.1 cm diam. pollination caused by insects, particularly the honey bees.

Fruit size much smaller than *X . granatum* Koen., less than 12.0 cm across; 10-20 seeded and the seed arrangement same as X. granatum Koen.

Seeds disperse through water current.

Flowering and fruiting time are dominant during March to June and mature fruits are seen to hang almost throughout the year.

Distributed in the mangrove forests and ecosystems of the east coast of India and Andaman & Nicobar Islands; an Old World mangroves.

Timber much used for cabinet works and as fuel wood. Bark astringent, used for dysentery, diarrhoea and other troubles.

FAMILY-VIII. *EUPHORBIACEAE*

The family-Euphorbiaceae is one of the largest Angiosperm family with more than 7,000 species, distributed throughout the tropical regions of the world. Among these diverse habitats, only three genera are considered by Tomlinson (1986) as the mangroves habitats or back mangals. The genus - *Excoecaria* L. and *Glochidion* J.R. & G. Forsst are distributed in the Old World Tropical mangals and the genus *Hippomane L. is* also reported from the New World Tropical mangal. In the Indian mangrove habitats, only the genus-*Excoecaria* L. is reported and the species *E. agallocha* L. is very common mangrove associate or back mangal in the intertidal and tidal areas throughout and also found in the above tidal zone of the coastal areas.

GENUS -12. *EXCOECARIA* L. Syst. Nat. (Ed.10) 1288.1759.

The genus - *Excoecaria* L. is having 35-40 species, of which *Excoecaria agallocha* L. Syst. Nat (Ed.10) 1288.1759. is the type species of the genus. Two species like, *E. agallocha* L. and *E. bicolar* Hassk. grow in the mangrove habitats of the Sundarbans. Though, *E. bicolar* Hassk. is very rare in distribution and the *E. agallocha* L. is very common, almost everywhere in the tidal and inter-tidal zones and also on the freshwater areas or the back mangal communities in the Old World Tropics. As such, this species *Excoecaria agallocha* L. is considered as the minor components of mangroves or mangrove associate or back mangal.

The species *E. agallocha* L. has no aerial root, but after soil erosion its horizontal spreading roots come out on the ground or surface soil and seems to be appeared as aerial roots. For exudation of excess salt, taken from the saline water and soil phase, the mature *Excoecaria agallocha* tree very often form or develop several perforated tumerous outgrowth on the base of the trunk. Besides the salt excretion, these perforated gall like outgrowths also perform the function of respiration in the tidal inundated, non-porous sedimented soil. Several salt glands are also present in the leaf, which perform the function of salt excretion.

Its milky poisonous latex and fleshy three loculae schizocarpus fruits are the characterisrtic features, similar to the other members of Euphorbiaceae.

Its seed germination is modified epigeal and viviparous germination is not conspicuous like other mangals.

Two species of *Excoecaria* L. in the Sundarbans mangals are compared with their key characters, though the common species *Excoecaria agallocha* L. is described in detail and *E. bicolor* Hassak is not collected during this course of studies.

Key to the Species of *Excoecaria* L.

1a. Handsome, halophytic tree, upto 15 m high, dioecious, with gall on the mature trunks, flowers catkin (male) and axillary (female), very common on the mangrove area in Sundarban. 1. *E. agallocha* L.

1b. Medium, halophytic shrub, upto 1.5 m high, monoecious, without any gall on the base of the plant; flowers very small, very occasional in the mangrove forest of Sundarbans. 2. *E.bicolar* Hassk.

Another species, *Excoecaria indica* (Willd.) Muel. Arg. (Syn. *Sapium indicum* Willd.), locally known as the 'Hurnabatul' is also reported by Naskar (1981) and Naskar & Guha Bakshi (1987); this collection was made from Rangabelia, the mangrove reclaimed area or back mangal of the Sundarbans. But in this treatise, this back mangal species is not dealt as this species is having very scarce distribution and insignificant in respect of the mangrove ecosystem.

12.1 *Excoecaria agallocha* L.Syst, Nat. (Ed.10) 1288.1759.

Chromosome No. 2n = 130 (Sidhu, 1961) and 2n = 140 (Datta, 1967)

Common Name : Blind-your-Eye-Mangrove or Blinding Tree

Local Name (Bengali) : Geoa or Gneoa

(Plate - 105, Plate- 106 and Fig. 22).

Medium dioecious tree, 15 m - 20 m high, deliquiscent, woody, perennial, partially deciduous, much branched, very dense with dark leaves and spreading branches. Bark grey, fissured, with poisonous milky latex, lenticelate, lenticel prominent on young twigs and also present in mature trunk base; mature trunk base with gall like tumerous outgrowths.

Roots generally without any aerial growth, but very often superficial spreading horizontal roots are exposed due to water erosion during tidal flashing; spreading surface or horizontal roots help to check soil erosion in this intertidal river banks and river flats or slopes.

Stem differentiated into nodes and internodes, internodes 0.7 cm in length, glabrous, green, round, solid, shoot orthotrophic with infrequently diffused branching, articulates, with almost an envelop of persistent stipules; stipules minute, ephemeral as lateral triangular scales on each side of the petiole. Leaves spirally arranged, somewhat cluster to the end of spirally arranged, erect shoot, simple, exstipulate, cauline, petiolate; petiole 1.5 cm to 2.0 cm in length, round, green, reddish occasionally on maturity, glabrous, lamina ovate elliptic, 4.0 cm to 7.5 cm in length and 2.4 cm - 3.5 cm broad, inconspicuously notched with a minute gland in the base of young leaves; basal glands 2 (-4) usually 1.0 cm diam. found each side of the blade at its insertion on the petiole; the glands usually circular often associated with additional outgrowths like the marginal glands. Each leaf 1(-2) axillary bud (s), with a series of minute decussately arranged, woody bud-scales, greenish, yellowish red at maturity, pulvinous, acute, angled, dorsiventral, coriaceous , somewhat fleshy, glabrous, odourless, unicostate reticulate venation. Mature leaves senescent during late winter or early summer and new leaves profusely grow during late summer or early monsoon months. Male inflorescence catkins, pendulous or erect,

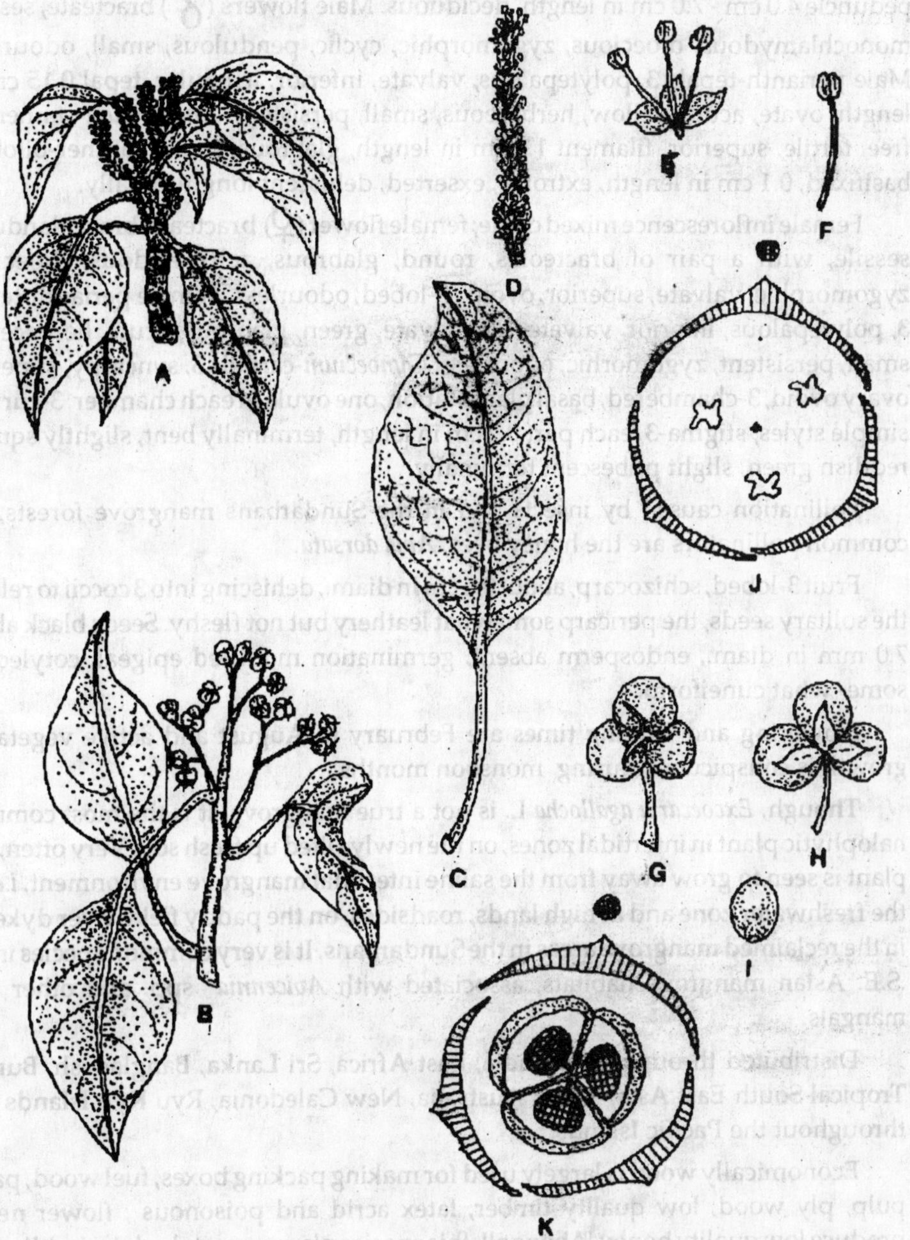

Fig. 22 : *Excoecaria agallocha* L .

A. Portion of flowering twig (male) x 1/3; B. Portion of flowerintg twig (female) x 1/3; C. Leaf : petiole - 1.8 cm long, lamina - 4.7 cm long x 2.4 cm width; D. Male inflorescence- 4.3 cm long; E. Male flower; F. Stamen: filament - 1.2 cm long; anther - 0.1 cm long; g. Female flower - 1.2 cm long; H. Female flower showing corolla; I. Seed- 0.5 cm long; J. Floral diagram (Male); K. Floral diagram (Female).

peduncle 4.0 cm - 7.0 cm in length, deciduous. Male flowers (♂) bracteate, sessile, monochlamydous, dioecious, zygomorphic, cyclic, pendulous, small, odourless. Male perianth-tepals 3, polytepalous, valvate, inferior, irregular, tepal 0.15 cm in length, ovate, acute, yellow, herbaceous, small, persistent. *Androecium*-stamens 3, free, fertile, superior, filament 1.2 cm in length, glabrous, round, anther-bilobed, basifixed, 0.1 cm in length, extrorse, exserted, dehiscent longitudinally.

Female inflorescence mixed cyme; female flower (♀) bracteate, bract glandular, sessile, with a pair of bracteoles, round, glabrous, achlamydous, dioecious zygomorphic, valvate, superior, ovoid, 3-lobed, odourless. Female perianth-tepals 3, polytepalous, inferior, valvate, tepal ovate, green, glabrous, acute, herbaceous, small, persistent, zygomorhic, odourless. *Gynoecium*-carpels 3, syncarpy, superior, ovary ovoid, 3-chambered, basal placentation, one ovule in each chamber. 3-curved, simple styles, stigma-3, each part 0.2 cm in length, terminally bent, slightly square, reddish green, slight pubescent terminally.

Pollination caused by insects and in the Sundarbans mangrove forests, the common pollinators are the honey bees, *Apis dorsata*.

Fruit 3-lobed, schizocarp, about 0.6 cm in diam., dehiscing into 3 coccii to release the solitary seeds, the pericarp somewhat leathery but not fleshy. Seeds black about 7.0 mm in diam., endosperm absent, germination modified epigeal, cotyledons some- what cuneiform.

Flowering and fruiting times are February to August and active vegetative growth is conspicuous during monsoon months.

Though, *Excoecaria agallocha* L. is not a true mangrove, it is the most common halophytic plant in intertidal zones, on the newly silted up fresh soil. Very often, this plant is seen to grow away from the saline intertidal mangrove environment, *i.e.*, in the freshwater zone and at high lands, roadsides, on the paddy field, river dykes or in the reclaimed mangrove areas in the Sundarbans. It is very common species in the S.E. Asian mangrove habitats, associated with *Avicennia* spp. and other true mangals.

Distributed throughout in India; East Africa, Sri Lanka, Bangladesh, Burma, Tropical South East Asia, North Australia, New Caledonia, Ryu Kyu Islands and throughout the Pacific Islands.

Economically wood is largely used for making packing boxes, fuel wood, paper pulp, ply wood, low quality timber; latex acrid and poisonous ; flower nectar produce low quality honey. Above all, this species play very vital role by holding the soil in the inter-tidal zones and add profuse biomass in the mangrove ecosystem.

Seed Dispersal, Seed-Germination and Seedling Morphology

Mature tri-loculer fruit has three seeds, dehisced longitudinally after complete drying. Three longitudinal cracks on the tri-locules of fruit make it to form 6 halves in each compartment, of which single seed is enclosed. Seeds are spread over on water surface, clay soil or elsewhere, depending the situation where the tree grow. Seeds may float on water and dispersed far away places; on reaching the soil phases, it response the germination after short dormancy within 15-20 days the seeds germination initiate.

Seed is round and flattened, attain 0.4 cm - 0.7 cm diam.; seed-coat is thin and black. During germination seeds gradually intake water and enlarge the size upto 1.0 cm diam. and short radicle comes out through seed-coat pore and penetrate into the soil layer. Seed-coat gradually ruptures and simultaneously radicle increases in length upwardly from the ground, upto 6.5 cm long. At this stage cotyledonary leaves become prominent and the seed-coat is pushed back at the tip of cotyledonary leaves, which attain the length upto 2.0 cm long and 1.5 cm broad. Cotyledonary leaves are green and enclosed short plumule. After 30 days, plumule completely flourishes two opposite foliage leaves, however the colyledonary leaves still remain attached at the junction of embryonic axis, until the seedling become mature.

FAMILY—IX. *MYRSINACEAE*

The family—Myrsinaceae has more than 1000 species under 30 genera; which are also distributed almost throughout the tropical and Sub-tropical zones (Tomlinson, 1986). This family—Myrsinaceae has only one minor element of mangrove genus, *viz., Aegiceras* Gaertner and two other mangrove associated or back mangal genus, *viz., Ardisia* Sw. and *Myrsine* L. Few species of these latter two genera are either grow as mangrove associates or as fringe mangal of the Old World Tropics; but these are not reported from the Indian mangrove ecosystems by the Indian Ecologists.

For having non-capsular, non-fleshy, single seeded fruit, this genus—*Aegiceras* Gaertner was treated under a distinct family — Aegiceraceae by de Candole in Prodromus 141 - 143.1844. But the other characters like, free central placentation, reginous elongated glands and elongated non-endospermous seeds resemble it as the tribe characters at *Ardisia* Sw. of Myrsinaceae. As such, in this present treatise this genus — *Aegiceras* Gaertner is considered as a member of Myrsinaceae.

GENUS - 13. *AEGICERAS* Gaertner, Fruct, et Sem 1 : 216. 1788.

Aegiceras Gaertner is the only minor mangrove genus of the family Myrsinaceae; it has only two minor mangrove species, *viz., A. corniculatum* (L.) Blanco and *A. floridum* Roem. & Schult. These species are distributed in the Old World Mangrove Regions, particularly in the S.E. Asian and Tropical Australian mangrove ecosystems, but not in the Pacific Islands. Only, *A. corniculatum* (L). Blanco is common but minor mangrove flora in the Indian mangals, including Andaman & Nicobar Islands. *A. floridum* Roem. & Schult. is reported only from N. Borneo, Java, Moluccas, Celebes and throughout the Philippines to Indochina (Tomlinson, 1986).

13.1. *Aegiceras corniculatum* (L.) Blanco *Fl.* Filip. 79.1837.

Rhizophora corniculatum L. Sp. Pl. 448.1753.

Aegiceras majus Gaertner Fruct, et Sem 1 : 216. 1788.

A. fragrans Konig, Konig & Sims Ann. Bot 1 : 131.1805.

Chromosome No. 2n = 36 (Subramanian, 1980); 2n = 46 (Faure, 1968) and 2n = 18 (Sidhu, 1961).

Common Name : River Mangrove.

Local Name (Bengali) : Khalsi

(Plate - 107 and Fig. 23)

Small tree or shurb, upto 4 m high, 16.0 cm - 20.0 cm trunk diam. at the base, spreading diffused branched, glabrous. Bark is almost smooth, pale or dark grey.

Tap roots are well developed, much branched, horizontally spreading, but without aerial roots; trunk base is much thick with the growth of basal roots.

Stem differentiated into nodes and internodes, glabrous, short apex, blunt and covered by leaf cluster and resting bud in terminally enclosed by scales.

Leaves simple, alternate or spirally arranged, exstipulates, cauline, petiolate; petiole 0.5 cm - 1.0 cm long, glabrous, pulvinous; lamina ovate-obovate, oblong or ovate-oblong or elliptic, entire, apex round to emerginate, cuneate base, shiny, margin curved downwardly, dorsiventral, coriaceous, mid-vein distinct in ventral side, 6.0 cm - 10.0 cm long and 3.0 cm - 5.0 cm broad.

Inflorescence simple umbels in the shoot apex or on the axillary branches, peduncle upto 0.5 cm long or occasionally absent; pedicellate, pedicel 0.8 cm - 1.0 cm long, glabrous, terete, bracts minute, bracteoles absent; *flower* pentamerous, pointed in bud condition, fragrant, 1.5 cm - 2.0 cm long and 0.5 cm - 1.0 cm across ; complete, bisexual, zygomorphic, erect; *sepals* - 5, polysepalous, twisted, free, oblong, entire, coriaceous, round apex, persistent, curved, 1.6 cm long and 0.2 cm broad; *petals* - 5, gamopetalous, twisted, slightly fused at base to form short tube, 0.95 cm long and 0.1 cm - 0.25 cm broad, reflexed at maturity, hairy at the mouth of petals and capitate hairs at the base, oblong - elliptic, entire, acuminate, herbaceous, white, not showy, inferior, deciduous; *stamens* - 5, opposite to petal, filaments 0.3 cm long, base united to form a tube, basifixed anthers, sagittate, longitudinally dehisced; inserted, introrse, bilobed, 0.25cm long, white, *ovary* 0.7 cm long, pyramidal, enlarged upto corolla mouth, ovary single loculed, several ovules in one chambered, but after fertilization one ovule survive, free central placentation; *style* -1, terminal, 0.3 cm long, glabrous, soft, gradually tapering, stigma absent.

Fruit capsule, 2.0 cm - 9.0 cm long, completely curved with pointed apex, calyx persistent at the base, capsule one seeded, 1.2 cm long, incipient viviparous germination and epigeal.

In the Sundarbans mangrove forest, this species start flowering during January and fruits are seen to hang on the mother plants upto September. The normal vegetative growth noted during summer and monsoon months. Active growth take place by terminal buds.

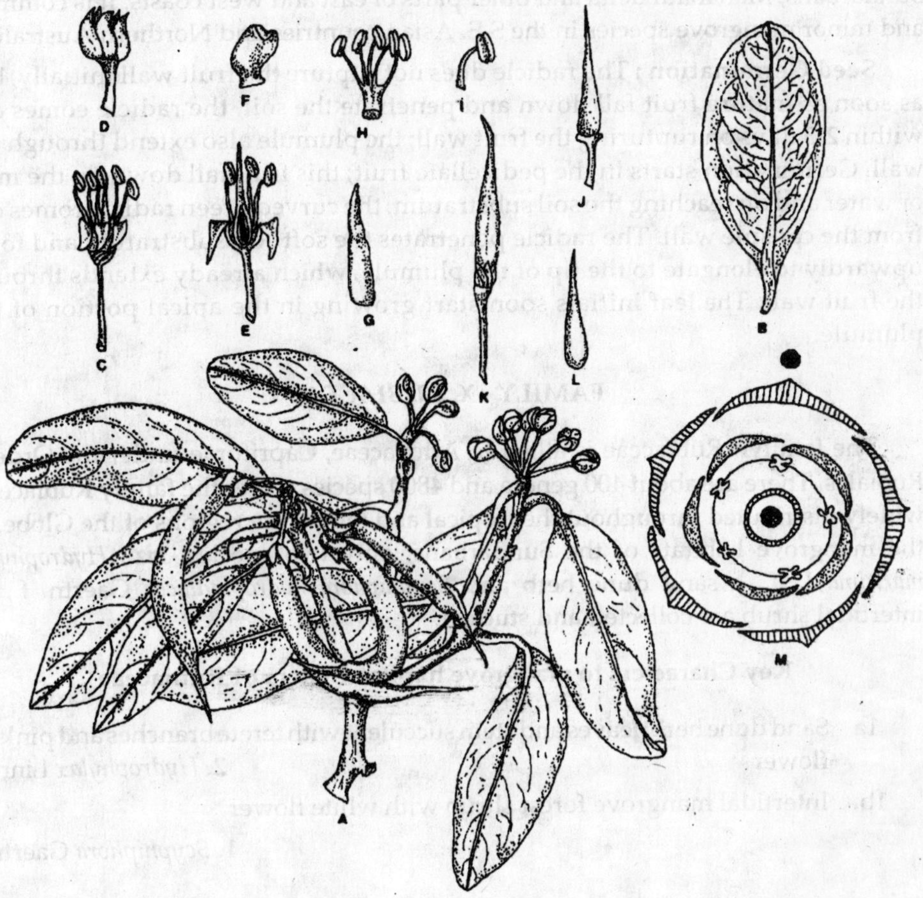

Fig. 23 : ***Aegiceras corniculatam*** **(L.) Blanco**

A. Portion of flowering twig x 1/3; B. Leaf : petiole - 1.0 cm long, lamina - 8.2 cm long x 3.0 cm width; C. Pedicellate flower : pedicel - 0.9 cm long, flower - 1.9 cm long x 0.5 cm across; D. Calyx - 1.3 cm long; E. L.S. of flower; F. Sepal; G. Petal - 0.9 cm long, x 0.25 m width; H. Androecium; I. Stamen : filament-0.25 cm long, anther - 0.25 cm long; . J. Gynoecium : carpel - 0.8 cm long; K. Fruit - 2.0 cm long; L. Seed - 1.2 cm long; M. Floral diagram.

A. corniculatum (L.) Blanco is very much dominant in the mid- and outer-estuaries, almost throughout the Sundarbans. This species often grow on the river slopes along with the members of Rhizophoraceae and Avicenniaceae.

Pollination caused by insects ; its sweet scented and nectariferous white flowers attract the insect, particularly the rock-bee, *Apis dorsata*. During the late winter and early summer months white coloured honey was collected by A. *dorsata* from *A. corniculatum* (L.) Blanco, which is considered the best quality honey of the Sundarbans.

A. corniculatum (L). Blanco is distributed throughout in India, particularly in

Sundarbans, Mahanadi delta and other parts of east and west coasts. It is common and minor mangrove species in the S.E. Asian countries and Northern Australia.

Seed Germination : The radicle does not rupture the fruit wall initially, but as soon as mature fruit fall down and penetrate the soil, the radicle comes out within 2-3 days on rupturing the fruit wall; the plumule also extend through the wall. Germination starts in the pedicellate fruit; this fruit fall down on the mud or water and on reaching the soil substratum, the curved green radicle comes out from the capsule wall. The radicle penetrates the soft soil substratum and force upwardly to elongate to the tip of the plumule, which already extends through the fruit wall. The leaf initials soon start growing in the apical portion of the plumule.

FAMILY - X. *RUBIACEAE*

The family - Rubiaceae is allied to Adoxaceae, Capriforae under the Order - Rubiales. There are about 400 genera and 4800 species under the family Rubiaceae, widely distributed throughout the tropical and temperate regions of the Globe. In the mangrove habitats of the Sundarbans, only two species, *viz.*, *Hydrophylax maritima* L. f., a sand dune herb and *Scyphiphora hydrophyllacea* Gaertn. f., an intertidal shrub are collected and studied.

Key Characters to mangrove habitat Genera of Rubiaceae

1a Sand dune herb; leaves and stem succulent with terete branches and pinkish flower 2. *Hydrophylax* Linn. f.

1b. Intertidal mangrove forest shrub with white flower 1. *Scyphiphora* Gaertn.f.

GENUS - 14. *SCYPHIPHORA* Gaertn. f. Fruct. Sem. Pl. 3: 91.1805

The genus - *Scyphiphora* Gaertn.f. is a monospecific genus and this plant — *S. hydrophyllacea* Gaerth. f. is distributed in the Mangals of Indo-Malayan Coasts and Australia.

14.1. *Scyphiphora hydrophyllacea* Gaertn. f. Fruct. Sem. Pl. 3 : 91.1805.

Ixora manila Blanco Fl. Filip. 635. 1837.

Local Name (Bengali) : Tagri Bani

(Plate - 108 and Fig. 24)

Shrub or small tree, attain a height upto 2.5 m; much branched, evergreen minor element of the mangroves or mangrove associate, in the intertidal zones of the outer estuary and on sandy soil of the Sundarbans.

Much spreading horizontal roots, underdeveloped tap roots and aerial roots wavy, not much conspicuous.

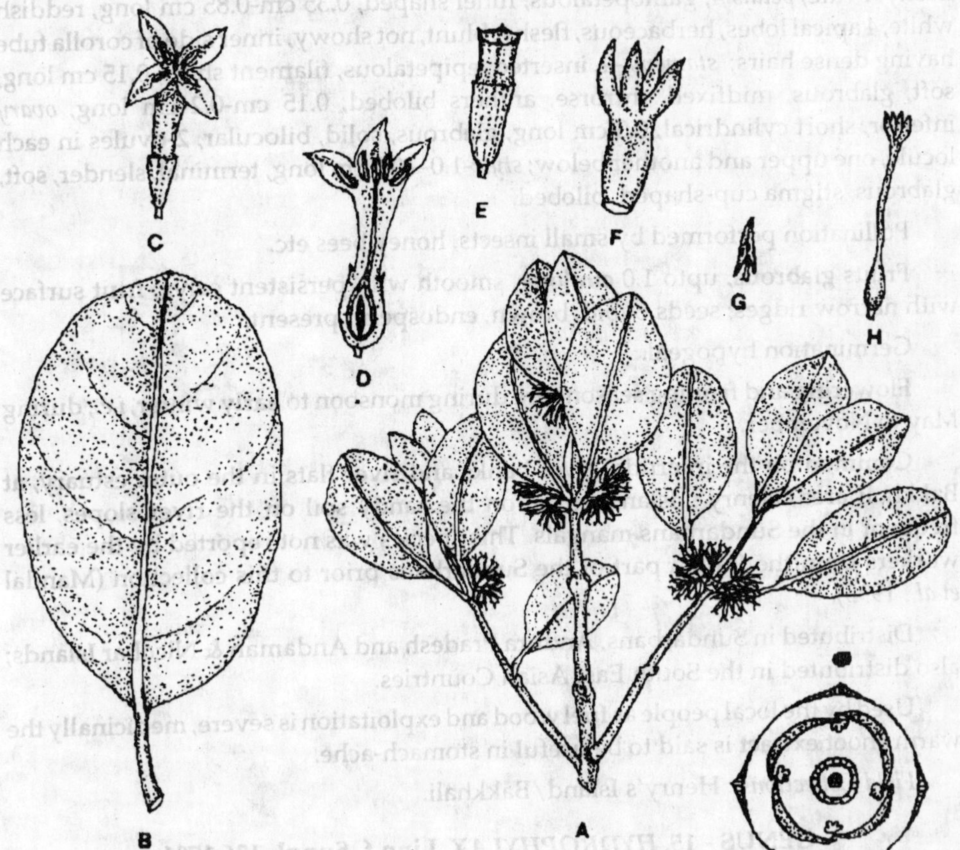

Fig. 24 : *Scyphiphora hydrohyllacea* Geartn. f.

A. Portion of flowering twig x 1/4; B. Leaf : petiole - 1.3 cm long, lamina- 7.0 cm long x 4.2 cm width; C. Flower - 1.0 cm long x 0.5 cm across; D. L.S. of flower; E. Calyx -0.6 cm long; F. Corolla: lobe-0.45 cm long, corolla tube-.0.35 cm width; G. Stamen-0.25 cm long; H. Gynoecium - 1.0 cm long; I. Floral diagram.

Internodes slightly ridged at the base of node, solid, terete, green, glabrous, flattened at the nodal portion.

Leaves simple, opposite decussate, interstipulate; stipule minute, 3 mm - 4 mm, cup-shaped, hairs at margin, petiolate; petiole 1.0 cm - 2.0 cm long, pulvinous, solid, green, glabrous, terete; lamina ovate-elliptic or ovate-oblong, entire, blunt, cuneate at base, dorsiventral, fleshy, green, glabrous, unicostate reticulate venation, 6.0 cm - 10.0 cm long and 3.0 cm - 5.5 cm width.

Inflorescence condensed cyme, having 4-7 flowers in aggregate, ebracteate, develop at node; *flower*-1.0 cm long × 0.5 cm across, tetramerous, complete, bisexual, erect, pedicellate; pedicel short, 0.15 cm long, solid, glabrous, terete; *sepals* 4, gamosepalous, persistent, 0.3 cm-0.6 cm long, glabrous, 4 apical short teeths, green, fleshy, ovate; *petals* 4, gamopetalous, funel shaped, 0.35 cm-0.85 cm long, reddish white, 4 apical lobes, herbaceous, fleshy, blunt, not showy, inner side of corolla tube having dense hairs; *stamens* -4, inserted, epipetalous, filament short, 0.15 cm long, soft, glabrous, midfixed, introrse, anthers bilobed, 0.15 cm-0.2 cm long, *ovary* inferior, short cylindrical, 0.3 cm long, glabrous, solid, bilocular, 2-ovules in each locule, one upper and another below; *style*-1.0—0.7 cm long, terminal, slender, soft, glabrous, stigma cup-shaped, bilobed.

Pollination performed by small insects, honey bees etc.

Fruits glabrous, upto 1.0 cm long, smooth with persistent calyx; fruit surface with narrow ridges; seeds ovoid, brown, endosperm present.

Germination hypogeal.

Flowering and fruiting seasons are during monsoon to early winter, *i.e.*, during May to November.

Common on the intertidal river banks and river flats in the outer estuary, at Bakkhali area/Henry's Island; grow on the sandy soil on the river slopes, less frequent in the Sundarbans mangals. This species was not reported by the earlier workers from the Indian part of the Sundarbans prior to this collection (Mandal *et al.*, 1995).

Distributed in Sundarbans, Andhra Pradesh and Andaman & Nicobar Islands; also distributed in the South East Asian Countries.

Used by the local people as fuel wood and exploitation is severe, medicinally the warm shoot extract is said to be useful in stomach-ache.

Field Collections : Henry's Island/Bakkhali.

GENUS - 15. *HYDROPHYLAX* Linn.f. Suppl. 126.1781

The genus - *Hydrophylax* Linn.f. is a sand dune flora, mainly distributed throughout the coastal area of India. It has only one species, *i.e., H. maritima* Linn.f. found in the sandy beach of coastal area.

15.1. *Hydrohylax maritima* Linn. f. Suppl. 126.1781.

(Plate - 109 and Fig. 25)

Prostrate much branched herb; creeping or twining on sandy beach.

Tap root system and adventatious roots develop from the nodes ; these profuse roots anchor the plant to the sandy substratum.

Internodes slightly ridged, easily differentiated, 1.2 cm long, fleshy, hollow, terete-flattened, glabrous, thick, succulent, dark green, poorly axile branches develop.

Simple, opposite decussate, interstipulate; sessile, shortly spathe covered the node, dark green, glabrous; lamina ovate elliptic, succulent, glabrous, dark green, 2.5 cm long and 1.7 cm width, acute, entire, main vein not distinct, isobilateral,

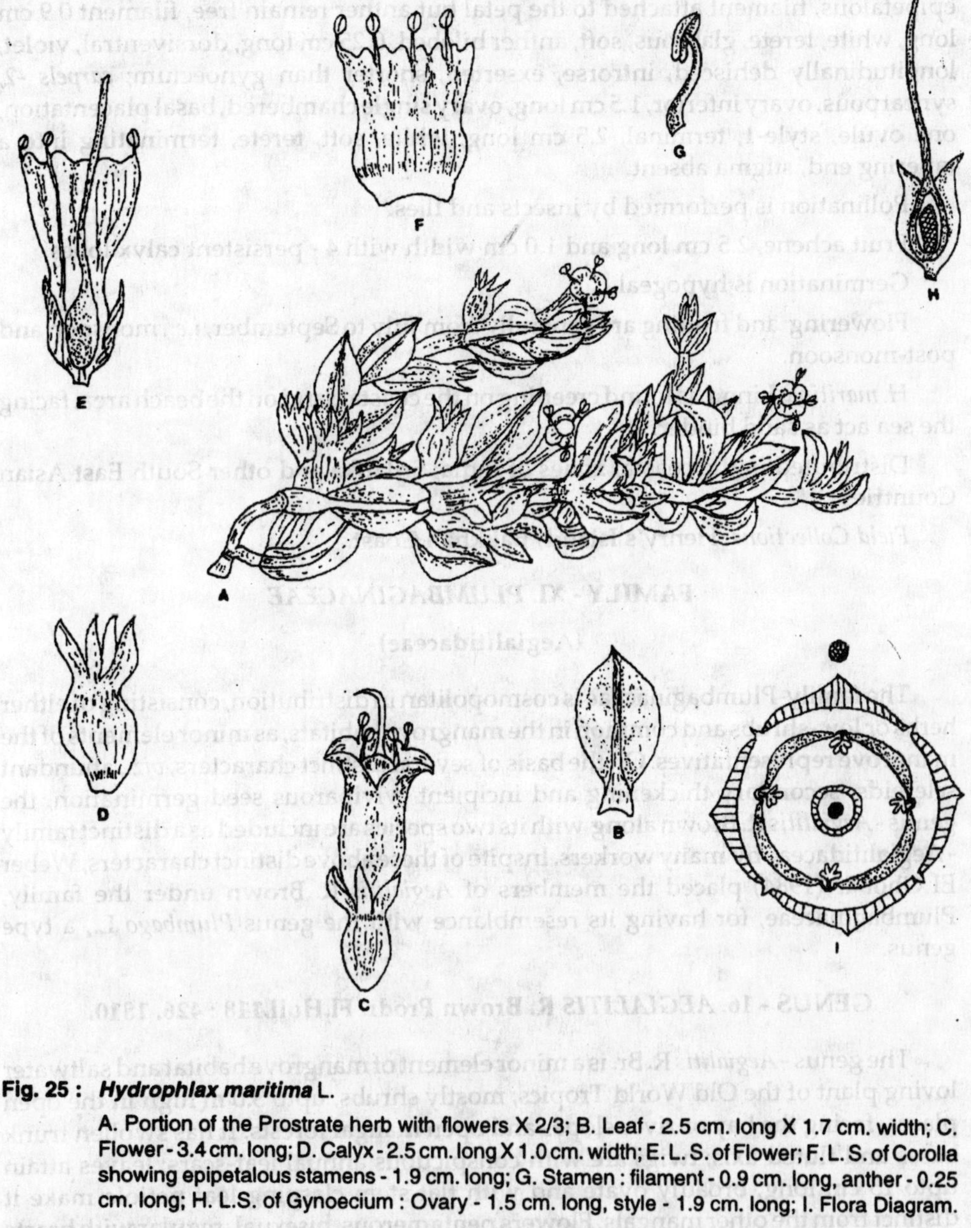

Fig. 25 : ***Hydrophlax maritima*** **L.**

A. Portion of the Prostrate herb with flowers X 2/3; B. Leaf - 2.5 cm. long X 1.7 cm. width; C. Flower - 3.4 cm. long; D. Calyx - 2.5 cm. long X 1.0 cm. width; E. L. S. of Flower; F. L. S. of Corolla showing epipetalous stamens - 1.9 cm. long; G. Stamen : filament - 0.9 cm. long, anther - 0.25 cm. long; H. L.S. of Gynoecium : Ovary - 1.5 cm. long, style - 1.9 cm. long; I. Flora Diagram.

unicostate, reticulate venation inconspicuously found. *Solitary flower* develop at the leaf base, *i.e.,* axile solitary flower; ebracteate, complete, bisexual, regular, erect, odourless, sessile, 3.4 cm long, hollow tube-shaped, tetramerous; *sepals*-4, gamosepalous 2.5 cm long × 1.0 width, superior, 4-lobes above, lobes 0.8 cm long, thick, fleshy, green, glabrous, valvate, alternate to petal, persistent; *petals* - 4, gamopetalous, hollow tube, 4-lobes above, lobes purple colour but tube white coloured, herbaceous, superior, valvate, glabrous, deciduous, 1.9 cm long; *stamens*-4, epipetalous, filament attached to the petal but anther remain free, filament 0.9 cm long, white, terete, glabrous, soft, anther bilobed, 0.25 cm long, dorsiventral, violet, longitudinally dehisced, introrse, exserted, shorter than gynoecium; *carpels* -2, syncarpous, ovary inferior, 1.5 cm long, ovary single chambered, basal placentation, one ovule, style-1, terminal, 2.5 cm long, white, soft, terete, terminating into a tapering end, stigma absent.

Pollination is performed by insects and flies.

Fruit achene, 2.5 cm long and 1.0 cm width with 4 - persistent calyx lobes.

Germination is hypogeal.

Flowering and fruiting are generally from July to September, *i.e.*, monsoon and post-monsoon.

H. maritima Linn.f. is found creeping on the coastal sand on the beach area, facing the sea act as sand binder.

Distributed in the coastal zones of India; Srilanka and other South East Asian Countries.

Field Collections: Henry's Island/Bakkhali Coast.

FAMILY - XI. *PLUMBAGINACEAE*

(Aegialitidaceae)

The family-Plumbaginaceae is cosmopolitan in distribution, consisting of either herbs or low-shrubs and common in the mangrove habitats, as minor elements of the mangrove representatives. On the basis of several distinct characters, *viz.*, abundant sclereids, secondary thickening and incipient viviparous seed germination, the genus - *Aegialitis* R. Brown along with its two species are included as a distinct family - Aegialitidaceae by many workers. Inspite of these above distinct characters, Weber El Ghobar (1984) placed the members of *Aegialitis* R. Brown under the family, Plumbaginaceae, for having its resemblance with the genus *Plumbago* L., a type genus.

GENUS - 16. *AEGIALITIS* R. Brown Prodr. Fl.Holl.118 : 426. 1810.

The genus - *Aegialitis* R. Br. is a minor element of mangrove habitat and saltwater loving plant of the Old World Tropics; mostly shrubs, upto 3.0 m high in the open places, facing the bay or river slopes and open mangal forests. It has swollen trunk bases and fluted axis; twigs are with conspicuous annual leaf-scars; leaves attain upto 15 cm long, broadly ovate and with flat stem clasping leaf petiole make it distinct from the other mangals. Flowers pentamerous, bisexual, regular with bracts and bracteoles, white; single seeded fruit and extended hypocotyle attain upto 7.0 cm long; germination epigeal and incipient viviparous.

There are two species under *Aegialitis* R. Brown, *viz.*, *A. annulata* R. Brown (the type species) and *A. rotundifolia* Roxb. *A. annulata* R. Br. is distributed in N. Australia and Malesia, while *A. rotundifolia* Roxb. is the common but minor elements of mangroves in Indian mangals and mangals of Bangladesh and Burma. These two species are never mixed with one other (Tomlinson, 1986). The distinctive features between these two species are (1) *A. annulata* R. Br. is having 9-10 corolla lobes, while *H. A. rotundifolia* Roxb. is having 5 corolla lobes.

In Indian mangroves, like Sundarbans of West Bengal, Mahanadi delta of Orissa and Andaman & Nicobar Islands, the species *A. rotundifolia* is common and here only the Indian species is described in respect of its morpho-anatomical peculiarities and ecological adaptabilities.

16.1. *Aegialitis rotundifolia* Roxb. Fl. Ind. 2 : 111.1824.

Common Name : Club Mangrove

Local Name (Bengali) : Tora or Satari

(Plate - 110 and Fig. 26)

Shrub, upto 3 m high, without any aerial roots, but trunk base broad due to basal fused upright roots; evergreen; stem swollen and fluted with spongy structure, about 20 cm diam. Bark dark grey or blackish grey and fissured.

Tap roots sunken, but not penetrate into deep; main roots from trunk base thick but gradually narrow and ultimately blunt tip, about 0.5 m long with many branched roots, which are spreading horizontally.

Stem straight, with inconspicuous leaf scars on shoot, branching only towards apex, but branching more or less parallel, arranged towards main shoot.

Leaf simple, alternate, cauline, exstipulate, petiolate, petiole 5.0 cm - 8.5 cm long, dilated at base, clasping the stem, tubular leaf-sheath, glabrous, solid but spongy, gradually narrow towards the lamina, terete, 0.8 cm diam. at base, green, lamina broadly ovate, 6.0 cm - 9.0 cm long and 6.0 cm - 9.0 cm broad, coriaceous, slightly fleshy, dark green, dorsiventral, dorsal side shiny, entire, obtuse, cuneate at base, mid-vein prominent but sub-costas, arranged parallely one another and terminated to margin.

Inflorescence raceme, branched, peduncle upto 6.0 cm long, developed from sheathing petiole. *Flower* bracteate, one bract and two bracteoles, bract ovate-lanceolate, entire, obtuse, curved, upto 1.0 cm long and 0.8 cm broad, green, bracteoles same as bract, but mucronate apex, 0.8 cm long covered by bract, pedicellate, pedicel 0.5 cm long, terete, greenish white, glabrous, erect, flower complete, pentamerous, regular, bisexual, odourless, erect, 1.5 cm long × 1.0 cm across. *Calyx* - sepals-5, gamosepalous, 5 lobes above, short adjoining at base, each lobe 1.0 cm long and 0.2 cm broad, entire, acute, green, glabrous, coriaceous, odourless, persistent, valvate, inferior, equal length. *Corolla*-petals 5 – 1.2 cm long × 0.15 cm width, polypetalous, linear oblong, white, imbricate, entire, obtuse, herbaceous, thin, mid-vein inconspicuouly found, odourless, alternate to sepals, equal length, not showy, glabrous. *Androecium*-stamens 5, 1.6 cm long, adhere, filament base adjoining to form short hollow tube, thin, 0.5 cm long, filament whitish, 1.3 cm long terete, glabrous soft, exserted, equal length, extrorse, basifixed, anther sagittate, bilobed, 0.2 cm long, yellowish-longitudinal dehisced, inferior. *Gynoecium*-carpels 5, syncarpy, ovary oblong, 0.4 cm long, one chambered, basal

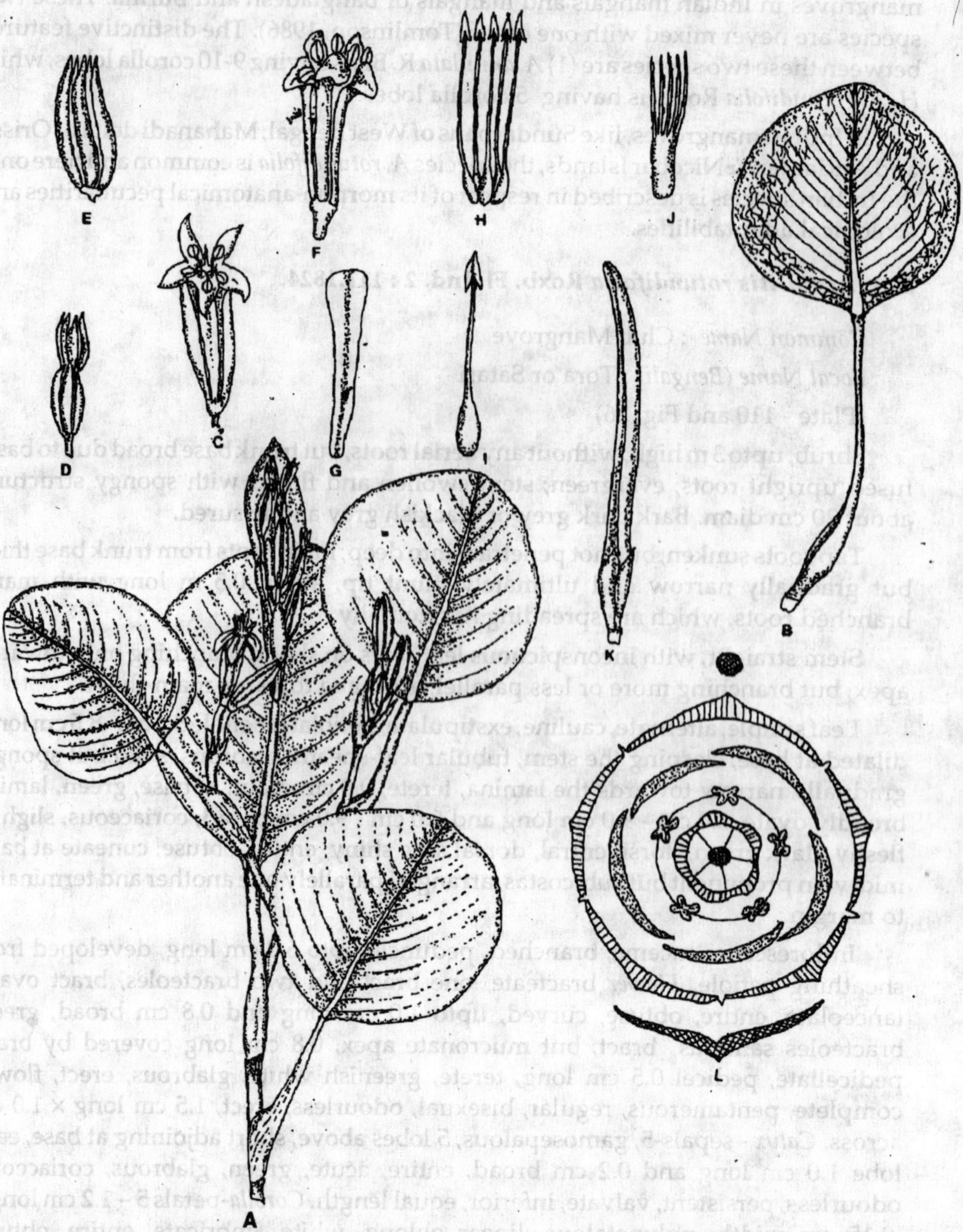

Fig. 26 : *Aegialitis rotundifolia* Roxb.

A. Portion of flowering twig x 1/3; B. Leaf : petiole - 7.0 cm long, lamina - 6.0 cm long x 6.0 cm width; C. Flower - 1.5 cm long; D. Flowering bud - 1.5 cm long; E. Calyx - 1.0 cm long; F. Corolla with androecium- 1.2 cm long; G. Petal - 1.2 cm long x 0.15 cm width, H. Androecium - 1.6 cm long; I. Stamen : filament - 1.3 cm long; anther - 0.2 cm long; J. Gynoecium : ovary - 0.4 cm long, style - 0.5 cm long; K. Fruit - 6.0 cm long L. Floral diagram.

placentation, styles 5, terminal, soft, glabrous, terete, 0.5 cm long, white, stigmas absent.

Pollination caused by insects and honey bee, *Apis dorsata.*

Fruit capsule, upto 7.0 cm long, persistent calyx, yellowish at young stage, become brown quickly after maturation. Seed single in each fruit, 7.0 cm long, black. Incipient viviparous seed germination epigeal.

Flowering and fruiting time are during the month of April to October; active vegetative growth conspicuous during monsoon months.

This species, *Aegialitis rotundifolia* Roxb. prefers to grow outer ridges or river slopes, away from the other species of the mangroves; grow mostly sea-facing and common on the outer estuaries. This species can tolerate high saline condition and grow in association with the other mangals, like *Lumnitzera racemosa, Ceriops decandra, Suaeda* spp.

Distributed in Sundarbans, Andaman & Nicobar Islands and Mahanadi delta; also common throughout the mangal habitats in the Asian tropics.

Seed Dispersal, Germination and Seedling Morphology

Mature fruits are deep brown and hard and dispersed by water current. During monsoon months, generally these mature fruits are abcissed from the pedicel, which are either settled on sticky silty mud or float through the water current. During this dormant stages, the pericarp along with seed absorbs water by imbibation. The floating fruit is found to reach by high tides and are placed gregariously on the sea shore or river flats.

During this stage, after settling the fruit respond to germination by absorbing water; the pericarp become soft and rupture longitudinally into 2-4 halves at the end and the yellowish fleshy radicle become exposed or open and penetrate slightly into the sticky soil through small curvature. After penetrating the soil, the tip of the radicle gradually comes up straight perpendicularly. Gradually several root initials are come out from the end of the radicle and immediately anchor the sticky mud. After 20-30 days latter, pericarps are completely detached from these seeds and the cotyledons become visible, covering short plumule. Gradually the cotyledonary leaves enlarge one after another and plumule gets exposed; short plumule gradually increases its length perpendicularly, foliage leaves develop one after another; the earlier leaf covers the latter leaf bud by the sheathing petiole.

FAMILY - XII. *PTERIDACEAE*

The family - Pteridaceae (Fern) is under the Order—Pteridales has about 35 genera and more than 1000 species (Tomlinson 1986). Among these the Genus - *Acrostichum* L. is generally grow in the back-mangal areas or mangrove communities and associated with several back-mangal species. Tomlinson (1986) has reported three species, *viz., Acrostichum aureum* L., *A. speciosum* Willd. and *A. danaeifolium* Langsd. & Fisch. from the mangals and back-mangal communities of the tropical and subtropical world. During the present investigations only *A. aureum* L. is collected from the mangals of the Sundarbans and described.

GENUS - 17. *ACROSTICHUM* L. Sp. Pl. 2 : 1067. 1753.

The Genus -*Acrostichum* L. is popularly known as 'mangrove fern', though it grow commonly in the mangrove cleared or back mangal zones. This genus is characterised by upper pinnal reduced and covered beneath with sporangia (acrostichoid condition). In the mangals of the Sundarbans, only the species *A. aureum* L. grow spontaneously.

17.1. *Acrostichum aureum* L. Sp. Pl. 2 : 1069. 1753.

Chrysodium aureum Mett. Fil. Hort. Lips. 21.1865.

Common Name : Mangrove Fern.

Local Name (Bengali) : Hodo

(Plate - 111 and Fig. 27)

Erect fern, attaining upto 2.0 m high ; having small bushy appearance; rhizomatous.

Fibrous typical fern like roots, but without any aerial roots.

Rhizome globose, horizontal, caudex, circinate vernation, young rachis covered by gregarious hair like structure, known as ramenta.

Frond simple, semipinnatifid, upto 1.0 m long, and 4.0 cm width, isobilateral, distinct mid-vein and reticulate vein present,entire, blunt end, green, and yellow on maturity with sorus, glabrous, coriaceous, unicostate reticulate venation.

Mature leaves or fronds become sporophyllous, diffuse sporangia at abaxial surface, mixed sporangia on both side of midvein, brown sporangia stalked, upper globose; sporangia formation is generally during winter months, *i.e.*, November to February.

Common back mangal fern throughout the mangals in India and also in the mangals of the Pantropical zones; grow within the intertidal and also in the above tidal zones, prefer with moderately freshwater inundation.

Tender fronds-leaves are edible as vegetable by the poor people. Medicinally the paste of rhizomes are said to used for treatment of boils.

FAMILY - XIII. *ACANTHACEAE*

Acanthaceae is a large tropical family, consisting large numbers of mesophytic herbs, shrubs and small trees. Among these, the genus - *Acanthus* L. is the type genus, having three species common in the mangrove habitats of the S.E. Asian regions and Tropical Africa and N. Australia.

GENUS - 18. *ACANTHUS* L. Sp. Pl. 639. 1753.

The genus - *Acanthus* L. is having about 50 species in the Old World Tropics, especially tropical Africa, Mediterranean region, North Australia and Tropical Asia. Among these, *A. ilicifolius* L. is the type species, common throughout the intertidal

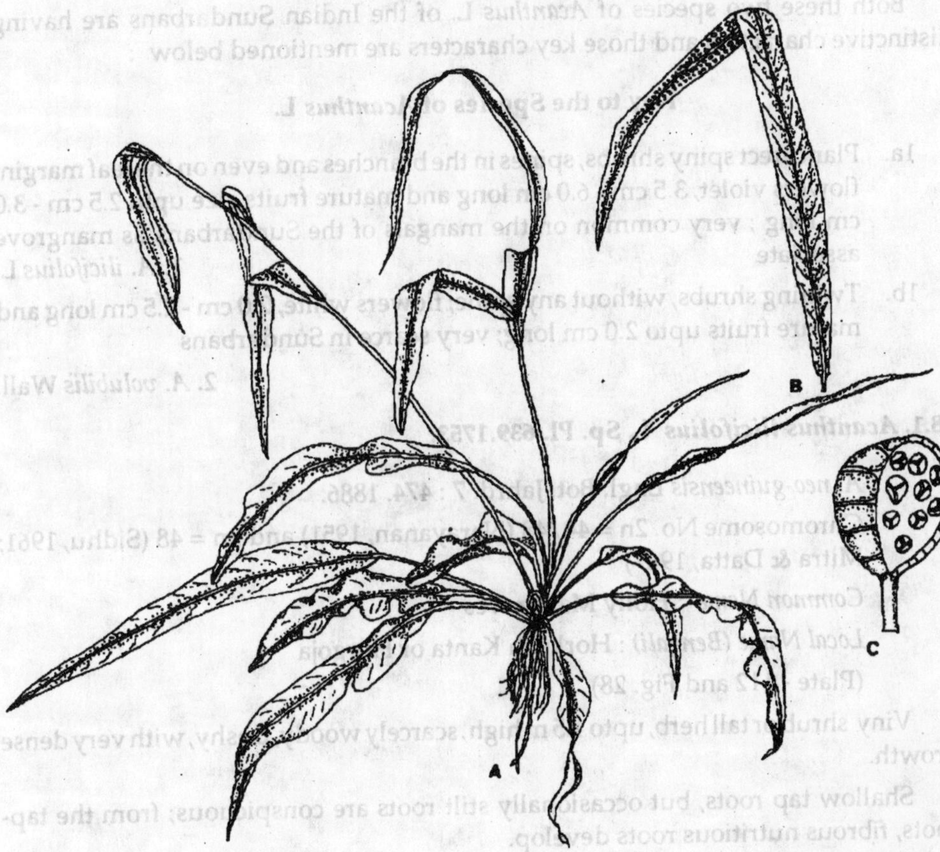

Fig. 27 : A. Entire plant B. Sporophyllous leaf C. Sporangium

and tidal mangals of S.E. Asian countries, Tropical Australia and Tropical Africa, as mangrove associate. Two other species, *viz., A. ebracteatus* Vahl and *A. volubilis* Wall. are inconspicuous in the mangal communities of S.E. Asia and elsewhere. Among these three species of *Acanthus* L., *A. ilicifolius* L. is having violate flower of 3.5 cm - 6.0 cm long and size of the fruit is 2.5 cm - 3.0 cm long; but the flowers of *A. ebracteatus* Vahl and *A. volubilis* Wall. are white and the length of flowers are less than 2.5 cm, fruit sizes are also less than 2.5 cm. Other vegetative characters have minor differences, *i.e., A. volubilis* Wall. is twinner and without any spines or arms.

In the Sundarbans mangals, *A. ilicifolius* L. is very much dominant, associated mangals; but *A. volubilis* Wall. is present very scarcely as twining shrub to other mangals. The third species, *A. ebracteatus* Vahl is not reported or collected from the Sundarbans mangals, though Tomlinson (1986) has reported its habitat similar to *Acanthus ilicifolius* L. But, *A. ebracteatus* is reported from Andaman Islands.

Both these two species of *Acanthus* L. of the Indian Sundarbans are having distinctive characters and those key characters are mentioned below :

Key to the Species of *Acanthus* L.

1a. Plant erect spiny shrubs, spines in the branches and even on the leaf margin; flowers violet, 3.5 cm - 6.0 cm long and mature fruits size upto 2.5 cm - 3.0 cm long ; very common on the mangals of the Sundarbans as mangrove associate 1. *A. ilicifolius* L.

1b. Twining shrubs, without any spine; flowers white, 2.0 cm - 2.5 cm long and mature fruits upto 2.0 cm long; very scarce in Sundarbans 2. *A. volubilis* Wall.

18.1. *Acanthus ilicifolius* L. Sp. Pl. 639.1753.

A. neo-guineensis Engl. Bot. Jahrd. 7 : 474. 1886.

Chromosome No. 2n = 44/48 (Narayanan, 1951) and 2n = 48 (Sidhu, 1961; Mitra & Datta, 1967)

Common Name : Holly Mangroves

Local Name (Bengali) : Horkoch Kanta or Horgoja

(Plate - 112 and Fig. 28)

Viny shrub or tall herb, upto 1.5 m high, scarcely woody, bushy, with very dense growth.

Shallow tap roots, but occasionally stilt roots are conspicuous; from the tap-roots, fibrous nutritious roots develop.

Stem differentiated into distinct nodes and internodes; internodes solid, glabrous, terete, greenish yellow, erect or spreading horizontally and sometime angled.

Leaf simple, opposite decussate, cauline, exstipulate, petiole short, 0.3 cm long, and 0.4 cm broad, flattened, glabrous, pulvinous to sheathing base, solid; lamina oblong, elliptic-oblong, spiny to very spiny margin and wavy; apex acute, basal angled, dorsiventral, green, odourless, unicostate reticulate venation conspicuously, upto 15.0 cm long and 6.0 cm broad.

Inflorescence racemose, spike, upto 16 cm long peduncled. *Flowers* bracteate, one bract, 0.8 cm long and 0.5 cm broad, curved, ovoid, entire, acute, basal cup-shaped, glabrous, greenish; two bracteoles partly covered by bract, same as bract in respect of shape but having less size than bract; flower upto 5.5 cm long and 3.8 cm across, bisexual, typically zygomorphic, complete, showy, erect, sessile, hypogynous, *calyx*-sepals 5, polysepalous, twisted, one cover completely other, outer one large, while inner one small, 1.7 cm - 2.0 cm long and 0.5 cm - 8.0 cm broad, ovate, margin folded, acute, basal curved, coriaceous, glabrous, greenish, persistent. *Corolla*-petals - 1, large, showy, violate, 3.5 cm - 4.0 cm long, × 3.4 cm - 3.6 cm broad, ovate, entire, obtuse, coriaceous, slightly fleshy, slightly curved, margin outerly

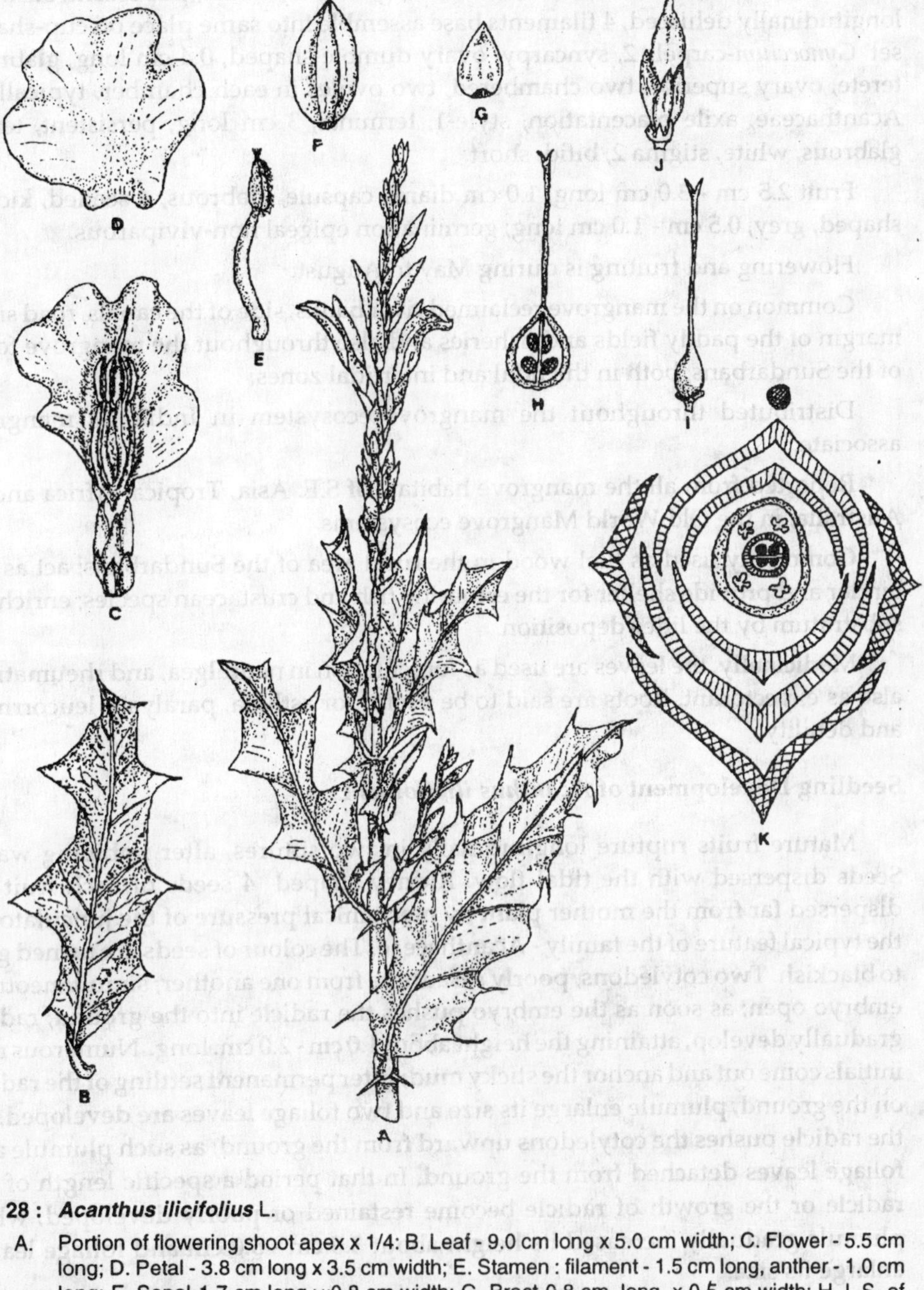

Fig. 28 : *Acanthus ilicifolius* L.

A. Portion of flowering shoot apex x 1/4: B. Leaf - 9.0 cm long x 5.0 cm width; C. Flower - 5.5 cm long; D. Petal - 3.8 cm long x 3.5 cm width; E. Stamen : filament - 1.5 cm long, anther - 1.0 cm long; F. Sepal-1.7 cm long × 0.8 cm width; G. Bract-0.8 cm. long x 0.5 cm width; H. L.S. of carpel; I. Gynoecium : ovary - 0.4 cm long, style - 3.0 cm long; J. Fruit with persistent calyx - 3.0 cm long; K. Floral diagram.

folded, deciduous. *Androecium*-stamens 4, free filament, but anther adhere loosely, filament 1.5 cm long, fleshy, solid, white, glabrous, 0.3 cm broad, flattened, erect, exserted, introrse, anther bilobed, 1.0 cm long, densely white, pubescent, basifixed longitudinally dehisced, 4 filaments base assemble into same place on cup-shaped set. *Gynoecium*-carpels 2, syncarpy, ovary dumble-shaped, 0.4 cm long, glabrous, terete, ovary superior, two chambered, two ovules in each chamber, typically as Acanthaceae, axile placentation, style-1, terminal, 3 cm long, persistent, terete, glabrous, white, stigma 2, bifid, short.

Fruit 2.5 cm - 3.0 cm long, 1.0 cm diam., capsule, glabrous, 4 seeded, kidney shaped, grey, 0.5 cm - 1.0 cm long; germination epigeal non-viviparous.

Flowering and fruiting is during May to August.

Common on the mangrove reclaimed river banks, side of the canals, road sides, margin of the paddy fields and fisheries and also throughout the mangrove forest of the Sundarbans; both in the tidal and intertidal zones.

Distributed throughout the mangrove ecosystem in India, as mangrove associate.

Reported from all the mangrove habitats of S.E. Asia, Tropical Africa and N. Australia, in the Old World Mangrove ecosystems.

Commonly used as fuel wood in the rural area of the Sundarbans; act as soil binder and provide shelter for the estuarine fish and crustacean species; enrich the substratum by the litter deposition.

Medicinally, the leaves are used as fomentation in neuralgea, and rheumatism, also as expectorant. Roots are said to be useful for asthma, paralysis, leucorrhoea and debility.

Seedling Development of *Acanthus ilicifolius* L.

Mature fruits rupture longitudinally in two sutures, after imbibing water. Seeds dispersed with the tidal flow. Kidney shaped 4 seeds of each fruit are dispersed far from the mother plant by mechanical pressure of the junculator as the typical feature of the family - Acanthaceae. The colour of seeds are turned grey to blackish. Two cotyledons, poorly disadhere from one another; simultaneously, embryo open; as soon as the embryo pushes the radicle into the ground; radicle gradually develop, attaining the height about 1.0 cm - 2.0 cm. long. Numerous root initials come out and anchor the sticky mud; after permanent settling of the radicle on the ground, plumule enlarge its size and two foliage leaves are developed. As the radicle pushes the cotyledons upward from the ground; as such plumule and foliage leaves detached from the ground. In that period a specific length of the radicle or the growth of radicle become restained or poorly developed, while plumule gradually increased its height about 3.0 cm - 5.0 cm and foliage leaves enlarge its size.

18.2. ***Acanthus volubilis*** **Wall. Pl. As. 2, 56. t.172. 1881**

Local Name (Bengali) : Lata Horgoja

(Plate - 113 and Fig. 29)

The plant is twining herb or shrub, attain the length, upto 4 m, much branched.

Tap root system developed and sometimes stilt roots are found. Node slightly ridged, internode variable in length, smooth surface, shiny, greenish, solid, terete.

Leaves simple, opposite decussate, exstipulate, petiolate; petiole upto 1.0 cm long, slightly flattened, glabrous, solid, pulvinous, green, lamina oblong-lanceolate, entire, blunt, angled, 8.0 cm long and 5.6 cm width, dorsiventral, succulent, green, glabrous, unicostate reticulate venation.

Inflorescence raceme, ebracteate, peduncle, upto 9.8 m long, erect, *flowers* ebracteate, sessile, complete, bisexual, irregular, erect, upto 2.5 cm long, *sepals* - 4, polysepalous, mixed size, 0.7 cm - 1.1 cm long and 0.3 cm - 0.6 cm width, green, glabrous, leathery, curved, entire, acute, superior. *Petals*-5, gamopetalous, fused to form single plate, base fused to form hollow short tube, while apex flattened to form 5 short lobes, 2.5 cm long and 1.6 cm width in flattened portion, white, fleshy, glabrous, entire; *stamens*-4, free, arranged in same whorle, equal size, filament-introrse, inserted, basifixed, flattened, glabrous, solid, hardy after maturation, 1. 0 cm long and 0.15 cm width, anther bilobed, brownish, 0.5 cm long and 0.2 cm width, gregarious hairs arising from ventral sides; *carpels*-2, syncarpous, ovary superior, dumble shaped, green, glabrous, 0.3 cm long and 0.15 cm width, two chambers, one ovule in each chamber, axile placentation, style-1, terminal, white, glabrous, terete, 1.3 cm long, stigma absent.

Pollination is caused by bumble bees and honey bees.

Fruit capsule, upto 2.0 cm long, persistent calyx, glabrous surface, shiny, seed with black seed coat.

Flowering and fruiting is during June to September.

Seed germination epigeal.

Very occasionally grow in the intertidal river bank thickets and twining on the other mangrove associated species, viz., *Excoecaria agallocha.*

This plant is collected from the degraded mangrove forest zone and intertidal areas of the Henry's Islands/near Kastala, within Bakkhali Forest range of the Western Sundarbans.

FAMILY - XIV. *TILIACEAE*

The tropical widely distributed family—Tiliaceae has about 50 genera and 400 species. These are mainly found in mesophytic habitats. Plants may be annual and perennial. Flowers are distinct and conspicuous, having numerous stamens. The fruits are generally capsule. Among these, the genus - *Brownlowia* Roxb. is the mangrove habitat plants.

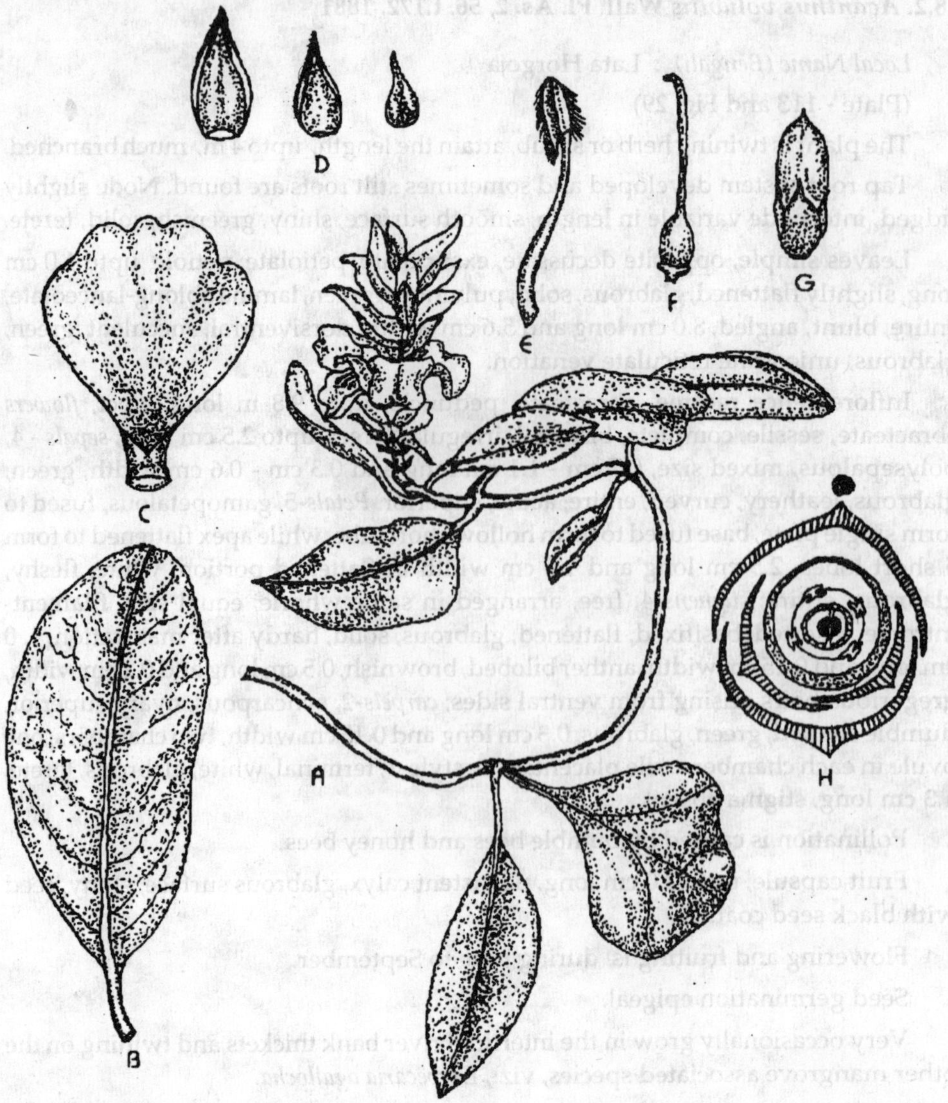

Fig. 29 : ***Acanthus volubilis*** **Wall.**

A. Part of Flowering twig x 1/3, B. petiole - 3.5 cm long, lamina 8 cm long × 5 cm width, C. Corolla, D. Calyx, E. Androecium, F. Gynoecium, G, Fruit, H. Floral diagram.

GENUS - 19. *BROWNLOWIA* Roxb. Pl. Corom. 3: 61, t. 265. 1819.

The genus - *Brownlowia* Roxb. having 30 species of which, most of these species are found in South East Asia. Two species of *Brownlowia, viz., B. argentata* Kurz and *B. lanceolata* (L.) Koesterm. are found in swampy areas. But, only the species *B. lanceolata* has been collected from the mangrove swamps of the Sundarbans.

19.1. *Brownlowia lanceolata* (L.) Kosterm. Reinw. 4 (4) : 73. 1959.

Local Name (Bengali) : Lata Sundari

(Plate - 114 and Fig. 30.)

Branched shrub attaining the height upto 3 m and the twining twig seems to be brownish, tap root system developed.

Bark brownish black, smooth, leaf simple, alternate, exstipulate, petiolate; petiole 0.5 cm - 2.0 cm long, brown, terete, not pulvinous, rough surface, solid, lamina 10.0-14.0 cm long and 2.5 - 4.0 cm width, lanceolate, entire, acute, round, coriaceous, dorsiventral, dorsal surface green, glabrous, shiny, ventral surface pale-green, leathery, unicostate reticulate venation. Mixed cyme, pendulous, pedunculate branched, ebracteate, *flower*; ebracteate, bisexual, complete, regular, pendulous, pedicellate; pedicel short - 0.15 cm long, terete, rough surface, brownish, flower 0.45 cm long and .4 cm across, not showy, *sepals*-4, polysepalous, ovate, entire, blunt, curved, outer surface rough having dark brown spots, inner surface glabrous, coriaceous, 0.3 cm long and 0.2 cm width at base, brownish, *petals* - 6, polypetalous, obovate, entire, curved, angled, narrow venation towards apex, herbaceous, yellow, 0.45 cm long and 0.2 cm width at apex. *Stamens* - indefinite, free, filament exserted, extrorse, glabrous, soft, terete, yellowish, basifixed, 0.2 cm long, anther-bilobed, 0.05 cm long, dark yellow, basifixed, *carpels* - 3, apocarpous,ovary superior, one chambered, two ovules in each chamber, axile placentation, 0.05 cm long, style - terminal, 0.2 cm long, glabrous, terete, greenish stigma absent.

Pollination caused by insect and honey bees.

Fruit capsule, small.

Flowering and fruiting periods are June to September.

The plant is treated as mangrove associate, as it mainly grow on the river slope, the intertidal river flats, banks, creeks and canals sides.

Collections were made from Henry's Island of Bakkhali range and Chhota Sudhanyakhali canal area of Pirkhali Block.

The plant is distributed throughout India and the Malayasia region.

FAMILY - XV. *MALVACEAE*

The family - Malvaceae is well known in the tropical and temperate region; it has about 80 genera, including 1500 species. Most of these members of Malvaceae are trees, shrubs and woody herbs (Tomlinson,1986). This family is characterised by mucillagenous juice, free of fused epicalyx, fused stamens to form a central staminal column; pentamerous flowers, yellow colour and ovary superior. Generally, few species under two genera, *viz., Hibiscus* L. and *Thespesia* Solander ex Correa. are distributed in the seashores, as mangrove associates or back-mangals.

Key Characters of the Mangrove Associates under Malvaceae

1a. Epicalyx present, calyx conspicuously 5-toothed, leaves purely cordate with usually 9-11 main veins; capsule dehisced into 5 separate valves 1. *Hibiscus* L.

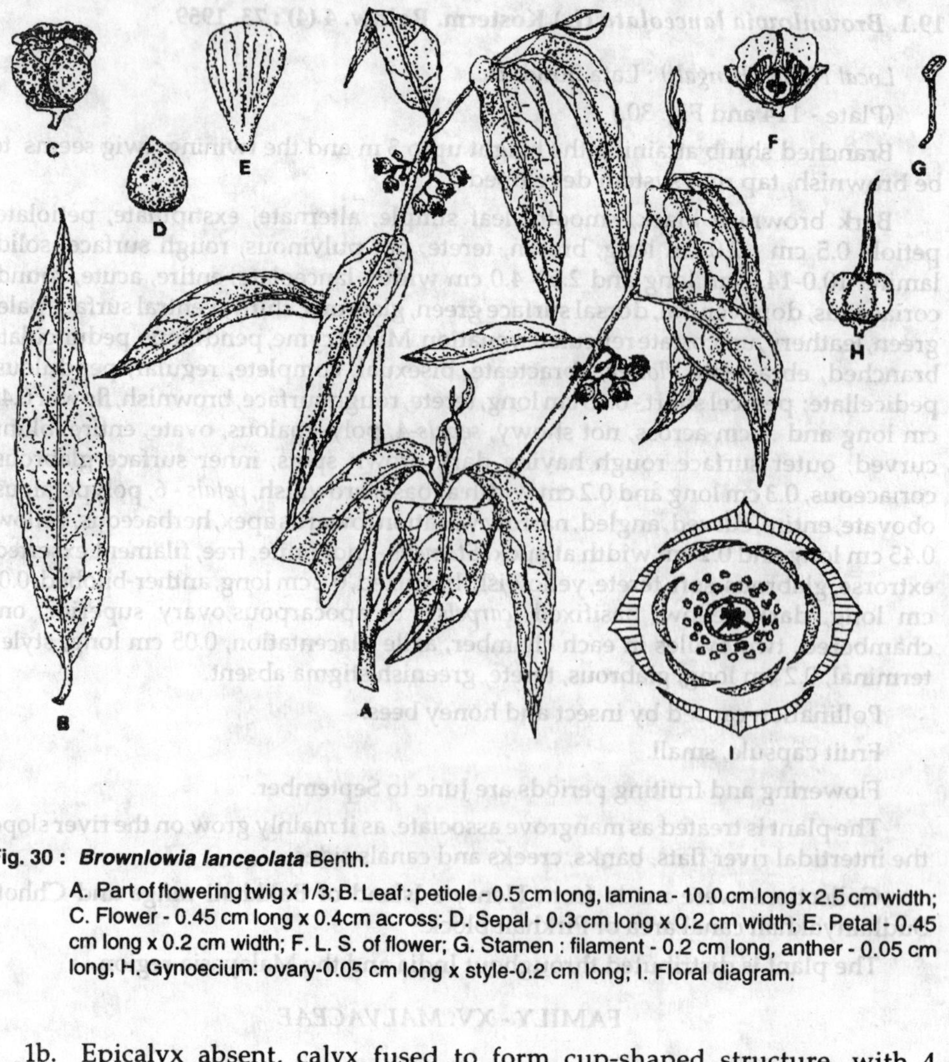

Fig. 30 : ***Brownlowia lanceolata*** **Benth.**

A. Part of flowering twig x 1/3; B. Leaf : petiole - 0.5 cm long, lamina - 10.0 cm long x 2.5 cm width; C. Flower - 0.45 cm long x 0.4cm across; D. Sepal - 0.3 cm long x 0.2 cm width; E. Petal - 0.45 cm long x 0.2 cm width; F. L. S. of flower; G. Stamen : filament - 0.2 cm long, anther - 0.05 cm long; H. Gynoecium: ovary-0.05 cm long x style-0.2 cm long; I. Floral diagram.

1b. Epicalyx absent, calyx fused to form cup-shaped structure, with 4 inconspicuously minute teeth; leaves cordate-lanceolate, with 7 main veins; capsule dehisced into 5 longitudinal sutures

2. *Thespesia* Soland. ex Correa

GENUS–20. *HIBISCUS* L. Gen. ed. I : 207.1737.

The genus–*Hibiscus* L. is characterised by 5 monadelphous stamens, representing by teeth at the top of the stamen tube.

This genus has about 300 species, which are distributed in the tropical and sub-tropical region. Some of these species are cultivated, especially *H. rosa-chinensis* for its showy flowers and many other medicinal properties.

But two species, *viz., Hibiscus tortuosus* Roxb. and *H. tiliaceous* L. are the common back mangals or mangrove associates, found in the coastal zones of the Sundarbans. In the present treaties these two species are collected and studied vividly.

Key Characters to the Mangrove Associated Species under–*Hibiscus* L.

1a. Petal yellow, longitudinal red veins upward parallely, deep red spot found in the base at inner side of flowers, stigma blackish-red

1. *H. tortuosus* Roxb.

1b. Petal light yellow, yellow-white, longitudinal veins upward parallely, crimson eye spot found in base at inner side of flowers, stigma pink yellow

2. *H. tiliaceous* L.

20.1. *Hibiscus tortuosus* Roxb. Hort. Beng. 51.1814 & Fl. Ind. iii : 92. 1832.

Common Name : Coast Cotton Tree/Wood.

Local Name (Bengali) : Bhola

(Plate–115 and Fig. 31)

Medium tree, 15 m high, deliquiscent branched, evergreen, perennial tree with mucilagenous latex.

Tap root system much spreading and deep sunken in the soil.

Internode 5 cm long, apical shoot not gregariously covered by leaves, green, terete, infrequently lenticels present, solid, woody, nodes not ridged; bark-grey black, longitudinally fissured but not conspicuously found.

Leaves simple, alternate, interstipular, stipule; ovate, acute, basal curved, green, pubescent, coriaceous, 1.5 cm long and 1.2 cm width, petiolate, petiole 5.5 cm long and 0.15 cm width, terete, solid, pulvinous, gregarious indistinct hairy, green petiolar terminal end slightly thick and cylinder, presence of axile dwarf branch, lamina 8.5 cm long and 9.5 cm width, purely cordate, slightly wavy, acuminate, coriaceous, dorsiventral, dorsal green, ventral whitish green, ventral surface rough, unicostate reticulate venation conspicuously found.

Cyme, peduncle having generally two flowers, one mature while other young, *i.e.*, unequal maturation, bracteate, bracts–2 on both side, 1.5 cm long and 1.2 cm width, ovate, entire, acute, leathery, brownish-green ; bracteate, pedicellate; pedicel 1.0-2.0 cm long and 0.4 cm width, solid, soft, green, terete, gregarious indistinct hairy, bisexual, complete, regular, 5.5 cm long and 4.5 cm across, showy, large, epicalyx–5, fused, cup-shaped, inferior, 5 apical lobes found at young stage, 1.0 cm long and 0.2 cm width but through maturation each lobe bifurcated, total 10 lobes formed, *calyx*-sepals-5, polysepalous, lanceolate, entire, acute, 1.5 cm long and 0.3 cm width, 2.0 cm across, hairy on both surfaces, inner surface shiny, inferior, imbricate, greenish, equal length, opposite to epicalyx shorter than corolla; *petals*–5, polypetalous, twisted, inferior, slightly touched at base, obovate, wrinkle, herbaceous, base thick, 5.2 cm long and 4.0 cm width in terminal part, while 0.8 cm width at base, deep red eye spot at basal inner side, glabrous, showy, dark yellow,

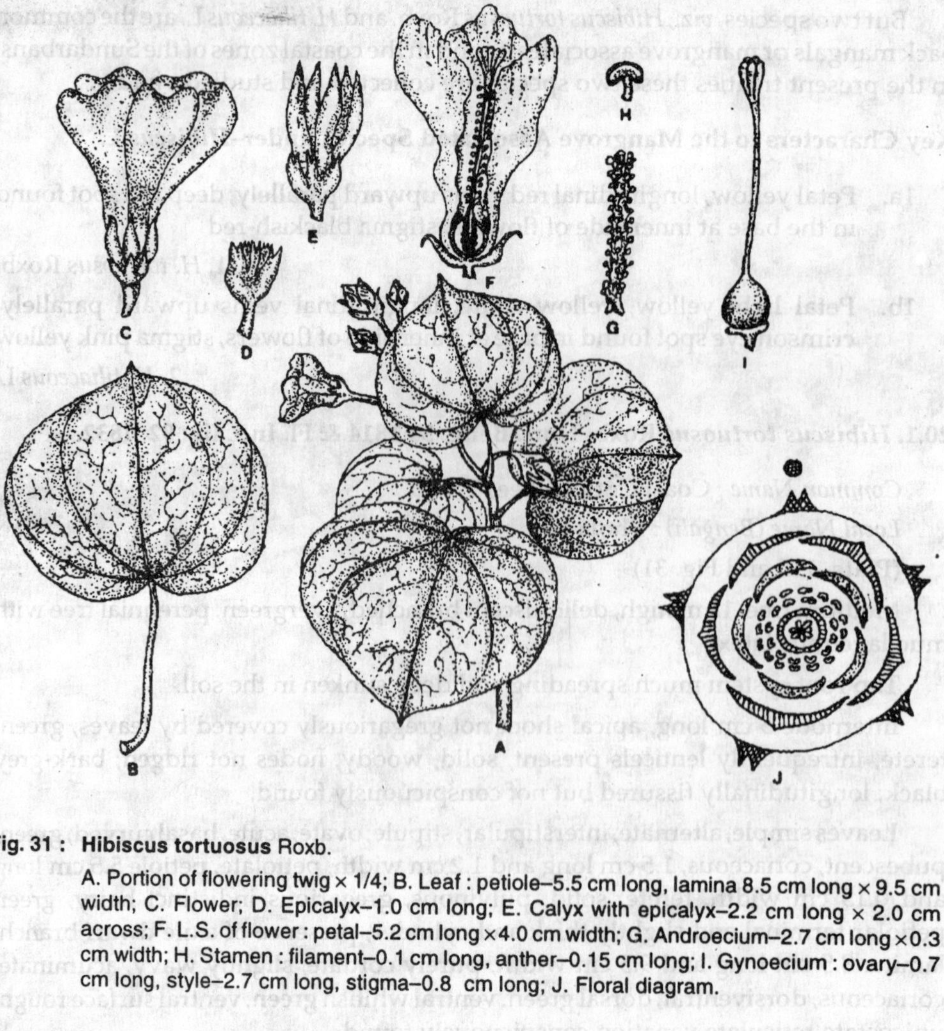

Fig. 31 : Hibiscus tortuosus Roxb.
A. Portion of flowering twig × 1/4; B. Leaf : petiole–5.5 cm long, lamina 8.5 cm long × 9.5 cm width; C. Flower; D. Epicalyx–1.0 cm long; E. Calyx with epicalyx–2.2 cm long × 2.0 cm across; F. L.S. of flower : petal–5.2 cm long × 4.0 cm width; G. Androecium–2.7 cm long × 0.3 cm width; H. Stamen : filament–0.1 cm long, anther–0.15 cm long; I. Gynoecium : ovary–0.7 cm long, style–2.7 cm long, stigma–0.8 cm long; J. Floral diagram.

deciduous, longitudinal 10.12 red veins upward parallely; *Androecium*, stamens-indefinite, monadelphous, filaments fuse to form staminal column surrounding style, 2.7 cm long, 0.3 cm width, hollow staminal column basally adhered to corolla, filamentous outgrowth, 0.1 cm long, white, glabrous, soft, inserted, anther unilobed transversely dehisced, kidney shaped, yellow, extrose, 0.15 cm width; *carpels*–5, syncarpous; superior, pyramidal shaped, 0.7 cm long and 0.8 cm diam., white greenish, glabrous, fleshy, 5-inconspicuous ridged found, 2 chambers, many ovules in two series in each chamber, axile placentation; style 1, terminal, soft, whitish, glabrous, 2.7 cm long, terete, stigmas–5, 0.8 cm long, each part distinctly separated, gradually thick towards terminal end, pubescent, blackish red distinct.

Pollination caused by bumble bees, honey bees and flies.

Fruits capsule, 2.0 cm–2.3 cm long, 2.1 cm width, cup-shaped, having persistent calyx, rough surface, 5 parts dehisced through longitudinal suture after maturation, brown.

Seeds black, flattened.

Germination is hypogeal.

Flowering and fruiting are mainly during November–March, but occasionally flowering and fruiting are seen throughout the year.

This is a mangrove associate and generally grow on the compact soil, above the frequent tidal reaches.

Distributed throughout the coastal states of India as mangrove associates or back mangals and in the tropical and sub-tropical regions of the Globe.

Bakkhali regions of the coastal area in Western Sundarbans.

20.2. *Hibiscus tiliaceous* L. Sp. Pl. 694.1753.

Chromosome No. 2n = 40 (Hsu, 1968); C. 80 (Skovsted, C 41; C. 92 (Skovsted, 1935); C. 92, C. 96 (Skottsbergc, 1955); C. 96 (Youngman, 1927; Krishnappa & Munirajappa, 1980).

Common Name : Native Hibiscus or Yellow Mallow Tree.

Local Name (Bengali) : Bhola

(Fig. 32)

Tall tree, upto 20 m high, deliquiscent branched, evergreen plant with mucilagenous latex.

Tap roots deep sunken, without any aerial roots.

Nodes slightly ridged, internodes 8.0 cm long, solid, woody, terete, green glabrous, whitish dust on young shoots, infrequently lenticellated; bark grey black, longitudinally fissured.

Leaves simple, alternate, interstipules, stipule ovate, entire, acute, basal curved, green, pubescent, coriaceous, 1.0 cm long and 0.5 cm width; petiolate, petiole, 10.0 cm long × 0.3 cm width, terete, green, pulvinous, lamina cordate, 17.5 cm long and 16.0 cm width, coriaceous, entire or slightly wavy, acuminate, dorsiventral, dorsal green glabrous, ventral whitish-green, rough, unicostate reticulate venation conspicuously.

Cyme, two flowers on each peduncle, one matured, while other young; bracteate, paired, bracts ovate, entire, acute, 1.0 cm long and 0.5 cm width, coriaceous, pubescent, green, pedicellate; pedicel upto 2.0 cm long × 0.3 cm width, bisexual, complete, regular, solid, soft, 4.0 cm long and 5.0 cm across, pentamerous, showy; epicalyx fused, 7-10 lobes, cup-shaped, persistent, lanceolate, entire, acute, outer surface green, inner surface whitish green, pubescent, 0.4 cm long and 0.3 cm width, valvate, inferior; *sepals*–5, gamosepalous, 5 long lobes, inferior, rosaceous, valvate, each lobe 1.2 cm long and 0.7 cm width, entire, acute, pubescent, outer whitish green with rough surface, inner shiny with smooth surface, persistent, alternate to corolla; *petals*–5, polypetalous, twisted, slightly fused at base, 3.8 cm long and 2.5 cm width, obovate, wrinkle, curved, basal thick, whitish-yellow, herbaceous, crimson eye spot

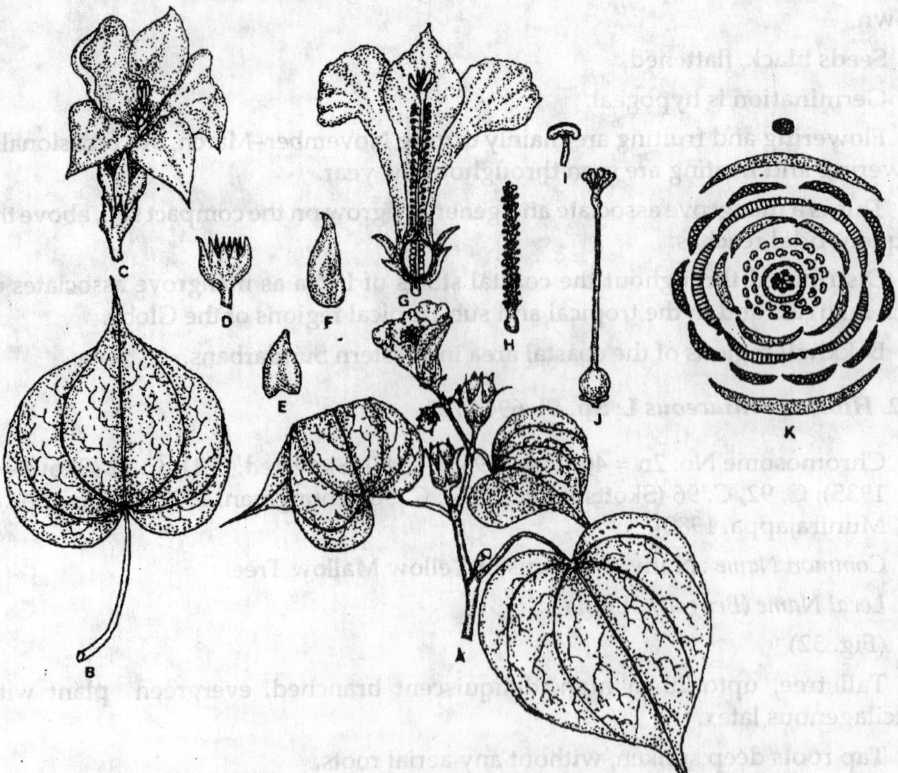

Fig. 32 : *Hibiscus tiliaceous* L.

A. Portion of flowering twig × 1/4; B. Leaf : petiole–9.5 cm long × 0.3 cm width, lamina –17.5 cm long × 16.0 cm width; C. Flower–4.0 cm long × 5.0 cm across; D. Epicalyx– 0.4 cm long; E. Bract –1.0 cm long × 0.5 cm width; F. Sepal –1.2 cm long × 0.7 cm width; G. L.S. of flower; H. Androecium–1.8 cm long × 0.3 cm width; I. Stamen–0.25 cm long; J. Gynoecium : Ovary– 0.5 cm long, style–2.0 cm long, stigma–0.5 long; K. Floral diagram.

found in base at inner side, glabrous, longitudinal, 10-12 yellow white veins upward parallely, deciduous, *androecium*, stamens–indefinite, monadelphous, filaments fuse to form staminal column surrounding style, 1.8 cm long and 0.3 cm width, filamentous outgrowth, 0.25 cm long, terete, white, soft, glabrous, anther unilobed, kidney-shaped, yellow, extrorse, basifixed; *carpels*–5, syncarpous, ovary superior, globose, 0.5 cm long and 0.4 cm width, whitish-green, glabrous, two locules, many ovules in each locule by series, axile placentation, style–1, terminal, white, soft, glabrous, 2 cm long, 0.05 cm width, terete, stigma–5, short, cylindrical, 0.5 cm long, pink yellow, gregarious, minute glands on the stigma.

Pollination caused by bumble-bees and honey bees.

Fruits capsule, 2.5 cm long and 1.5 cm width, globose, with 5-persistent calyx and epicalyx, dehisced into 5 parts, through 5-longitudinal sutures.

Seeds numerous, black, flattened.

Seed germination is hypogeal.

Flowering and fruiting are mainly during winter season, but occasionally throughout the year.

This species is considered as mangrove associate or back mangals; occasionally planted as avenue tree.

Distributed throughout the coastal regions and also in the world tropics.

Collections were made from the roadside at Ghusighata/Kulti area of the Sundarbans.

Medicinally, the leaves are laxative and resolvent; infusion used as a vulnerary; mucilage of the bark is said to be useful in dysentery; roots are used for the preparation of embrocations for the treatment of rheumatism and lumbago sees infusion is used as an emetic.

Table 61 : Comparative Morphology between Two Species of *Hibiscus* L.

Characters	*H. tortuosus* Roxb.	*H. tiliaceous* L.
Habit	Medium tree, upto 15 m high.	Tall tree, upto 20-25 m high.
Stem	internode upto 5 cm long.	Internode upto 8 cm long.
Leaf	Petiole 5.5 cm long and 0.15 cm width, lamina 8.5 cm long and 9.5 cm broad.	Petiole 10 cm long and 0.3 cm width, lamina upto 17.5 cm long and 16 cm broad.
Flower	Pedical 1-2 cm long and 0.4 cm width, flower 5.5 cm long and 4.5 cm across, sepal 1.5 cm long, 0.3 cm width, petal 5.2 cm long and 4.0 cm width, dark red spot at basal interior side; staminal column 2.7 cm long and 0.3 cm width ovary 0.7 cm long and 0.8 cm diam.	Pedical 0.8 cm–2.0 cm long and 0.3 cm width, flower 4.0 cm long and 5.0 cm across ; sepals 1.2 cm long and 0.7 cm width; petal 3.8 cm long and 2.5 cm width, crimson eye spot at basal interior side; staminal column 1.8 cm long and 0.3 cm width, ovary 0.5 cm long and 0.4 cm diam.
Fruit	2.3 cm long and 2.1 cm width.	2.5 cm long and 1.5 cm width.

GENUS–21. *THESPESIA* Soland ex Correa, Ann. Mus. IX : 290.t.8.f. 2.1807.

The genus–*Thespesia* Soland ex Correa consists of 15 species, which are distributed in the tropical countries; out of these 15 species, *T. populnea* (L.) Soland ex Correa, *T. populneoides* (Roxb.) Kostel and *T. lampus*(Cav.) Dalz. & Gibs. are reported from the mangrove habitats of the Sundarbans and other Indian coasts, as back mangals or mangrove associates.

In the present treaties only *Thespesia populnea* (L.) Soland ex Correa is described in detail and *T. populneoides* (Roxb.) Kostel is mentioned in brief.

21.1. *Thespesia populnea* (L.) Soland ex Correa, Ann. Mus. Par. IX : 290.1807.

Chromosome No. $2n = 24$ (Hazra & Sharma, 1971) and $2n = 26$ (Youngman, 1927).

Common Name : Portia Tree or Umbrella Tree or Indian Tulip Tree or Parespipal.

Local Name (Bengali) : Paras or Habal

(Plate–116 and Fig. 33)

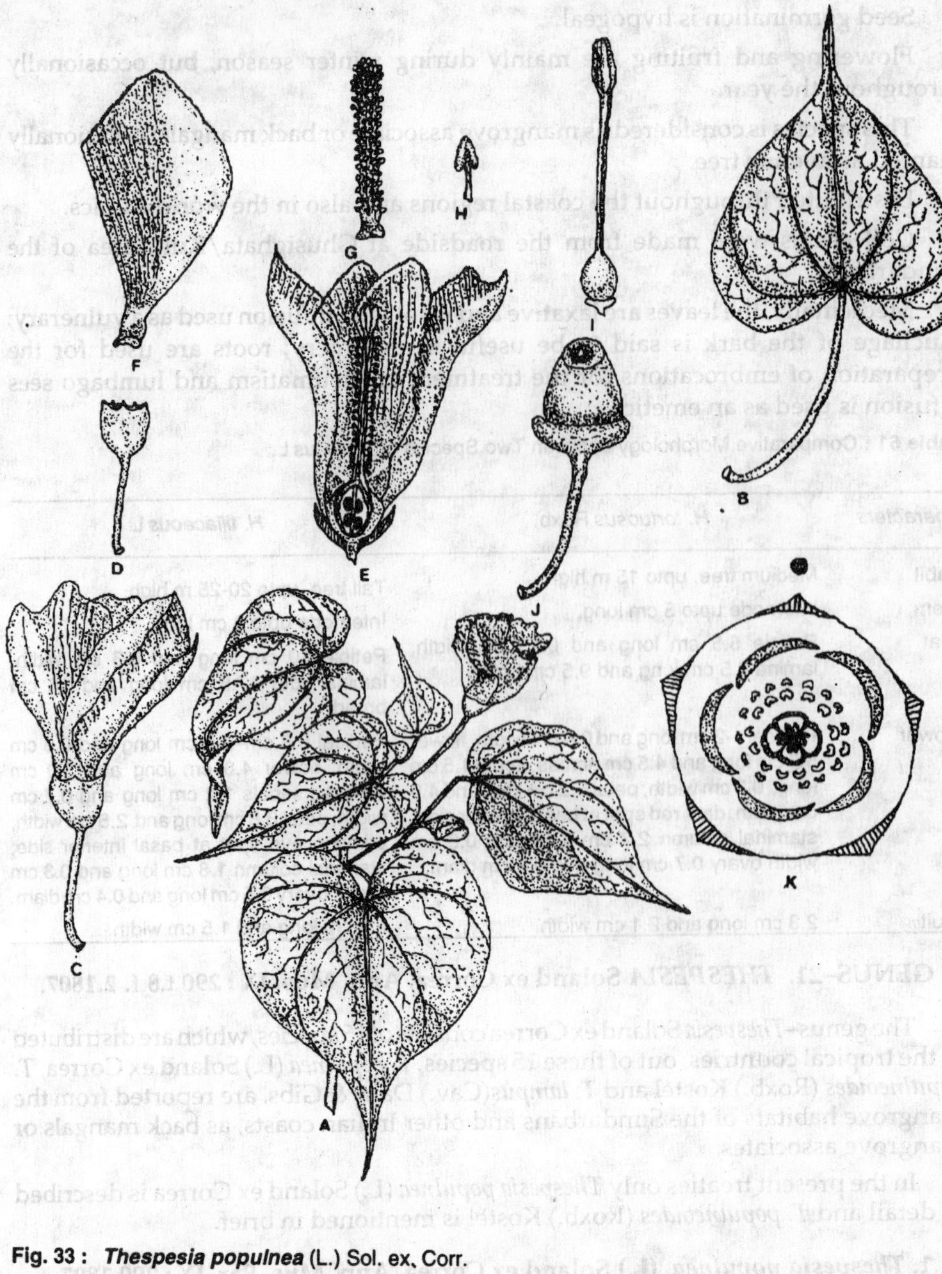

Fig. 33 : ***Thespesia populnea*** (L.) Sol. ex. Corr.

A. Portion of flowering twig × 1/4; B. Leaf : petiole–10.5 cm long, lamina 15.0 cm long × 11.7 cm width; C. Flower–5.2 cm long × 4.5 cm across; D. Calyx–1.2 cm long; E. L.S. of flower; F. Petal–5.0 cm long × 4.0 cm width; G. Androecium–2.8 cm long × 0.13 cm width; H. Stamen : filament– 0.6 cm long, anther 0.15 cm long; I. Gynoecium : ovary - 0.5 cm long, style–2.5 cm long, stigma–0.5 cm long; J. Fruit–2.5 cm long × 2.5 cm width; K. Floral diagram.

Medium tree, upto 15 m high, deliquiscent branched, evergreen, presence of mucilagenous latex.

Tap roots well developed and much spreading into deep horizon of soil; occasionally hang from the branches.

Nodes not ridged, internodes short, green, terete, infrequently lenticellate, solid, woody; bark-grey black, longitudinally fissured but not conspicuously.

Leaves simple, alternate, intrastipulate, stipule present at both the sides of petiolar base, lanceolate, entire, acute, green, glabrous, coriaceous, mid-vein present, after maturation it become ceased; petiolate, petiole, long, pulvinous, gregarious indistinct hairy, green, solid, terete, 10.5 cm long and 2.0 cm width, slightly twining, lamina cordate, 15 cm long × 11.7 cm width, entire, acute, coriaceous, dorsiventral, dorsal green with smooth surface, ventral whitish-green with rough surface, unicostate reticulate venation, whitish yellow mid-vein distinct.

Flower solitary but raceme, mature flower at base, ebracteate, pedicellate, pedicel 5.0 cm–12.0 cm long × 0.3 cm width, solid, green, terete, bisexual, regular, complete 5.5 cm long and 4.5 cm across, showy, large, slightly fragrant, *sepals*–5, gamosepalous, cup-shaped, inferior, valvate, 5 minute lobes, green, rough surface, 1.2 cm long and 1.5 cm diam., persistent, shorter than corolla; *petal*-5, polypetalous, twisted, inferior, slightly touched at base, obovate, herbaceous, wrinkle, 5 cm long × 4 cm width at terminal part, while 0.8 cm width at base, dark red spot at basal interior side, base thick while terminal side thin, glabrous, showy yellow but turned into pink through gradual maturation, deciduous, longitudinal veins upward parallely *androecium*, *stamens*–indefinite, monadelphous, filaments fuse to form staminal column surrounding style ; 2.8 cm long and 0.13 cm width, basally attached to corolla, filamentous outgrowth, 0.6 cm long, branched or unbranched inserted, soft, white yellow, glabrous, anther unilobed, kidney shaped, yellow, transversely dehisced, extrorse, 0.15 cm width ; *carpels*–5, syncarpy, superior, cup-shaped, 0.5 cm long and 0.8 cm diam., white yellow, glabrous, fleshy, 5-chambered, two ovules in each chambers, axile placentation, style-1, terminal, soft, white-yellow, glabrous, fleshy 2.5 cm long, terete, stigma-5, 0.5 cm long, 5 parts, each part longitudinally flattened, glabrous white yellow, distinct.

Pollination caused by bumble bees and honey bees.

Fruit–capsule, 2.5 cm long and 2.5 cm width, cup-shaped with partly covered by persistent calyx, 5 halves inconspicuously found, longitudinally dehisced, brown after maturation.

Seeds flattened, black.

Germination is hypogeal.

Flowering and fruiting are during winter but occasionally throughout the year.

This species is considered as back mangals or mangrove associates, commonly grow in the above tidal zones of the inner estuary and occasionally planted as avenue tree.

Timber producing plants; distributed throughout the coastal states in India and also in the tropical South East Asian coasts.

Medicinally, barks, leaves, flowers and fruits are used in cutaneus affection; seed oil is said to be used in skin troubles; barks, roots and fruits are astringent and said to be used in dysentery and hemorrhoids. Seeds are purgative.

Henry's Islands of the Western Sundarbans.

FAMILY–XVI. *RUTACEAE*

The family–Rutaceae is cosmoplitan and it has over 1300 species under 140 genera. This family is characterised with scented leaves and aromatic fruits with sour taste. The genus *Atalantia* Corr. (Syn. *Merope* Roem.) is occasionally grow in the back-mangal communities or canal and river sides, in the mangrove habitats.

GENUS–22. *ATALANTIA* Corr. (Syn *Merope* Roem. Syn. Monogr. 1 : 44.1846)

This is monotypic genus and the only species *Atalantia correa* M. Roem. [Syn. *Merope angulata* (Willd.) Swingle] is occasionally grow as mangrove associated species and its suitable habitat is the back-mangal zones or river banks; frequently inundated with tidal water. It has also reported by Swingle (1943) as a salt tolerant species.

22.1. *Atalantia correa* M. Roem.

Merope angulata (Willd.) Swingle Bull. Wash. Bio. Soc. 2 : 22.1915.

Common Name : Mangrove Lime.

Local Name (Bengali) : Ban Lebu

(Plate–117 and Fig. 34)

Occasional in the back-mangal, river flats, in the low saline zone; shrub, 2.5 m–3.0 m tall, evergreen, much branched, stout stem base. Tap roots, no aerial roots, but trunk or stem base swollen, flat. Bark blackish, longitudinally fissured. Stem glabrous, shiny, green, round, solid, two spines at each node; lemon like scented; nodes not ridged.

Leaves simple, alternate, stipulate (stipule as two spines), petiolate, petiole 0.3 cm long and 0.2 cm broad, green, glabrous, pulvinous, lamina lanceolate, oblong, entire, acute, dorsiventral, green, ventral side pale green, glabrous, coriaceous, 7 cm long and 3.0 cm–4.0 cm width, aromatic, unicostate reticulate venation.

Inflorescence solitary, *flower* bisexual, complete, actinomorphic, medium 1.8 cm long × 0.5 cm across, not showy, erect, pentamerous; *sepals*–5, gamosepalous, green, 5-teeth above, cup-shaped, 0.2 cm long, and 0:25 cm across, valvate ; *petals*–5, polypetalous, each petal lanceolate, white, fleshy, acute entire, 0.9 cm long and 0.4 cm width, not showy, twisted; *stamens*–10, in single whorle, filament long, upto 0.25 cm and 0.08 cm width, fleshy, white, one third of the total stamen, anther introrse, inserted, longitudinal dehisced, white, 0.4 cm long and 0.1 cm width; *carpels*-syncarpy, ovary superior, triangular, green, 0.4 cm long, many ovules in each chamber, axile placentation; style–1, terminal, solid, green, glabrous, 0.3 cm long, cylindrical, stigma absent.

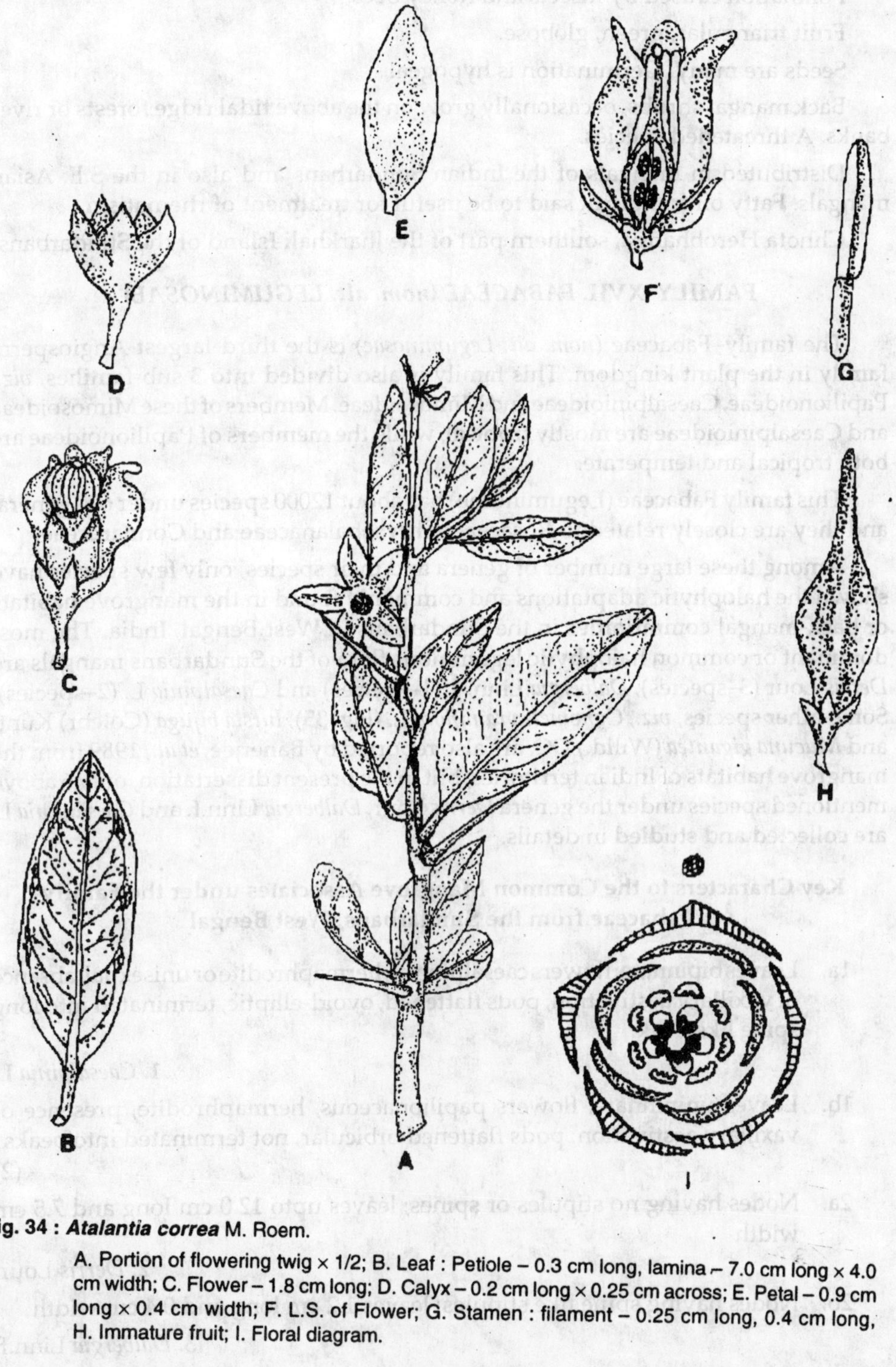

Fig. 34 : *Atalantia correa* M. Roem.

A. Portion of flowering twig × 1/2; B. Leaf : Petiole – 0.3 cm long, lamina – 7.0 cm long × 4.0 cm width; C. Flower – 1.8 cm long; D. Calyx – 0.2 cm long × 0.25 cm across; E. Petal – 0.9 cm long × 0.4 cm width; F. L.S. of Flower; G. Stamen : filament – 0.25 cm long, 0.4 cm long, H. Immature fruit; I. Floral diagram.

Pollination caused by insects and honey bees.

Fruit triangular, green, globose.

Seeds are many. Germination is hypogeal.

Back mangal shrubs, occasionally grow on the above tidal ridge forests or river banks. A threatened species.

Distributed in mangals of the Indian Sundarbans and also in the S.E. Asian mangals. Fatty oil from fruit said to be useful for treatment of rhematism.

Chhota Herobhanga, southern part of the Jharkhali Island of the Sundarbans.

FAMILY–XVII. *FABACEAE* (*nom. alt. LEGUMINOSAE*)

The family–Fabaceae (*nom. alt. Leguminosae*) is the third largest Angiosperm family in the plant kingdom. This family is also divided into 3 sub-families, *viz.*, Papilionoideae, Caesalpinioideae and Mimosoideae. Members of these Mimosoideae and Caesalpinioideae are mostly tropical, while the members of Papilionoideae are both tropical and temperate.

This family Fabaceae (Leguminosae) has about 12000 species under 600 genera; and they are closely related to Rosaceae, Chrysobalanaceae and Connaraceae.

Among these large number of genera and their species, only few species have shown the halophytic adaptations and commonly found in the mangrove habitats or back-mangal communities in the Sundarbans in West Bengal, India. The most dominant or common halophytic leguminous flora of the Sundarbans mangals are *Derris* Lour (3–species), *Dalbergia* Linn. f. (1–species) and *Caesalpinia* L. (2–species). Some other species, *viz.*, *Cynometra ramiflora* L. (Fig. 35), *Intsia bijuga* (Colebr) Kunt. and *Mucuna gigantea* (Willd.) DC. are also reported by Banerjee, *et al.*, 1989 from the mangrove habitats of Indian territories; but in the present dissertation, only 6 above mentioned species under the genera *Derris* Lour, *Dalbergia* Linn.f. and *Caesalpinia* L. are collected and studied in details.

Key Characters to the Common Mangrove Associates under the Family-Fabaceae from the Sundarbans, West Bengal

1a. Leaves bipinnate; flowers caesalpinous, hermaphrodite or unisexual, absence of vaxillary aestivation; pods flattened, ovoid-elliptic, terminated into long spine like beak.

1. *Caesalpinia* L.

1b. Leaves unipinnate; flowers papilionaceous, hermaphrodite, presence of vaxillary aestivation; pods flattened orbicular, not terminated into beaks (2)

2a. Nodes having no stipules or spines; leaves upto 12.0 cm long and 7.5 cm width

2. *Derris* Lour.

2b. Nodes having spine like stipules; leaves 1.2 cm long and 0.4 cm width

3. *Dalbergia* Linn.f.

GENUS–23. *CYNOMETRA* L. 1753 (Sp. Pl. : 382.1753 Gem, Pl. Ed. 5.179.1754).

The genus-*Cynometra* L. (Caesalpiniaceae) having about 70 species which are mainly distributed in the lowland forest of tropical zone. The genus is characterised by unijugate or bijugate leaves, small bud scales, 4 calyx segments, 5 petals, stamens–10, 1 seeded fruit. Among these species, *Cynometra ramiflora* L. occasionally grows in the mangrove habitats of the Sundarbans and elsewhere. During this studies a germinted seed is collected and planted in the non-saline soil and irrigated with tap water. The seedling plant is studied only.

23.1. *Cynometra ramiflora* Linn. Sp.Pl. : 382.1753

Fig. 35.

Common Name : Wrinkle Mangrove.

Local Name (Bengali) : Shinger.

Cynometra ramiflora Linn. is a small tree, attain the height upto 8-15 m tall, deliquiscent branched, evergreen perennial tree. Tap root system well developed, but no aerial roots. Bark-glabrous, not fissured, greyish, internode variable in length.

Leaves unijugate or bijugate, compound, exstipulate, paripinnate, rachis upto 8.0 cm long, terete, solid, glabrous; leaflets 2.5 cm–4.0 cm long and 1.0 cm–1.5 cm width, lower pairs of leaflet comparatively small, mid costa obliquely placed, elliptic, entire, obtuse, angular, coriaceous, dorsiventral, glabrous, green, unicostate reticulate venation.

Flowers 3.5 cm–5 cm in diameter, in axillary racemes, reddish or purple; pods, 5–10 in a raceme, 2.5 cm long, elliplic-oblique, wrinkled and 1–seeded; seed flat, smooth.

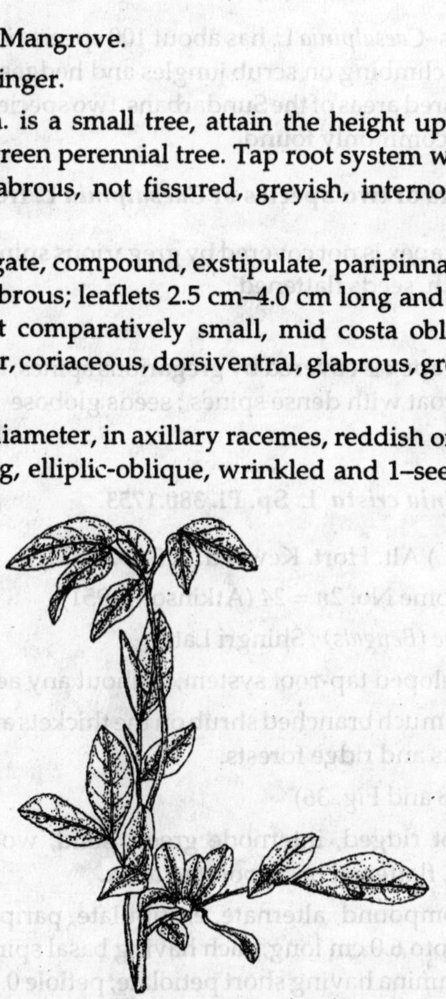

A

Fig. 35 : ***Cynometra ramiflora*** **L**

A. Portion of Vegetative twig X 1/2

Flowering and fruiting time is October to March. Distributed in the mangrove forest of Sundarbans, Mahanadi Delta and Andaman & Nicobar Islands.

This plant is treated as back mangal; commonly distributed in the ridge forests of the intertidal zones.

The plant is distributed in Old World, *i.e.*, Malayasia, Phillipines, Sri Lanka and North Australia. Wood is used for fuel and timber and also used for building construction purposes (Fig. 35).

Root decoction is purgative; seed oil is said to be used for leprosy, scabies and other cutaneous diseases.

GENUS–24. *CAESALPINIA* L. Sp. Pl. 380.1753 (Caesalpiniaceae)

The Genus–*Caesalpinia* L. has about 100 species in the tropical and sub-tropical zones; mostly climbing on scrub jungles and hedges. In the back-mangal zones or mangrove cleared areas of the Sundarbans, two species, *viz.*, *C. crista* L. and *C. bonduc* (L.) Roxb. are commonly found.

Key Characters of two Species of *Caesalpinia* L. from Sundarbans

1a. Shoot apex is not covered by gregarious spines; flowers bisexual; fruit coat smooth, seeds flattened

1. *C. crista* L.

1b. Shoot apex is covered by gregarious spines; flowers unisexual by abortion; fruit coat with dense spines ; seeds globose

2. *C. bonduc* (L.) Roxb.

24.1. *Caesalpinia crista* L. Sp. Pl. 380.1753.

C. *nuga* (L.) Alt. Hort. Kew. ed. 2, 3 : 32.1811.

Chromosome No. $2n = 24$ (Atkinson, 1951)

Local Name (Bengals) : Shingri Lata

Well developed tap-root system, without any aerial roots.

Climbing much branched shrub on the thickets and hedges, from dense bush of the river banks and ridge forests.

(Plate–118 and Fig. 36)

Nodes not ridged, internode green, solid, woody, having frequent spines, spines brown, flattened base, pointed apex.

Leaves compound, alternate, exstipulate, paripinnate, leaf axis upto 25.0 cm long, leflets upto 6.0 cm long, each having basal spines, dark green, axis glabrous, terete, solid, lamina having short petiolate; petiole 0.2 cm long, terete, soild, green, glabrous, pulvinous, lamina ovate-elliptic, entire, acute, angled, 3.5 cm long and 1.5 cm width, coriaceous, green, dorsiventral, ventral whitish green, unicostate reticulate venation conspicuously.

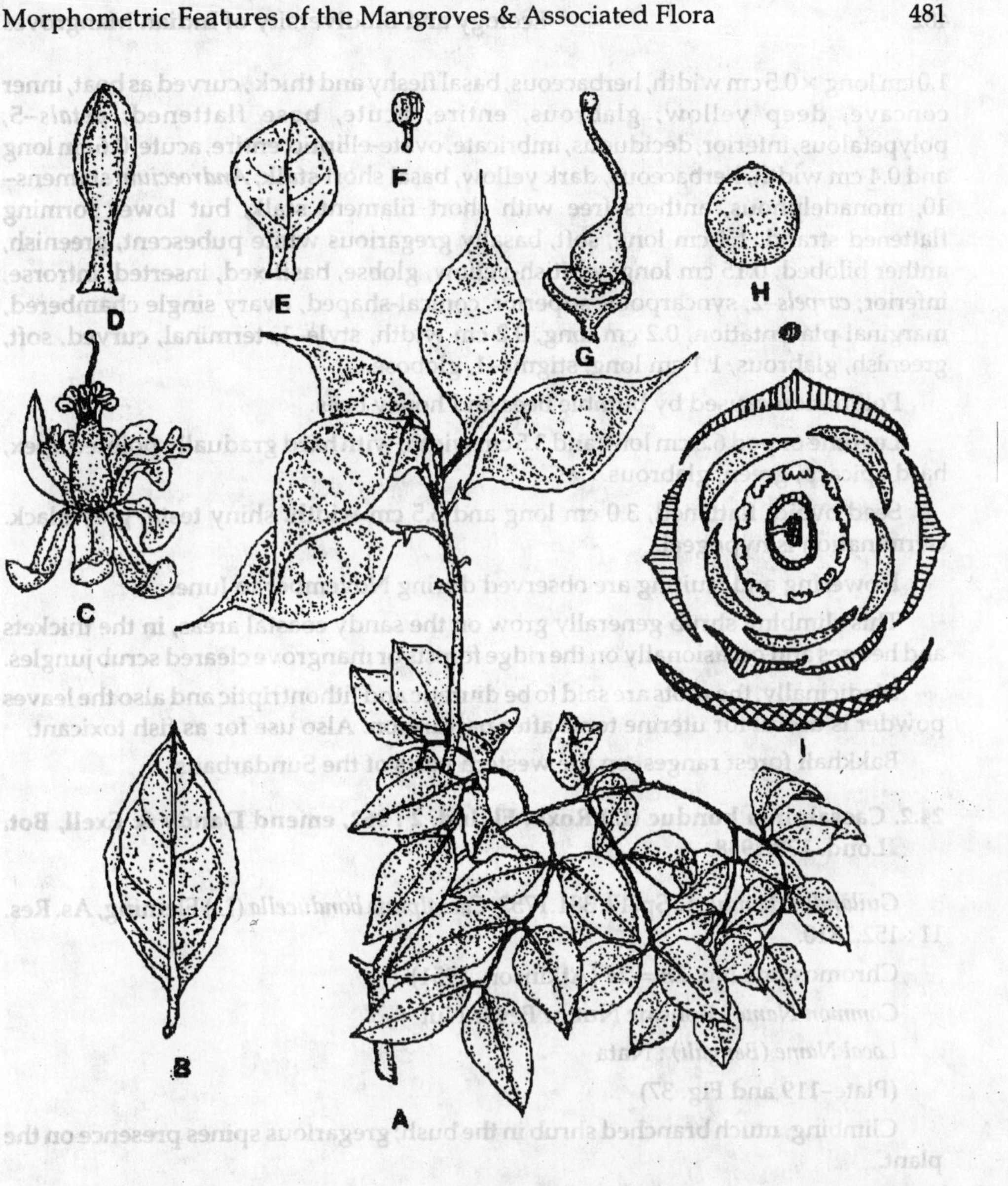

Fig. 36 : ***Caesalpinia crista*** **L.**

A. Twig with fruits × 1/2; B. Leaflet : stalk–0.2 cm long, lamina–3.5 cm long ×1.5 cm width; C.Flower–1.5 cm long; D. Sepal–1.0 cm long × 0.5 cm width; E. Petal–0.6 cm long × 0.4 cm width; F. Stamen–0.3 cm long; G. Carpel–1.3 cm long; H. Seed–2.5 cm long × 3.0 cm width; I. Floral diagram.

Raceme, peduncle branched, upto 15.0 cm long, solid, green, glabrous; complete, bisexual, irregular, hypogynous, showy, flower; pedicellate, pedical 1.2 cm long, solid, green, soft, glabrous, pendulous, 1.5 cm long and 1.0 cm across; thalamus flattened, green, soft, glabrous; *sepals*–5, polysepalous, imbricate, inferior, deciduous,

1.0 cm long × 0.5 cm width, herbaceous, basal fleshy and thick, curved as boat, inner concave, deep yellow, glabrous, entire, acute, base flattened; *petals*–5, polypetalous, inferior, deciduous, imbricate, ovate-elliptic, entire, acute, 0.6 cm long and 0.4 cm width, herbaceous, dark yellow, basal short stalk; *Androecium* stamens–10, monadelphous, anthers free with short filament stalk, but lower forming flattened strand, 1.4 cm long, soft, basally gregarious white pubescent, greenish, anther bilobed, 0.15 cm long, whitish-yellow, globse, basifixed, inserted, introrse, inferior; *carpels*–2, syncarpous, superior, conical-shaped, ovary single chambered, marginal placentation, 0.2 cm long, 0.2 cm width, style–1, terminal, curved, soft, greenish, glabrous, 1.1 cm long, stigma–1, globose.

Pollination caused by bumble bees and honey bees.

Legume or pod 6.5 cm long and 3.5 cm width, with hard gradually pointed apex, hard epicarp, green, glabrous.

Seed ovoid, flattened, 3.0 cm long and 3.5 cm width, shiny testa, grey black. Germination is hypogeal.

Flowering and fruiting are observed during November to June.

This climbing shrub generally grow on the sandy coastal areas, in the thickets and hedges and occasionally on the ridge forests or mangrove cleared scrub jungles.

Medicinally, the roots are said to be diuretic and lithontriptic and also the leaves powder is useful for uterine tonic after parturition. Also use for as fish toxicant.

Bakkhali forest ranges, on the western parts of the Sundarbans.

24.2. Caesalpinia bonduc (L.) Roxb. Fl. Ind. 2 : 362, emend Dandy & Exell, Bot. Lond. 179.1938.

Guilandina bonduc L. Sp. Pl. 381.1753. *Caesalpinia bonducella* (L.) Fleming, As. Res. 11 : 152.1810.

Chromosome No. $2n = 24$ (Atkinson, 1951)

Common Name : Bonduc Nut or Fever Nut

Local Name (Bengali) : Nata

(Plate–119 and Fig. 37)

Climbing, much branched shrub in the bush; gregarious spines presence on the plant.

Tap root system developed, but without any aerial roots.

Nodes not ridged, internodes having gregariously spines, solid, woody, mature branches brown, while immature shoot brownish yellow, having dense pubescent; spines 0.2 cm long, brown, stipule pointed.

Leaves compound, alternate, stipulate, paripinnate, leaf axis upto 25.0 cm long, leaflet upto 8.0 cm long, each having gregarious spines, axis glabrous, terete, solid, lamina having short petiolate; petiole 0.1 cm long and 0.05 cm width, glabrous, green, pulvinous, lamina elliptic-oblong, entire, mucronate, 2.6 cm long and 1.3 cm width, coriaceous, glabrous, yellowish green, dorsiventral, unicostate reticulate venation conspicuously.

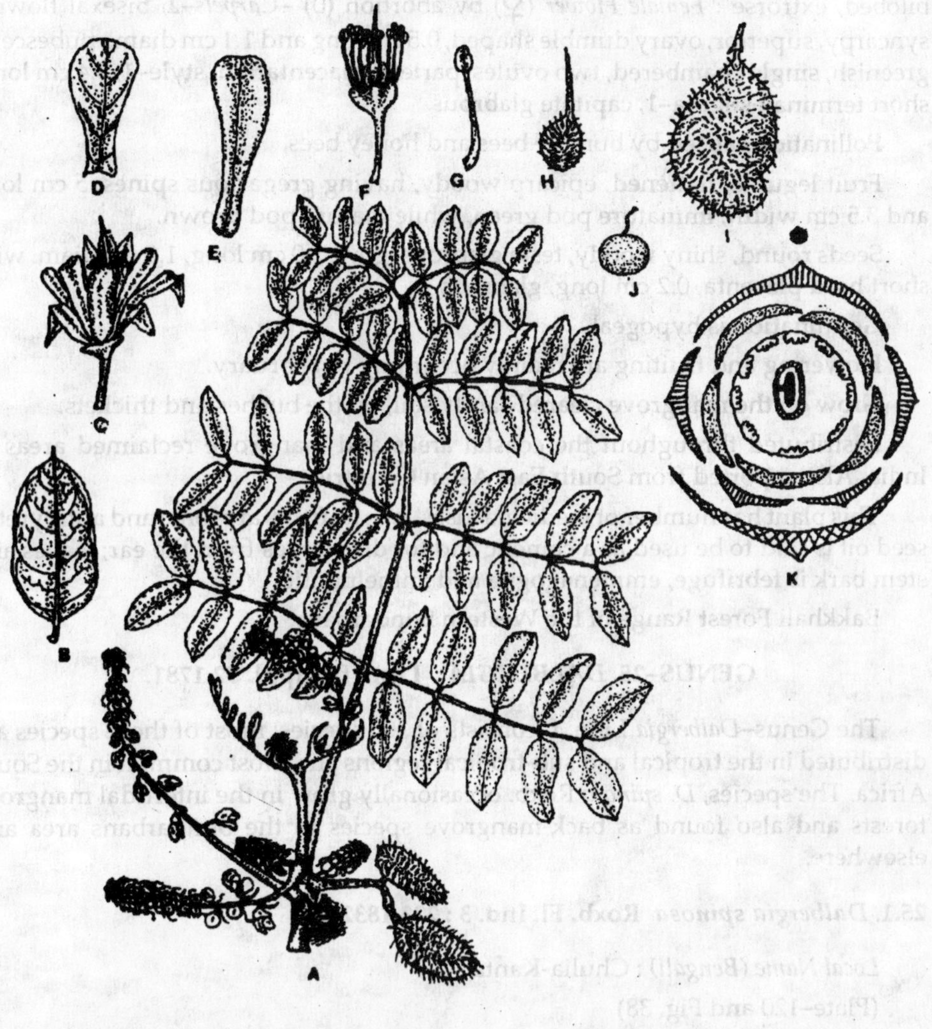

Fig. 37 : ***Caesalpinia bonduc*** **(L.) Roxb.**

A. Twig with Inflorescence & Fruits X 1/2, B. Leaflet. Stalk - 0.1 cm. long, Lamina - 2.6 cm. long × 1.3 cm. width; flower-1.0 cm. long X 1.8 cm. across; D. Sepal - 1.0 cm. long X 0.5 cm. width E. Petal - 1.2 cm. long X 0.3 cm. width; F. Pedicel with androecium and gynoecium; G. Stamen-0.8 cm. long; H. Carpel : Ovary - 0.4 cm long, Style - 0.4 cm long; I. Fruit - 5.0 cm. long; J. Seed - 1.0 cm long; K. Floral Diagram.

Receme lateral or axillary, upto 20.0 cm long, peduncle, solid, green; *flower* unisexual, irregular, hypogynous, showy. bracteate. *Male Flowers* (♂) 1.0 cm long × 1.8 cm across, *sepals*–5, polysepalous, ovate-oblique, entire, pubescent, blunt, tufted, curved, upto 1.0 cm long and 0.5 cm width; *petals*–5, polypetalous, imbricate, yellow, oblong, 1.5 cm long and 0.6 cm width, alternate to sepals; *stamens*–10, free, filaments hairy, terete, soft, solid, upto 0.8 cm long, dorsifixed, exerted, anther

bilobed, extrorse : *Female Flower* (♀) by abortion (0) –*Carpels*–2, bisexal flowers syncarpy, superior, ovary dumble shaped, 0.5 cm long and 1.1 cm diam., pubescent, greenish, single chambered, two ovules, parietal placentation, style–1, 0.4 cm long, short terminal; stigma–1, capitate glabrous.

Pollination caused by bumble bees and honey bees.

Fruit legume, flattened, epicarp woody, having gregarious spines, 5 cm long and 3.5 cm width, immature pod green, while mature pod brown.

Seeds round, shiny woody, testa glabrous, grey, 1.0 cm long, 1.0 cm diam. with short hard placenta, 0.2 cm long, glabrous.

Germination is hypogeal.

Flowering and fruiting are mainly occur July to February.

Grow on the mangrove cleared sandy soil, on the bushes and thickets.

Distributed throughout the coastal areas and mangrove reclaimed areas in India. Also reported from South East Asian Countries.

This plant has number of medicinal uses, *viz.*, its fruits are tonic and antipyretic; seed oil is said to be used as a cometic and for discharges from the ear; leaves and stem bark is febrifuge, emmenagogue and anthelmintic.

Bakkhali Forest Range of the Western Sundarbans.

GENUS–25. *DALBERGIA* Linn. f. Suppl. 52.1781.

The Genus–*Dalbergia* Linn. f. consists of 300 species; most of these species are distributed in the tropical and sub-tropical regions and most common in the South Africa. The species, *D. spinosa* Roxb. occasionally grow in the intertidal mangrove forests and also found as back-mangrove species in the Sundarbans area and elsewhere.

25.1. *Dalbergia spinosa* Roxb. Fl. Ind. 3 : 223.1832.

Local Name (Bengali) : Chulia-Kanta

(Plate–120 and Fig. 38)

Much branched shrub or small tree with twining habitat; attain a height upto 8 m. Tap-roots deep sunken but without any aerial roots.

Nodes not ridged, internodes solid, woody, glabrous, terete, brown-black.

Leaves compound, imparipinnate, alternate, stipulate, 2 woody spines at each node, upto 4 cm long, gradually tapering end, rachis upto 8.0 cm long, pulvinous, glabrous, woody, leaflets petiolate ; petiole 0.2 cm long, soft, greenish, glabrous, leaflets upto 1.2 cm long and 0.4 cm broad, coriaceous, unicostate reticulate venation.

Receme, panicles, peduncles upto 4.0 cm long; flower ebracteate, pedicellate, pedicel 0.1 cm long, soft, green, glabrous, complete, bisexual, zygomorphic, 0.45 cm long, erect, small, not showy; *sepals*–5, gamosepalous, 5 lobes above, 0.25 cm long, glabrous, green, fleshy, soft, vaxillary; *petals*-5,

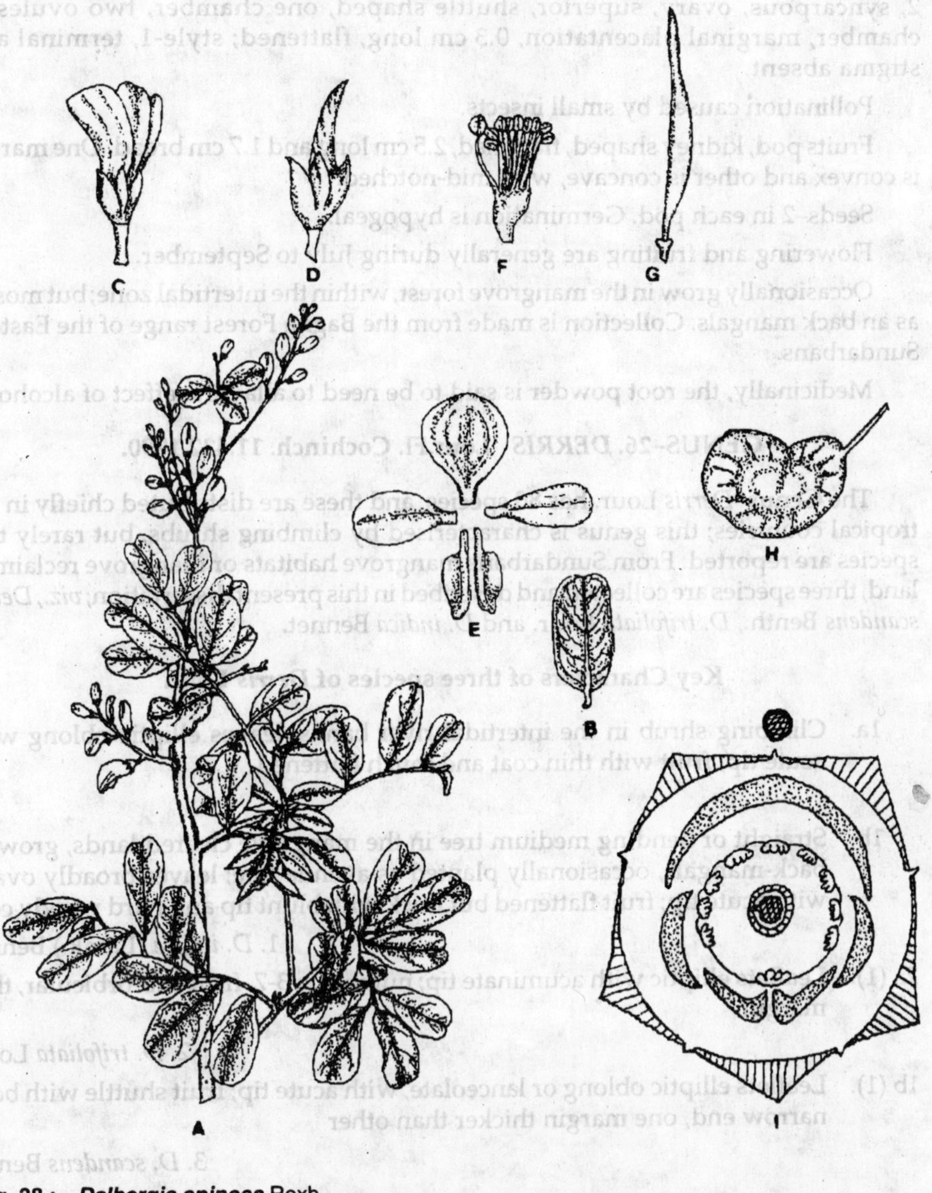

Fig. 38 : ***Dalbergia spinosa*** **Roxb.**

A. Portion of shoot with flowers X 1/2; B. Leaflet : stalk - 0.2 cm. long, lamina - 1.0 cm. long ; C. Flower - 0.45 cm. long; D. Calyx with Gynoecium; E. Corolla : standard - 0.25 cm long X 0.25 cm. width, wing - 0.25 cm. long, keel - 0.2 cm. long; F. Androecium - 0.2 cm. long; G. Carpel 0.3 cm. long; H. Fruit - 2.3 cm. long X 1.7 cm. width; I. Floral Diagram.

polypetalous, inferior, white, glabrous, herbaceous, odourless, vaxillary, standard–1, large, 0.25 cm long and 0.25 cm broad, two wings, 0.25 cm long, two keels–0.25 cm long; *stamens* 10 (9 + 1), diadelphous, filaments fuse to form short column, anthers 0.2 cm long, anthers free, soft, white, glabrous; *carpels*–

2, syncarpous, ovary, superior, shuttle shaped, one chamber, two ovules in chamber, marginal placentation, 0.3 cm long, flattened; style-1, terminal and stigma absent.

Pollination caused by small insects.

Fruits pod, kidney shaped, flattened, 2.5 cm long and 1.7 cm broad. One margin is convex and other is concave, with mid-notched.

Seeds–2 in each pod. Germination is hypogeal.

Flowering and fruiting are generally during July to September.

Occasionally grow in the mangrove forest, within the intertidal zone; but mostly as an back mangals. Collection is made from the Bagna Forest range of the Eastern Sundarbans.

Medicinally, the root powder is said to be need to allay the effect of alcohol.

GENUS–26. *DERRIS* Lour.Fl. Cochinch. 11.432.1790.

The Genus–*Derris* Lour. has 80 species and these are distributed chiefly in the tropical countries; this genus is characterised by climbing shrubs, but rarely tree species are reported. From Sundarbans mangrove habitats or mangrove reclaimed land, three species are collected and described in this present dissertation; *viz.*, *Derris scandens* Benth., *D. trifoliata* Lour. and *D. indica* Bennet.

Key Characters of three species of *Derris* Lour.

1a. Climbing shrub in the intertidal river banks; leaves elliptic oblong with acute tip; fruit with thin coat and much flattened

(2)

1b. Straight or bending medium tree in the mangrove cleared lands, grow as back-mangals, occasionally planted as avenue tree; leaves broadly ovate, with acute tip; fruit flattened but thick with blunt tip and hard woody coat

1. *D. indica* (Lamk.) Bennet

2a (1). Leaflets elliptic with acuminate tip; number is 3-7, fruit sub-orbicular, thin margin

2. *D. trifoliata* Lour.

1b (1). Leaflets elliptic oblong or lanceolate, with acute tip; fruit shuttle with both narrow end, one margin thicker than other

3. *D. scandens* Benth.

26.1. *Derris trifoliata* Lour. Fl. Cochinch. 433.1790.

D. uliginosa Benth. Miq. Pl. Jungh. 1 : 252.1852.

Pongamia uliginosa DC. Prodr. 2 : 416.1824.

Galedupe uliginosa Roxb. Hort. Beng. 53.1814 & Fl. Indica 3 : 243.1831.

Local Name (Bengali) : Pan-Lata

(Plate–121 and Fig. 39)

Climbing much branched shrub or twining habit, evergreen, perennial.

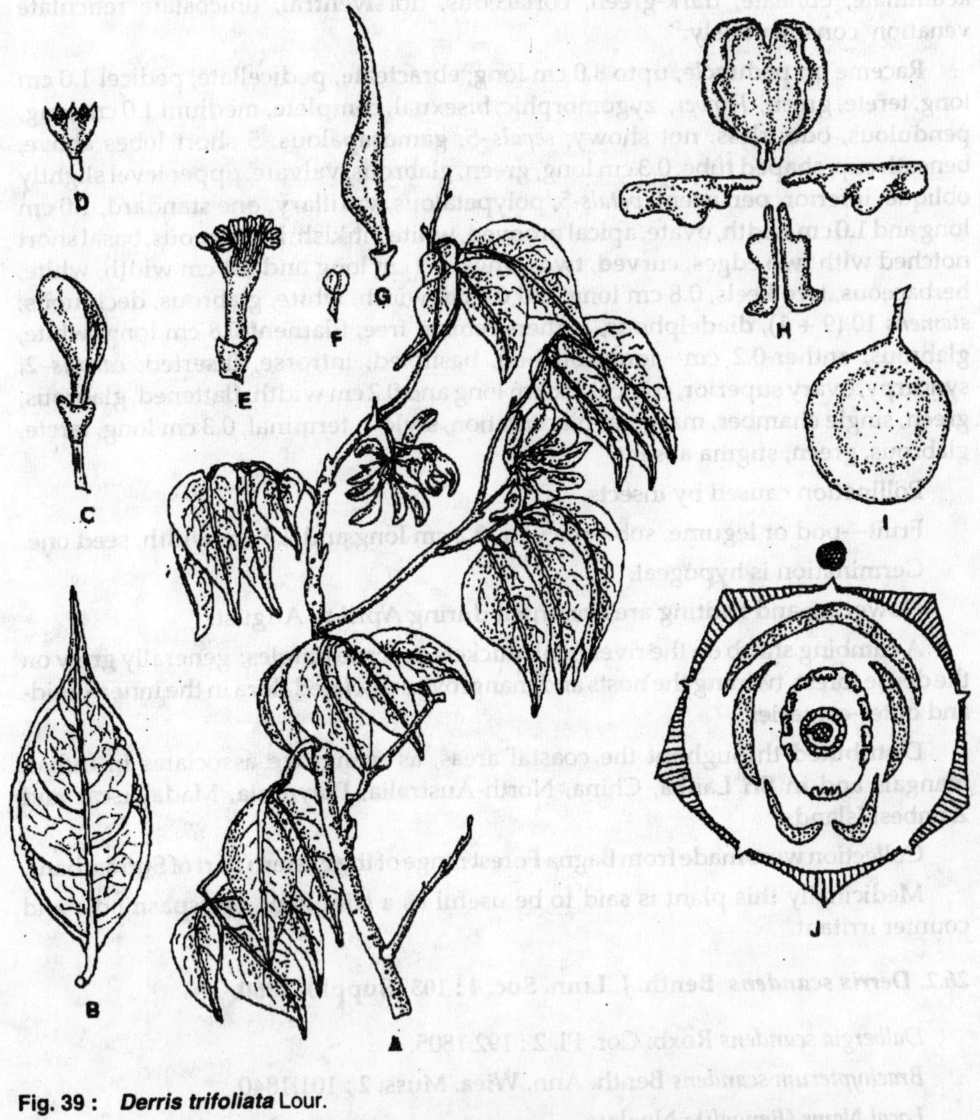

Fig. 39 : ***Derris trifoliata*** **Lour.**

A. Portion of shoot with flowers × 1/2; B. Leaflet : stalk-0.3 cm long, lamina–6.8 cm long × 3.1 cm width; C. Flower–1.0 cm long; D. Calyx–0.3 cm long; E. Androecium–1.0 cm long; F. Stamen; G. Carpel–0.8 cm long; H. Corolla : standard–1.0 cm long × 0.9 cm width, wing-0.9 cm long, keel-0.8 cm long; I. Fruit–3.2 cm long × 2.5 cm width; J. Floral diagram.

Tap roots deep sunken and without any aerial outgrowth.

Nodes not rigid, internodes upto 5.0 cm long, 0.3 cm diam., glabrous, green, rounded, twining.

Compound leaves, leaf axis upto 15.0 cm long, pulvinous, terete, green alternate, imparipinnate, leaflets upto 8.0 cm long and upto 3.1 cm broad, 3-7 numbers, petiolate; petiole 0.5 cm long, terete, glabrous, green, leaflets elliptic-oblong, entire, acuminate, cuneate, dark-green, coriaceous, dorsiventral, unicostate reticulate venation conspicuously.

Raceme on peduncle, upto 8.0 cm long; ebracteate, pedicellate, pedicel 1.0 cm long, terete, green, *Flower*, zygomorphic, bisexual, complete, medium 1.0 cm long, pendulous, odourless, not showy; *sepals*–5, gamosepalous, 5 short lobes above, beneath cup-shaped tube, 0.3 cm long, green, glabrous, valvate, upper level slightly oblique, inferior, persistent; *petals*-5, polypetalous, vaxillary, one standard, 1.0 cm long and 1.0 cm width, ovate, apical grooved, white pinkish, herbaceous, basal short notched with two edges, curved, two wings, 0.9 cm long and 0.4 cm width, white, herbaceous, two keels, 0.8 cm long and 0.4 cm width, white, glabrous, deciduous; *stamens* 10 (9 + 1), diadelphous, anthers remain free, filament 0.8 cm long, white, glabrous, anther-0.2 cm long, bilobed, basifixed, introrse, inserted; *carpels*–2, syncarpy, ovary superior, conical, 0.6 cm long and 0.2 cm width, flattened, glabrous, green, single chamber, marginal placentation, style-1, terminal, 0.3 cm long, terete, glabrous, green, stigma absent.

Pollination caused by insects.

Fruit—pod or legume, sub-orbicular, 3.2 cm long and 2.5 cm width, seed one.

Germination is hypogeal.

Flowering and fruiting are dominant during April to August.

A climbing shrub on the river bank thickets or scrub jungles; generally grow on the dense forest, twining the hosts and mangrove associated flora in the inner–, mid- and outer-estuaries.

Distributed throughout the coastal areas, as mangrove associates in Indian mangals and in Sri Lanka, China, North-Australia, Polynesia, Madagascar and Zambesi Islands.

Collection were made from Bagna Forest range of the Eastern part of Sundarbans.

Medicinally this plant is said to be useful as a stimulant, antispasmodic and counter irritant.

26.2. *Derris scandens* Benth. J. Linn. Soc. 4 : 103 (Suppl.), 1860.

Dalbergia scandens Roxb. Cor. Pl. 2 : 192.1805.

Brachypterum scandens Benth. Ann. Wiea. Muss. 2 : 101.1840.

Local Name (Bengali) : Noalata

(Plate–122 and Fig. 40)

Climbing, much branched shrub having twining habit, evergreen, perennial plant.

Tap-root system having deep sunken, but without aerial roots.

Nodes not ridged, internode solid, woody, glabrous, dark brown, terete, grooved ridged, rough surface, frequently small dot shaped lenticellated.

Leaves compound, alternate, exstipulate, imparipinnate, leaf axis, upto 19.0 cm long, solid, woody, green, abaxial, curved, while adaxial narrow grooved, pulvinous, solid, soft, , lamina elliptic-lanceolate, entire, blunt, glabrous, upto 6.3 cm long and

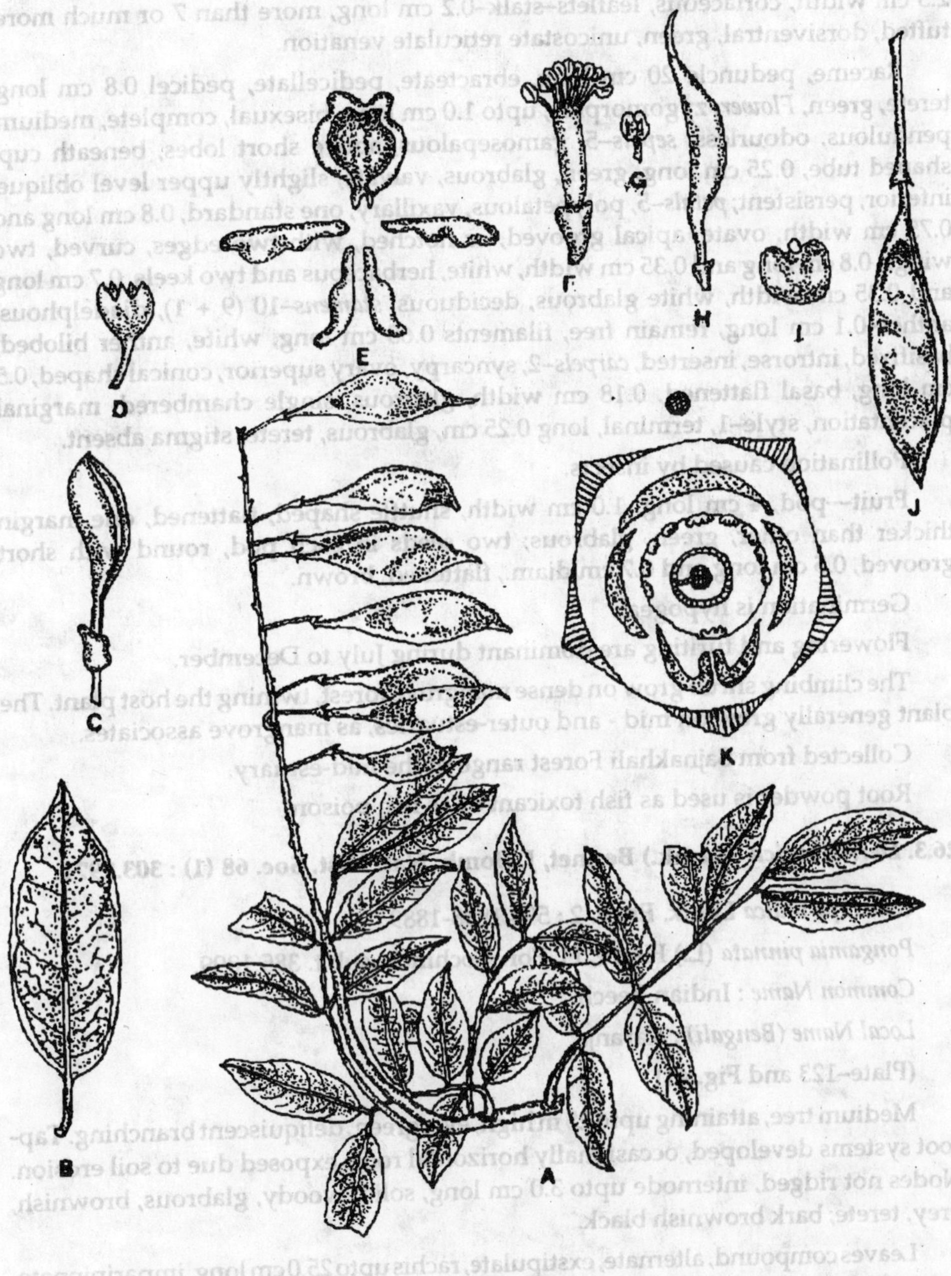

Fig. 40 : ***Derris scandans*** **Benth.**

A. Portion of shoot with fruits × 1/2; B. Leaflet : stalk-0.2 cm long, lamina–6.3 cm long × 2.5 cm width; C. Flower–1.0 cm long; D. Calyx–0.25 cm long; E. Corolla; standard–0.8 cm long × 0.75 cm width, wing–0.8 cm long × 0.35 cm width, keel-0.7 cm long × 0.35 cm width; F. Androecium–0.75 cm long; G. Stamen; H. Carpel–0.6 cm long; I. Seed-0.5 cm long × 0.7 cm diam.; J. Fruit with stalk : stalk-1.0 cm long & fruit–4.0 cm long; K. Floral diagram.

2.5 cm width, coriaceous, leaflets–stalk–0.2 cm long, more than 7 or much more; tufted, dorsiventral, green, unicostate reticulate venation.

Raceme, peduncle 20 cm long, ebracteate, pedicellate, pedicel 0.8 cm long, terete, green, *Flower;* zygomorphic, upto 1.0 cm long, bisexual, complete, medium, pendulous, odourless; *sepals*–5, gamosepalous, above short lobes, beneath cup-shaped tube, 0.25 cm long, green, glabrous, valvate, slightly upper level oblique, inferior, persistent; *petals*–5, polypetalous, vaxillary, one standard, 0.8 cm long and 0.75 cm width, ovate, apical grooved, or notched, with two edges, curved, two wings, 0.8 cm long and 0.35 cm width, white, herbaceous and two keels, 0.7 cm long and 0.35 cm width, white glabrous, deciduous; *stamens*–10 (9 + 1), diadelphous, anther–0.1 cm long, remain free, filaments 0.65 cm long, white, anther bilobed, basifixed, introrse, inserted, *carpels*–2, syncarpy, ovary superior, conical shaped, 0.5 cm long, basal flattened, 0.18 cm width, glabrous, single chambered, marginal placentation, style–1, terminal, long 0.25 cm, glabrous, terete, stigma absent.

Pollination caused by insects.

Fruit—pod, 4 cm long, 1.0 cm width, shuttle shaped, flattened, one margin thicker than other, green, glabrous; two seeds in each pod, round with short grooved, 0.5 cm long and 0.7 cm diam., flattened, brown.

Germination is hypogeal.

Flowering and furiting are dominant during July to December.

The climbing shrub grow on dense mangrove forest, twining the host plant. The plant generally grow on mid - and outer-estuaries, as mangrove associates.

Collected from Sajnakhali Forest range of the mid-estuary.

Root powder is used as fish toxicant or as fish poison.

26.3. *Derris indica* (Lamk.) Bennet, J. Bomb. Nat. Hist. Soc. 68 (1) : 303.1971.

Galedupa indica Lamk. Ency. 2 : 594.1888-1889.

Pongamia pinnata (L.) Pierre, Fl. For. Cochinch. sub.t. 385.1899.

Common Name : Indian Beech.

Local Name (Bengali) : Karanja

(Plate–123 and Fig. 41)

Medium tree, attaining upto 25 m high, evergreen, deliquiscent branching. Tap-root systems developed, occasionally horizontal roots exposed due to soil erosion. Nodes not ridged, internode upto 3.0 cm long, solid, woody, glabrous, brownish, grey, terete; bark brownish black.

Leaves compound, alternate, exstipulate, rachis upto 25.0 cm long, imparipinnate, green, pulvinous, 0.1 cm width, leaflets petiolate; petiole 1.0 cm long, 0.1 cm width, ovate, entire, acuminate, angled, green, dorsiventral, glabrous, unicostate reticulate venation conspicuously, coriaceous, 10.0 cm long and 5.0 cm width.

Raceme, peduncle, upto 18.0 cm long; ebracteate, pedicellate; pedicel 1.0 cm long, terete, green, *flower* zygomorphic, 1.2 cm long, bisexual, complete, pendulous,

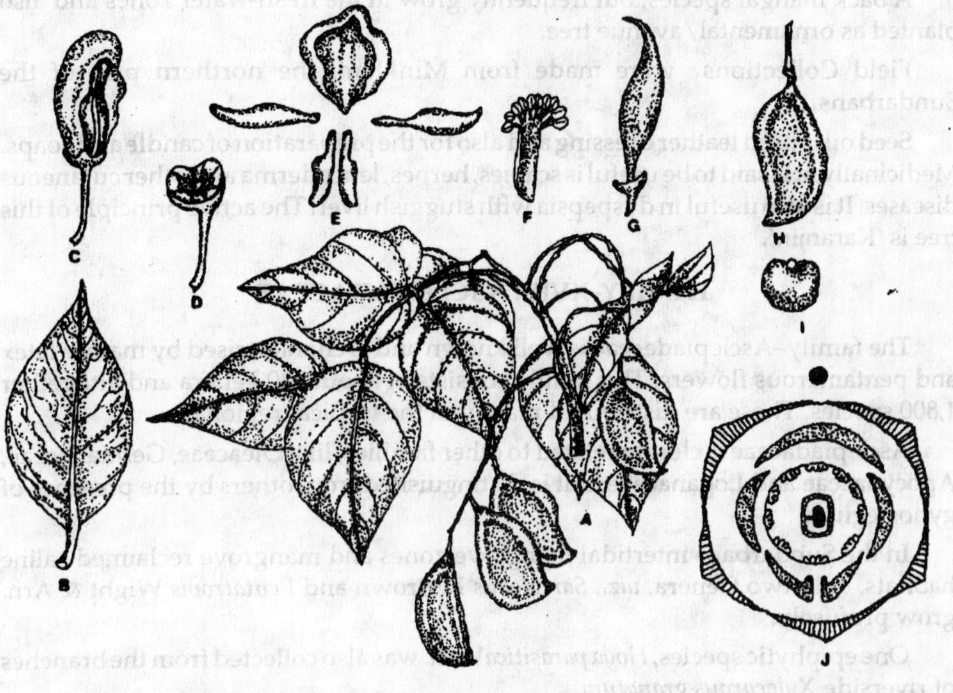

Fig. 41 : ***Derris indica*** **Bennet**

A. Portion of shoot with fruits × 1/3; B. Leaflet : stalk–1.0 cm long × 0.1 cm width, lamina–10.0 cm long × 5.0 cm width; C. Flower–1.2 cm long; D. Calyx–0.6 cm across; E. Corolla : standard–1.0 cm long × 0.9 cm width; wing–1.2 long × 0.35 cm width, keel–1.0 cm long × 0.3 cm width; F. Androecium–1.0 cm long × 0.2 cm width; G. Carpel–1.3 cm long, 0.15 cm width; H. Fruit–5.8 cm long × 3.5 width; I. Seed; J. Floral diagram.

odourless; *sepals*–5, gamosepalous, cup-shaped, green, glabrous, valvate, inferior, persistent, 0.6 cm across; *petals*–5, polypetalous, vaxillary, one standard, 1 cm long and 0.9 cm width, apical notched, white pinkish, herbaceous, basal short appendages, curved, two wings, 1.2 cm long and 0.35 cm width, white, herbaceous and two keels, 1.0 cm long, 0.3 cm width, white, glabrous, deciduous; *stamens*–10 (9 + 1), diadelphous, staminal column upto 1.0 cm long and 0.2 cm width, anthers remain free, anther bilobed, basifixed, introrse, inserted; *carpels*–2, syncarpous, conical, ovary superior, 1.3 cm long and 0.15 cm width, mid-region flattened, two ovules in single chamber, marginal placentation, style–1, terminal, 0.25 cm long, glabrous, stigma absent.

Pollination is cuased by insects and flies.

Fruits-pod, nut like, 5.8 cm long and 3.5 cm broad with persistent calyx, elliptic, glabrous, green, but black brown on maturity and woody on drying, thick upto 5 cm. Seeds 2, more or less square, brown seed coat.

Germination is hypogeal.

Flowering and fruiting are during April to August, but fruit hang upto December in the mother tree.

A back mangal species, but frequently grow in the fresh-water zones and also planted as ornamental/avenue tree.

Field Collections were made from Minkhan, the northern part of the Sundarbans.

Seed oil is used leather dressing and also for the preparation of candle and soaps. Medicinally, it is said to be useful is scabies, herpes, leucoderma and other cutaneous diseases. It is also useful in dyspepsia with stuggish liver. The active principle of this tree is 'Karanjin'.

FAMILY–XVIII. ASCLEPIADACEAE

The family–Asclepiadaceae is well known and is characterised by mainly latex and pentamerous flowers. This family consists of about 280 genera and altogether 1,800 species. These are distributed mainly in the tropical regions.

Asclepiadaceae is closely related to other families, like Oleaceae, Gentianaceae, Apocynaceae and Loganaceae; but is distinguished from others by the presence of gynostegium.

In the Sundarbans intertidal mangrove zones and mangrove reclaimed saline habitats, only two Genera, *viz.*, *Sarcolobus* R. Brown and *Pentatropis* Wight & Arn. grow profusely.

One epiphytic species, *Hoya parasitica* Wall. was also collected from the branches of riverside *Xylocarpus granatum.*

Key Characters to the Common Genera of Asclepiadaceae

1a. Prostrate or climbing herbs or shrubs; leaves large to medium, upto 8.5 cm long and 4.0 cm broad, pedicel stout, not pendulous, flowers having no corona, but single ciliated like structure at the base of inner side of petal, fruits long and broad

1. *Sarcolobus* R. Brown

1b. Twining herb, leaves small to medium, upto 2.5 cm long and 1.8 cm broad, pedicel weak and pendulous; flower with corona, fruits long

2. *Pentatropis* W. & A.

GENUS–27. *SARCOLOBUS* R. Brown, Mem. Wern. Soc. i : 34.1809.

The genus–*Sarcolobus* R. Brown consists of 15 species, which are distributed chiefly in South East Asia and Malaysia.

Prostrate or climbing herb, poorly branched, presence of milky latex. Tap root system, root sometime exposed to the above ground due to soil erosion.

Key Characters to Different Species of *Sarcolobus* R. Brown

1a. Leaves thick and coriaceous, upto 8.5 cm long and 4.0 cm width, flowers ebracteate, rosaceous, petal white-pink coloured, having no brown spots, inflorescence branched cyme

1. *S. globosus* Wall.

1b. Leaves coriaceous, upto 6.5 cm long and 2.7 cm width, flower bracteate, slightly cup-shaped with (below) hollow tube, petal white-pink coloured and scattered brown spots on petal at outer surface, inflorescence corymb 2. *S. carinatus* Wall.

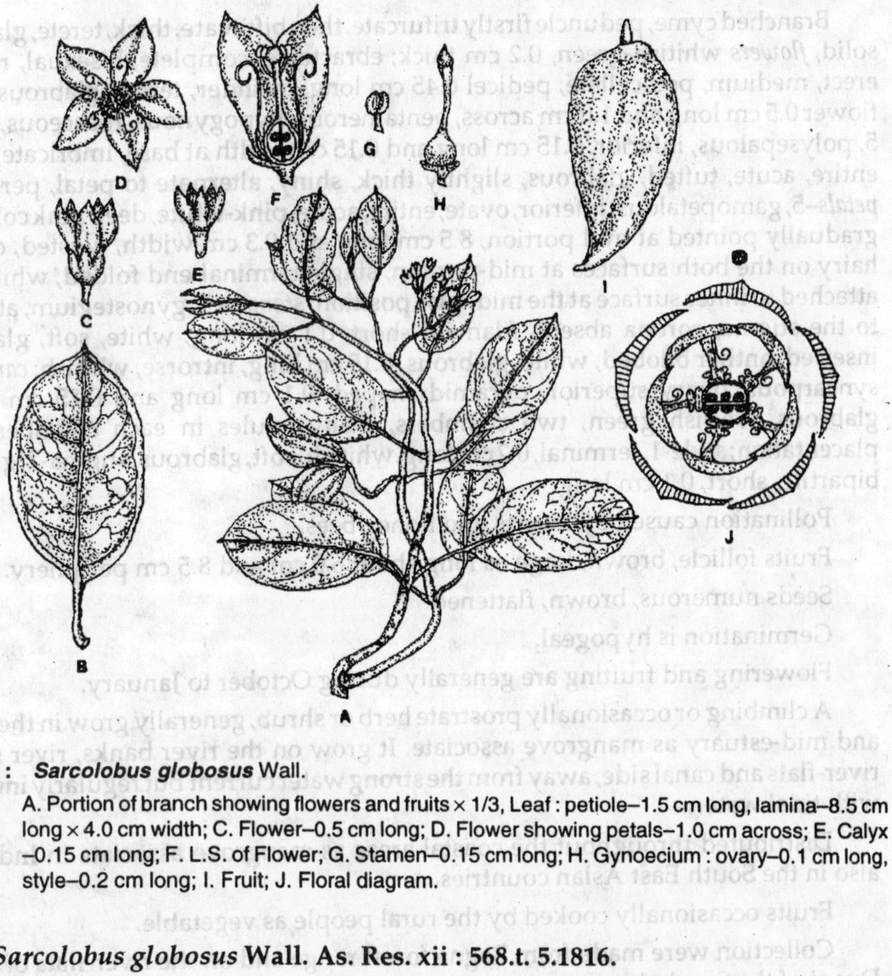

Fig. 42 : ***Sarcolobus globosus*** **Wall.**
A. Portion of branch showing flowers and fruits × 1/3, Leaf : petiole–1.5 cm long, lamina–8.5 cm long × 4.0 cm width; C. Flower–0.5 cm long; D. Flower showing petals–1.0 cm across; E. Calyx –0.15 cm long; F. L.S. of Flower; G. Stamen–0.15 cm long; H. Gynoecium : ovary–0.1 cm long, style–0.2 cm long; I. Fruit; J. Floral diagram.

27.1. *Sarcolobus globosus* Wall. As. Res. xii : 568.t. 5.1816.

Local Name (Bengali) : Baole Lata

(Plate–124 and Fig. 42)

Prostrate and climbing herb or shrub, poorly branched, presence of milky latex.

Tap-roots deep sunken, often exposed due to soil erosion but without any aerial roots.

Nodes slightly ridged, but not jointed, internode upto 8.0 cm long, solid, glabrous, semi-terete, green, 0.2 cm–0.3 cm-diam.; in climbing condition branches become weak and long but in prostrate condition branches become thick, stout and showing secondary thickening.

Leave simple, opposite decussate, exstipulate, petiolate, petiole 1.5 cm long and 0.2 cm width, whitish green, terete-flattened, in adaxial slight mid-grooved, glabrous, solid, pulvinous; lamina eliptic-oblong, entire, acute, angled, coriaceous, 8.5 cm long and 4.0 cm broad, glabrous, dorsiventral, green, distinct medium, unicostate reticulate venation conspicuous.

Branched cyme, peduncle firstly trifurcate, then bifurcate, thick, terete, glabrous, solid, *flowers* whitish-green, 0.2 cm thick; ebracteate, complete, bisexual, regular, erect, medium, pedicellate; pedicel 0.45 cm long, cylinder, terete, glabrous, solid, flower 0.5 cm long and 1.0 cm across, pentamerous, hypogynous, rosaceous, *sepals*–5, polysepalous, inferior, 0.15 cm long and 0.15 cm width at base, imbricate, ovate, entire, acute, tufted, glabrous, slightly thick, shiny, alternate to petal, persistent; *petals*–5, gamopetalous, inferior, ovate, entire, acute, pink-white, deep pink coloured, gradually pointed at mid portion, 8.5 cm long and 0.3 cm width, twisted, distinct hairy on the both surfaces at mid-poriton, single terminal end folded, white cilia, attached to inner surface at the mid-base position; *stamens* 5, gynostegium, attached to the stigma, corona absent, filament short, 0.1 cm long, white, soft, glabrous, inserted, anther bilobed, white, glabrous, 0.15 cm long, introrse, whitish; *carpels*–2, syncarpous, ovary superior, pyramid shaped, 0.1 cm long and 0.15 cm width, glabrous, whitish green, two chambers, many ovules in each chamber, axile placentation; style-1, terminal, 0.2 cm long, whitish, soft, glabrous, stigma-2, globose, bipartite, short, 0.1 cm long.

Pollination caused by insects and honey bees.

Fruits follicle, brown, large as long about 6.0 cm and 8.5 cm periphery.

Seeds numerous, brown, flattened.

Germination is hypogeal.

Flowering and fruiting are generally during October to January.

A climbing or occasionally prostrate herb or shrub, generally grow in the inner- and mid-estuary as mangrove associate. It grow on the river banks, river slopes, river-flats and canal side, away from the strong water current but regularly inundate with tidal water.

Distributed throughout the coastal areas as mangrove associate in India and also in the South East Asian countries.

Fruits occasionally cooked by the rural people as vegetable.

Collection were made from Bagna forest range and on the river-flats of Durga Duani of the Sundarbans.

27.2. *Sarcolobus carinatus* Wall. As. Res XII : 570. t. 4.1816.

Local Name (Bengali) : Baole Lata

(Plate–125 and Fig. 43)

Climbing herb or shrub, branched, presence of milky latex.

Tap-roots deep sunken, sometime prostrate roots are exposed due to soil erosion; without any aerial roots.

Nodes slightly ridged, but not jointed, internodes upto 12.0 cm long, solid, glabrous, terete, green grey, 0.15 cm–0.25 cm diam.

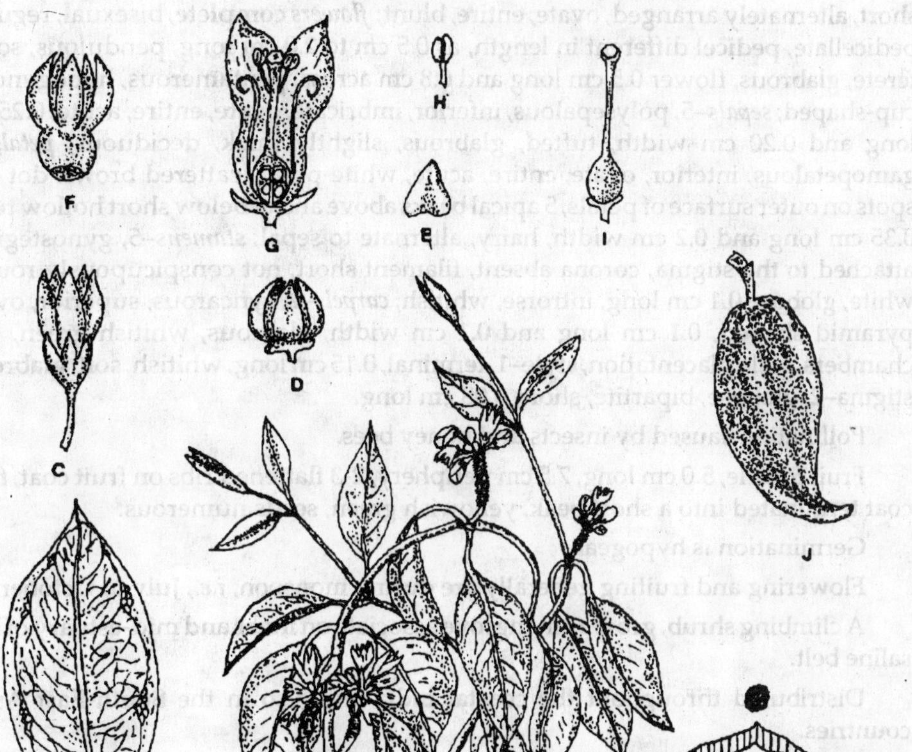

Fig. 43 : *Sarcolobus carinatus* Wall.

A. Portion of branch showing inflorescences × 1/2; B. Leaf; petiole–2.7 cm long, lamina-6.5 cm long × 2.7 cm width; C. Flower–0.5 cm long; D. Calyx–0.25 cm long; E. Sepal–0.25 cm long × 0.2 cm width; F. Corolla: corolla tube–0.15 cm long, corolla lobe–0.2 cm long; G. L.S. of Flower; H. Stamen–0.1 cm long; I. Gynoecium : ovary–0.7 cm long, style–0.15 cm long, stigma–0.8 cm long; J. Fruit–5.0 cm long ; K. Floral diagram.

Leaves simple, opposite decussate, exstipulate, petiolate; petiole 2.7 cm long and 0.15 cm width, yellowish-green, terete, glabrous, solid, pulvinous, lamina elliptic-oblong, entire, acute, cup-shaped or angled, coriaceous, 6.5 cm long and 2.7 cm width, glabrous, green, dorsiventral, distinct mid-vein, unicostate reticulate venation conspicuous.

Corymbose; peduncle unbranched, 2 cm long and 0.15 cm width, glabrous, solid, arise at the nodal axis, bract scars present on the peduncle; bracteate, bract short, alternately arranged, ovate, entire, blunt; *flowers* complete, bisexual, regular, pedicellate, pedicel different in length, as 0.5 cm to 2.0 cm long, pendulous, solid, terete, glabrous, flower 0.5 cm long and 0.8 cm across, pentamerous, hypogynous, cup-shaped; *sepals*–5, polysepalous, inferior, imbricate, ovate, entire, acute, 0.25 cm long and 0.20 cm width, tufted, glabrous, slightly thick, deciduous; *petals*–5, gamopetalous, interior, ovate, entire, acute, white-pink, scattered brown dot like spots on outer surface of petals, 5 apical beaks above and at below short hollow tube, 0.35 cm long and 0.2 cm width, hairy, alternate to sepal; *stamens*–5, gynostegium attached to the stigma, corona absent, filament short, not conspicupously round, white, globose 0.1 cm long, introrse, whitish; *carpels*–2, syncarous, superior, ovary pyramid shaped, 0.1 cm long and 0.1 cm width, glabrous, whitish green, two chambers, axile placentation, style–1, terminal, 0.15 cm long, whitish, soft, glabrous, stigma–2, globose, bipartite, short, 0.08 cm long.

Pollination caused by insects and honey bees.

Fruit follicle, 5.0 cm long, 7.5 cm periphery, 2-3 flattened ribs on fruit coat, fruit coat terminated into a short beak, yellowish green, seeds numerous.

Germination is hypogeal.

Flowering and fruiting generally are during monsoon, *i.e.*, July to October.

A climbing shrub, grow as mangrove associate on inner and mid-estuary of low saline belt.

Distributed throughout the coastal India and also in the South-East Asian countries.

Collection was made from the Gosaba area of the Sundarbans.

Table 62 : Comparative Morphology of the Two Species of *Sarcolobus* spp.

Characters	*Sarcolobus globosus*	*Sarcolobus carinatus*
Habit	Prostrate and climbing shrub or herb	Generally climbing shrub or herb.
Stem	Internode, upto 8.0 cm long, 0.2 cm–0.3 cm diam., green or green-brownish.	Internode upto 12.0 cm long, 0.15 cm–0.25 cm diam., green grey.
Leaf	Petiole upto 1.5 cm long and 0.2 cm width, lamina upto 8.5 cm long and 4.0 cm width, thick and coriaceous.	Petiole upto 2.7 cm long and 0.15 cm width, lamina upto 6.5 cm long and 2.7 cm width, coriaceous.
Inflorescence	Branched cyme.	Unbranched corymb.
Flowers	Ebracteate, pedicel 0.45 cm long, flower 0.50 cm long and 1.0 cm across; sepal upto 0.15 cm long and 0.15 cm width; petal-0.5 cm long and 0.3 cm width; white pink anther 0.15 cm long; ovary upto 0.10 cm long and 0.15 cm width, style 0.2 cm long.	Bracteate, pedicel 0.5 cm to 2.0 cm long, flower 0.5 cm long and 0.8 cm across; sepals 0.25 cm long and 0.2 cm width; petals 0.35 cm long and 0.2 cm width, scattered brown spots on outer petal surface; anther 0.1 cm long; ovary upto 0.1 cm long and 0.1 cm across, style 0.15 cm long.
Fruit	6.0 cm long and 8.5 cm periphery.	5.0 cm long and 7.5 cm periphery.
Ecology	Generally grow on inner-estuary of low-saline belt.	Generally grow on inner and mid-estuary of low-saline belt.

GENUS–28. *PENTATROPIS* Wight & Arnold

The species *P. capensis* (Linn.f.) Bullock is the only species found in the Sundarbans.

28.1. *Pentatropis capensis* (Linn. f.) Bullock, Kew Bull. 1955 : 284.1995.

P. microphylla (Roxb.) W. & A. Wight Contri. 52.1834

Chromosome No. $2n = 22$ (Biswas, 1955)

Hindi Name : Ambarvel

(Plate–126 and Fig. 44)

Climbing herb, having twining habitat, branched, presence of milky latex.

Tap-roots, but no aerial roots.

Node ridged, not jointed, internode upto 8.0 cm long and upto 0.2 cm width, terete, solid, woody, rough surface, mature branches grey coloured with rough surface, while young twig green colour with glabrous surface.

Simple, opposite decussate, exstipulate, petiolate; petiole 0.8 cm long and 0.15 cm width, solid, fleshy and soft, green, glabrous, pulvinous; lamina ovate-elliptic, entire, mucronate, cup-shaped, green, fleshy, glabrous, 2.5 cm long and 1.8 cm width, dorsiventral, unicostate reticulate venation inconspicuously.

Cyme, pedunculate, peduncle 0.25 cm long and 0.05 cm width, terete, green, glabrous, group buds on each peduncle stalk, at base presence of numerous hairy bracts; bracts lanceolate, entire, acute, tufted; *flower* bracteate, bisexual, complete, regular, pedicellate, pedicel upto 2.6 cm long, terete, narrow but tufted, glabrous, brownish, flower hypogynous, generally one flower arise from each peduncle, 0.5 cm long and 0.3 cm across; *sepals* 5, polysepalous, rosaceous, imbricate, persistent, greenish, herbaceous, pubescent, lanceolate, 0.15 cm long, entire, acute, alternate to petals; *petals*-5, gamopetalous, inferior, outer upper surface brownish-white and glabrous, imbricate, 0.4 cm long and 0.2 cm width, coriaceous, not showy; *stamens*-5, gynostegium, corona-5, flattened, 0.15 cm long and 0.05 cm width, fleshy, soft, brownish-white, forming hollow column surrounding style; anther bilobed, brown, introrse, filament absent or short, attached with stigma; *carpels*-2, syncarpous, superior, ovary conical shaped, greenish-white, fleshy, soft, 0.1 cm long, two chambered, many ovules in each chamber, axile placentation; style–1, terminal, 0.08 cm long, soft, terete, greenish white, stigma-2, globose, soft, bifurcate, distinctly, 0.05 cm diam.

Pollination caused by small insects.

Fruits follicle, 3-4 cm long, cylindrical, gradually tapering end, rigid, glabrous, green, ruptured in one suture. Seeds numerous, brown, 0.2 cm diam.

Seed germination is hypogeal.

Flowering and fruiting are generally concentrated during November to January.

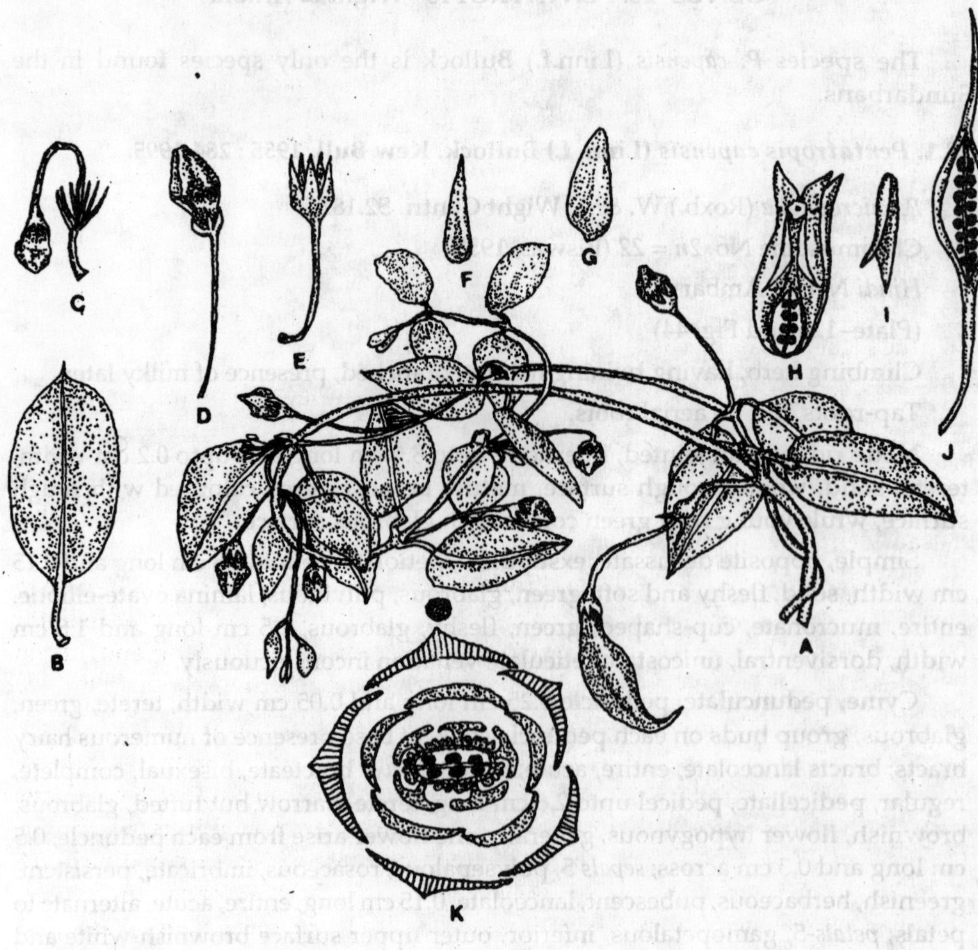

Fig. 44 : ***Pentatropis capensis*** **Bull.**

A. Portion of twining branch showing flowers and fruit × 1/2; B. Leaf; petiole–0.8 cm long, lamina–2.5 cm long × 1.8 cm width; C. Inflorescence with single flower; D. Pedicellate flower; pedicel–1.3 cm long, flower-0.5 cm long; E. Calyx; F. Sepal–0.1 cm long; G. Petal–0.4 cm long × 0.2 cm width; H. L.S. of flower; I. Corona–0.1 cm long; J. Fruit with stalk : stalk-2.5 cm long, fruit-4.0 cm long; K. Floral diagram.

Climbing herb, twin the host and generally grow on mid-and outer-estuaries, on tidal inundated plain lands or river flats.

Distributed throughout the coastal regions of India and also in the Tropical coasts of the Old World.

Collection was made from Kakdwip area of the Western Sundarbans.

Medicinally, this plant is said to be useful as alternative and refrigerant.

FAMILY–XIX. *AMARYLLIDACEAE*

The family–Amaryllidaceae is characterised by rhizomatous stem, leafy only in spring or in rains; inflorescence usually on scape, with spathe, cymose but often umbel or head like condensed, solitary flower occasionally.

The family consists of about 85 genera, including 1100 species, which are usually distributed in the Tropical or Sub-Tropical countries. Out of these, the Genus–*Crinum* L. is a moist loving group of plants and *Crinum defixum* Ker-Gawler grow in the inner-estuary, within the intertidal limits, river-flats and river slopes.

GENUS–29. *CRINUM* L. Gen. ed. I : 97.1737.

The Genus–*Crinum* L. has about 100-110 species, which are distributed in the Tropical and Sub-Tropical regions, especially on the sea coasts. The species *Crinum defixum* Ker-Gawler is collected from the Sakha Bidyadhari river flats at Kulti, in the Sundarbans area.

29.1. *Crinum defixum* Ker-Gawler Quart. J. Sci. 3 : 105.1817.

Chromosome No. 19, 20, 20 + B, 24, 30, 50, 60 (Subramanian, 1979) and $2n = 22$ (Tandan & Mathur, 1966 and Raina, 1978)

Local Name (Bengali) : Sukha Darsan

(Plate–127 and Fig. 45)

Stout, bulbous herb, attaining the height upto 1.5 m.

Fibrous roots, without any aerial out growth of roots.

Rhizome bulb, ovoid or globose, caudex.

Simple in rosette, lamina upto 1.0 m long and 12.0 cm breadth, slightly sheathing at base, linear, entire, acute, adaxially convex, dark green, fleshy, glabrous, isobilateral, unicostate, parallel venation.

Umbel, scape axillary, erect, cylindrical, upto 1.0 m long and 3.0 cm broad, triangular, fleshy, solid, spongy, glabrous, green, 5-10 flowers on each peduncle, a common thin white layer cover the buds at young stage; *flower* ebracteate, bisexual, regular, erect, epigynous, upto 5.5 cm long and 5.0 cm across, showy; *tepals*–5, gamotepalous, 5 lobes above, linear, mucronate, herbaceous, imbricate, 3.0 cm long and 0.3 cm broad, white, convex, slightly folded, glabrous, deciduous, beneath fuse to form hollow tube, fleshy, spongy, upto 1.8 cm long and 0.4 cm width; *stamens*-6, free, filaments jointed at base by white thin layer, filament upto 2.2 cm long and 0.1 cm width, terete, soft, glabrous, white, mid position curved, anther bilobed, 0.15 cm width and 1.8 cm long, dark yellow with mid-narrow evergreen linear line, exserted, extrorse; *carpels*–2, syncarpy, ovary inferior, globose, 0.5 cm long, 4 chambered, one ovule in each chamber, axile placentation ; style-1, terminal, upto 4.5 cm long, terete, soft, glabrous, white below, navy green upper, stigma-1, globose.

Pollination caused by wind and insect. Fruit is sub-globose. Seeds 1-2, rugose. Germination is hypogeal.

Flowering and fruiting period ranges from June to September.

This bulbous herb occasionally grow on the silted up sedimented soil in the intertidal river banks and river flats of the river Sakha Bidyadhari at Kulti area.

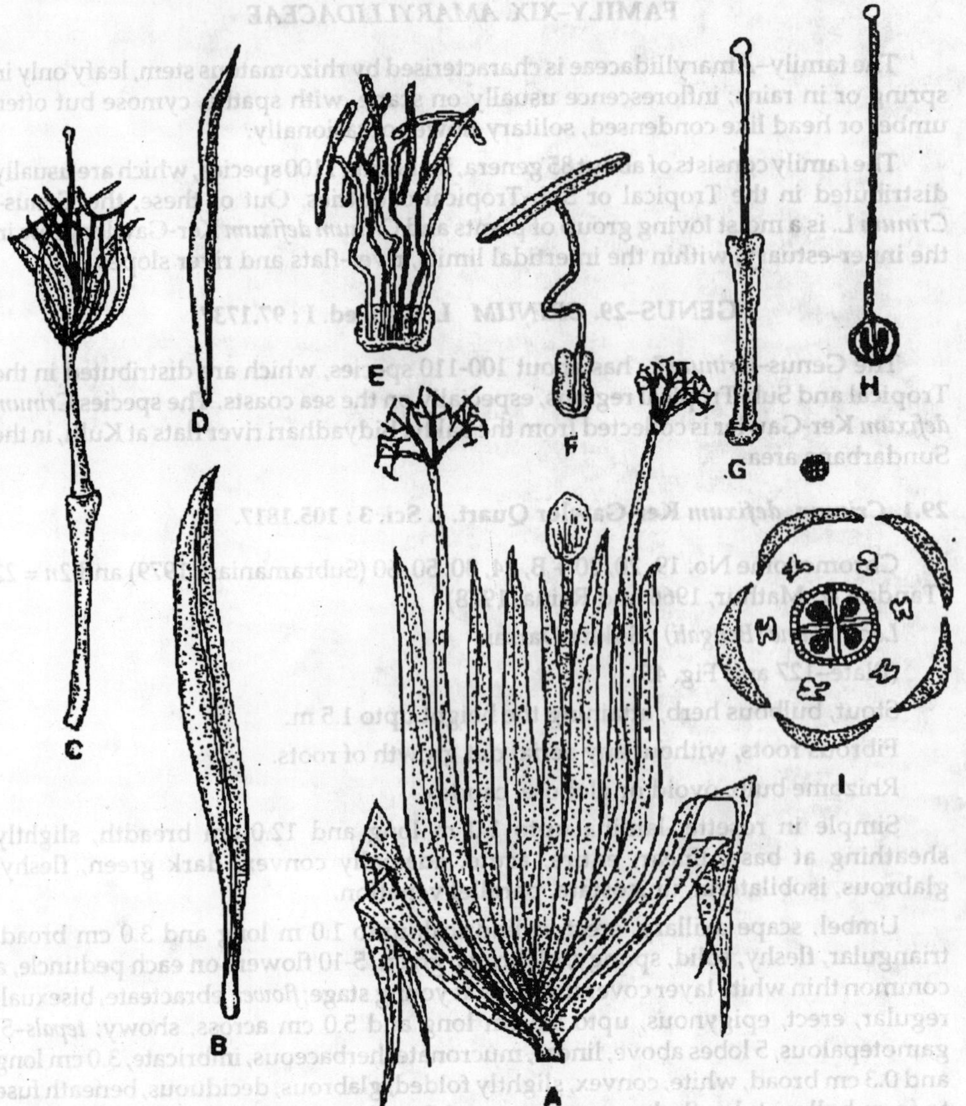

Fig. 45 : ***Crinum defixum Ker - Gawler***

A Entire plant showing habit with inflorescence X 1/150; B. Leaf - 100 cm. long X 12.0 cm. width; C. Inflorescence; D. Tepal - 3.0 cm. long X 0.3 cm. width; E. Androecium; F. Stamen : filament - 2.1 cm. long, anther - 1.8 cm. long; G. Gynoecium : ovary - 0.5 cm. long style - 4.5 cm. long; H. L. S. of carpel; I. Floral Diagram.

Distributed in the inner estuaries in the S.E. Asian mangals.

Collections were made from Kulti/Ghusighata.

Medicinally, this plant is said to be useful as diaphoretic and emollient and used for the treatment of burns, whitlow and curbuncles.

FAMILY–XX. CHENOPODIACEAE

The family–Chenopodiaceae is characterised by having fleshy habit, 2-3 carpellate and one ovuled ovary. Chenopodiaceae is closely related to Amaranthaceae–on the basis of monochlamydous flower, uniseriate stamens and 1-celled ovary. It is also related to Phytolaccaceae–on the plan of floral structure.

This family has about 102 genera and 1,400 species, distributed throughout the World. The common herbaceous genus–*Suaeda* Forsk. ex Scop. alongwith its two species, *i.e., S. maritima* Dumort. and *S. nudiflora* Roxb. are dominant in the intertidal river flats, river beds and forest floores of the Sundarbans. Another halophytic species, *Suaeda monoica* Forsk. ex. Gmel. is also common in the Tamil Nadu and Andhra Pradesh coasts, but not reported from the coastal zones or mangrove forest areas of the Sundarbans. The genus–*Salicornia* L. along with its single species. *S. brachiata* Roxb. is also reported by Naskar & Guha Bakshi (1987) and Naskar (1993) from the mangrove reclaimed coastal areas or river banks of the Sundarbans in West Bengal. But, during this field investigation this *S. brachiata* Roxb. was not collected and studied in details.

GENUS–30. *SUAEDA* Forsk. ex Scop. Fl. Aegypt. Arab. 69.1775.

The Genus–*Suaeda* Forsk. ex Scop. is cosmopolitan on the coastal zones and salty areas. These are characterised by halophytic herbs with fleshy leaves and dense cymes.

This genus consists of 110 species which have distributed all over the coastal tropics and sub-tropics.

The members under the genus–*Suaeda* Forsk. ex Scop. are prostrate or erect herb, upto 1.0 m tall, much branched, green or reddish in appearance. Tap root systems and without any aerial roots, though some tap roots are exposed due to soil erosion. Internodes not rigid, solid, woody, slightly, rough, terete, grey green. Leaves simple, alternate, exstipulate, sessile, upto 3.0 cm long and 0.4 cm width, fleshy, glabrous, green or reddish after maturation, ovate-oblong, isobilateral. Spike at leaf axis; *flowers* small, upto 0.1 cm long and 0.25 cm across, bisexual, complete, regular, ebracteate, sessile, each flower at leaf axis, erect or horizontal, hypogynous; *tepals*–5, polytepalous, inferior, imbricate, ovate, acute, entire, fleshy, upto 0.08 cm long and 0.08 cm width, persistent; *stamens*-5, free, filament white, soft, exerted, anther bilobed, dark yellow, extrorse; *carpels*-3, syncarpous, ovary superior, 0.5 cm long and 0.5 cm width, whitish green, single chambered with single ovule; style 2-3, terminal, 0.08 cm long. Drupe (included utricle), 0.15 cm long and 0.15 cm across, with persistent perianth. Seeds erect or horizontal.

Key to the Commonly Available Species under *Suaeda* Forsk. ex. Scop.

1a. Leaves broad, green, styles–2, seeds usually horizontal
1. *S. maritima* Dumort.

1b. Leaves semi-terete, green, but reddish after maturation, styles-3, seeds erect
2. *S. nudiflora* Roxb.

30.1. *Suaeda maritima* Dumort. Fl. Belg. 22.1827.

Chromosome No. *n = 18* (Jinno. 1956)

Common Name : Common Indian Saltwort.

Local Name (Bengali) : Giria Sak/Gire Sak

(Plate–128, Fig. 46)

Erect much branched herb, 0.30 m–0.65 m tall, green.

Tap-roots, occasionally exposed due to soil erosion during tidal current.

Internodes not rigid, solid, woody, rough, terete, green grey.

Leaves simple, alternate, exstipulate, sessile, 1.0 cm long and 0.4 cm broad, succulent, glabrous, 0.05 cm thick, green, ovate-oblong, entire, round, isobilateral.

Spikes at leaf axis; ebracteate, small flowered, *flower*–0.1 cm long and 0.15 cm across; bisexual, complete, regular, sessile, not showy, erect, horizontal, hypogynous; *tepals*-5, polytepalous, rosaceous, inferior, imbricate, ovate, acute, entire, succulent, green, thick 0.08 cm long and 0.08 cm width, glabrous, equal length, persistent; *stamens*-5, free, filament white, soft, glabrous, terete, 0.1 cm long, basifixed, exserted, anther bilobed, dark yellow, globose, extrorse, 0.08 cm long; *carpels*–3, syncarpous, ovary superior, 0.05 cm long and 0.05 cm width, whitish green, succulent, glabrous, single chambered with single ovule, basal placentation, style-2, 0.08 cm long, soft, white, glabrous, terminal, stigma-0.

Pollination caused by wind and water currents and flies etc.

Fruits drupe, 0.15 cm width and 0.10 cm long with persistent perianths. Germination is hypogeal.

Flowering and fruiting are generally from October to December.

A common halophytic and herbaceous plant, grow on the mid–and–outer estuaries, on the muds and river flats. Grow gregariously and cover large scale in the frequently inundated river flat lands. Distributed throughout the coastal areas and in the salt marshes in India and South East Asian coasts.

Occasionally cooked as vegetable by the rural poor people; cattles occasionally grazed on the herbaceous flora; acted as soil binder.

Collection was made from Pakhiralay of the central parts of Indian Sundarbans.

30.2. *Suaeda nudiflora* Roxb. Moq. Ann. Sci. Nat. 23.1831.

Chromosome No. 2n = 36 (Sidhu 1964)

Common Name : Common Indian Saltwort.

Local Name (Bengali) : Gire Sak/Giria Sak

(Plate–129 and Fig. 47)

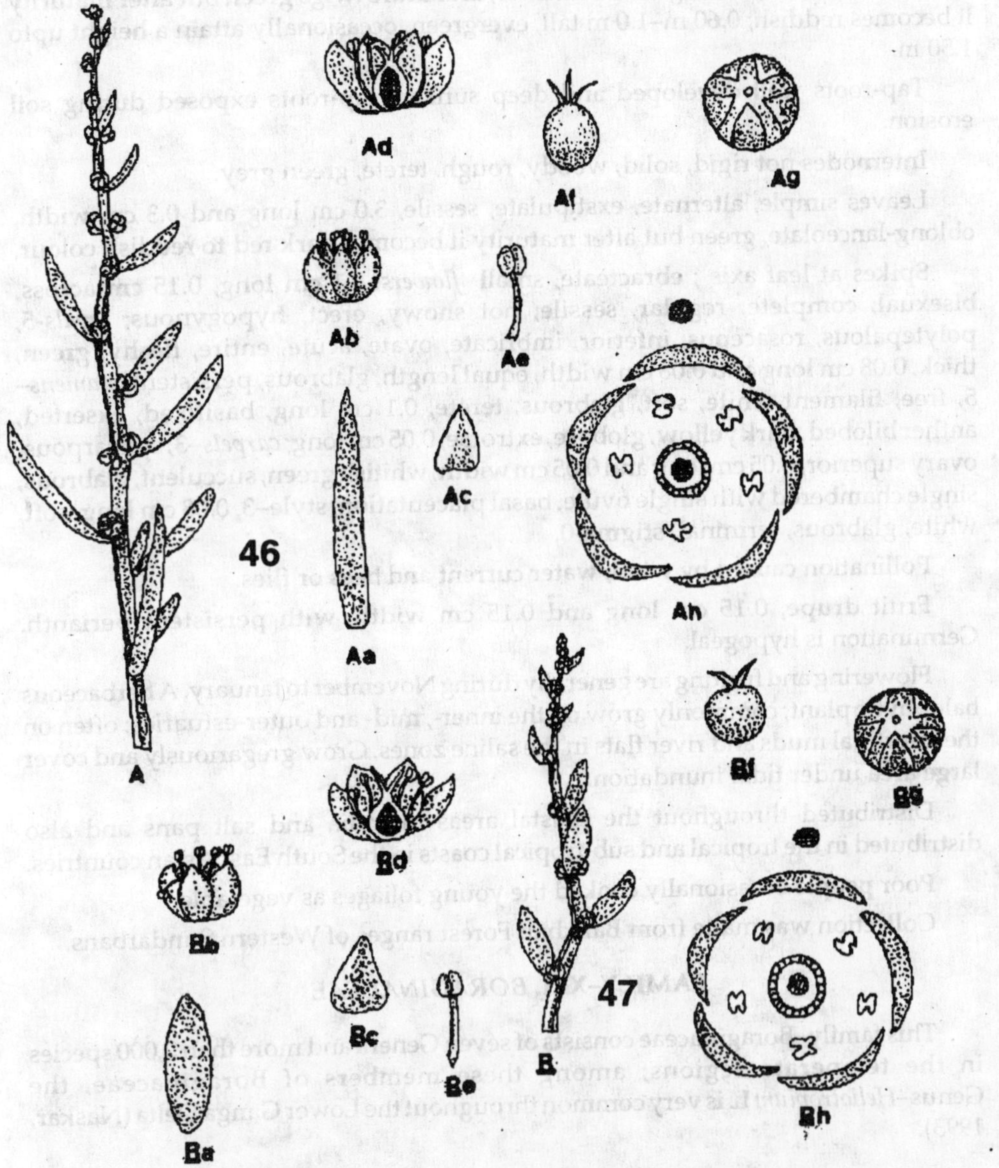

Fig. 46 : ***Suaeda maritima*** **Dumort.**

B. Portion of flowering twig × 2/3; Ba. Leaf–1.0 cm long × 0.4 cm width; Bb. Flower–0.1 cm long; Bc. Tepal–0.08 cm long; Bd. L.S. of flower; Be. Stamen : filament–0.1 cm long, anther–0.05 cm long; Bf. Gynoceium : 0.1 cm long; Bg. Fruit–0.15 cm long × 0.15 cm across, Bh. Floral diagram.

Fig. 47 : ***Suaeda nudiflora*** **Roxb.**

A. Portion of flowering twig × 2/3; Aa. Leaf–3.0 cm long × 0.3 cm width; Ab. Flower–0.1 cm long × 0.18 cm across; Ac. Tepal; Ad. L.S. of flower; Ae. Stamen : filament–0.1 cm long, anther–0.05 cm long; Af. Gynoecium; 0.1 cm long; Ag. Fruit–0.15 cm long × 0.15 cm width; Ah. Floral diagram.

Prostrate herb, having erect branches; immature twigs green but after maturity it becomes reddish; 0.60 m–1.0 m tall, evergreen, occasionally attain a height upto 1.50 m.

Tap-roots well developed and deep sunken; tap-roots exposed during soil erosion.

Internodes not rigid, solid, woody, rough, terete, green grey.

Leaves simple, alternate, exstipulate, sessile, 3.0 cm long and 0.3 cm width, oblong-lanceolate, green but after maturity it becomes dark red to reddish colour.

Spikes at leaf axis ; ebracteate, small *flowers*, 0.1 cm long, 0.15 cm across, bisexual, complete, regular, sessile, not showy, erect, hypogynous; *tepals*-5, polytepalous, rosaceous, inferior, imbricate, ovate, acute, entire, fleshy, green, thick, 0.08 cm long and 0.08 cm width, equal length, glabrous, persistent; *stamens*–5, free, filament white, soft, glabrous, terete, 0.1 cm long, basifixed, exserted, anther bilobed, dark yellow, globose, extrorse, 0.05 cm long; *carpels*–3, syncarpous, ovary superior, 0.05 cm long and 0.05 cm width, whitish green, succulent, glabrous, single chambered with single ovule, basal placentation, style–3, 0.08 cm long, soft, white, glabrous, terminal, stigma-0.

Pollination caused by wind, water current and bees or flies.

Fruit drupe, 0.15 cm long and 0.15 cm width, with persistent perianth. Germination is hypogeal.

Flowering and fruiting are generally during November to January. A herbaceous halophytic plant; commonly grow on the inner-, mid- and outer-estuaries; often on the intertidal muds and river flats in the saline zones. Grow gregariously and cover large area under tidal inundation.

Distributed throughout the coastal areas in India and salt pans and also distributed in the tropical and sub-tropical coasts in the South East Asian countries.

Poor people occasionally cooked the young foliages as vegetable.

Collection was made from Bakkhali Forest ranges of Western Sundarbans.

FAMILY–XXI. *BORAGINACEAE*

This family–Boraginaceae consists of seven Genera and more than 2,000 species in the temperate regions; among these members of Boraginaceae, the Genus–*Heliotropium* L. is very common throughout the Lower Ganga Delta (Naskar, 1993).

GENUS–31. *HELIOTROPIUM* L. Sp. Pl. 130-1753.

In the Lower Ganga Delta, *Heliotropium* L. is having four common species, *viz.*, *H. indicum* L., *H. curassavicum* L., *H. ovalifolium* Forsk. and *H. strigosum* Willd. (Naskar, 1993). Out of these four species, *Heliotropium curassavicum* L. is very much common in the mangrove reclaimed lands, river banks and intertidal flat river slopes of the Sundarbans. This species are noted to grow in the salt crusted soil and very often grow in the highly saline soil and frequently saline water inundated zones. For having its high salt tolerance ability and abundance growth in the

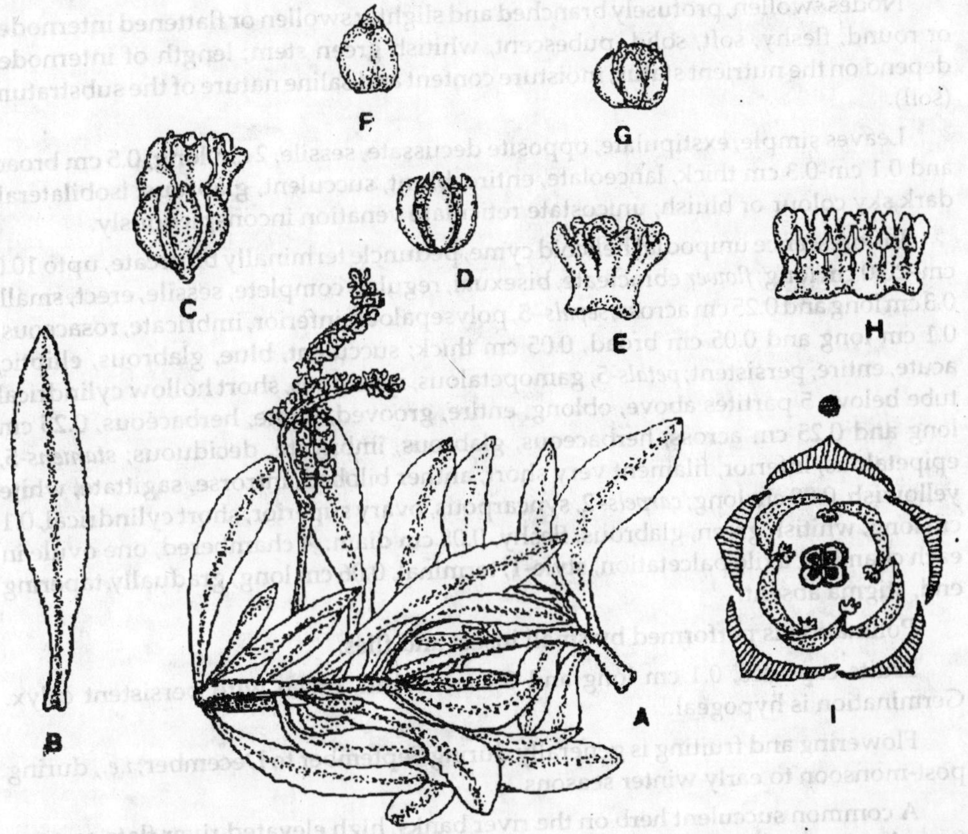

Fig. 48 : ***Heliotropium curassavicum*** **L.**

A. Portion of prostrate herb with inflorescence × 2/3; B. Leaf-2.0 cm long × 0.5 cm width; C. Flower–0.3 cm long; D. Calyx : sepal–0.1 cm long ×0.05 cm width; E. Corolla–0.25 cm long × 0.25 cm across; F. Carpel–0.1 cm long; G. Fruit with persistent calyx–0.1 cm long; H. L.S. of corolla with epipetalous stamens; I. Floral diagram.

mangrove cleared forest zones, *H. curassavicum* L. is considered as mangrove associate; this is much succulent halophytic herb and an exotic species.

31.1. *Heliotropium curassavicum* L. Sp. Pl. 130.1753.

Chromosome No. $2n = 24$ (Subramanyan & Kamble, 1967); $2n = 26$ (Heiser & Whiteker, 1948; Britton, 1951 and Sarkar *et al.*, 1973) and $2n = 28$ (Pal, 1957).

Local Name (Bengali) : Nona Hatisnur
(Plate–130 and Fig. 48)

Prostrate, much branched and spreading herb; on maturity form sub-erect branches; branches and foliages are dark-bluish.

Tap root much branched; deep rooted–about 0.3 m depth.

Nodes swollen, profusely branched and slightly swollen or flattened internodes or round, fleshy, soft, solid, pubescent, whitish green stem; length of internodes depend on the nutrient status, moisture content and saline nature of the substratum (soil).

Leaves simple, exstipulate, opposite decussate, sessile, 2 cm long, 0.5 cm broad and 0.1 cm-0.3 cm thick, lanceolate, entire, blunt, succulent, glabrous, isobilateral, dark sky colour or bluish; unicostate reticulate venation inconspicuously.

Inflorescence unipodial helicoid cyme, peduncle terminally bifurcate, upto 10.0 cm–20.0 cm long; *flower* ebracteate, bisexual, regular, complete, sessile, erect, small, 0.3 cm long and 0.25 cm across; *sepals*–5, polysepalous, inferior, imbricate, rosaceous, 0.1 cm long and 0.05 cm broad, 0.05 cm thick; succulent, blue, glabrous, elliptic, acute, entire, persistent; *petals*-5, gamopetalous, rosaceous, short hollow cylindrical tube below, 5 partites above, oblong, entire, grooved, white, herbaceous, 0.25 cm long and 0.25 cm across, herbaceous, glabrous, imbricate, deciduous; *stamens*-5, epipetalous, inferior, filament very short, anther bilobed, introrse, sagittate, white yellowish, 0.08 cm long; *carpels*–2, syncarpous, ovary superior, short cylindrical, 0.1 cm long, whitish green, glabrous, fleshy, 0.08 cm diam., 4 chambered, one ovule in each chamber, axile palcetation, style-1, terminal, 0.06 cm long; gradually tapering end, stigma absent.

Pollination is performed by small insect and flies.

Fruits capsules, 0.1 cm long and 0.1 cm broad, green with persistent calyx. Germination is hypogeal.

Flowering and fruiting is generally during September to December; *i.e.*, during post-monsoon to early winter seasons.

A common succulent herb on the river banks, high elevated river flats or open forest lands; above the frequent tidal reaches, especially in the saline salt crusted soil.

The fleshy, much succulent leaves and stems provide morphological support to adapt this halophytic herb in the highly saline soil. Though, the surface soil in the habitat of *H. curassavicum* L. are very often noted with salt crust, especially during late winter and summer months, its sunken root may absorb less saline water from the deeper zones. Its sunken root profusely spreading from tap roots, help to intake the less saline water and the ultra-filtration mechanism may filtrate the salt molecule during absorption of solutes from the substratum (Chapman 1976). Its sunken stomata also reduces the water loss during transpiration also.

Heliotropium curassavicum is distributed almost throughout India, especially from the coastal saline habitats and in dry conditions. It is said to be introduced from West Indies and Prain (1903a, 1903b, 1905) has not reported this taxon from the lower parts of Bengal and Sundarbans, as well.

Medicinally, the root powder is said to be important for sores and wounds.

FAMILY–XXII. *CONVOLVULACEAE*

The family–Convolvulaceae is closely allied to the families, Solanaceae, Nolanaceae and Scrophulariaceae and included under the Order–Solanales, due to presence of false septum of ovary, persistent calyx and bilobed structures of corolla, intra-xylary phloem. This family–Convolvulaceae is also allied to the family–Boraginaceae. This family-Convolvulaceae consists of about 50 genera and 1,200 species, which are mostly distributed in the tropical and sub-tropical reigon of the World. *Ipomoea* L. is one of the distinct genus of the family–Convolvulaceae and Naskar (1993) has reported 13 species under this genus–*Ipomoea* L. from the Lower Ganga Delta.

GENUS–32. *IPOMOEA* L. Sp. Pl. 159.1753.

The Genus–*Ipomoea* L. is generally climbing herbs and preferred aquatic or semi-aquatic habitats. This genus has 500 species, which are mostly distributed to the Tropical and warmer temperate regions of the Globe. Some of these species are cultivated in warmer countries for its tubers or vegetative parts for vegetables. The species *Ipomoea pes-caprae* (L.) R. Brown mostly grow in the sandy beaches and sandy soils of the coastal areas in Sundarbans. Banerjee, *et al.* (1989) have also reported *Ipomoea tuba* (Schl.) G. Don as the hinterland species along river banks and throughout near the tidal swamps of India, Indo-China, Malesia.

32.1. *Ipomoea pes-caprae* (L.) R. Brown, Tuckery, Narr. exped. Zaire 1816, (Mart. 1818); Sweet, Hort. Suburb, Lond. 35 (Jul. 1818); Stern, Taxon X. 238.

Convolvulus pes-caprae L. Sp. Pl. 159.1753. *Ipomoea biloba* Forsk. Fl. Aegypt. Arab. 44.1776.

Chromosome No. *2n = 30* (Watanabe, 1939; Miege, 1960; Chuang *et al.* 1963 & Jones, 1964)

Local Name (Bengali) : Chagal Kuri

(Plate–131 and Fig. 49)

Prostrate, creeping, herb; much branched; branching with fast growth during monsoon months; branches slightly erect, while main stalk horizontal to the ground.

Tap roots and several adventatious roots develop from the nodes. These adventitious roots are anchored the branches with the sand, as well as main stalk to the sandy soil and also act as sand binder in the naked sand-dune of coastal regions.

Weak stem, internodes upto 11.0 cm long, terete, green, solid, slightly fleshy, 0.3 cm diam., glabrous.

Alternate, cauline, upright, exstipulate, simple, petiolate; petiole upto 12.0 cm long, terete, glabrous, base brownish green, fleshy, solid, basal portion conspicuously twisted, pulvinous; lamina apically bilobed, slightly folded, each fold broadly ovate, entire, straight, common notch of both folded lamina about 0.3 cm deep, succulent, dorsiventral, glabrous, unicostate reticulate venation; sub-costas longer than the main costa, green in ventral surface, main costa along with subcostas distinctly found, each fold upto 5.5 cm long and 4.5 cm width.

Solitary (cyme), peduncle long about 7.0 cm long solid, glabrous, green, terete, erect, peduncle branched, long stalk bears large flower, while dwarf stalk having bud. *Flower*–5.5 cm long × 3.0 cm across, ebracteate, bisexual, complete, regular,

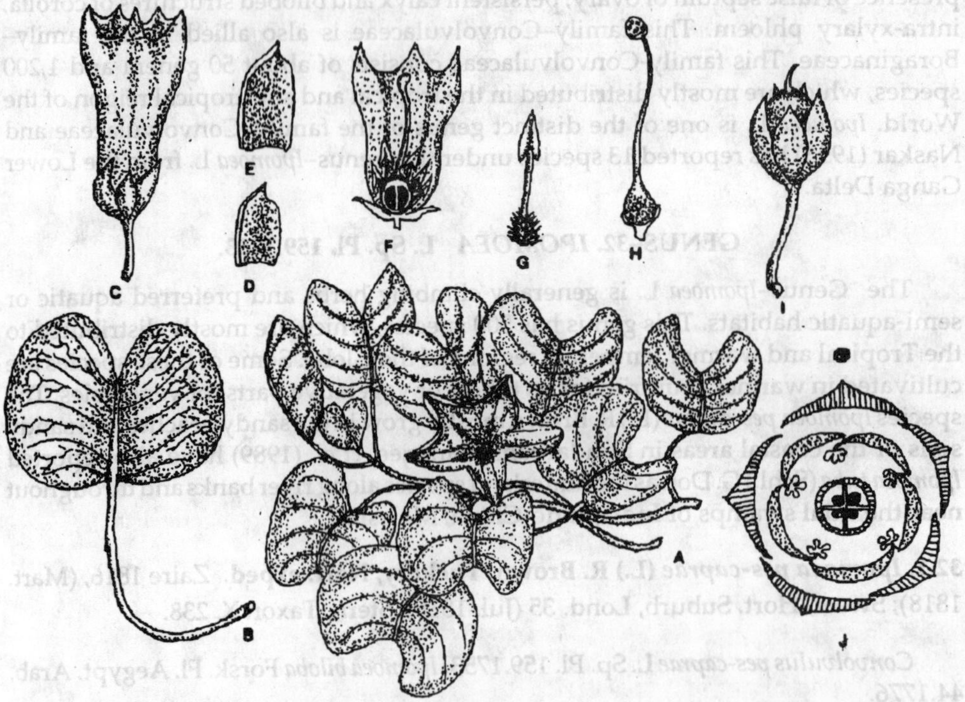

Fig. 49 : ***Ipomoea pes-caprae*** **(L.) Sweet**

A Part of the flowering twig ×1/2; B. Leaf; petiole–12.0 cm long, lamina-5.5 cm long × 4.5 cm width; C. Flower-5.5 cm long × 3.0 cm across; D. Small sepal-0.9 cm long; E. Large sepal-1.2 cm long × 0.6 cm width; F. L.S. of the flower; G. Stamen; spine base of the filament–0.2 cm long, smooth filament-0.8 cm long, anther-0.5 cm long; H. Gynoecium : ovary–0.2 cm long, style–1.7 cm long, stigma-0.1 cm long; I. Fruit with stalk : stalk-2.5 cm long, fruit-1.0 cm long ×1.3 cm periphery. J. Floral diagram.

erect, pedicellate, pedicel, 2.5 cm long, solid, terete, green, erect, odourless, pentamerous, perigynous, *sepals*-5, polysepalous; 3 inner large, 2 outer small, inferior, imbricate, 1.2 cm long—0.9 cm long, respectively and 0.6 cm width, ovate, oblong, entire, mucronate, glabrous, herbaceous, green, inconspicuous, parallel margin from base to apex, shorter than corolla; *petals*-5, gamopetalous, twisted, funnel shaped, 5 apical pointed beaks present, which are inwardly bending, fleshy, inferior, herbaceous, basally thick, violet at base gradually colour become fade with maturation, showy, equal length, alternate to sepal; *stamens*-5, epipetalous, filament base slightly attached to the corolla base, 2 large and 3 small filaments, 2 cm long and 1 cm long, respectively and 0.15 cm width at base, solid, base of the filament covered by gregarious white velvet hairs, filament white, solid, glabrous, inserted, anther bilobed, introrse, shorter than gynoecium, longitudinally dehisced, 0.5 cm long and 0.15 cm width, sagittate; *carpels*–2, syncarpous, ovary superior, conical, short, 0.2 cm long, ovary 4-chambered, each chamber having one ovule, axile placentation, style-1, terminal, 1.7 cm long, glabrous, solid, terete, white; stigmas–2, bifid, globose, finely ornamented.

Pollination is performed by butterflies and bumble bees.

Fruit capsule, with apical beak, 1.0 cm long and 1.3 cm periphery, with persistent large calyx, reddish at magin.

Germination is hypogeal.

Flowering and fruiting are during post-monsoon and winter seasons, *i.e.*, October to February. This plant mostly grow on the sandy beach of the sea facing, beyond the tidal reach.

Distributed throughout the sandy coast and sandy soil of India and tropical zones; this is actually a sand dune flora and plenty in the coastals sand of the Sundarbans.

Leaves used in rheumatism, stomachic tonic and the extract of the leaves have the astringent, diauretic, laxative properties, occasionally cattle gazed on these green leaves, a prominent sand binder.

FAMILY–XXIII. *AIZOACEAE* (*nom. alt. Ficoidae* Benth. & Hook.f.)

The family–Aizoaceae is allied to Phytolaccaceae, Caryophyllaceae and also to Cactaceae. The family is characterised by Xerophytic or halophytic herbs or shrubs, with opposite or alternate stipules or exstipulate; leaves often fleshy and with cymes, regular flower. The family–Aizoaceae has about 130 genera, including 1,200 species; which are mostly distributed to South Africa, but also in Tropical Africa, Asia, Australia and California.

The Genus–*Sesuvium* L. is the only representative of the family–Aizoaceae from the intertidal mangrove habitats of the Sundarbans.

GENUS–33. *SESUVIUM* L. Syst. ex. X.1058.1759.

The genus–*Sesuvium* L. consist of about 8 species in the tropical and sub-tropical coasts. Most of these species are having halophytic adaptability; but from the Lower Ganga Delta region, only *S. portulacastrum* L. is collected and reported by Naskar & Guha Bakshi (1987) and Naskar (1990 & 1993).

33.1. *Sesuvium portulacastrum* L. Syst. X. 1058.1759.

Chromosome No. *2n = 16* (Di Fulvio, 1967); *2n = 36* (Sharma & Bhattacharya (1956); *2n = 48* (Rughaban & Srinivasan, 1940) and C. *48* (Carr, 1978)

Local Name (Bengali) : Gada Bani

(Plate–132 and Fig. 50)

Prostrate, much branched herb, occasionally creeping on the muddy frequently tidal inundated slopes of saline habitats.

Tap roots are poorly developed, adventatious roots develop from the nodes, 2.5 cm – 4.0 cm long, glabrous, fleshy, terete, shiny, spongy, 0.2 cm–0.4 cm diam., nodes jointed and easily separated, pink, short erect branching, mid zone of the internodes slightly swollen, while node indistinctly grooved.

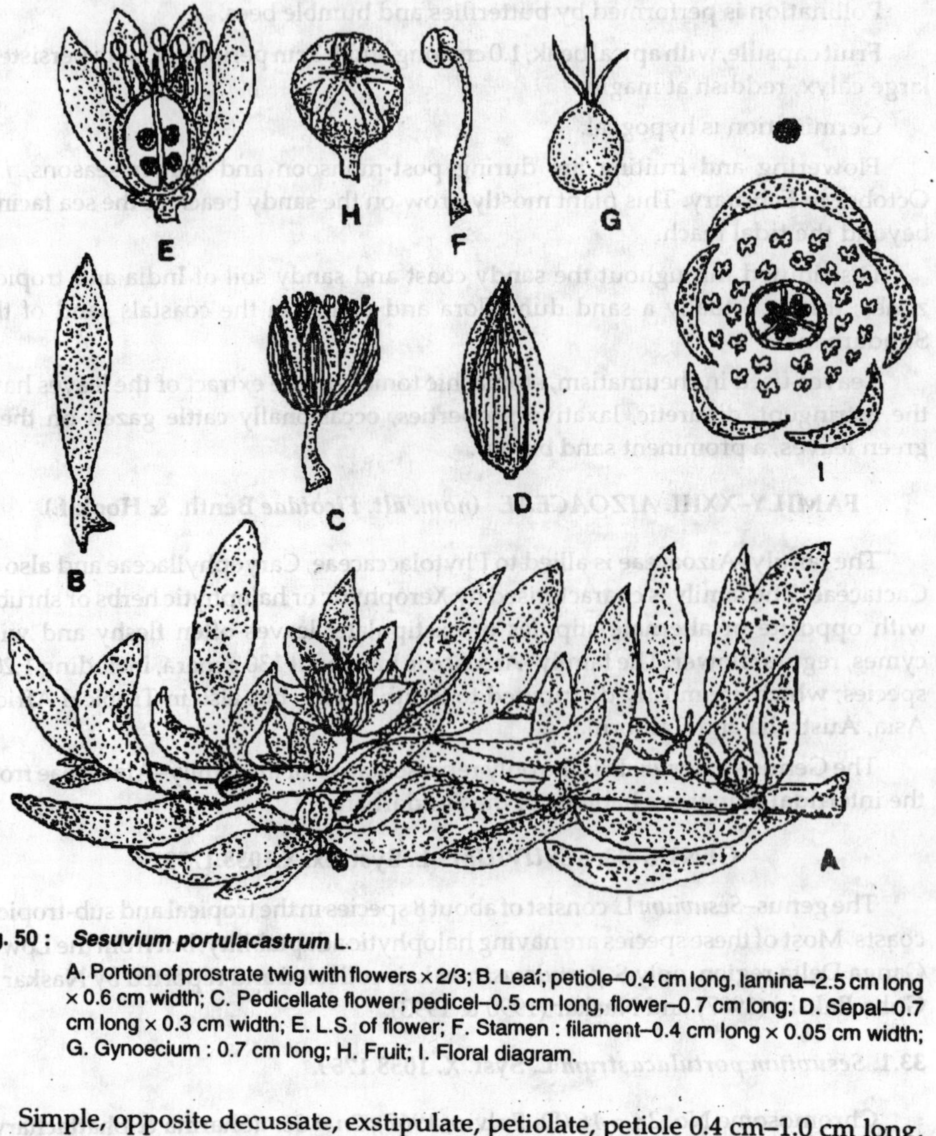

Fig. 50 : ***Sesuvium portulacastrum*** **L.**

A. Portion of prostrate twig with flowers × 2/3; B. Leaf; petiole–0.4 cm long, lamina–2.5 cm long × 0.6 cm width; C. Pedicellate flower; pedicel–0.5 cm long, flower–0.7 cm long; D. Sepal–0.7 cm long × 0.3 cm width; E. L.S. of flower; F. Stamen : filament–0.4 cm long × 0.05 cm width; G. Gynoecium : 0.7 cm long; H. Fruit; I. Floral diagram.

Simple, opposite decussate, exstipulate, petiolate, petiole 0.4 cm–1.0 cm long, fleshy, flattened, glabrous, pulvinous, spongy, lamina lanceolate, 4.5 cm long and 0.8 cm width, fleshy, glabrous, entire, acute, angled, isobilateral, inconspicuously unicostate reticulate venation, distinct mid-vein.

Solitary cyme at axis; *flower* ebracteate, complete, bisexual, regular, pedicellate, pedicel 0.5 cm long and 0.1 cm width, fleshy, terete, pink, glabrous, soft, solid, erect, medium, 0.7 cm long and 0.7 cm across, hypogynous; *sepals*-5, polysepalous, rosaceous, inferior, imbricate, persistent, elliptic, entire, acute, fleshy, thick, glabrous, mid portion thick, terminated into mucronate tip, margin white pink, herbaceous, glabrous, 0.7 cm long and 0.3 cm width, inner portion whitish; *corolla* absent; stamens numerous, free, filament 0.4 cm long and 0.05 cm width, flattened at the base, basifixed, exserted, pink violet, soft, glabrous, anther bilobed, 0.05 cm long, white violate, introrse, inferior, globose; *carpels*-3, syncarpous, ovary superior, ovoid, 0.3 cm long, 0.2 cm width, whitish green, soft, glabrous, 3-chambered, style-3, 0.4 cm long, terminal, soft, whitish, glabrous, stigma-0.

Pollination is performed by small insects and flies.

Fruit–capsule, 1 cm long and 1.5 cm diam. with persistent calyx.

Flowering and fruiting are generally from the post-monsoon to winter seasons, *i.e.*, October to January.

Germination is hypogeal.

Creeping herbs, densely covered the bare flat land of saline coasts, reclaimed area etc.

Distributed throughout the coastal areas in India; a Pantropic species.

FAMILY–XXIV. *VERBENACEAE*

There are altogether 400 tropical and sub-tropical species under the Family-Verbenaceae. In the mangroves and back-mangal communities two genera are conspicuous; in Sundarbans two genera, *viz.*, *Clerodendron* L. and *Premna* L. are reported as the back-mangals or as mangrove associates. But here only the Species *C. inerme* (L.) Gaertn. is described as the dominant mangrove associate and common flora of the present day's Sundarbans. This *C. inerme* (L.) Gaertn. has some positive role in the ecosystem of mangroves, in the tidal and intertidal estuarine ecosystems of the Sundarbans.

GENUS–34. *CLERODENDRON* L.

Four species are common in the Lower Ganga Delta of West Bengal, *viz.*, *Clerodendron inerme* (L.) Garten., *C. indicum* (L.) Kuntze, *C. splandens* G. Don. and *C. viscosum* Vent. Out of these four species *C. inerme* (L.) Gaertn. is very common and abundantly grow on the river banks and intertidal flat river slopes as mangrove associate; this species can also grow in the freshwater zone and largely planted as hedge plants in the boundary of the gardens, away from the mangrove environment.

34.1. *Clerodendron inerme* (L.) Gaertn. Fruct. 1 : 271.1788.

Chromosome No. $2n = 46$ (Raman & Kesavan, 1963; Sharma & Mukhopadhyay (1963) and $2n = 48$ (Sobti & Singh, 1961).

Local Name (Bengali) : Ban-Jnui or Batraj

(Plate–133 and Fig. 51)

Straggling much branched shrub, upto 3.0 m long and terminal branch very often twining, slender, twiggy, dark evergreen and form dense bush in the river banks and river slopes, occasionally inundate with the saline tidal water; as such, it is grouped as mangrove associate or back-mangal; it has no characters similar to the mangroves.

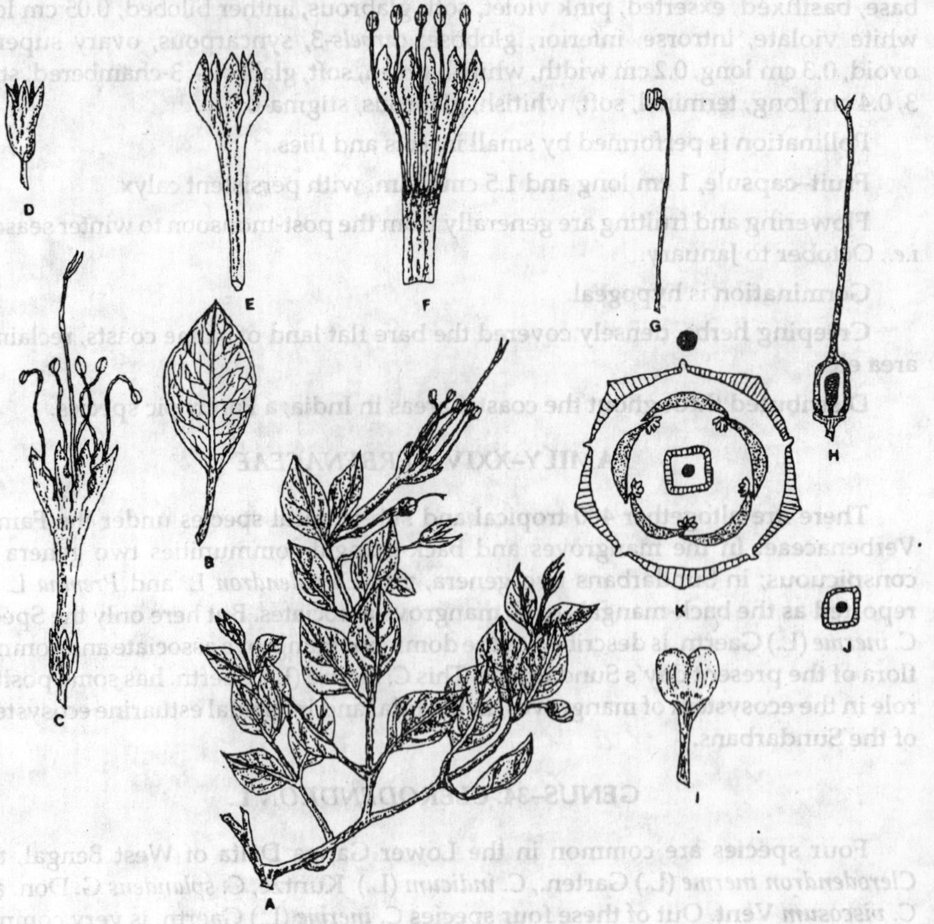

Fig. 51 : ***Clerodendron inerme*** **(L.) Gaertn.**

A. Portion of the flowering twig × 1/2; B. Leaf; petiole–0.8 cm long, lamina-3.8 cm long ×1.8 cm width; C. Flower–5.6 cm long; D. Calyx–0.5 cm long; E. Corolla : corolla tube-2.0 cm long, corolla lobe–0.5 cm long × 0.35 cm width; F. L.S. of corolla showing epipetalous stamens; G. Stamen : filament–3.5 cm long, anther-0.1 cm long; H. L.S. of Gynoecium : ovary–0.2 cm long, style-4.5 cm long; I. Fruit-1.5 cm long ×1.0 cm width; J. T.S. of ovary; K. Floral diagram.

Tap roots underground, but not much developed and deep sunken.

Stem differentiated into nodes and internodes, internodes variable in length, 2.0 cm–8.0 cm long, terete, glabrous, yellowish green, solid, upto 0.3 cm diam., nodes slightly ridged.

Leaf simple, opposite decussate, cauline, exstipulate, petiolate, petiole 0.6 cm–1.0 cm long, terete, glabrous, solid, soft, brownish green, pulvinous, lamina, ovate, elliptic, entire, acute, angled, dorsiventral, green, fibrous, herbaceous, 3.5 cm–4.5 cm long and 1.5 cm–2.0 cm broad, unicostate reticulate venation, odourless.

Inflorescence cymes, axillary penduculate. *Flower*–5.6 cm long × 1.0 cm across ebracteate, pedicellate, pedicel upto 1.0 cm long, terete, soft, greenish white, solid, glabrous; complete, bisexual, zygomorphic, erect, odourless, medium hypogynous; *sepals*-5, gamosepalous, 5 short lobes above, 0.15 cm long, entire, beneath cup-shaped tube, 0.5 cm long, green, glabrous, thin, persistent, valvate; *corolla*-petals-5, gamopetalous, 5 *petal* lobes above, equal length, imbricate, each lobe 0.5 cm long and 0.35 cm broad, elliptic, entire, acute, herbaceous, white, glabrous, beneath cylindrical long tube attaining 2.0 cm height and 0.2 cm diam., glabrous, white ; *androecium*-stamens 4, epipetalous, anthers remain free, filament long, 3.5 cm long, white, soft, terete, glabrous, exserted, extrorse, dorsiventral, anther bilobed, longitudinal dehisced, 0.1 cm long, yellowish ; *gynoecium*-carpels 2, syncarpy, ovary 0.2 cm long, cylindrical glabrous, greenish, one chambered, single, ovate, basal placentation; style-1, terminal, long 4.5 cm, terete, soft, glabrous, white, stigmas 2, bifid.

Pollination caused by small insects and honey bees are the specific pollinator in the Sundarbans mangrove areas.

Fruits-pyriform drupe, woody, 1.5 cm–2.0 cm long; seeds four in each fruit.

Germination is hypogeal, grow on the river banks and river flats and the fruits or the seeds are dispersed through water current.

Very often this plant is also seen in the sides of freshwater canals or in the mesophytic zones and also planted as dense dark green hedge plant in the peripheries of ornamental garden.

Medicianlly, the leaves are used as febrifuge; it contain an amorphous bitter principle, resembling the chiretta (Swertia chirayita). Leaves is said to be used as poultice to resolve buboes.

FAMILY–XXV. *LORANTHACEAE*

Plants under the family–Loranthaceae are either epiphytic or semi-parasitic habitats and there are several genera and species; these are reported both from tropical and temperate regions of the globe. In the mangrove forests of the Sundarbans, the two genera, *viz.*, *Viscum* L. and *Dendrophthoe* Mart. *Macrosolen* (Bl.) Reichenb. are very often grow on the branches of the mangrove trees, like *Avicennia* spp., *Excoecaria* sp. and *Xylocarpus* sp. The species under the above genera *Viscum* L., *Dendrophthoe* Mart. and have not any halophytic adaptations or characteristics similar to the halophytes; but as their close association on mangrove hosts these species are discussed and studied.

These two genera can be distinctly identified with the following key and field characters :

1a. Leaves broad, upto 10.0 cm long and 6.5 cm broad; flowers large, red and showy and bisexual. 1. *Dendrophthoe* Mart.

1b. Leaves narrow, upto 6.0 cm long and 2.0 cm broad ; flowers small, greenish and not showy, unisexual 2. *Viscum* L.

GENUS–35. *VISCUM* L.

Two species under the genus–*Viscum* L. are reported by Naskar & Guha Bakshi (1987) from the mangrove forest of the Sundarbans; these two species can be identified distinctly with the following key characters–

1a. Leaves obliquely ovate or fulcate, acute 1. *Viscum monoicum* Roxb.

1b. Leaves equally elliptic or elliptic lanceolate, obtuse 2. *Viscum orientale* Willd.

During this studies *Viscum monoicum* Roxb. is not collected and studies cannot be made.

35.1. *Viscum orientale* Willd. Sp. Pl. 4 (2) : 737.1806.

Chromosome No. *2n = 20* (Schaeppi & Steindle, 1942 & Barlow, 1963) and 2n = 26 (Wiens & Barlow, 1971).

Local Name (Bengali) : Mandá

(Plate–134 and Fig. 52)

Semi-parasitic or epiphytic evergreen shrub; very common on the branches of the *Excoecaria agallocha* tree in the Sundarbans.

Not developed any normal roots, but for its parasitic or epiphytic nature it form haustorium and anchor the branch of the host trees.

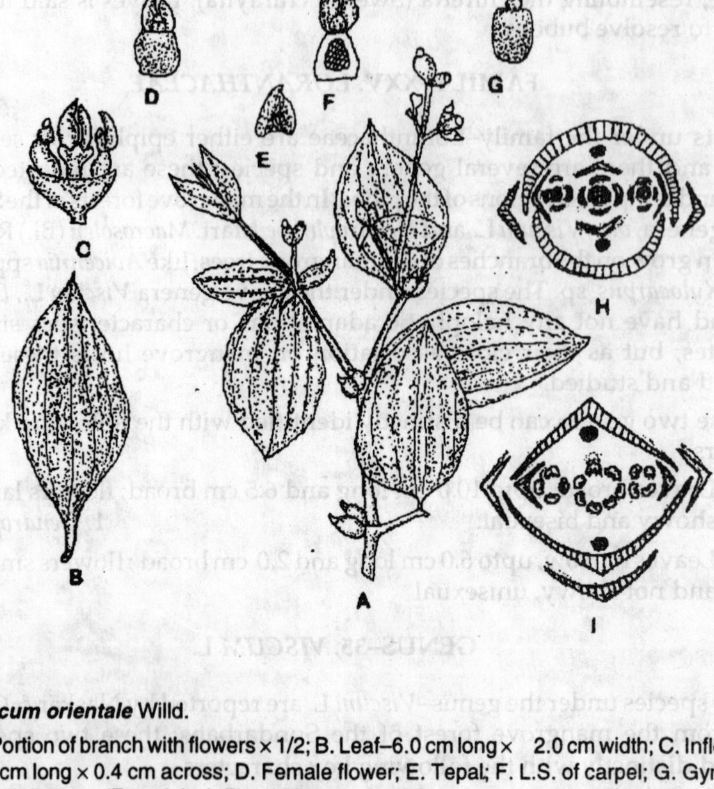

Fig. 52 : *Viscum orientale* Willd.

A. Portion of branch with flowers × 1/2; B. Leaf–6.0 cm long × 2.0 cm width; C. Inflorescence–0.4 cm long × 0.4 cm across; D. Female flower; E. Tepal; F. L.S. of carpel; G. Gynoecium; H. Floral diagram (Female) ; I. Floral diagram (male).

Stem much differentiated into nodes and internodes, much branched and each internodes separate on drying; branches both pendulous and / or erect, horizontal to the host plant; nodes ridged, easily dissociated, internodes 2.4 cm long, flattened, glabrous, solid, dark green. Leaf opposite decussate, simple, exstipulate, petiolate, petiole not prominent but continued of lamina, glabrous, yellowish, lamina various shape, elliptic–lanceolate, lanceolate, reniform; having different length, ranges from 4.0 cm to 6.0 cm and 0.8 cm–2.0 cm broad, entire, acute, tapering, thick, fleshy, glabrous, multicostate, parallel convergent venation (3 costa), vein inconspicuously present, not dorsiventral, dark green.

Inflorescence cyme–0.4 cm long × 0.4 cm across, cupshaped, capitulum head, green, glabrous, having 3 flowers of different maturation, middle flower more mature than the others two comparatively less mature. *Flower* bracteate, sessile, erect, odourless, minute, 0.1 cm long, regular, incomplete, unisexual; *perianth*-tepals-4, polytepalous, imbricate, deciduous, green, ovate, entire, acute, glabrous, fleshy, thick, superior, enclose the androecium; *gynoecium*-carpel-1, ovary inferior, rod-shaped, 0.1 cm long, one chambered, one ovule, basal placentation, style absent, stigma-1, spherical, globose, minute.

Pollination caused by honey bees, *Apis dorsata* and insects.

Fruits green, fleshy. Seed dispersed by wind and insect. Commonly grow on the *Excoecaria agallocha* and other mangrove trees, as epiphytes.

GENUS–36. *DENDROPHTHOE* Mart.

36.1. ***Dendrophthoe falcata*** (L.f.) Etting.

Local Name (Bengali) : Bara Manda

Chromosome No. $2n = 18$ (Barlow, 1966; Barlow & Wiens, 1971; Johri & Nag, 1974 and Sarkar *et al.* 1978)

(Plate–135 and Fig. 53)

Semi-parasitic or epiphytic plant on the mangrove trees, like *Avicennia officinalis, A. alba, Xylocarpus granatum, Excoecaria agallocha* and others.

Normal roots absent but haustoria formed to attach on the host trees.

Stem well differentiated into nodes and internodes, branched, deliquiscent, nodes ridged, prominently found, not easily separated like *Viscum* spp.; internodes 2.0 cm–3.0 cm long, terete, brown, solid, lenticellate, woody, cauline. Leaf opposite, decussate, exstipulate, simple, petiolate, petiole 0.5 cm long and 0.4 cm broad, flattened, glabrous, pulvinous, red, lamina–10.0 cm long × 6.5 cm width, ovate, elliptic, wavy, obtuse, tapering, dorsiventral, light green, fleshy, slightly succulent, glabrous, mid-vein red, unicostate, reticulate venation, mid-vein conspicuously found, but sub-veins inconspiousously found, sweet scented.

Inflorescence raceme, simple, ebracteate, peduncle unbranched. *Flowers* bisexual, complete, regular, pendulous, showy, deep red, 5.7 cm long, pedicellate, pedicel 0.6 cm long, glabrous, terete, green, pulvinous, slightly curved soft, sweet scented, pentamerous; *calyx*-sepals 5, gamosepalous, cup-shaped, grooved, oblique in upper surface, glabrous, green, not apical lobe or beak; *corolla*-petals 5, gamopetalous, forming a holo tube, 3.5 cm long, basal larger than the upper portion,

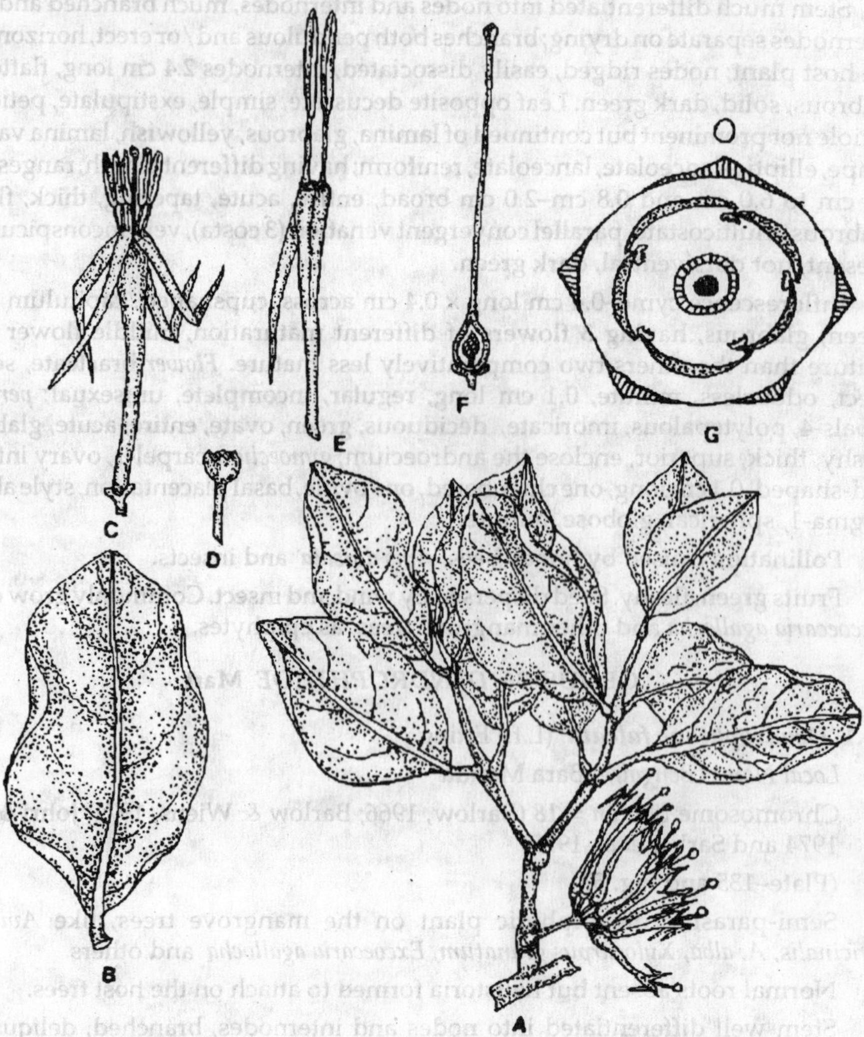

Fig. 53 : ***Dendrophthoe falcata*** **(L. f.) Etting.**
A. Portion of branch with inflorescence ×1/3; B. Leaf : petiole–0.5 cm long; lamina-10.0 cm long × 6.2 cm width; C. Flower–5.7 cm long; D. Calyx; E. Petal with epipetalous stamen; F. L.S. of carpel : ovary–0.4 cm long, style–4.7 cm long; G. Floral diagram.

apical 5 lobes of equal length, lobe 1.5 cm long and 0.1 cm broad, thick, glabrous, acicular, acute, entire, red, pendulous, terminally bent, total length of the corolla is about 5.0 cm, superior, imbricate; *androecium*-stamens 5, epipetalous, filament attached to the apex of the corolla tube from where apical lobe furcated, filament 0.5 cm long, round, soft, glabrous, white, basifixed, anther bilobed, globose, 0.8 cm long, longitudinal dehisced, red coloured, inserted, extrorse, same level surface to stigma, anthers equal length, and upper equal levelled, superior; *gynoecium*-carpels 2, syncarpous, ovary inferior, cup-shaped, 0.4 cm long and 0.25 cm diam., single chambered, single ovule, basal placentation; style terminal, 4.7 cm long, gradually tapering apex, terete, soft, glabrous, white, stigma-1, minute, globose, apical slightly grooved.

Pollination caused by insects and honey bees, *Apis dorsata*.

This plant grows commonly on the top branches of the mangrove trees but also grow on the freshwater or mesophytic trees throughout, *viz.*, tree like *Mangifera indica*.

The other epiphytic shrub, *Macrosolen cochinchinensis* (Lour.) van Teig. (Syn. *Loranthus cochinchinensis* (Lour.) is also reported from the mangrove trees of the Sundarbans.

FAMILY–XXVI. *POACEAE*
(*Nom. alt.* Gramineae)

GENUS–37. PORTERESIA Takeoka, Bull. Nat. Sci Mus. Tokyo 8 : 406. 1965.

37.1. *Porteresia coaractata* (Roxb.) Takeoka, Bull. Nat. Sci. Mus. Tokyo 8 : 406. 1965. *Oryza coarctata* Roxb. Fl. Ind. (2nd ed.) 2 : 206.1832. *O. triticoides* Griff. Nat. 3 : 8.1851. *Sclerophyllum coarctatum* (Roxb.) Griff. Nat. 3 : 8.1851.

Local Name (Bengal) : Dhani Ghash

(Plate–136, 137 and Fig. 54)

Culm standing, erect from a stout creeping rhizome or root stock, internode 8.0 cm–14.0 cm long, terete, greenish yellow, glabrous, nodes swollen, hollow internodes. Rhizome or root stock much branched and polished, pseudo—tap roots formed in the loosely silted up soil layer, roots upto 1.5 m long penetrated upto the deep consolidated soil; leaf blades margin unequals, mid-vein obscure, coriaceous, linear, acuminate, serrate or dented, leaf sheath long, undulately reticulate, polished, ligules very short, lamina upto 30.0 cm long, 0.5 cm broad, parallel venation, isobilateral.

Spikelets 15.0 cm-20.0 cm long, arranged on stem, stalked, each spikelet consisting of special bracts, the so called glumes; the 1st and 2nd glumes do not subtained flower and 0.35 cm and 0.20 cm long, respectively, 3rd glume having flower, known as lemma, 1.3 cm long and 0.4 cm width, teeth acuminate, hard, entire, rough outer surface, while inner surface glabrous, shiny, folded, slight longitudinal ridged row on outer surface; over-lap the edges of the opposite bracteoles, known as palea; 1.0 cm long and 0.3 cm width, bifolded; upper rough surface, inner surface glabrous, shiny, hard, entire, acute. *Flowers* hermaphrodite, zygomorphic, complete, erect, 1.3 cm long and 0.4 cm width; *tepals* are known as lodicules–2, free, herbaceous, glabrous, white, soft, 0.15 cm long and 0.15 cm width, ovoid, entire, obtuse; *stamens*-6, free, filament 0.5 cm long, white, glabrous, soft, versatile, exerted, anther bi-lobed, 0.45 cm long, extrorse, yellow, longitudinal dehisced, *carpel*-1, ovary superior, conical flax shaped, glabrous, shiny, I chambered, white, 0.15 cm long, 1 ovuled, style 2, terminal white, soft, 0.1 cm long, stigmas 2, 0.3 cm long, white, feather.

Pollination is performed by wind.

Fruit caryopsis, 1.4 cm long and 0.45 cm width. Seed brown coloured, thin 1.0 cm long.

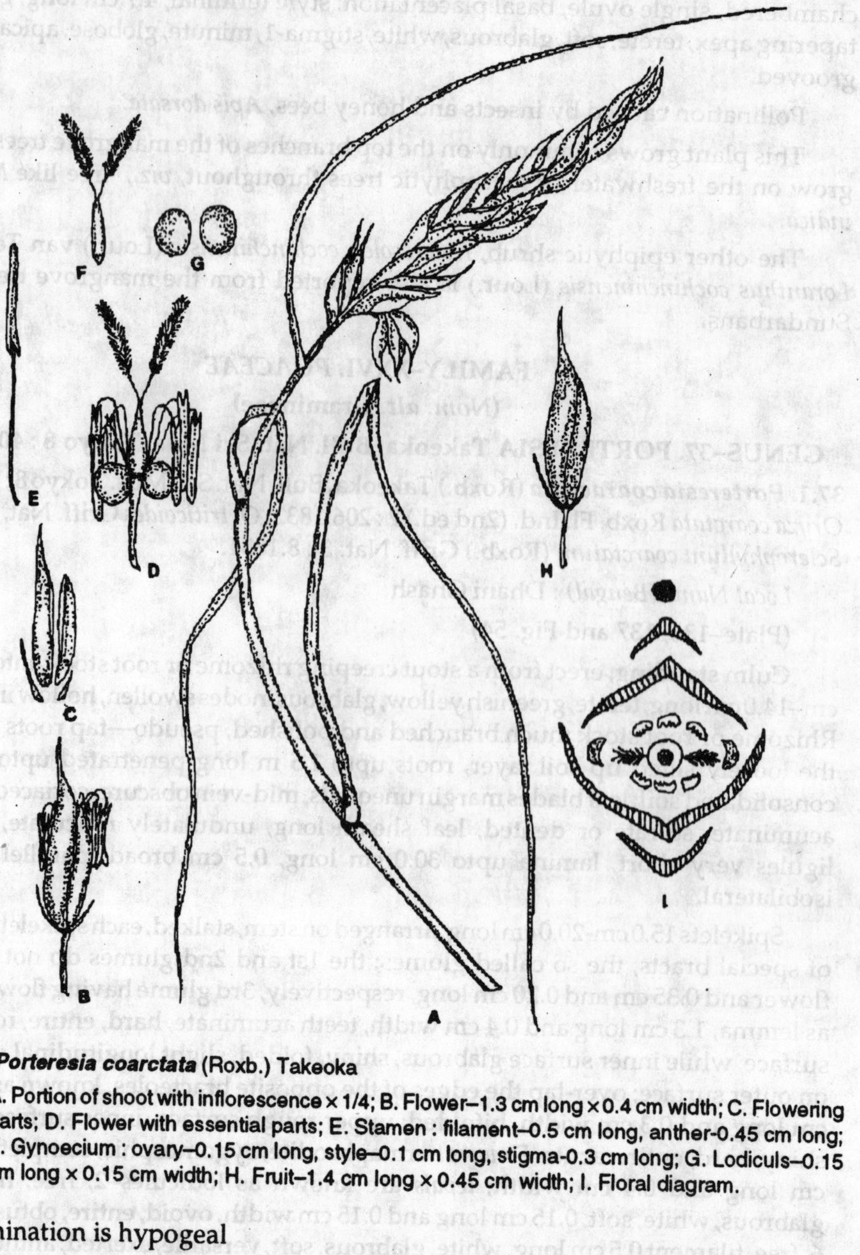

Fig. 54 : ***Porteresia coarctata*** **(Roxb.) Takeoka**

A. Portion of shoot with inflorescence × 1/4; B. Flower–1.3 cm long × 0.4 cm width; C. Flowering parts; D. Flower with essential parts; E. Stamen : filament–0.5 cm long, anther-0.45 cm long; F. Gynoecium : ovary–0.15 cm long, style–0.1 cm long, stigma-0.3 cm long; G. Lodiculs–0.15 cm long × 0.15 cm width; H. Fruit–1.4 cm long × 0.45 cm width; I. Floral diagram.

Germination is hypogeal

Flowering and fruiting period extend from the middle of June to December.

Distributed throughout the coastal silt and newly deposited soil. In the S.E. Asian mangals this species is considered as the pioneer species in the succession processes of the mangals in the frequent intertidal zone.

Ripe seeds are cooked like rice in the rural areas of the Sundarbans areas at Sagar Islands.

FAMILY- XXVII. *TAMARICACEAE*

The family- Tamaricaceae is characterised by shrubs or herbs with alternate, simple inter-exstipulate leaves; often hedge like; flowers solitary or in raceme inflorescence, basal parietal placentation, seeds hairy.

This family- Tamaricaceae consists of 120 species, under 4 genera; most of these species are distributed in the temperate and sub-tropical desert and mostly to the shore lines.

GENUS - 38. TAMARIX L. Sp. Pl. 279.1753

The genus- *Tamarix* L. has about 90 species, which are distributed to West Europe, Mediterannean to Indian and North China coasts. Four species of the genus - *Tamarix* L., *viz.*, *T. gallica* L., *T. dioica* Roxb., *T. troupii* Hole and *T. aphylla* (L.) Kanza have been reported by Naskar (1993) from the Lower Ganga Delta in Indian sub-continent. Most of these species are distributed in the coastal areas. Only the commonly growing species, *i.e.*, *Tamarix gallica* L. is studied and described.

38.1. *Tamarix gallica* L. Sp. Pl. 270. 1753.

Chromsome No. $2n = 24$

Local Name (Bengali) : Ban Jhau/Nona Jhau

(Plate-138 and Fig. 55)

Shrubs with deliquiscent branching, occasionally small tree, upto 10 m high but sometime bushy appearance.

Tap roots much developed but without any aerial roots.

Nodes and internodes not distinctly distinguished, presence of leaf scars; older trunks black brown, longitudinally fissured, while in young - twiggy shoot covered with leaves or leaf scars; woody, solid, terete, pubescent, green.

Leaves simple, exstipulate, opposite, decussate, sessile, lamina, triangular, 0.15 cm long and 0.15 cm width at base, green, fleshy, thick, entire, tapering end, covered.

Catkins, peduncle pendulous, 12.0 cm long, branched, bracteate, bracts acicular, 0.15 cm long, tapering long end, *Flower*–pink violet, complete, bisexual, regular, pedicellate; pedicel short, 0.05 cm long, terete, soft, pink violet, pubescent, erect, 0.10 cm long and 0.05 cm across, hypogynous, *sepals*- 5, gamosepalous; cup-shaped inferior, valvate, 0.05 cm long, × 0.10 cm across, *petals*-5, apical acicular lobe, coriaceous, glabrous, pink red, persistent, valvate, inferior, 0.05 cm long, herbaceous, thin, white ovate elliptic, entire, blunt, equal length in one whorl, deciduous not showy; *stamens* - 5, free, filament white, 0.05 cm long, soft, terete, glabrous, basifixed, inserted, anther bilobed, sagittate, introrse, 0.04 cm long, white greenish, inferior; *carpels*- 3, syncarpous, ovary superior, 0.05 cm long, one chambered, basal placentation, style short, stigma short, globose, glabrous.

Pollination is caused by small insects and honey bees or flies.

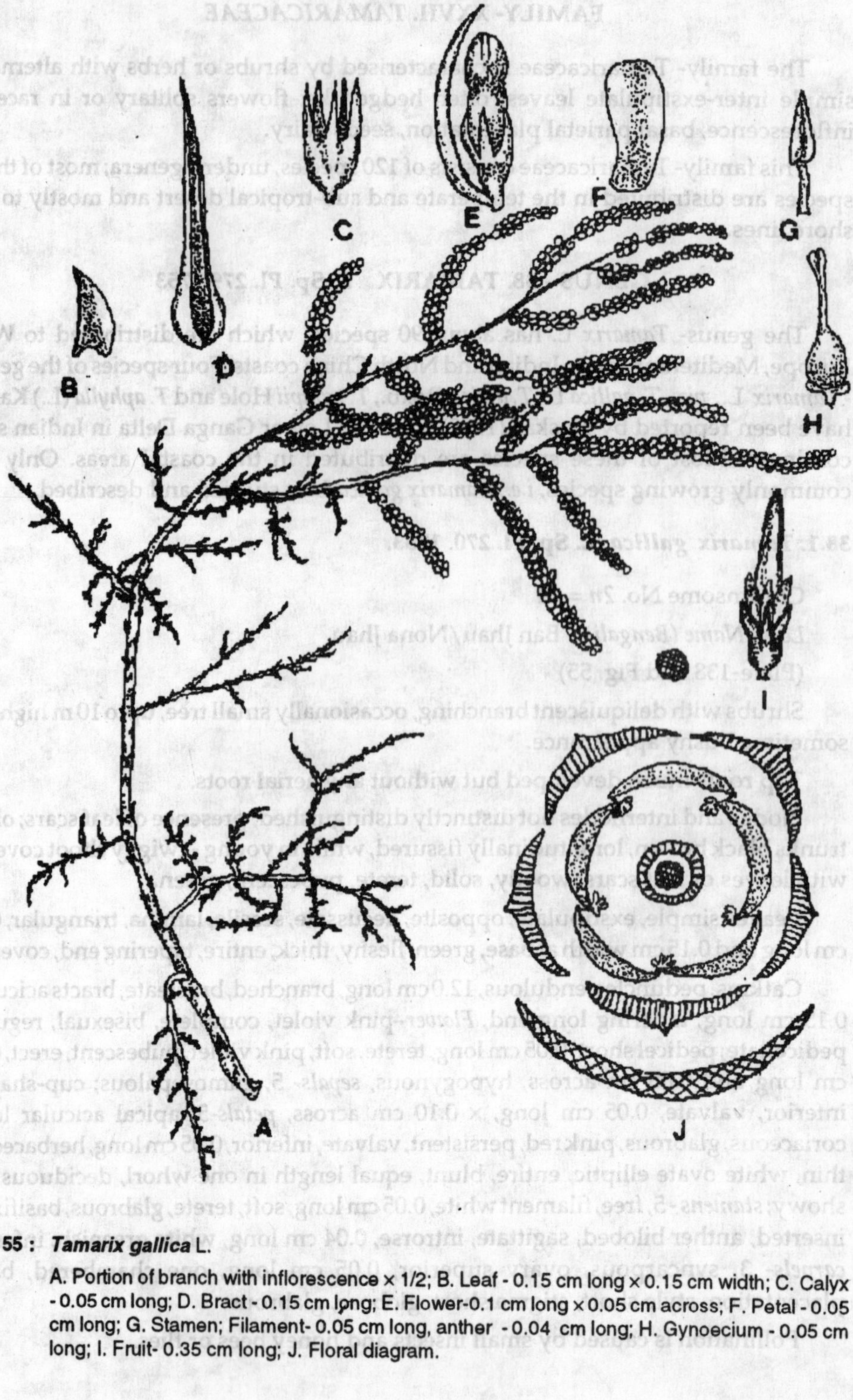

Fig. 55 : ***Tamarix gallica*** **L.**

A. Portion of branch with inflorescence × 1/2; B. Leaf - 0.15 cm long × 0.15 cm width; C. Calyx - 0.05 cm long; D. Bract - 0.15 cm long; E. Flower-0.1 cm long × 0.05 cm across; F. Petal - 0.05 cm long; G. Stamen; Filament- 0.05 cm long, anther - 0.04 cm long; H. Gynoecium - 0.05 cm long; I. Fruit- 0.35 cm long; J. Floral diagram.

Capsule gradually tapering, 0.35 cm long with persistent calyx, red violet coloured, dehiscent into 3 parts.

Seed germination is hypogeal.

Flowering and fruiting are generally from November to March.

A halophytic shrub or small tree, distributed in the mangrove reclaimed or mangove clear river banks or river flats in the Sundarbans; more dominant in the sandy beaches.

Distributed in the Indian coastal sandy soil and throughout the tropics.

Commonly used as fuel wood by the local people of the Sundarbans.

FAMILY- XXVIII. *SOLANACEAE*

The family-Solanaceae was placed under the Order-Tubiflorae, on the basis of their floral structure. Engler (1892) placed this family under the Tubiflorae and latter on Wettstein (1901) has supported this view but he considered that this family is closely related to Convolvulaceae. According to Randle (1904-1925) the family-Solanaceae is placed under Tubiflorae. Hallier considered this family as primitive member of the Tubiflorae, but this family has derived from Linaceae.

On the basis of these above views of the different Taxonomists, the family is related to Convolvulaceae but differ from it in having numerous seeds.

Solanaceae has about 85 genera, including 2,200 species; which are mostly distributed throughout the tropical and temperate regions (Mukherjee & Ganguli, 1964). The Genus-*Solanum* L. is the 'type-genus' of this family- Solanaceae.

GENUS-39. SOLANUM L. Sp. Pl. 187.1753.

The Genus - *Solanum* L. is having 10 species in the Lower Ganga Delta (Naskar, 1993); among these 10 species of the *Solanum* L., only *Solanum trilobatum* L. and *S. surattense* Burm. f. are common in the mangrove reclaimed areas of the Sundarbans and coastal sands of the Southern parts of the Sundarbans. As such, these species are considered in this present studies as mangrove associate.

These two species can be separated from their distinctive habits and habitats as follows :

1a. Weak stemmed, diffused herb, grow on the sandy beach.
1. *S. surattense* Burm. f.

1b. Scandent or trailing shrubs on thickets as back mangal.
2. *S. trilobatum* L.

39.1. *Solanum trilobatum* L. Sp. Pl. 188.1753.

Chromosome No. $2n = 24$ (Bhaduri, 1933; Rao, 1962; Mitra, 1967; Krishnappa, 1968; Madhavan 1968 and Sanjappa & Satyananda, 1979.

Local Name (Bengali) : Lata Gurbegun

(Plate-139 and Fig. 56).

Fig. 56 : ***Solanum trilobatum*** **L.**

A. Portion of climbing branch with flowers and fruits × 1/2; B. Leaf: petiole - 1.0 cm long, lamina - 3.0 cm long × 1.2 cm width; C. Flower - 2.5 cm long × 1.5 cm across; D. L. S. of flower; E. Petal with epipetalous stamen - 1.0 cm long; F. Stamen : filament - 0-.1 cm long, anther - 0.4 cm long × 0.2 cm width; G. Gynoecium : ovary - 0.2 cm long, style - 1.0 cm long; H. Fruit - 1.0 cm long; I. Floral diagram.

Trailing with prickle hooked, shrub with weak branches, grow on bushy dry soil. Root poorly developed but sunken upto a depth of 0.3 m.

Nodes and inter-nodes not rigid; internodes covered by infrequently flattened prickles, gradually tapering towards end, yellowish internodes terete, dark violet, green, glabrous, solid, one side green and other side dark violet-green, woody.

Simple, exstipulate, alternate, petiolate; petiole 1.0 cm long and 0.1 cm broad, terete, dark violet, glabrous, soft, solid, pulvinous, covered by infrequently prickles, lamina trilobate but not distinctly serrated, coriaceous, 3.0 cm long and 1.0 cm - 2.0 cm broad, entire, round, dorsiventral, dorsal surface dark violet, ventral surface green, unicostate, reticulate venation conspicuous.

Corymbose raceme; *flowers* ebracteate, bisexual, regular, complete, pedicellate; pedicel 0.5 cm - 1.5 cm long, terete, solid, soft, pubescent, violet green, erect, medium 2.5 cm long and 1.5 cm across, showy, hypogynous; *sepals*- 5, polysepalous, inferior, ovate, entire, acute, fleshy, pubescent outer surface, inner surface shiny green, 0.2 cm long and 0.08 cm broad, persistent, rosaceous, imbricate; *petals*- 5, gamopetalous, inferior, bell shaped, ovate, entire, acute, thin herbaceous, violet, distinct venation on petals, slightly adhere at base, imbricate, showy, 1.0 cm long and 0.5 cm broad; *stamens*- 5, epipetalous, filament 0.1 cm long, white, soft, glabrous, basifixed, anther 0.4 cm long and 0.2 cm broad, yellow, extrorse, inserted, globose, inferior; *carpels*- 2, syncarpous, superior, short cylindrical, 0.2 cm long and 0.1 cm diam, white, glabrous, fleshy, 4 lobes obliquely partites, axile placentation, style - 1, terminal, terete, soft, glabrous, 1.0 cm long, stigma absent.

Flowering and fruiting are mainly concentrated during postmonsoon and winter, i.e., October- March; occasionally seen throughout the year.

Pollination is caused by insect and honey bees.

Berry globose, 1.0 cm diam. with persistent calyx, greenish with white soft at apex in unripe stage, but deep red on maturity.

Seeds numerous, flat, yellowish, albuminous.

Germination is hypogeal.

A halophytic climber of trailing shrub on the river bank bushes or hedges of the mangrove reclaimed areas of the Sundarbans; especially thickets and hedges.

Distributed throughout the coastal area of West Bengal, Orissa and in the Deccan peninsula. In the Tropical World, particularly in Sri Lanka and Malay Peninsula are its common habitats.

Leaves and fruits occasionally cooked as vegetable.

The other species *Solanum surattense* Burm. f., the beach flora is not collected during the studies and not described in this present treaties.

Mangrove Associate and Less Known Species of Sundarbans

Several other back mangrove species or/and the mangrove associate Angiosperm species are occasionally or very scarcely grow and distributed in these mangrove habitats or beyond the intertidal zones at Sundarbans, West Bengal and elsewhere. Some of these species were also collected and studied. These less known species belonging to different genera and families are also mentioned herein brief.

3. Other Frequent Back Mangal Families

FAMILY-1. *BARRINGTONIACEAE* (*nom. alt.* Lecythidaceae)

GENUS–1 : *Barringtonia* Forster nom. cons.

1. ***Barringtonia acutangula*** (L.) Gaertn., Fruct. 2 : 97, t. 101. 1791.

Local Name (Bengali) : Hijal

Medium sized, deliquiscent much branched, leaves dense in terminal part of branches, lamina oblong–elliptic, petiole 6.0-7.0 cm long, inflorescence racemes with pink coloured or cream coloured flower, pendulous, fruits dumble shaped, quadrangular.

Flowering and fruiting- May to September.

This plant is found in river banks of the Sundarbans and are also found throughout the coastal region of India, South -East Asia and Polynesia.

The woods are used for fuel and building construction purposes. Root barks are employed as fish poison. Seed powder is said to be used as an emetic and expectorant.

2. ***Barringtonia racemosa*** Roxb. (Hort. Beng. 52. 1814), Fl. Ind. 2 : 634.1924.

Local Name (Bengali) : Sumudra

Medium sized tree upto15 m tall, bark grey, leaves 10.0-25.0 cm long × 6.0-10.0 cm width, elliptic - oblong, glabrous, membranous, main costa 10-20 pairs. Flowers cream coloured with pendulous raceme terminal, stamens numerous, filaments exserted, fruits rounded, seeds solitary.

Flowering and fruiting- April to October.

This plant is found in river banks of Sundarbans and is also found throughout the coastal area of India; distributed in Malaya, Sri Lanka, Bangladesh and Polynesia.

The woods are used for fuel and fruits are used for medical purposes. Seed powder contain saponia and used as fish poison.

FAMILY-2. *BIGNONIACEAE*

GENUS–2 : *Dolichandrone* Seem.

3. ***Dolichandrone spathacea*** (L. f.) K. Schum., Fl. Kais. Wilh. Land 123.1889.

Chromosome No. $2n = 40$ **(Venkatasubhan, 1944) (Plate-140)**

Local Name (Bengali) : Gorshingiah

Medium tree, leaves pinnately compound; leaflets 5.0-7.5 cm long, flowers large, showy, capsule 25.0-35.0 cm long, seeds rectangular with corky wings.

Flowering and fruiting - November to March. This plant is infrequently found in Sundarbans, found in Orissa coast, Andaman & Nicobar Islands, distributed in Bangladesh, Sri Lanka, Burma and Malaya.

The woods are used for fuel. Seed powder is said to be antiseptic and used in spasmodic affections.

FAMILY-3. *MALVACEAE*

Genus–3 : *Thespesia* Soland ex Corr.

4. ***Thespesia populneoides*** (Roxb.) Kostel, Allg. Med. Pharm. Fl. 5.1861.

Local Name (Bengali) : Paras

Shrub or small tree, upto 7 m tall, branched, leaves ovate, entire, acuminate, petiole 6.0-10.0 cm long, flowers solitary, axillary, 6.0-7.0 cm in diam. yellow, peduncle 8.0-10.0 cm long, capsule pyriform.

Flowering and fruiting are throughout the year.

This plant is found in the Sudarbans; also found in plains and coastal area of India.

Wood and barks are used for the production of red colouring matter; all parts of the plants are used for healing the skin diseases.

FAMILY-4. *ASCLEPIADACEAE*

Genus–4 : *Tylophora* R. Br.

5. ***Tylophora tenuis*** Bl. Bijdr. 1062. 1826;

Local Name (Bengali) : Antomul

Slender twining herb, leaves linear-lanceolate, puberulous above, acute; cyme lateral, dense branching, peduncles about 2.0 cm long, corolla pink, globose.

Flowering and fruiting- November to March.

This plant is found in the river banks and side of creeks of Sundarbans; also found in coastal area of India and distributed in S. E. Asia.

Plants are used for excessive perspiration, bilious swellings and small pox.

FAMILY-5. *PANDANACEAE*

Genus–5 : *Pandanus* Parkins

6. ***Pandanus odoratissimus*** Soland ex Parkinson, J. Voy. H.M.S. Endeav. 46. 1773.

Local Name (Bengali) : Keya Kanta

Small tree with dichotomously branched, upto 5 m tall, stemless, palm-like, sub-erect, supported by aerial roots; leaves upto 2 m long., serrated margin. sword shaped, acuminate, spadix long, inflorescence sweet scented, carpels confluents, stigma short, staminodes absent, fruits oblong, drupes numerous.

Flowering and fruiting- March to November.

This plant is found in waste land of Sundarbans and also found in all over the Indian coastal area and distributed in Burma and Sri Lanka.

The leaves are used for making cordage and obtain strong white fibres; also used for the remedy of small pox, leprosy etc.

FAMILY-6. *MELIACEAE*

Genus–6 : *Amoora* Roxb.

7. *Amoora (Aglaia) cucullata* Roxb., Corr. Pl. 3 : 54. t. 258. 1819.

   Local Name (Bengali) : Amur (Plate-141)

Medium tree upto 15 m tall, leaves compound, imparipinnate 25-38 cm long, leaflets 2-4 pairs and on old one opposite, 7.5- 10.0 cm long x 3-6 cm width, oblong elliptic, sub-acute, glabrous, pedicel short sepals rounded, petals - longer, staminal tube shorter than the petals, ovoid, ovary stalked, ovoid-oblong.

Flowering and fruiting - October to January.

Infrequently found in the mangrove forests and coastal area of India and distributed in Sri Lanka and Burma.

Woods are used for fuel and timber.

FAMILY-7. *TAMARICACEAE*

Genus–7 : *Tamarix* L.

8. ***Tamarix dioica*** Roxb., (Hort. Beng. 22. 1814) nom. nud. & Fl. Ind. 2 : 101- 1824. Chromosome No. *2n = 24* Malik, 1960).

   Local Name : Lal Jhau

Small tree upto 5 m tall, leaves minute, adpressed sheathing, sheaths tubular, flowers pink, cylindrical, loosely panicle spikes, capsules oblong, tapering.

Flowering and fruiting July to October.

This Plant is found in the river banks of Sundarbans also found in coastal area of India and distributed in Burma, Baluchistan Eygyt.

Barks are used for remedy of skin diseases.

FAMILY-8. *CHENOPODIACEAE*

Genus–8 : *Salicornia* L.

9. ***Salicornia brachiata*** Roxb. Fl. Ind. I. 82 : 1820.

   Chromosome No. *2n = 36* (Wulff, 1932) (Plate-132)

Stem fleshy, node ridged, easily separated, leaves simple, fleshy, elliptic, green, flower bisexual, petals yellow, showy, stamens - 1-2, seeds with testa containing hooked hairs.

Flowering and fruiting — July to November.

This plant is found in inundated area of narrow creeks and river sides of Sundarbans and adjacent area, also found in coastal area of W. Peninsula and distributed in Sri Lanka and Bangladesh.

Sometimes it is used as vegetables.

FAMILY-9. *VERBENACEAE*

Genus–9 : *Premna* L.

10. ***Premna corymbosa* (Burm. f.) Rottl. & Wild.**, Greeala. Nat. Freude Neuc. Schr. 4 : 187. 1803.

Local Name (Bengali) : Gamiari or Bhui Biravi

Small tree, young shoot tuberulous, branched, leaves oblong-ovate, slightly acuminate, entire, truncate at base, flowers foetid, greenish-yellow, pubescent, corymbose cyme, fruits globose, black.

Flowering and fruiting—April to July. Infrequently found in mangrove Sundarbans forests, also found in Andaman & Nicobar Islands and distributed in Sri Laka and Malay.

Leaves are used for remedy of piles and tumours.

FAMILY-10. *CACTACEAE*

Genus–10 : *Opuntia* L.

11. ***Opuntia dillenii* (Ker. Gawler) Haw.**, Suppl. Pl. Succ. 79.1819.

Local Name (*Bengali*) : Nag Phara (Plate-142)

Purely xerophytic shrub, jointed and phylloclades, flowers yellow, berries purple, pear shaped, tubercled.

Flowering and fruiting — March to August. Found in waste land and jungles, also found in the desert area of India and distributed in the desert area of the all over the world.

The fruits and phylloclades are said to be edible.

FAMILY-11. *LAURACEAE*

Genus–11 : *Cassytha* L.

12. *Cassytha filiformis* L. Sp. Pl. 35.1753.

Chromosome No. $2n = 24$ (Sobha & Ramchandran) and $2n = 48$ (Okada & Tanaka, 1975)

Local Name (Bengali) : Akas-bel

Leafless twining parasitic herb, stem solid, cord like twined, branched, flower small, sessile, bracteoles short, fruits upto 0.7 cm long, white, globose, smooth, enclosed in fleshy perianth tube.

Flowering and fruiting - September to December. Infrequently found on the *Excoecaria* (Gyano) tree, as parasite and distributed in the coastlines of India and several countries of world.

The flora is used for the remedy of chronic dysentery, skin diseases etc.

FAMILY-12. *POACEAE*

Genus–12 : *Phragmites* Trin.

13. ***Phragmites kakra*** (Retz.) Trib, ex Steud., Nom. (ed.2) 2 : 324 : 1814.
Local Name (Bengali) : Nalor Dharma

Culms erect, stout, creeping, leafblades flat, lanceolate, ligule narrow, inflorescence panicles, silky hair, spikelets bi-sexual, 3-4 fid, pedicel and glumes glabrous, fruits grain free, lossely enclosed, by the lemna and palea, oblong semi-terete, hilum oblong short.

Flowering and fruiting — December to March.

Found in the marshy places, found throughout India and distributed in Afganistan, Japan, Asia and Australia.

Used for paper pulp.

Genus–13 : *Myriostachya* Hook . f.

14. ***Myriostachya wigtiana* (Nees ex Steud) Hook. f.** Fl. Brit. India 7 : 327. 1896.
Local Name (Bengali) : Nalai

Culms tufted, erect, stout, sheathed, sometimes floating with long branched roots, smooth, simple, leafblade broad, serrated, finely acuminate, sheath very long, inflorescence panicles, rachis smooth, crowded, facicled and whorled flowering, spikelets 4-8 flowered, compressed, pedicels short. Fruits grain obliquely ovoid, sub-terete.

Flowering and fruiting — June to January.

Frequently found in the riverbeds of Sundarbans, also found in the coastal area of India and distributed in Sri Lanka and Tenasserim.

Genus–14 : *Aeluropus* L.

15. ***Aeluropus lagopoides* (L.,) Trin. ex Thw.** Enum. pl. Zeyl. 374. 1864.
Chromosome No. 20 (Murin *et al*. 1970 and Banguar *et al*., 1969)
Local Name (Bengali) : Nona Durba

Prostrate, creeping, erect, grass, stem solid, leafblade linear, panicles pinkish.
Flowering and fruiting — March to July.

Frequently found in the sand dune and beach area of the Sundarbans, also found throughout India and distributed in the Sri Lanka, Baluchistan etc.

FAMILY -13. *CYPERACEAE*

Genus–15 : *Scirpus* L.

16. ***Scirpus littoralis* Schrad.**, Fl. Germ. 1.142. t. 5.1806.

Root stock stout, hard, with fibrous roots, leaves membranous, acuminate, sheath soft, fleshy, long, inflorescence at the top of the stem, spikelets pale brown, oblong, terete, pedicellate, nuts broad, obvoid, glabrous with long apiculation, dull-brown.

Flowering and fruiting — June to Novembr.

Frequently found in the river bank and river flats and also found in reclaimed area of Sundarbans. It is also found in Punjab, W. Pensula and Central India. It is distributed in Sri Lanka, Persia, Central Asia, Africa and Mediterranean region.

FAMILY-14. *RUPPIACEAE* (*nom alt.* NAJADACEAE)

Genus -16. *Ruppia* L.

17. ***Ruppia maritima*** **L. (Syn. rosetelle Koch.**

Local Name (Bengali) : Nona Jhangi (Plate-143)

Submerged, delicate, soft, green much branched herb in the brackishwater fisheries or stagnant brackishwater pools of the Sundarbans.

Identified a nutritious food for number of brackishwater fish and Prawn species.

FAMILY-15. *SAPOTACEAE*

Genus -17 : *Manilkara* Andans.

18. ***Manikara hexandra*** **(Roxb.) Dub.** Ann. Mus. Col. Marseille (ser. 3) 3: 9.1915.

Chromosome No. $2n = 24$ (Mehra & Bawa 1969)

Local Name : Bilati Bakul or Khirkhejur

(Plate - 144)

Tall evergreen tree with dense leafy crown.

Introduced and or naturalised in the reclaimed mangrove areas Netidhopani of the Sundarbans. Bank is febrifuge; seed produce edible oil.

FAMILY-16. *ARACEAE*

Genus -18 : *Cryptocoryne* Fisch. ex Wydler

19. ***Cryptocoryne ciliata*** **(Roxb.) Fisch. ex Wydler,** Linnaea 5 : 428. 1830.

Chromosome No. $2n = 22$ (Marchant, 1972)

Vern. Name : Kerali

(Plate - 145)

Erect herb on the intertidal river flats or canal sides, in the marshy or moist localities.

FAMILY-17. *CUSCUTACEAE* (Splitted out from Convolvulaceae)

Genus-19 : *Cuscuta* L.

20. ***Cuscuta reflexa*** **Roxb.** Pl.Cor. 2:3, t. 104. 1799.

Local Name : Swarnalata

Twining yellow leafless fleshy stem parasitic herb.

Common as parasitic and forming dense mass on low trees and shrubs.

Medicinally, used as purgative in flatulence and liver complaints.

21

ANATOMICAL FEATURES OF THE MANGROVES

FAMILY–I. *RHIZOPHORACEAE*

Cuticle very thick, hypodermal layers towards epidermis, consisting of different layers. Bowman (1916) interpreted that the number of hypodermal layers may vary on the basis of salt concentration. Stomata confined to the lower surface. Mesophyll-palisade tissue consisting of 1-4 layers, spongy tissue with distinct inter-cellular spaces. In the axis-cork arising superficially in the hypodermis, wood-vessels moderately small, scalariform perforation, rays generally 3-5 cells wide.

Anatomical Key Characters to the Genera of Rhizophoraceae

A. Corkwart found as black spots on the lamina — 1. *Rhizophora* L.

AA. Corkwart absent — B.

B. Epidermis two layered thick in upper surface — 4. *Kandelia* (DC) W. & A.

BB. Epidermis one layered thick in upper surface — C.

C. Petiolar stele showing crescent-shaped, consisting of 5 solitary bundles — 2. *Bruguiera* Lamk.

CC. Petiolar stele showing ring of bundles surrounding single or two isolated bundles in centre — 3. *Ceriops* Arn.

GENUS–1. *RHIZOPHORA* L.

Corkwarts frequently occur as black spots in the surfaces of lamina; epidermis multilayered thick in the upper surface; hypodermis 2-3 layers thick Mesophyll–palisade tissue consisting of one layered thick, while spongy tissue multilayered thick. *Petiole*-stele showing ring of bundles surrounding solitary or net like medullary strands. In the axis, stele-circular cylinder, open collateral, ray—mostly homogeneous.

Anatomical Key Characters to the Species of *Rhizophora* L.

A. 'H' shaped idioblasts present in mesophyll tissue; corkwart only in lower surface 1. *R. mucronata* Lamk.

AA. 'H' shaped idioblasts absent in mesophyll tissue; corkwart occur on both surface 2. *R. apiculata* Blume

1.1. *Rhizophora mucronata* Lamk.

(Plate–146 and Fig. 57)

Leaf : dorsiventral, petiolate.

Lamina : Corkwarts frequently occur as black circular spot in the lower surface. *Cuticle*–very thick and waxy. *Epidermis*-3 layered thick on upper surface and 1-2 layered thick on lower surface; *Hypodermis*–3 layered thick below the upper surface, while hypodermal layers absent in lower surface; *Stomata*–anomocytic, 40.5 ±1 μ in length and 27.8 ± 1μ in width, confined to the lower surface, deeply sunken; *Mesophyll*–one layer thick palisade tissue, long, vertically placed, spongy tissue consisting of 10-12 layers thick, with large inter-cellular spaces. *Idioblasts*–'H' shaped idioblasts occur in the mesophyll tissue; *Veins*–embedded in the mesophyll tissue: *Petiole; epidermis*–one layer thick; *Cortex*–consisting of multilayered thick parenchymatous tissue with large air cavities in the middle *Crystals*–star shaped, solitary, secretory elements–oil cells found in cortical cells. *Endodermis*–one layered thick. *Stele*–transverse section in distal end showing ring of bundles surrounding single medullary strand. *Pericycle*–one layer thick; *vascular strand*–closed collateral, phloem in outer, xylem in inner, xylem endarch.

Axis : *Epidermis*–one layered thick in young axis, while in mature axis, epidermal layer replaced by cork, cork consisting of two types of horizons, upper horizon having multilayered thick, small cells, while lower horizon having multi-layered thick, rectangular shaped, vertically placed. *Cortex* multilayered thick, parenchymatous, star shaped-crystals and brachysclereids scattered in cortex; *endodermis*; one layered thick; *stele* transverse section showing circular cylinder; *pericycle*-one layered thick; *vascular strand*-open collateral, phloem in outer, associated with parenchymatous cells; *cambium*–multilayered thick, xylem / wood-vessels small, diffuse porous, scalariform perforation, boarder pitted, xylem traversed by fibres. *Parenchyma*-paratracheal, towards centre, rays-multiseriate, 2-3 celled width, mostly homogeneous, upright cells; parenchymatous cells, compactly arranged, brachysclereids scattered in pith.

Stilt Roots : *Epidermis*–one layered thick, ruptured by lenticels; *Hypodermis*–multilayered thick, parenchymatous cells, compactly arranged, distinct, one celled thick circular layer in hypodermis; *cortex* consisting of multi-layered thick, parenchymatous cells, infrequently cavities, solitary sclereids; *endodermis*-one layered thick; *stele* circular; *pericycle* one layered thick; vascular strand, amphi-cribral, solitary strands scattered in the peripheral surface of stele; *pith*-consisting of parenchymatous cells, compactly arranged.

Hypocotyle : *Epidermis*–one layered thick ruptured by frequently corkwarts, *cortex*-consisting of parenchymatous cells, large air cavities in circular series below

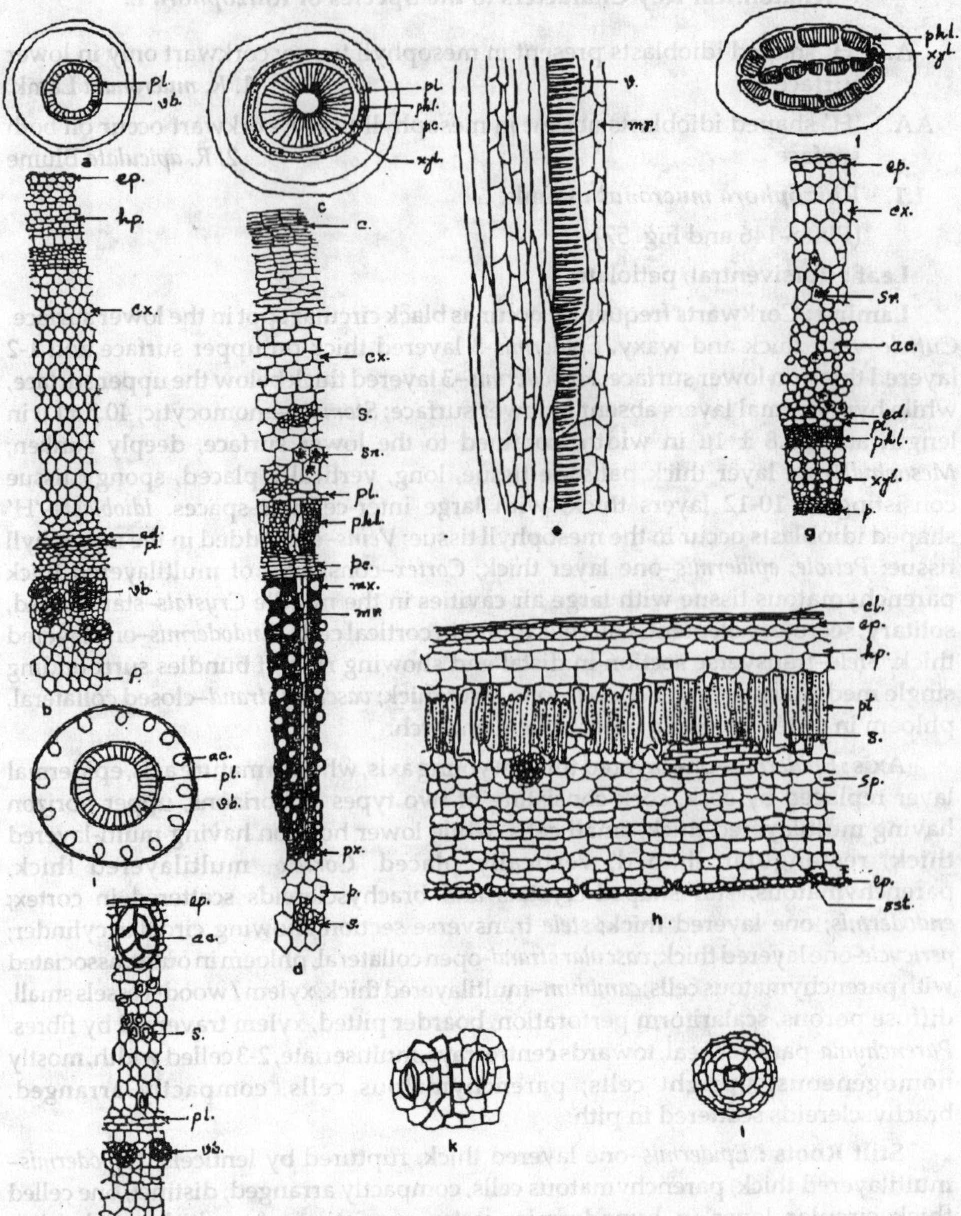

Fig. 57 : ***Rhizophora mucronata*** **Lamk.**

(a) T.S. of stilt root (diagrammatic); (b) T.S. of a portion of stilt root; (c) T.S. of stem (diagrammatic); (d) T.S. of a portion of stem; (e) L.S. of stem. ; (f) T.S. of petiole (diagrammatic); (g) T.S. of a portion of petiole; (h) T.S. of lamina; (i) T.S. of hypocotyle (diagrammatic); (j) T.S. of a portion of hypocotyle; (k) Stomata; (l) Corkwart.

the epidermal layer, sclereids scattered in single or gathered; *endodermis*-one layered thick, *Stele*-circular ring, *pericycle*-one layered thick, *vascular strand*, amphi-cribral, placed below the pericycle layer, *pith*-parenchymatous cells, compactly arranged.

1.2. *Rhizophora apiculata* Blume

(Plate–147 and Fig. 58)

Leaf : dorsiventral, petiolate.

Lamina : Corkwarts occur as black circular spot in both the upper and lower surfaces; *cuticle*-very thick and waxy; *Epidermis*-4 layered thick, while one layer thick in lower surface; *Hypodermis*–2 layered thick below the upper surface, while hypodermal layers absent in lower surface; *Stomata*-anomocytic, 41.1 ± 1 μ in length and 22.8 ± 1 μ in width, confined to the lower surface, deeply sunken. *Mesophyll*—one layered thick palisade tissue, long, vertically placed, spongy tissue consisting of 5-6 layered thick, with inter-cellular spaces, veins embedded in the mesophyll tissue; *Petioles;–epidermis*-one layered thick. *Cortex*-consisting of multi-layered thick, parenchymatous tissue with large air cavities in middle; *endodermis*-one layered thick; *stele*-transverse section in distal end showing ring of bundles, surrounding reticulate medullary strand; *pericycle*-one layered thick; *vascular strand*–open collateral, phloem in outer, associated with sclereids, 2-3 layered thick cambial ring in middle and xylem in inner, xylem mesarch.

Axis : *Epidermis*–one layered thick, in young axis, while in mature axis epidermal layer replaced by *cork*, cork-consisting of two types of horizon, upper horizon having multilayered thick and lower horizon having also multilayered thick, *cortex*-multilayered thick, parenchymatous, few brachy; sclereids scattered in cortex *stele*-transverse section showing circular cylinder; *pericycle*-one layered thick, *vascular strand*-open collateral phloem in outer associated with fibres and sclereids, middle cambial ring, xylem/wood-vessels small, diffuse porous, scalariform perforation, parenchyma-paratracheal, rays-multiseriate, 3 celled width, mostly homogeneous, most of the cells upright, few procumbent cells scattered, *pith*-parenchymatous cells, compactly arranged, brachysclereids scattered in pith.

Stilt Root : *Epidermis*–one layer thick, ruptured by lenticels; *Hypodermis*–multilayered thick, parenchymatous cells, compactly arranged, consisting of two types of horizon, upper horizon having large celled thick, lower horizon having cells, horizontally placed, *cortex*-consisting of multilayered thick, parenchymatous cells, infrequently cavities, solitary sclereids and brachy sclereids; *Endodermis*–one layer thick; *Vascular strand*–amphi-cribral, solitary strands scattered in the peripheral surface of stele, *pith*-consisting of parenchymatous cells, compactly arranged.

Hypocotyle : Anatomy same as *R. mucronata*, except epidermis–not ruptured, *i.e.*, smooth walled surface.

GENUS–2. *BRUGUIERA* Lamk.

Lamina : *Epidermis*–one layer thick; *Hypodermis*–1-2 layers thick; *Stomata*–confined to the lower surface; *Mesophyll*–palisade tissue consisting of multilayered thick, spongy tissue–consisting of multilayered thick; *Petiole*–stele showing crescent-

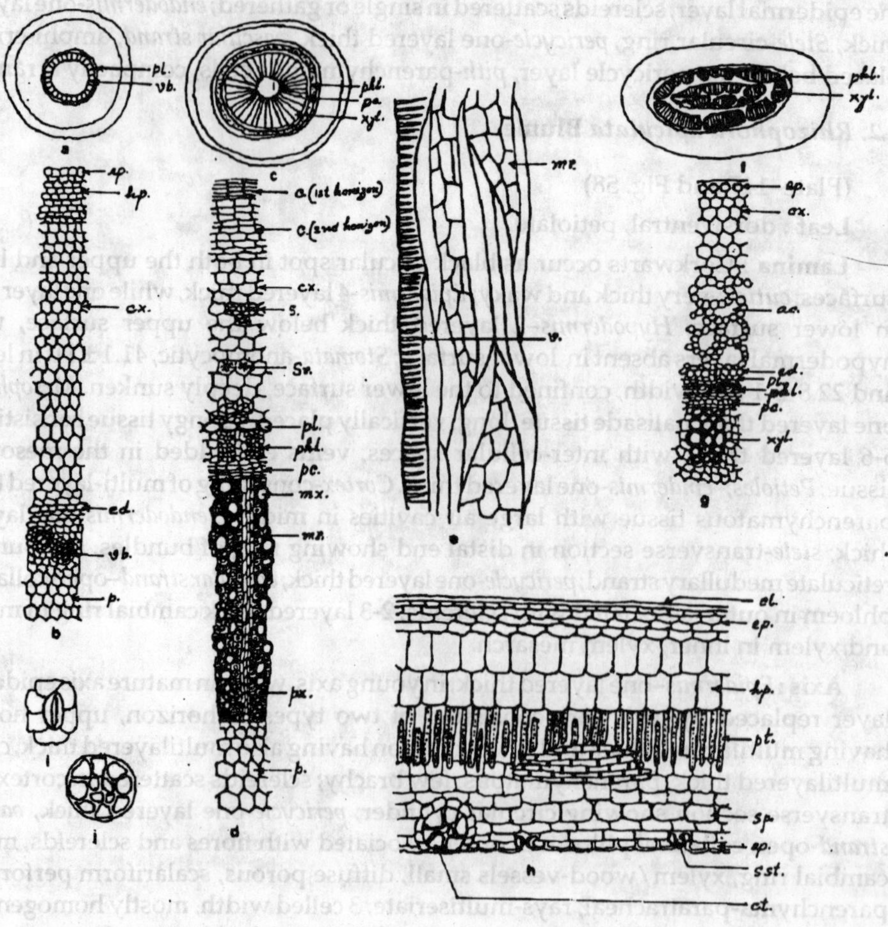

Fig. 58 : *Rhizophora apiculata* Blume

(a) T.S. of stilt root (diagrammatic); (b) T.S. of a portion of stilt root; (c) T.S of stem (diagrammatic); (d) T.S. of a portion of stem.; (e) L.S. of stem; (f) T.S. of petiole (diagrammatic); (g) T.S. of a portion of petiole; (h) T.S. of lamina; (i) Stomata; (j) Corkwart.

shaped, consisting of 5 solitary bundles. In the axis, *stele* circular cylinder, open collateral, ray-homogeneous.

Anatomical Key Characters to the Species of *Bruguiera* Lamk.

A. Extra–stelar xylem absent in cortex — B

AA. Extra–stelar xylem present in cortex — C

B. Hypodermis 1–layer thick below the upper epidermis

1. *B. gymnorrhiza* Lamk.

BB. Hypodermis 2-layer thick below the upper epidermis

2. *B. sexangula* Poir.

C. Hypodermis 1 layered thick below the upper epidermis

3. *B. cylindrica* Blume

CC. Hypodermis 2 layered thick below the upper epidermis

4. *B. parviflora* W. & A.

2.1. *Bruguiera gymnorrhiza* Lamk.

(Plate–148 and Fig. 59)

Leaf : dorsiventral, petiolate.

Lamina : *Cuticle*–very thick and waxy; *Epidermis*–one layered thick; *Hypodermis*–one layered thick below the upper epidermis, while 2-layered thick below the lowered epidermis; *Stomata*-anomocytic, 20.2 ± 1 μ in length and 13.3 ± 1 μ in width, confined to the lower surface, deeply sunken; *Mesophyll*–palisade tissue 4 layered; thick, rectangular, compactly arranged, vertically placed, spongy tissue consisting of 10-12 layered thick, with inter-cellular spaces, *veins* embededd in the mesophyll tissue; *Petiole–Epidermis*–one layer thick, *cortex*-consisting of multilayered thick, parenchymatous tissue, with large air cavities in middle, *endodermis*-one layered thick, *stele*-transverse section in distal end showing crescent-shaped strand, consisting of 5 solitary bundles, *pericycle*-one layered thick, *vascular strand*-closed collateral, phloem i n outer, xylem in inner, xylem endarch.

Axis : *Epidermis*–one layered thick in young axis, while in mature axis, epidermal cells replaced by cork, *cork*-consisting of multilayered thick, rectangular shaped, cortex-multilayered thick, having large air cavities in immature axis, while air cavities absent in mature axis, *crystals*–sphaeraphides are frequently occur in immature cortex, while absent in mature cortex, oil cells present in cortex *endodermis*-one layered thick *stele*-transverse section showing circular cylinder, pericycle-one layered thick; associated with sclereids cells, while large celled sclereids ring interrupted by parenchymatous cells in mature axis; *vascular strand*-open collateral, phloem in outer, cambium-2-3 layered thick, xylem/wood-vessels-small, diffuse porous, scalariform perforation, border pitted, xylem traversed by fibres; *parenchyma*–paratracheal, ray–multiseriate, 2-4 celled width, homogeneous, procumbent cells, *pith*-parenchymatous cells, compactly arranged.

Hypocotyle : *Epidermis*–one layered thick, *cortex*-consisting of parenchymatous cells, having no air cavities, *pericycle*–multilayered thick, *vascular strand*-amphicribral, occur towards the peripheral region as a circular ring, *pith*-consisting of parenchymatous tissue, compactly arranged.

2.2. *Bruguiera sexangula* Poir.

(Plate–149 and Fig. 60)

Leaf : Dorsiventral, petiolate.

Lamina : *Cuticle*-very thick and waxy, *epidermis*-one layered-thick, below the upper

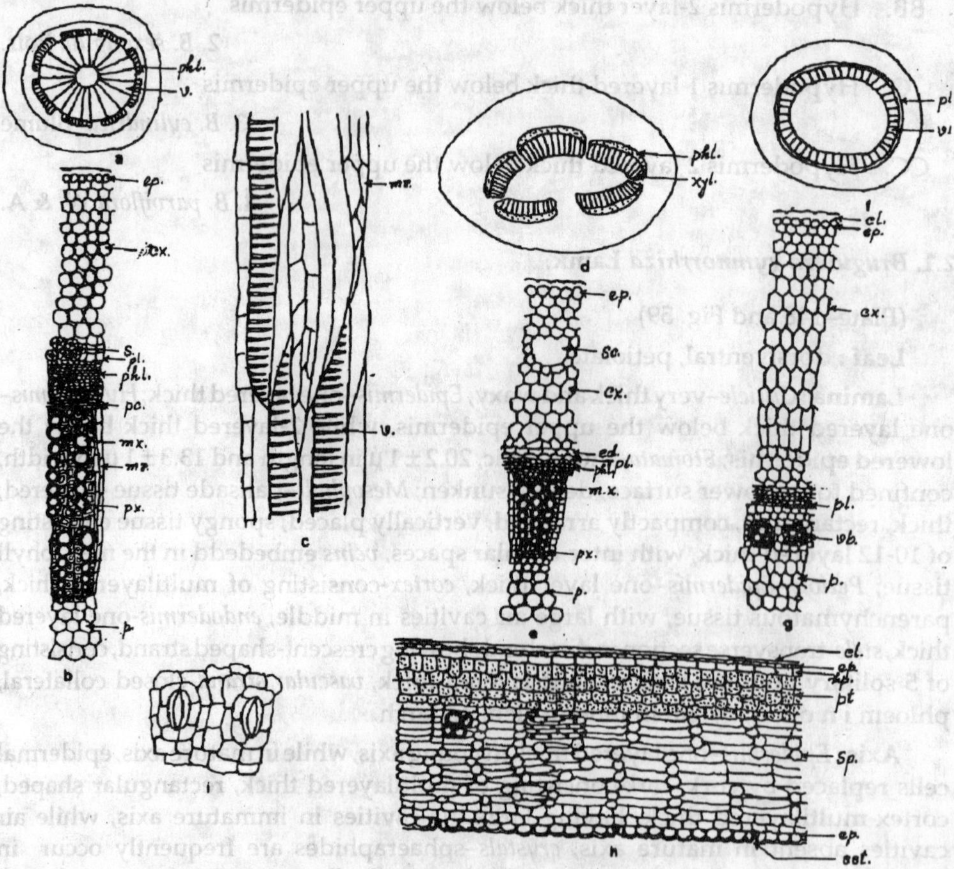

Fig. 59 : ***Bruguiera gymnorrhiza*** **(L.) Lamk.**

(a) T.S. of stem (diagrammatic) ; (b) T.S. of a portion of stem.; (c) L.S. of stem.; (d) T.S. of petiole (diagrammatic); (e) T.S of a portion of petiole; (f) T.S. of hypocotyle (diagrammatic); (g) T.S of a portion of hypocotyle; (h) T.S. of lamina; (i) Stomata.

surface, *hypodermis*–2 layered thick below the upper epidermis and two layered thick in the lower epidermis *stomata*–anomocytic, 21.5 ± 1 μ in length and 10.7 ± 1 μ in width, confined to the lower surface, deeply sunken. Mesophyll–palisade–3 layered thick, rectangular, compactly arranged, vertically placed, spongy tissue consisting of 8-9 layered thick, with large inter-cellular spaces, veins embedded in the mesophyll, *petiole-epidermis*-one layered thick, cortex-consisting of multilayered thick, parenchymatous tissue with large air cavities in middle, endodermis–one layered thick, *stele*-transverse section in distal end showing crescent shaped strand, consisting of 5 solitary bundles, *pericycle*–one layered thick, *vascular strand*–open collateral, phloem in outer 3 layered thick, cambial ring in middle, xylem in inner, endarch.

Axis : *Epidermis*–one layered thick in young axis, while in mature axis epidermal

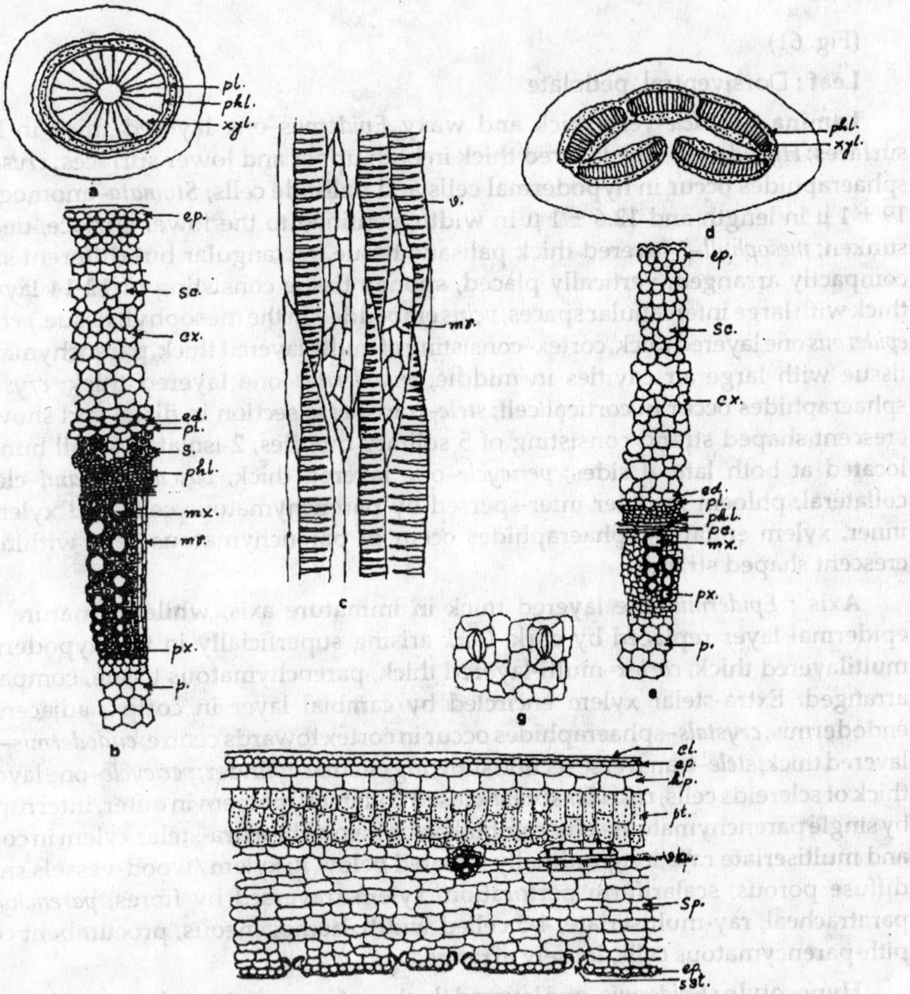

Fig. 60 : ***Bruguiera sexangula*** (Lour.) Poir.

(a) T.S. of stem (diagrammatic); (b) T.S. of a portion of stem; (c) L.S. of stem.; (d) T.S. of petiole (diagrammatic); (e) T.S. of a portion of petiole; (f) T.S. of lamina ; (g) Stomata.

layer replaced by cork, *cork*–consisting of multilayered, *cortex*–multilayered thick, with large air cavities in immature axis and in mature axis no air cavities found, *endodermis*–one layered thick, distinctly found, stele–transverse section showing circular cylinder, *pericycle*-multilayered thick associated with sclereids, *vascular strand*-closed collateral. *Phloem* in outer, associated with fibres and sclereids, xylem / wood-vessels small, diffuse porous, scalariform perforation, border pitted, xylem traversed by fibres, parenchyma paratracheal, *ray*-multiseriate, 2-layered thick, homogeneous, upright cells; *pith*-parenchymatous cells, compactly arranged.

Hypocotyle : Same as *B. gymnorrhiza* Lamk.

2.3. *Bruguiera cylindrica* Blume

(Fig. 61)

Leaf : Dorsiventral, petiolate.

Lamina : *Cuticle*-very thick and waxy *Epidermis*–one layered thick in both surfaces; *Hypodermis*–one layered thick in both upper and lower surfaces, *crystals*–sphaeraphides occur in hypodermal cells and palisade cells; *Stomata*–anomocytic, 19 ± 1 μ in length and 12.6 ± 1 μ in width, confined to the lower surface, deeply sunken; *mesophyll*–3 layered thick palisade tissue, rectangular but different sizes, compactly arranged, vertically placed, spongy tissue consisting of 12-14 layered thick with large intercellular spaces; *veins* embedded in the mesophyll tissue, *petiole*–*epidermis* one layered thick, cortex–consisting of multi-layered thick, parenchymatous tissue with large air cavities in middle; *endodermis*–one layered thick; *crystals*–sphaeraphides occur in cortical cell; *stele*–transverse section in distal end showing crescent shaped strand consisting of 5 solitary bundles, 2 isolated small bundles located at both lateral sides; *pericycle*–one layered thick, *vascular strand*–closed collateral, phloem in outer inter-spersed by parenchymatous cells and xylem in inner, xylem endarch, sphaeraphides occur in parenchymatous cells within the crescent shaped strand.

Axis : *Epidermis*–one layered thick in immature axis, while in mature axis epidermal layer replaced by cork. *Cork* arising superficially in the hypodermis, multilayered thick, *cortex*–multi-layered thick, parenchymatous tissue, compactly arranged. Extra-stelar xylem encircled by cambial layer in cortex, adjacent to endodermis, *crystals*–sphaeraphides occur in cortex towards centre; *endodermis*–one layered thick, *stele*–transverse section showing circular cylinder; *pericycle*–one layered thick of sclereids cells, *vascular strand*-closed collateral, phloem in outer, interrupted by single parenchymatous cells, cambial patch below the extra-stelar xylem in cortex and multiseriate rays longitudinally situated below it; xylem/wood-vessels small, diffuse porous, scalariform perforation, xylem traversed by fibres, *parenchyma*-paratracheal, ray-multiseriate, 4-5 celled width, homogeneous, procumbent cells; pith-parencymatous cells, loosely arranged.

Hypocotyle : *Epidermis*–one layered thick, *cortex*-consisting of parenchymatous cells, compactly arranged, oil cells occur; *Endodermis*–one layered thick, not distinctly separated from cortical layers, *stele*–circular ring, *pericycle*–multi-layered thick, *vascular strand*–amphicribral occur towards the peripheral region, *pith*-consisting of parenchymatous tissue, compactly arranged.

2.4. *Bruguiera parviflora* W. & A.

(Fig. 62)

Leaf : Dorsiventral, petiolate.

Lamina : Cuticle–very thick and waxy; *Epidermis*–one layered thick in both surfaces; *Hypodermis*–2-3 layered thick below the upper epidermis and one layered thick in lower surface; *stomata*-anomocytic, 20.2 ± 1 μ in length and 11.8 ± 1 μ in width confined to the lower surfaces, deeply sunken. *Mesophyll*–4 layered thick, palisade

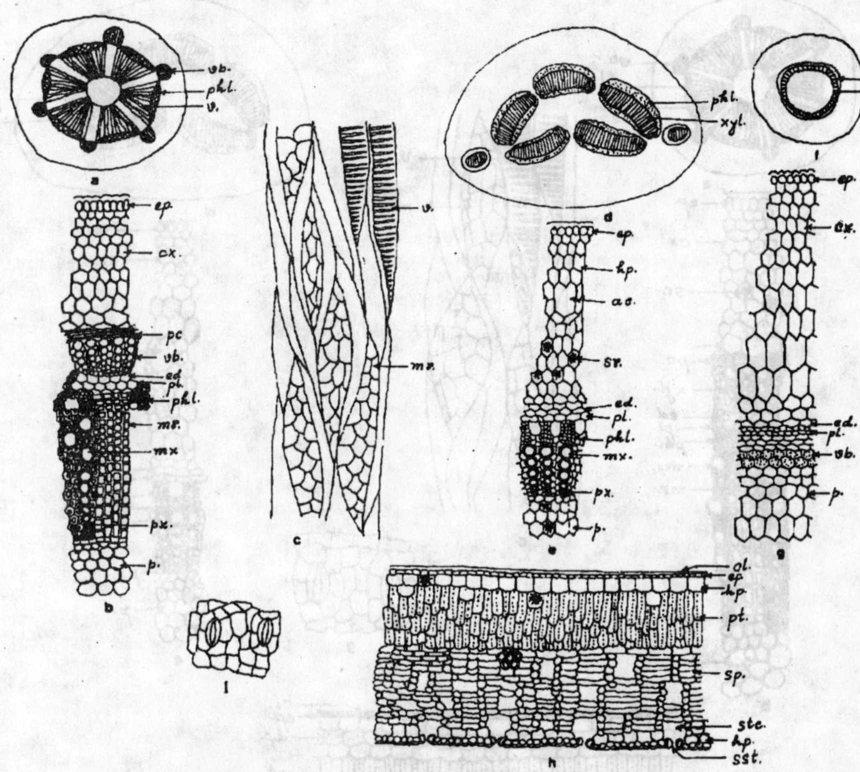

Fig. 61 : *Bruguiera cylindrica* (L.) Blume

(a) T.S. of stem (diagrammatic); (b) T.S. of a portion of stem.; (c) L.S. of stem.; (d) T.S of petiole (diagrammatic); (e) T.S. of a portion of petiole; (f) T.S. of hypocotyle (diagrammatic); (g) T.S. of a portion of hypocotyle; (h) T.S. of lamina; (i) Stomata.

tissue, rectangular, compactly arranged, vertically placed, spongy tissue consisting of 6-8 layered thick with large inter-cellular spaces, *veins* embedded in the mesophyll tissue; *petiole-epidermis* one layer thick, *cortex*-consisting of multi-layered thick, parenchymatous tissue with large air cavities in middle; *endodermis*-one layered thick, *stele*-transverse section in distal end showing crescent shaped strand consisting of 5 solitary bundles, *pericycle*–multi-layered thick, *vascular strand*–open collateral, phloem in outer, cambial ring in middle and xylem in inner, xylem endarch.

Axis : *Epidermis*–one layered thick in immature axis, while in mature axis, epidermal layer replaced by cork. *Cork*-arising superficially in the hypodermis, multilayered thick; *cortex*-multilayered thick, parenchymatous tissue, compactly arranged, extra stelar xylem encircled by cambial layer in cortex, adjacent to endodermis; *Endodermis*–one layered thick, *stele*-transverse section showing circular cylinder, *pericycle*–multi-layered thick transversed with sclereids, *vascular strand*; closed collateral-phloem in outer extra cambial patch below the extra stelar strand in cortex; xylem / wood-vessels small, diffuse porous, scalariform perforation, xylem

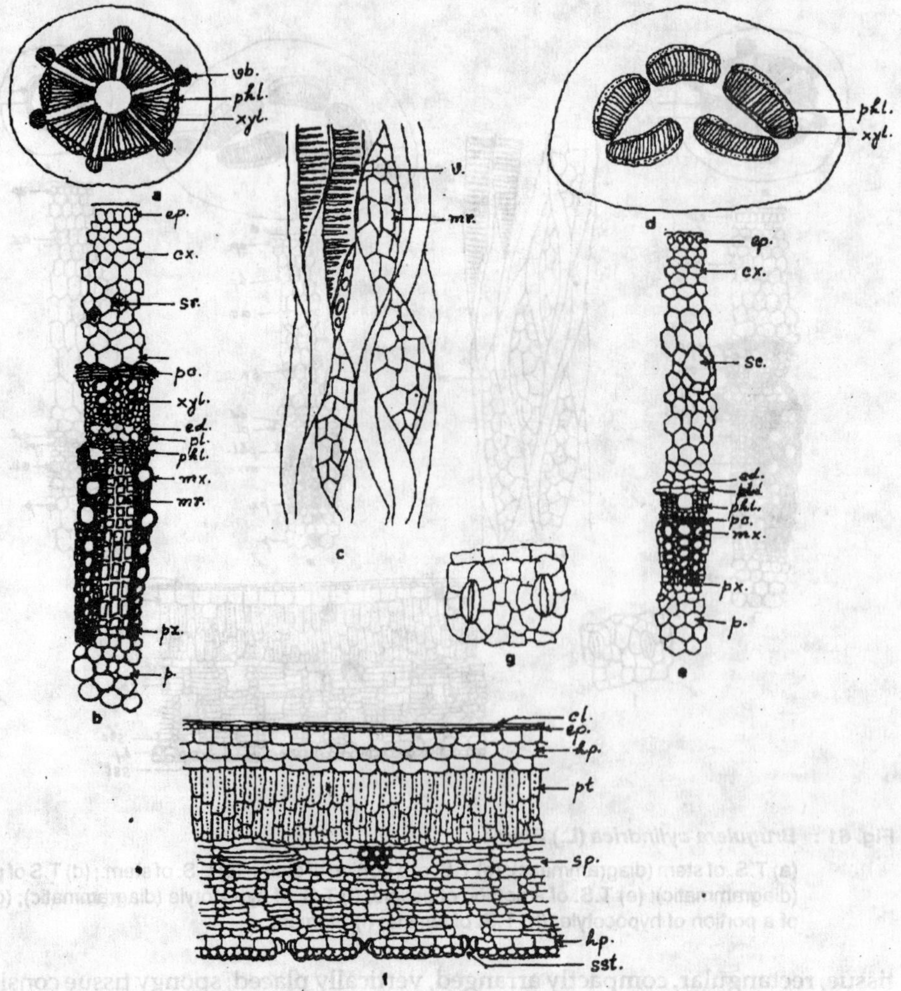

Fig. 62 : ***Bruguiera parviflora*** **W&A. ex Griff**

(a) T.S. of stem. (diagrammatic); (b) T.S. of a portion of stem.; (c) L.S. of stem.; (d) T.S. of petiole (diagrammatic); (e) T.S. of a portion of petiole; (f) T.S. of lamina; (g) Stomata.

traversed by fibres, *parenchyma*-paratracheal, ray–multiseriate. 6-7 celled width, homogeneous, procumbent cells; *pith*-parenchymatous cells, compactly arranged.

Hypocotyle : Same as *B. cylindrica* Blume

GENUS–3. *CERIOPS* Arn.

Lamina : *Epidermis*–one layered thick; *Hypodermis*–2 layered thick below the upper epidermis and also in the lower surface; *stomata*–confined to the lower surface. *Mesophyll*–palisade tissue–1 layered thick, spongy tissue consisting of multilayer thick with large inter-cellular spaces, *petiole*–stele; transverse section showing semilunar shaped strand. In the axis, *cortex* having air cavities *stele-vascular strand* closed collateral, ray-homogeneous, mostly upright cells.

Anatomical Key Characters to the Species of *Ceriops* Arn.

A. Petiolar stele showing semilunar strand, consisting of ring of bundles surrounding single medullary bundle

1. C. *decandra* (Griff.) Ding Hou

AA. Petiolar stele showing semilunar strand, consisting of ring of bundles surrounding two solitary medullary bundles

2. C. *tagal* (Perr.) Robinson

3.1. *Ceriops decandra* (Griff.) Ding Hou

(Fig. 63)

Leaf : Dorsiventral, petiolate.

Lemina : *Cuticle*–very thick and waxy; *Epidermis*-one layered thick both in upper and lower surfaces; *Hypodermis*–two layered thick below the upper epidermis and in the lower epidermis; *stomata*–anomocytic, 26.6 ± 1 μ in length and 15.8 ± 1 μ in width, confined to the lower surface, deeply sunken. *Mesophyll*–2 layered thick palisade tissue, in a single longitudinal row, one celled long, and other dwarf, spongy tissue consisting of 12-15 celled thick, parenchymatous cells, with large inter-cellular spaces, *veins* embedded in the mesophyll tissue; *petiole-epidermis* one layered thick, *cortex*-consisting of multilayered thick, parenchymatous tissue with large inter-cellular spaces in middle, sphaeraphides and few oil cells occur in cortex; *endodermis*-one layered thick; *stele*-transverse section showing semilunar shaped strand consisting of ring of bundles surrounding solitary medullary bundle; *pericycle*-2 layered thick, sclereids ring inter-spersed by parenchymatous cells *vascular strand*-closed collateral pholem in outer, xylem in inner, xylem endarch.

Axis : *Epidermis*–one layered thick in young axis, while in mature axis epidermal cell replaced by cork, *cork*-consisting of multi-layered thick, rectangular shaped, *cortex*-multilayered thick, having large air cavities in immature axis, while air cavities absent in mature axis; *Endodermis*-one layered thick, *stele*-transverse section showing circular cylinder, *pericycle*-multi-layered thick, consisting of sclereids cells, *vascular strand*-closed collateral, phloem in outer; xylem / wood-vessels small, diffuse porous, scalariform perforation, xylem traversed by fibres, *parenchyma*-paratracheal, ray-multiseriate, 2 celled width, homogeneous, upright cells, *pith*-parenchymatous cells, compactly arranged, few cells occur in pith.

Stilt Root : *Epidermis*–one layered thick, *cork*-consisting of multi-layered thick, compactly arranged, longitudinally placed. *Hypodermis*–multi-layered thick, compactly arranged, horizontally placed, *cortex*–consisting of air cavities inter-spersed by tracheids; annular perforation, *endodermis*-one layered thick, distinctly separated from cortex, *stele*-circular cylinder, *pericycle*-multi-layered thick associated with frequent brachy–sclereids in groups, *vascular strand*-open collateral, phloem in outer, multi-layered thick, cambial ring in middle, xylem / wood-vessels medium,

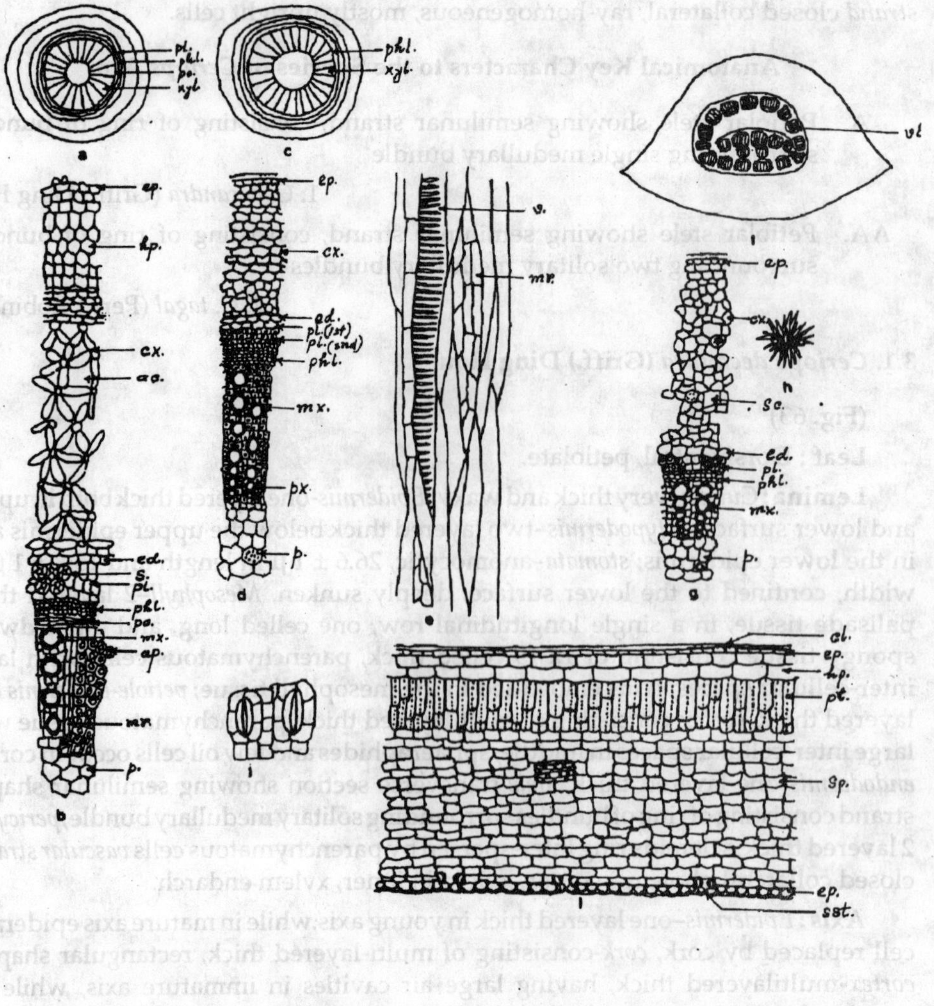

Fig. 63 : ***Ceriops decandra* (Griff.) Ding Hou**

(a) T.S. of root (diagrammatic); (b) T.S. of a portion of root; (c) T.S. of stem. (diagrammatic); (d) T.S. of a portion of stem.; (e) L.S. of stem.; (f) T.S. of petiole (diagrammatic); (g) T.S. of a portion of petiole; (h) Sphaeraphide; (i) T.S. of lamina; (j) Stomata.

diffuse porous, *parenchyma*–apotracheal, transversed by few fibres, ray-multiseriate, fibres in group; *pith*-parenchymatous cells, compactly arranged.

3.2. *Ceriops tagal* (Perr.) Robinson

(Plate–150 and Fig. 64)

Leaf : Dorsiventral, petiolate.

Lamina : Cuticle–very thick and waxy; *Epidermis*–one layered thick in both upper and lower surfaces. *Hypodermis*–2 layered thick below the upper epidermis

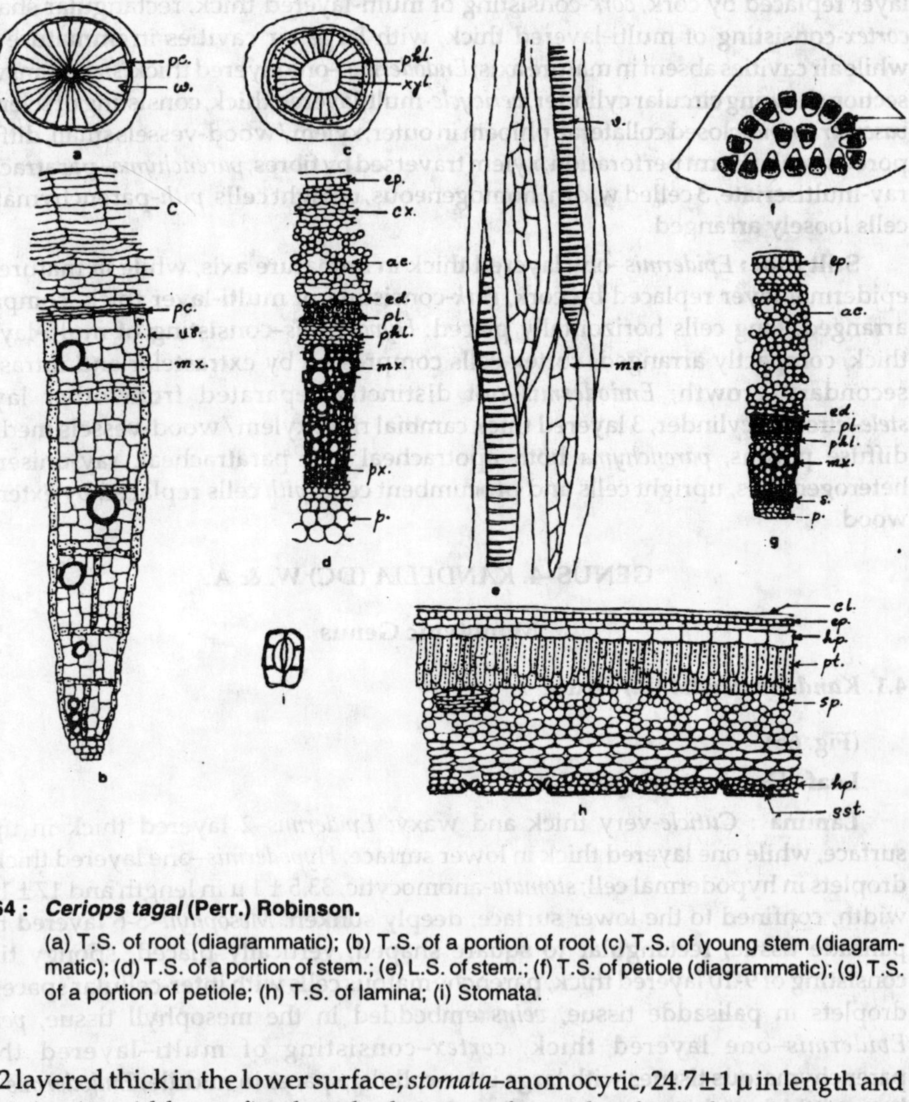

Fig. 64 : ***Ceriops tagal*** **(Perr.) Robinson.**

(a) T.S. of root (diagrammatic); (b) T.S. of a portion of root (c) T.S. of young stem (diagrammatic); (d) T.S. of a portion of stem.; (e) L.S. of stem.; (f) T.S. of petiole (diagrammatic); (g) T.S. of a portion of petiole; (h) T.S. of lamina; (i) Stomata.

and 2 layered thick in the lower surface; *stomata*–anomocytic, 24.7 ± 1 μ in length and 13.3 ± 1 μ in width, confined to the lower surfaces, deeply sunken. *Mesophyll*–2 layered thick palisade tissue in a single longitudinal row, one cell long while other dwarf, spongy tissue consisting of 10-12 layered thick, parenchymatous cells, with large inter-cellular spaces, *veins* embedded in the mesophyll tissue, *petiole-epidermis*-one lyered thick, *cortex*-consisting of multilayered thick, parenchymatous tissue with large inter-cellular spaces in middle, *endodermis*–one layered thick, *stele*-transverse section showing semilunar shaped strand consisting of ring of bundles surrounding two isolated medullary bundles, *pericycle*–2 layered thick-consisting of sclereids cells, *vascular strand*–closed collateral, phloem in outer and xylem in inner, xylem endarch.

Axis : *Epidermis*–one layered thick in young axis, while in mature axis epidermal layer replaced by cork, *cork*-consisting of multi-layered thick, rectangular shaped, *cortex*-consisting of multi-layered thick, with large air cavities in immature axis, while air cavities absent in mature axis; *Endodermis*–one layered thick, *stele*-transverse section showing circular cylinder, *pericycle*-multilayered thick, consisting of sclereids, *vascular strand*-closed collateral, phloem in outer; xylem / wood-vessels small, diffused porous, scalariform perforation, xylem traversed by fibres, *parenchyma*–paratracheal, ray-multiseriate, 3 celled width, homogeneous, upright cells, *pith*-parenchymatous, cells loosely arranged.

Stilt root : *Epidermis*–one layered thick in immature axis, while in mature axis epidermal layer replaced by cork, *cork*-consisting of multi-layer thick, compactly arranged, long cells horizontally placed; *Hypodermis*–consisting of multi-layered thick, compactly arranged, cortex-cells compressed by extrastelar and intrastelar secondary growth; *Endodermis*–not distinctly separated from other layers, *stele*–circular cylinder, 3 layered thick cambial ring; xylem / wood-vessels medium, diffuse porous, *parenchyma* both apotracheal and paratracheal, ray-uniseriate, heterogeneous, upright cells and procumbent cells, *pith* cells replaced by extended wood.

GENUS–4. *KANDELIA* (DC) W. & A.

Monotypic Genus

4.1. *Kandelia candel* (L.) Druce

(Fig. 65)

Leaf : Dorsiventral, petiolate.

Lamina : *Cuticle*-very thick and waxy; *Epidermis*–2 layered thick in upper surface, while one layered thick in lower surface. *Hypodermis*–one layered thick, oil droplets in hypodermal cell; *stomata*-anomocytic, 33.5 ± 1 μ in length and 17 ± 1 μ in width, confined to the lower surface, deeply sunken. *Mesophyll*–5-6 layered thick palisade tissue, rectangular to square shaped, vertically placed, spongy tissue consisting of 9-10 layered thick, parenchymatous cells with inter-cellular spaces, oil droplets in palisadde tissue, *veins* embedded in the mesophyll tissue, *petiole*; *Epidermis*–one layered thick, *cortex*–consisting of multi–layered thick, parenchymatous tissues with large inter-cellular spaces in middle; *Endodermis*–one layered thick, *stele*–transverse section showing ovoid shaped strand, consisting of ring of bundles, surrounding linear medullary strand having several bundles, *vascular strand*–closed collateral, phloem in outer and xylem in inner, xylem mesarch.

Axis : *Epidermis*–one layered thick in young axis, while in mature axis epidermal cells replaced by cork, *cork*-consisting of multi-layered thick, *cortex*-mature cortical cells are of two types, upper horizon having small cells, while lower horizon have medium cells, with sphaeraphides and oil droplets, one type of cells in mature axis; *Endodermis*–consisting of one celled thick, *stele*–transverse section showing circular strand, *pericycle*–consisting of wavy ring of sclereids cells, *vascular strand*-closed collateral, phloem in outer; xylem / wood-vessels small, diffuse porous, scalariform

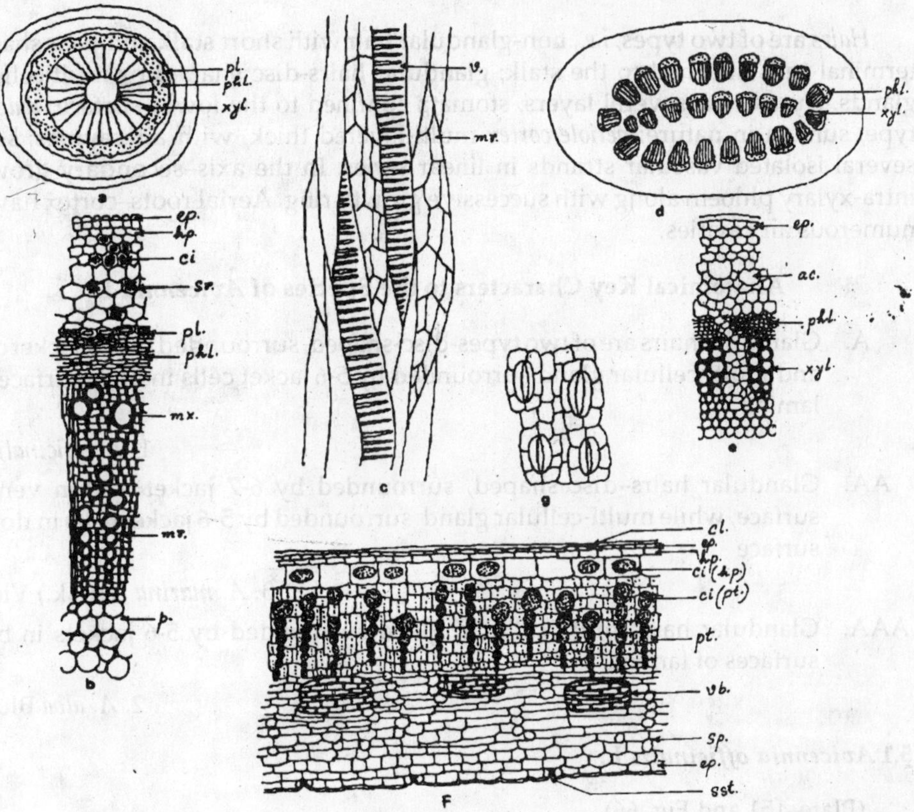

Fig. 65 : ***Kandelia candel*** **(L.) Druce**
(a) T.S. of stem. (diagrammatic); (b) T.S. of a portion of stem.; (c) L.S. of a portion of stem.; (d) T.S. of petiole (diagrammatic); (e) T.S. of a portion of petiole; (f) T.S. of lamina; (g) Stomata.

perforation, border pitted, parenchyma-square, *paratracheal*, ray-multiseriate, 2-3 celled width, homogeneous, upright cells, *pith*-consisting of parenchymatous cells, loosely arranged.

FAMILY–II. *AVICENNIACEAE*

Leaf hairs are of two types–glandular hair and non-glandular hair; non-glandular hair giving a white velvet appearance to the lower surface of the leaf; stomata confined to the lower surface, generally *diacytic* type, sunken in nature; *petiole-cortex*, multi-layered thick, in axis, *cork* originating superficially, *stele*-secondary growth of anomalous type, intra-xylary phloem along with successive growth ring.

GENUS–5. *AVICENNIA* L.

Hairs are of two types, *i.e.*, non-glandular hair with short stalk and pear shaped terminal cells attached to the stalk; glandular hairs-disc shaped or multicellular glands. *Hypodermis* several layers, stomata confined to the lower surface, *diacytic* type, sunken in nature; *petiole-cortex*-multi-layered thick, with air cavities; stele-several isolated vascular strands in linear series. In the axis–secondary growth, intra-xylary phloem along with successive growth ring. Aerial roots–cortex having numerous air cavities.

Anatomical Key Characters to the Species of *Avicennia* L.

A. Glandular hairs are of two types-disc-shaped, surrounded by 6-7 jacket cells and multi-cellular gland surrounded by 5-6 jacket cells in both surfaces of lamina

1. *A. officinalis* L.

AA. Glandular hairs–disc shaped, surrounded by 6-7 jacket within ventral surface, while multi-cellular gland surrounded by 5-6 jacket cells in dorsal surface

3. *A. marina* (Forsk.) Vierh.

AAA. Glandular hairs–multi-cellular, gland surrounded by 5-6 jackets in both surfaces of lamina

2. *A. alba* Blume

5.1 *Avicennia officinalis* L.

(Plate–151 and Fig. 66)

Leaf : Dorsiventral, petiolate.

Lamina : *Cuticle*–waxy on upper surface, hairs–non-glandular,uniseriate hairs with short stalk and pear shaped terminal cells attached on it, providing white velvet appearance to the lower surface; glandular hairs–two types : disc shaped gland bounded by 6-7 jacket cells and multi-cellular gland bounded by 5-6 jackets cells in both the surfaces, frequency of disc-shaped glands comparatively more than multi-cellular glands, *epidermis*-one layered thick in both upper and lower surfaces; *Hypodermis*–4 layered thick beneath the uper epidermis, rectangular, horizontally placed, compactly arranged; *stomata*-diacytic, 13.7 ± 0.5 μ in length and 10.4 ± 0.5 μ in width, confined to the lower surface, deeply sunken. *Mesophyll*–3 layered thick, palisade tissue rectangular shaped, vertically placed, 3-4 layered thick spongy parenchyma, compactly arranged, sclerenchymatous cells surrounding vascular strands, few sclereids and fibres scattered in the spongy parenchymatous tissue, *veins* embedded in the mesophyll tissue. *Petiole*–uniseriate hairs occur in grooved of mature petiole, *epidermis*-one layered thick, wall deposited by suberin, *hypodermis*–2 layered thick, collenchymatous tissue, *cortex*-consisting of parenchymatous cells along with numerous air cavities towards endodermis, single sclereids scattered in cortex, one layered thick, *stele*-transverse section showing crescent shaped vascular strand and two isolated strands on both curved ends. *Pericycle*-one layered thick,

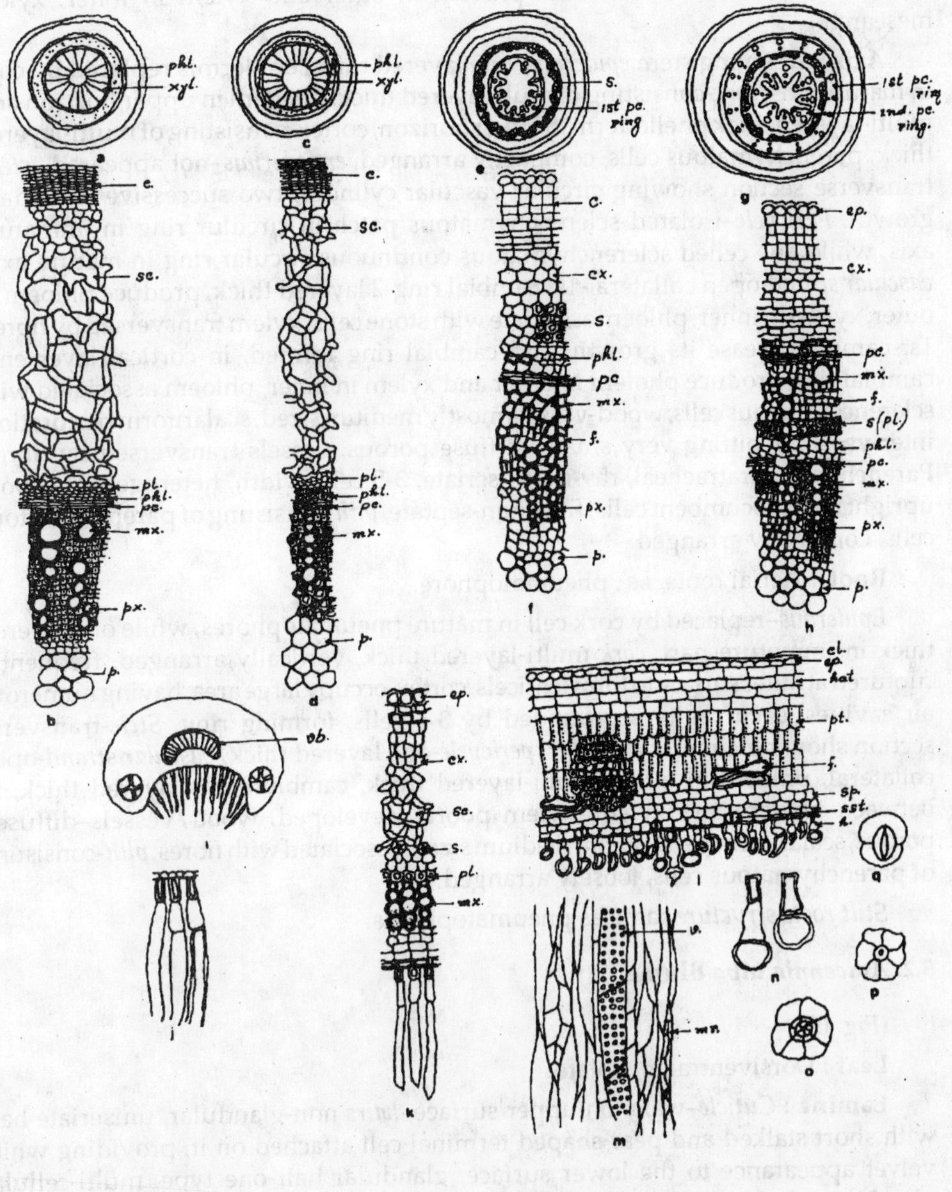

Fig. 66 : *Avicennia officinalis* L.

(a) T.S. of pneumatophore (diagrammatic); (b) T.S. of a portion of pneumatophore; (c) T.S. of stilt root (diagrammatic); (d) T.S. of a portion of stilt root; (e) T.S. of young stem. (diagrammatic) showing one vascular ring; (f) T.S. of a portion of young stem; (g) T.S. of mature stem. (diagrammatic) showing 2 rings of vascular bundles; (h) T.S. of a portion of mature stem; (i) T.S. of petiole (diagrammatic); (j) Non-glandular hairs; (k) T.S. of a portion of petiole; (l) T.S of lamina; (m) L.S. of a portion of stem.; (n) Pear shaped-non-glandular hairs; (o) Multicellular gland; (p) One celled gland; (q) Stomata.

vascular strand : closed collateral-phloem in outer and xylem in inner, xylem mesearch.

Axis : In young stem *epidermis*–one layered thick, epidermis replaced by cork in mature axis, *cork*-consisting of multilayered thick, phellem in upper horizon and multi-layered thick phelloderm in lower horizon, cortex-consisting of multi-layered thick, parenchymatous cells, compactly arranged, *endodermis*–not appeared, *stele*-transverse section showing circular vascular cylinder, two successive secondary growth. *Pericycle*–isolated sclerenchymatous patch in circulur ring in immature axis, while one celled sclerenchymatous continuous circular ring in mature axis, *vascular strand*-open collateral-1st cambial ring-2 layered thick, produce phloem in outer, xylem in inner, phloem associate with stone cells, xylem transversed by fibres, 1st cambium cease its growth, 2nd cambial ring formed, in cortical layer end, cambial ring produce pholem in outer and xylem in inner, phloem associated with sclerenchymatous cells; wood-vessels mostly medium sized, scalariform perforation, inter-vascular pitting very small, diffuse porous, vessels transversed by fibres. Parenchyma-paratracheal, ray-multi-seriate, 3-5 cells width, heterogeneous, both upright and procumbent cells, fibre non-septate, *pith*-consisting of parenchymatous cells, compactly arranged.

Root : Aerial roots, *i.e.*, pneumatophore.

Epidermis–replaced by cork cell in mature pneumatophores, while one layered thick in immature part, *cork* multi-layered thick, vertically arranged, frequently ruptured at outer surface to form lenticels, cortex-occupy large area, having numerous air cavities, each cavity surrounded by 3-4 cells, forming ring. *Stele*-transverse section showing circular cylinder, *pericycle*-one layered thick, *vascular strand*-open collateral, phloem in outer, multi-layered thick, cambium 2-3 layered thick, in between phloem and xylem, xylem poorly developed, wood/vessels–diffused porous, scalariform perforation, medium sized, associated with fibres, *pith*-consisting of parenchymatous cells, loosely arranged.

Stilt roots structure same as pneumatophores.

5.2. *Avicennia alba* Blume

(Fig. 67)

Leaf : Dorsiventral, petiolate.

Lamina : *Cuticle*–waxy on upper surface, *hairs* non-glandular, uniseriate hair with short stalked and pear shaped terminal cell attached on it, providing white velvet appearance to the lower surface, glandular hair-one type, multi-cellular gland bounded by 4-5 jacket cells in both surfaces, frequency of glands in ventral surface more than glands of dorsal surface; *epidermis* one layered thick in both upper and lower surfaces; *hypodermis*-5 layered thick beneath the upper epidermis, rectangular, horizontally placed, compactly arranged; *stomata*-diacytic, 16.4 ± 1 μ in length and 13.3 ± 1 μ in width, confined to the lower surfaces, deeply sunken, *mesophyll*; 2 layered thick-palisade tissue, rectangular shaped, vertically placed, 4-5 layered thick spongy parenchymatous tissue, compactly arranged, sclerenchymatous cells surrounding vascular strands, few sclereids and fibres

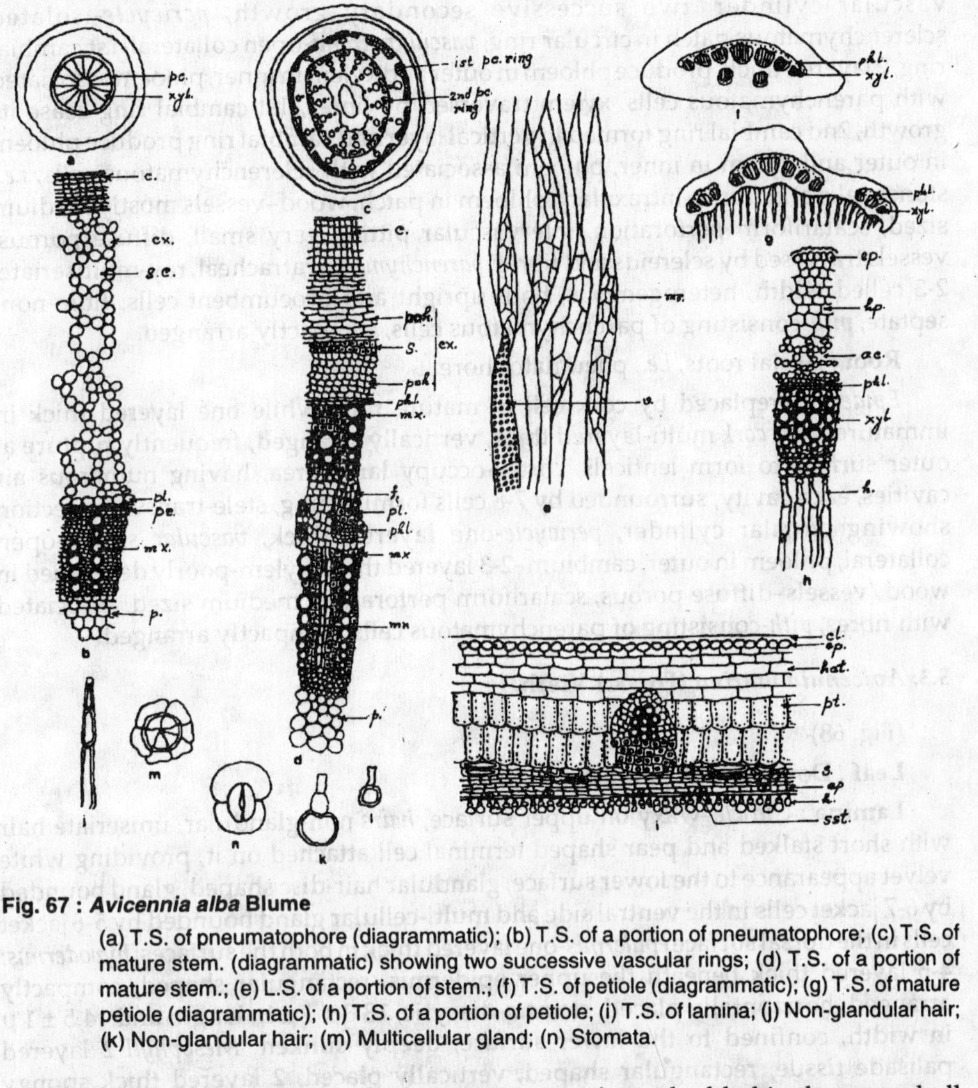

Fig. 67 : *Avicennia alba* Blume

(a) T.S. of pneumatophore (diagrammatic); (b) T.S. of a portion of pneumatophore; (c) T.S. of mature stem. (diagrammatic) showing two successive vascular rings; (d) T.S. of a portion of mature stem.; (e) L.S. of a portion of stem.; (f) T.S. of petiole (diagrammatic); (g) T.S. of mature petiole (diagrammatic); (h) T.S. of a portion of petiole; (i) T.S. of lamina; (j) Non-glandular hair; (k) Non-glandular hair; (m) Multicellular gland; (n) Stomata.

scattered in the spongy parenchymatous tissue, *veins* embedded in the mesophyll tissue. *Petiole* : *hairs*-uniseriate hairs present in grooved of mature petiole, *epidermis*-one layered thick, wall deposited by suberin, *cortex*-consisting of parenchymatous cells, multi-layered thick, along with numerous air cavities towards endodermis; *endodermis*-one layered thick, *stele*-transverse section showing numerous solitary strands in a linear line, *pericycle*-one layered thick, *vascular strand*-closed collateral. Phloem in outer and xylem in inner, xylem mesarch.

Axis : In young stem, *epidermis*-one layered thick, epidermis replaced by cork in mature axis; *cork*-consisting of multilayered thick, vertically placed; *cortex*-consisting of multilayered thick, two layered horizon, upper layered horizon having collenchymatous cells, lower horizon having parenchymatous cells, compactly arranged, *endodermis*-one layered thick; *stele*-transverse section showing circular vascular cylinder, two successive secondary growth, *pericycle*-isolated sclerenchymatous patch in circular ring, *vascular strand*-open collateral-1st cambial ring, 2 layered thick, produce phloem in outer and xylem in inner; phloem associated with parenchymatous cells, xylem traversed by fibres, 1st cambial ring cease its growth, 2nd cambial ring formed in vertical layer and cambial ring produce phloem in outer and xylem in inner, phloem associated with sclerenchymatous cells, *i.e.*, stones cells and fibres, intraxylary phloem in patch, wood–vessels mostly medium sized, scalariform perforation, intervascular pitting very small, diffuse porous, vessels traversed by sclereids and fibres, *parenchyma*-paratracheal, ray-multiseriate, 2-3 celled width, heterogeneous, both upright and procumbent cells, fibre-non-septate, *pith*-consisting of parenchymatous cells, compactly arranged.

Root : Aerial roots, *i.e.*, pneumatophore.

Epidermis–replaced by cork cell in mature root, while one layered thick in immature root, *cork* multi-layered thick, vertically arranged, frequently rupture at outer surface to form lenticels, cortex-occupy large area, having numerous air cavities, each cavity, surrounded by 7-8 cells forming ring, stele-transverse section showing circular cylinder, *pericycle*-one layered thick, *vascular strand*-open collateral, phloem in outer, cambium–2-3 layered thick, xylem-poorly developed in wood/vessels–diffuse porous, scalariform perforation, medium sized, associated with fibres, *pith*-consisting of parenchymatous cells, compactly arranged.

5.3. *Avicennia marina* (Forsk.) Vierh.

(Fig. 68)

Leaf : Dorsiventral, petiolate.

Lamina : *Cuticle*–waxy on upper surface, *hairs* non-glandular, uniseriate hair with short stalked and pear shaped terminal cell attached on it, providing white velvet appearance to the lower surface, glandular hair-disc shaped, gland bounded by 6-7 jacket cells in the ventral side and multi-cellular gland bounded by 5-6 jacket cells in the dorsal surface *epidermis*-one layered thick in both the surfaces; *hypodermis*; 4-5 layered thick beneath the upper epidermis, rectangular shaped, compactly arranged, horizontally placed, *stomata*-diacytic, 18.3 ± 1 μ in length and 14.5 ± 1 μ in width, confined to the lower surface, deeply sunken. *Mesophyll*-2 layered palisade tissue, rectangular shaped, vertically placed, 2 layered thick spongy parenchymtous tissue, compactly arranged, sclerenchymatous cells-surrounding vascular strands, few sclereids and fibres scatterd in the spongy parenchymatous tissue, *veins* embedded in the mesophyll. *Petiole*; hairs uniseriate, present in the grooved of mature petiole, *epidermis*-one layered thick, *hypodermis*-one layered thick, parenchymatous cells. *Cortex*-consisting of parenchymatous tissues, along with numerous air cavities towards endodermis, one layered thick, *stele*–transverse section showing numerous solitary bundles in a linear line; *pericycle*-one layered thick, vascular strand-closed collateral, phloem in outer and xylem in inner, xylem mesarch.

Axis : In young stem hair 2-3 celled, uniseriate, *epidermis*–one layered thick,

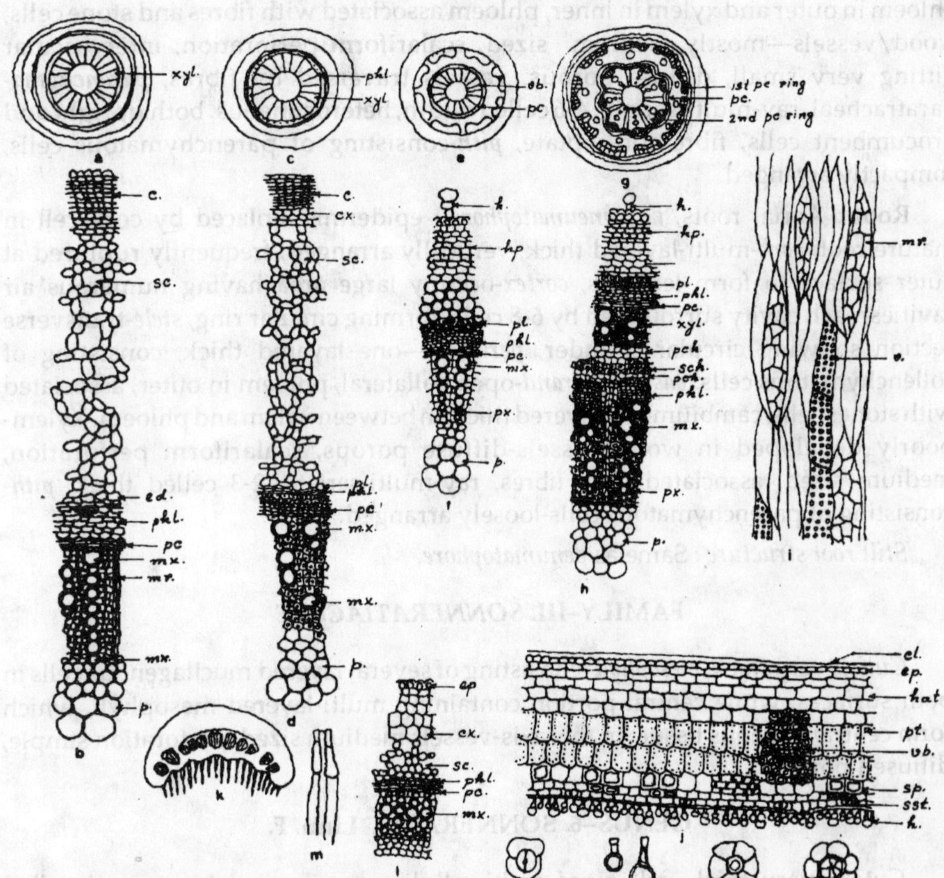

Fig. 68 : ***Avicennia marina*** (Forsk.) Vierh.

(a) T.S. of Pneumatophore (diagrammatic); (b) T.S. of a portion of Pneumatophore; (c) T.S. of stilt root (diagrammatic); (d) T.S of a portion of stilt root; (e) T.S. of young shoot (diagrammatic) showing one successive vascular ring; (f) T.S. of a portion of young shoot; (g) T.S. of mature stem (diagrammatic) showing two successive vascular rings; (h) T.S. of a portion of mature stem; (i) L.S. of a portion of mature stem; (j) T.S. of lamina; (k) T.S. of petiole (diagrammatic); (l) T.S. of a portion of petiole; (m) Non-glandular hair; (n) One-celled gland; (o) Multi-celled gland; (p) Non-glandular pear shaped hairs; (q) Stomata.

epidermis replaced by cork in mature axis, *cork*-consisting of multilayered thick, square shaped cells, vertically placed, *cortex*-consisting of two type of layers, upper horizon having multi-layered thick, collenchymatous cells, while lower horizon consisting of parenchymatous cells, *endodermis*-not appeared, *stele*-transverse section showing circular vascular cylinder-two successive growth rings appeared, *pericycle*-continuous circular sclerenchymatous ring, *vascular strand*-open collateral, 1st cambial ring-produce phloem in outer and xylem in inner, xylem transversed by fibres, 1st cambial ring ceases its growth, 2nd cambial ring develops in the cortex, produces phloem in outer and xylem in inner, phloem associated with fibres and stone cells, wood/vessels—mostly medium sized, scalariform perforation, intervascular pitting very small, diffuse porous, vessels traversed by fibres, *parenchyma*-paratracheal, ray-multi-seriate, 4-6 celled width, heterogeneous, both upright and procumbent cells, fibre-non-septate, *pith*-consisting of parenchymatous cells, compactly arranged.

Root : Aerial roots, *i.e., pneumatophore* : epidermis-replaced by cork cell in mature root, *cork*-multi-layered thick, vertically arranged, frequently ruptured at outer surface to form lenticels, *cortex*-occupy large area having numerous air cavities, each cavity surrounded by 6-8 cells, forming circular ring, *stele*-transverse section showing circular cylinder, *pericycle*—one layered thick, consisting of collenchymatous cells, *vascular strand*-open collateral, phloem in outer, associated with stones cells, cambium–3-4 layered thick, in between xylem and phloem. Xylem-poorly developed in wood vessels-diffuse porous, scalariform perforation, medium sized, associated with fibres, ray-multi-seriate, 2-3 celled thick, *pith*-consisting of parenchymatous cells-loosely arranged.

Stilt root structure : Same as *penumatophore.*

FAMILY–III. *SONNERATIACEAE*

Cuticle very thick, *mesophyll*–consisting of several layered mucilagenous cells in both surfaces, while central portion containing multi-layered mesophyll, which form central aqueous tissue. In this axis-vessels medium sized, perforation simple, diffused porous.

GENUS–6. SONNERATIA Linn. F.

Cuticle very thick, *salt gland*-multi-cellular, surrounded by several celled jacket cells; *stomata*-confined to both the surfaces, *anomocytic,* deeply sunken. *Mesophyll*-consisting of several layered mucilagenous cells below the hypodermal layer in both these surfaces, while central portion containing multi-layered mesophyll, which form central aqueous-tissue, *petiole*-vascular bundle-hadrocentric. In the axis, vessels medium sized, diffuse porous, xylem traversed by fibres, rays-uniseriate. *Stilt root*-cortical layer consisting of numerous air cavities.

Anatomical Key Characters to the Species of *Sonneratia* Linn. f.

(In Petiole)

A. Straight mid flattened solitary vascular strand with both terminal bent ends — 3. *S. caseolaris* Engler

AA. Slightly curved solitary vascular strand with both terminal bent ends — 1. *S. apetala* Buch.-Ham.

AAA. Semilunar solitary vascular strand — 2. *S. griffithii* Kurz.

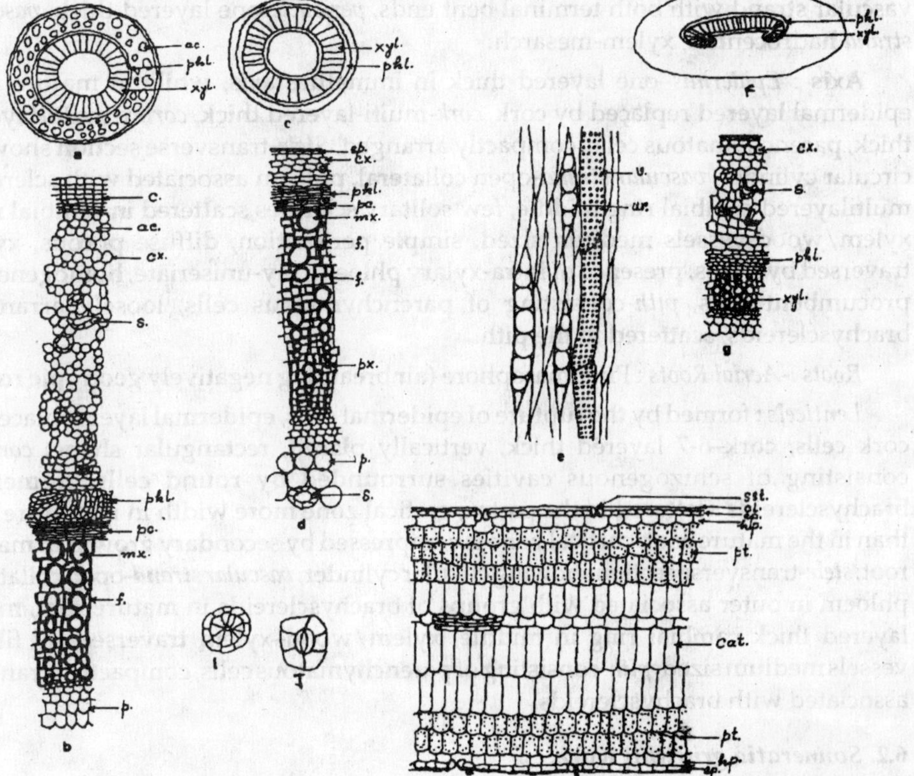

Fig. 69 : *Sonneratia apetala* Buch. Ham.

(a) T.S. of pneumatophore (diagrammatic); (b) T.S. of a portion of pneumatophore; (c) T.S. of stem. (diagrammatic) ; (d) T.S. of a portion of stem; (e) L.S. of a portion of stem.; (f) T.S. of petiole (diagrammatic); (g) T.S of a portion of petiole; (h) T.S. of lamina; (i) Multi-celled gland; (j) Stomata.

6.1. *Sonneratia apetala* Buch.–Ham.

(Fig. 69).

Leaf : Isobilateral, petiolate.

Lamina : *Cuticle*–one layered thick, *epidermis*-one layered thick in both upper and lower surfaces, *salt gland* multi-cellular, surrounded by several jacket cells, confined to the dorsal surface; *stomata*-anomocytic, confined to both the surfaces, deeply sunken, 19 ± 1 μ in length and 10.4 ± 0.5 μ in width in ventral surface, while 18 ± 1 μ in length and 16 ± 0.5 μ in width in dorsal surface; *Mesophyll*-consisting of 4 layered and 3 layered thick mucilaginous palisade cells below the upper and lower surfaces, respectively; while central portion includes 3 layered thick tissue, compactly arranged, vertically placed, rod-shaped, *idioblasts*-scattered scleren chymatous cells in mesophyll tissue, *veins* embedded in the mesophyll tissue; *Petiole-epidermis*-one layered thick, *cortex*-consisting of parenchymatous tissue, multi-layered thick, scattered sclereids in cortex; *stele*-transverse section showing slightly curved solitary vascular strand with both terminal bent ends, *pericycle*-one layered thick, *vascular strand* hadrocentric, xylem-mesarch.

Axis : *Epidermis*–one layered thick in immature axis, while in mature axis, epidermal layered replaced by cork, *cork*-multi-layered thick, *cortex*-multi-layered thick, parenchymatous cells, compactly arranged. *Stele*-transverse section showing circular cylinder, *vascular strand*-open collateral, phloem associated with sclereids, multilayered cambial ring middle, few solitary sclerides scattered in cambial ring, xylem/wood-vessels medium sized, simple perforation, diffuse porous, xylem traversed by fibres, presence of intra-xylary phloem, ray-uniseriate, homogeneous, procumbent cells, *pith*-consisting of parenchymatous cells, loosely arranged, brachysclereids, scattered in the pith.

Roots :-Aerial Roots : Pneumatophore (air breathing negatively geotropic roots).

Lenticels : formed by the rupture of epidermal cells, epidermal layer replaced by cork cells, cork–6-7 layered thick, vertically placed, rectangular shape; *cortex* : consisting of schizogenous cavities surrounded by round cells, numerous brachysclereids scattered in the cortex, cortical zone more width in immature root than in the mature root, cortical zone is compressed by secondary growth in mature root; *stele*-transverse section showing circular cylinder, *vascular strand*-open collateral, phloem in outer associated with groups of brachysclereids in mature root, multi-layered thick cambial ring in middle, xylem/wood–xylem traversed by fibres, vessels medium sized, *pith*-consisting of parenchymatous cells, compactly arranged, associated with brachysclereids.

6.2. *Sonneratia griffithii* Kurz.

(Plate–152 and Fig. 70)

Leaf : Isobilateral, petiolate.

Lamina : *Cuticle*–one layered thick, *epidermis*–one layered thick in both upper and lower surfaces, salt gland multi-cellular, surrounded by several jacket cells, confined to the dorsal surface, *stomata*-anomocytic, confined to both the surfaces, deeply sunken, 24 ± 1 μ in length and 12.6 ± 0.5 μ in width in ventral surface, while 29.4 ± 1 μ in length and 22.8 ± 0.5 μ in width in dorsal surface. *Mesophyll*-consisting of 4 layered thick and 3 layered thick mucilage palisade cells below the upper and lower surfaces respectively. While, central portion includes 7-8 layered thick aqueous tissue, compactly arranged, angular cells; *Idioblasts*-scattered sclerenchymatous in mesophyll tissue, *vines* embedded in the mesophyll tissue. *Petiole* : *epidermis*-one layered thick, *cortex*-consisting of parenchymatous tissue, multi-layered thick, compactly arranged. *Stele*-transverse section showing semilunar solitary vascular strand, *pericycle*-one layered thick, *vascular strand*-hadrocentric, xylem-mesarch.

Axis : *Epidermis*–one layered thick in immature axis, while in mature axis, epidermal layered replaced by cork, *cork*-multi-layered thick, compactly arranged, vertically placed, *cortex*–multi-layered thick, parenchymatous cells, compactly arranged; *Stele*-transverse section showing circular cylinder; *Vascular* strand-open collateral, phloem associated with sclereids, multilayered cambial ring in middle,

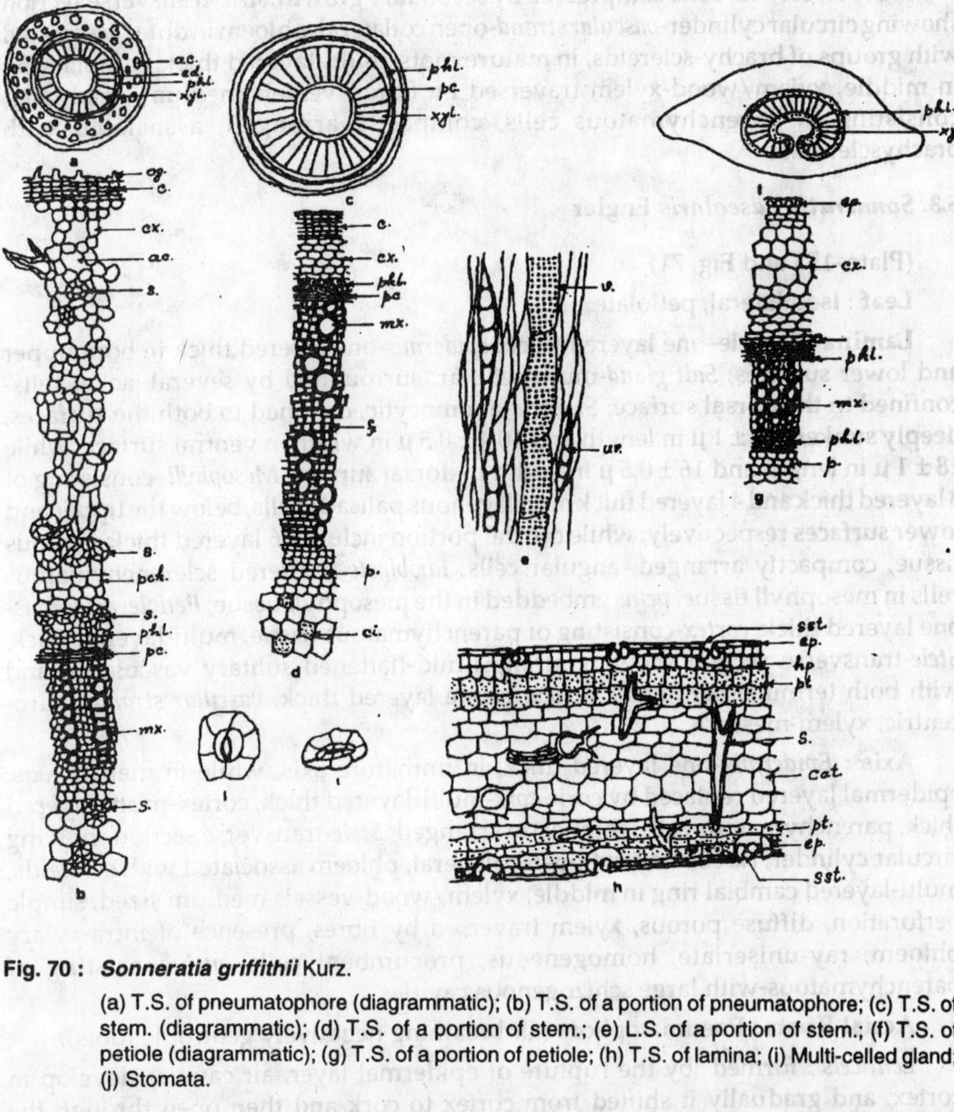

Fig. 70 : ***Sonneratia griffithii*** Kurz.

(a) T.S. of pneumatophore (diagrammatic); (b) T.S. of a portion of pneumatophore; (c) T.S. of stem. (diagrammatic); (d) T.S. of a portion of stem.; (e) L.S. of a portion of stem.; (f) T.S. of petiole (diagrammatic); (g) T.S. of a portion of petiole; (h) T.S. of lamina; (i) Multi-celled gland; (j) Stomata.

xylem/wood-vessels medium sized, simple perforation, diffuse porous, xylem traversed by fibres, presence of intra-xylary phloem, ray-uniseriate, homogeneous, procumbent cells, *pith*-consisting of parenchymatous cells, few oil cells scattered in pith.

Roots : *Aerial Roots* : Pneumatophore (air breathing negatively geotropic roots).

Lenticels : formed by the rupture of epidermal layer, epidermal layer replaced by cork layer, cork-consisting of 4-5 layered thick, compactly arranged, vertically placed, square shaped, *cortex*-consisting of schizogenous cavities surrounded by ovoid shaped cells, numerous brachy-sclereids and branched-sclereids scattered in the cortex, cortical zone more width in immature root than in the mature root, in mature root, cortical zone compressed by secondary growth, *stele*-transverse section showing circular cylinder-*vascular strand*-open collateral, phloem in outer associated with groups of brachy-sclereids, in mature roots, multi-layered thick, cambial ring in middle, xylem/wood-xylem traversed by fibres, vessels medium sized, *pith*-consisting of parenchymatous cells, compactly arranged, associated with brachysclereids.

6.3. *Sonneratia caseolaris* Engler

(Plate–153 and Fig. 71)

Leaf : Isobilateral, petiolate.

Lamina : Cuticle–one layered thick, *epidermis*-one layered thick in both upper and lower surfaces; *Salt gland*-multi-cellular, surrounded by several jacket cells-confined to the dorsal surface; *Stomata*-anomocytic, confined to both the surfaces, deeply sunken; 19 ± 1 μ in length and 10.4 ± 0.5 μ in width in ventral surface, while 18 ± 1 μ in length and 16 ± 0.5 μ in width in dorsal surface; *Mesophyll*–consisting of 3 layered thick and 4 layered thick mucilagenous palisade cells, below the upper and lower surfaces respectively; while central portion includes 6 layered thick aqueous tissue, compactly arranged, angular cells; *Idioblasts*–scattered sclerenchymatous cells in mesophyll tissue, *veins* embedded in the mesophyll tissue; *Petiole*; *epidermis*-one layered thick, *cortex*-consisting of parenchymatous tissue, multi-layered thick; *Stele*-transverse section showing straight mid-flattened solitary vascular strand with both terminal bent ends; *pericycle*–one layered thick, *vascular strand*-hadrocentric, xylem-mesarch.

Axis : *Epidermis*–one layered thick, in immature axis, while in mature axis, epidermal layered replaced by cork, *cork*-multi-layered thick, cortex-multi-layered thick, parenchymatous cells, compactly arranged; *Stele*-transverse section showing circular cylinder, *vascular strand*-open collateral, phloem associated with sclereids, multi-layered cambial ring in middle, xylem/wood-vessels medium sized, simple perforation, diffuse porous, xylem traversed by fibres, presence of intra-xylary phloem, ray-uniseriate, homogeneous, procumbent cells, *pith*-consisting of parenchymatous-with large schizogenous cavities.

Aerial Roots : Pneumatophore (air breathing negatively geotropic roots).

Lenticels : formed by the rupture of epidermal layer, air cavities develop in cortex, and gradually it shifted from cortex to cork and then open through the epidermal layer; epidermal layer replaced by cork layer, *cork*–5-6 layered thick, vertically placed, rectangular shaped, *cortex*-consisting of schizogenous cavities surrounded by ovoid cells, numerous branched sclereids and brachy-sclereids scattered in the cortex, cortical zone more width in immature root than in the mature root, in mature root, cortical zone compressed by secondary growth, *stele*-transverse section showing circular cylinder, *vascular strand*-open collateral, phloem in outer associated with groups of brachysclereids in mature root, multi-layered thick cambial ring in middle, xylem/wood, xylem traversed by fibres, vessels–medium sized, *pith*-consisting of parenchymatous cells, compactly arranged, associated with brachy-sclereids.

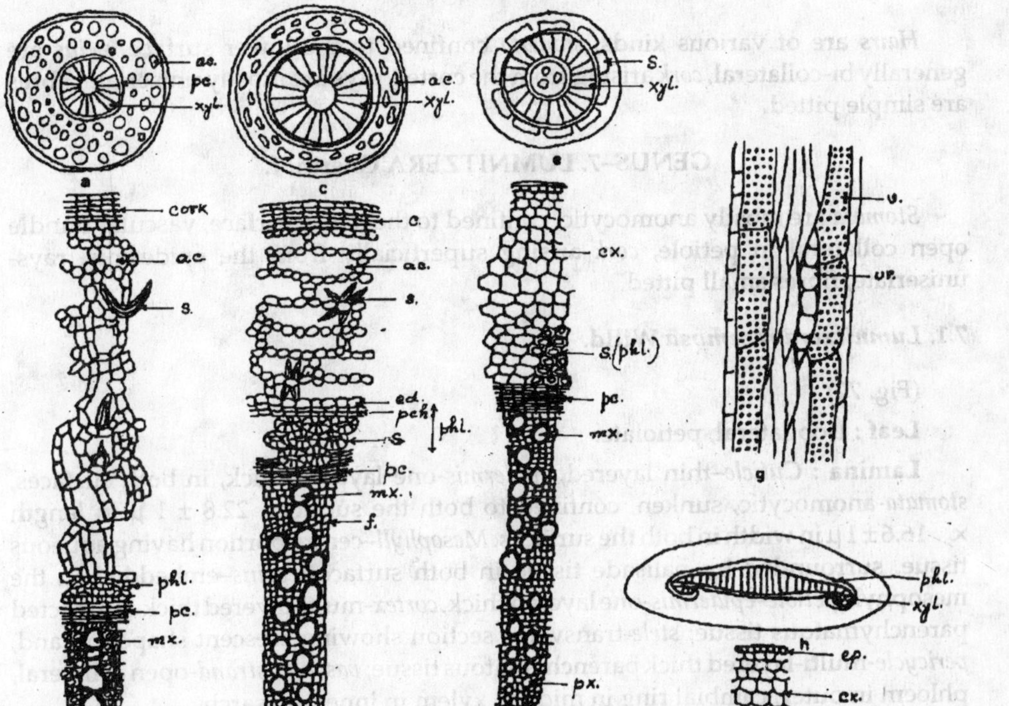

Fig. 71 : ***Sonneratia caseolaris*** **(L.) Engler**

(a) T.S. of young pneumatophore (diagrammatic); (b) T.S. of a portion of young pneumatophore; (c) T.S. of mature pneumatophore (diagramamtic); (d) T.S. of a portion of mature pneumatophore; (e) T.S. of stem (diagrammatic); (f) T.S. of a portion of stem; (g) L.S. of a portion of stem.; (h) T.S. of petiole (diagrammatic); (i) T.S. of a portion of petiole; (j) T.S. of lamina; (k) Multi-celled gland; (l) Stomata.

FAMILY–IV. *COMBRETACEAE*

Hairs are of various kinds, *stomata* confined to the lower surface, *veins* are generally bi-collateral, *cork* arising from the cortex, rays are mostly uniseriate, fibres are simple pitted.

GENUS–7. LUMNITZERA Gaertn. f.

Stomata are mostly anomocytic, confined to the lower surface, vascular bundle open collateral in petiole, *cork* arising superficially from the epidermis, rays-uniseriate, fibres small pitted.

7.1. *Lumnitzera racemosa* Willd.

(Fig. 72)

Leaf : Isobilateral, petiolate.

Lamina : *Cuticle*–thin layered, *epidermis*-one layered thick, in both surfaces, *stomata*-anomocytic, sunken, confined to both the surfaces, $22.8 \pm 1\ \mu$ in length $\times$ $16.6 \pm 1\ \mu$ in width in both the surfaces. *Mesophyll*–centre portion having aqueous tissue, surrounded by palisade tissue in both surfaces, *veins*–embedded in the mesophyll; *petiole–epidermis*-one layered thick, *cortex*-multi-layered thick compacted parenchymatous tissue; *stele*-transverse section showing crescent shaped strand, *pericycle*-multi-layered thick parenchymatous tissue; *vascular strand*-open collateral, phloem in outer, cambial ring in middle, xylem in inner, mesarch.

Axis : Epidermal layer replaced by cork in mature axis, cork-multi-layered thick walled cells, *cortex*-multi-layered thick compacted parenchymatous tissue; stele-transverse section showing circular cylinder, *pericycle*-multi-layered thick parenchymatous tissue, *vascular strand*-closed collateral, phloem traversed by parenchyma and solitary sclereids, xylem/wood-vessels medium sized, diffuse porous, scalariform perforation, traversed by fibres, fibres-simple pitted, parenchyma-paratracheal, ray-uniseriate, mostly homogeneous, *pith*-compacted parenchymatous tissue having large solitary sclereids.

FAMILY–V. *ARECACEAE*

Epidermal layer thickly cutinized, *mesophyll* having sclereids patches, *veins* surrounded by sclereids.

Key to the Genera of Arecaceae

A.	Stomata confined to both the surfaces	2. *Phoenix* Roxb.
AA.	Stomata confined to the lower surface only	1. *Nypa* Steck

GENUS–8. *NYPA* Steck

Stomata confined to the lower surface, generally *Pandanus* type, sunken, vascular bundle surrounded by sclereides, bundles scattered in the cortex in leaf axis, leaf axis showing sunken stomata of its outer surface.

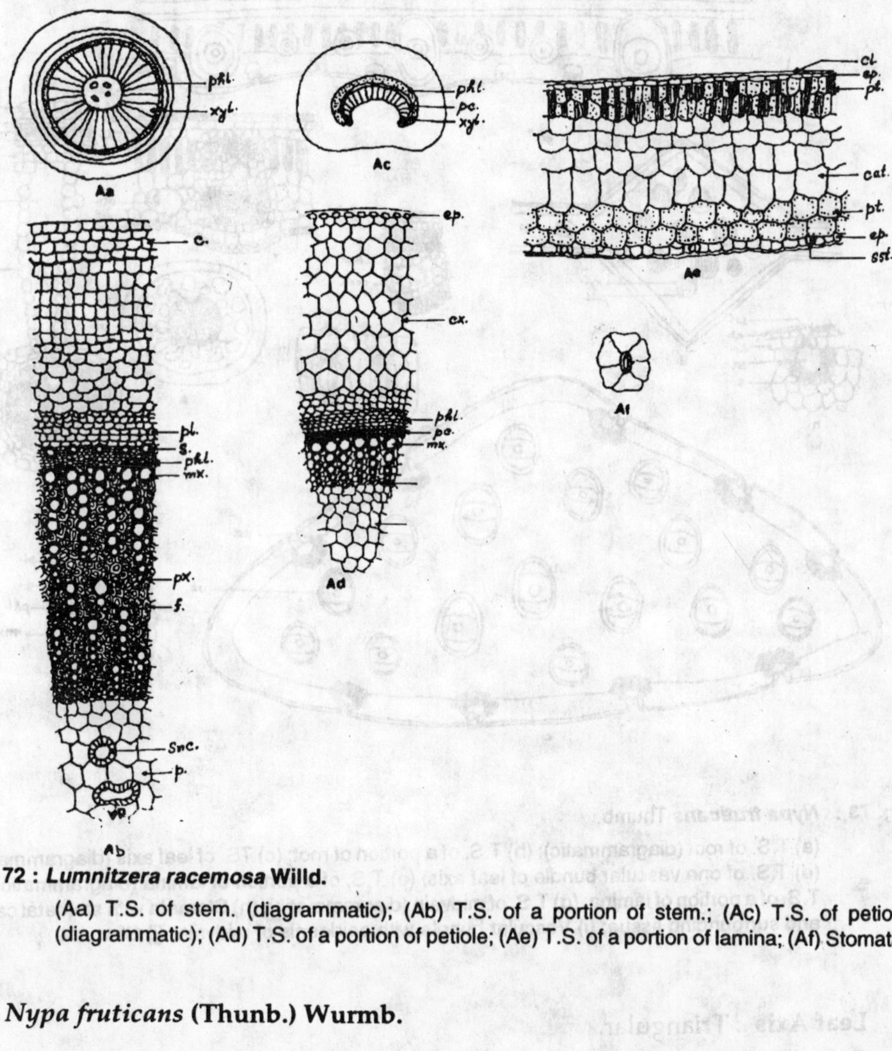

Fig. 72 : *Lumnitzera racemosa* Willd.

(Aa) T.S. of stem. (diagrammatic); (Ab) T.S. of a portion of stem.; (Ac) T.S. of petiole (diagrammatic); (Ad) T.S. of a portion of petiole; (Ae) T.S. of a portion of lamina; (Af) Stomata.

8.1. *Nypa fruticans* (Thunb.) Wurmb.

(Fig. 73)

Leaf : Dorsiventral.

Lamina : *Cuticle*–one layered thick, epidermis–one layered thick, cells thickly cutinized; *stomata–Pandanus* type, confined to the lower surface, deeply sunken, 30.4 ± 1 μ in length × 20 ± 1 μ in width. *Hypodermis*–2 layered thick on upper surface, *mesophyll*–palisade tissue-one layered thick, vertically placed, spongy tissue one layered thick, vertically placed, spongy tissue consisting of compacted parenchymatous tissue, *veins*–midvein rhombose shaped, several bundles gathering, surrounded by sclereids, bundles of various size arranged in linear row in mesophyll, each bundle surrounded by sclereids and each bundle surrounded by one layered thick pericycle.

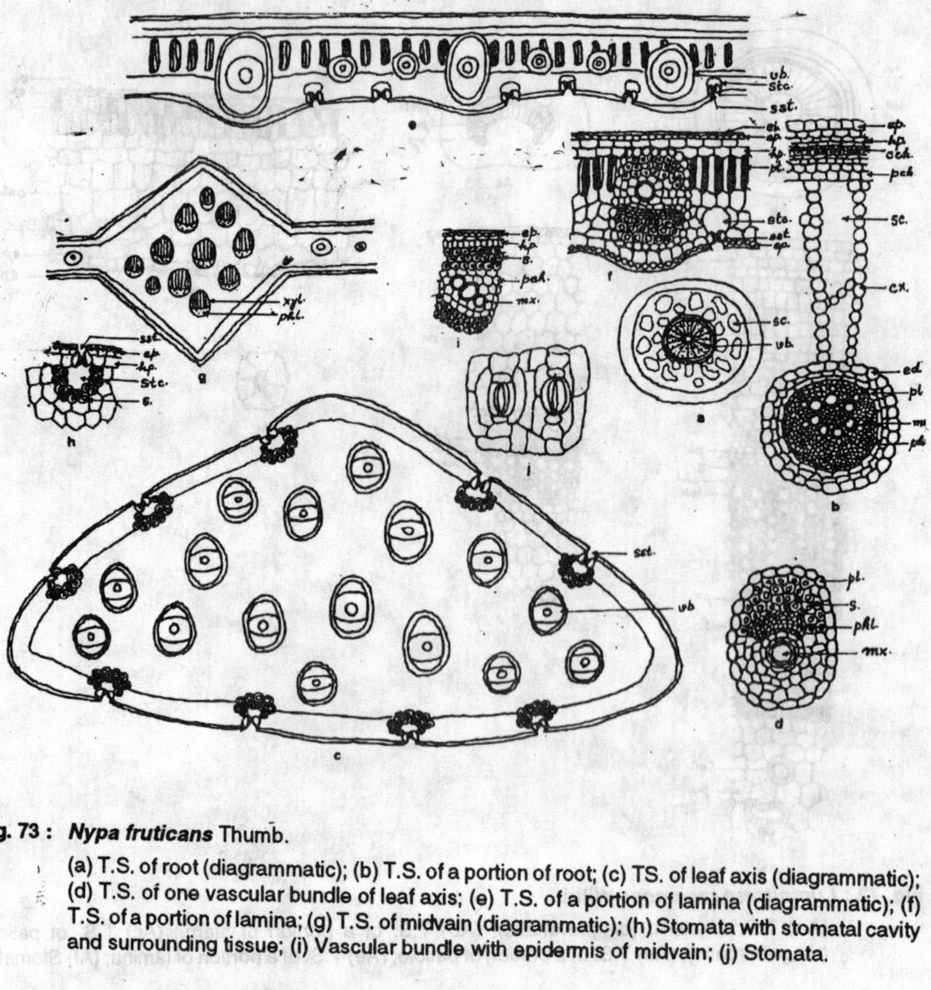

Fig. 73 : ***Nypa fruticans*** Thumb.

(a) T.S. of root (diagrammatic); (b) T.S. of a portion of root; (c) TS. of leaf axis (diagrammatic); (d) T.S. of one vascular bundle of leaf axis; (e) T.S. of a portion of lamina (diagrammatic); (f) T.S. of a portion of lamina; (g) T.S. of midvain (diagrammatic); (h) Stomata with stomatal cavity and surrounding tissue; (i) Vascular bundle with epidermis of midvain; (j) Stomata.

Leaf Axis : Triangular.

Epidermis–one layered thick, cutinized, *stomata* deeply sunken, 30.4 ± 1 μ in length × 20 ± 1 μ in width, sub-stomatal cavity surrounded by sclereids, *hypodermis*-one layered thick, cortex-consisting of compacted parenchymatous tissue, vascular bundle-numerous bundles scattered in cortex, each bundle surrounded by one layered thick pericycle layer, *vascular bundles*–closed collateral, traversed by sclereids.

Root : Underground : *epidermis*–one layered thick, *hypodermis* multi-layered thick parenchymatous tissue including one layered thick collenchymatous tissue, *cortex*-large vertical schizogenous cavities formed by isometric cells, *endodermis*–one layered thick, *vascular bundles*-radial, condensed parenchymatous tissue, thickly cutinized, xylem-exarch, *pith*-cutinized parenchymatous tissue.

GENUS–9. *PHOENIX* Roxb

Epidermal layer thickly cutinized, *stomata* mostly sunken, *mesophyll* having idioblasts, bundles surrounded by sclereids.

9.1. *Phoenix paludosa* Roxb.

(Fig. 74)

Leaf : Isobilateral.

Lamina : *Cuticle*–thin, *eipdermis*-one layered thick in both surfaces, cells rectangular. *Stomata*-*Pandanus* type, confined to both the surfaces, deeply sunken, 20.2 ± 1 μ in length × 12 ± 0.5 μ in width, *hypodermis*-one layered thick in both surfaces, *mesophyll*-palisade tissue chloren-chymatous, extended to both the surfaces, sclereids patches below the upper hypodermal layer, sub-stomatal chamber surrounded by fibres in lower hypodermal layer, veins–embedded in the mesophyll tissue, one large bundles situated after few bundles alternately, each bundles surrounded by sclereids.

Aerial Root : *Pneumatothode*

Epidermis–one layered thick, thickly cutinized, *hypodermis*-multi-layered thick, compacted parenchymatous tissue, *cortex*-consisting of large schizogenous cavities, each cavity surrounded by isodiametric cells, each cavities become enlarged in mature root, each cell having thickly well deposition as fungal hyphal association, each inter-cavities portion strengthened by sclereids patches, *endodermis*–one layered thick; *stele*-circular strand, *pericycle*-one layered thick, vascular bundle-radial, xylem-exarch traversed by sclereids, phloem associated with sclereids, pith-compacted parenchymatous cells.

FAMILY - VI. *STERCULIACEAE*

Hairs : stellate hairs in lower surface, *stomata* generally in lower surface, petiole showing cylindric stele, *cork* arising from outer surface of cortex, crystals are scattered.

GENUS - 10. *HERITIERA* Aiton

Stellate hairs in lower side make the lamina tufted leathery form; *stomata* confined to the lower side, diacytic, *petiole* showing cylindric stele, cork arising outer surface of the cortex, crystals are scattered, mucilagenous cavities scattered in petiole and in stem.

10.1. *Heritiera fomes* Buch. -Ham.

(Plate - 154, Fig. 75)

Leaf : Dorsiventral, petiolate.

Lamina : *Stellate* hairs in lower side make the lamina tufted leathery form; cuticle - one layered thick; *epidermis* - one layered thick in both surfaces, lower epidermal cells cutinized; *stomata* - diacytic, deeply sunken, confined to the lower surfaces, 12.6 ± 0.5 μ in length × 14.5 ± 0.5 μ in width, *hypodermis*-one layered thick in both the surfaces; *mesophyll* - palisade tissue-one layered thick, spongy tissue-loose

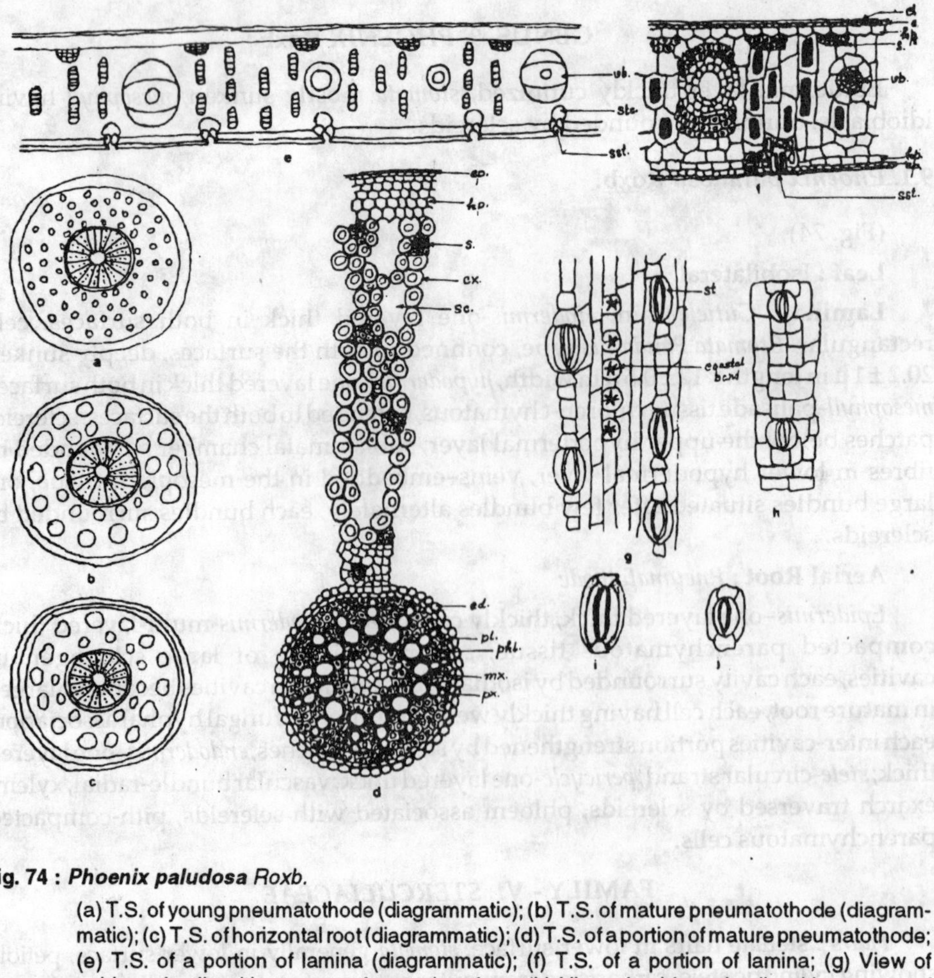

Fig. 74 : *Phoenix paludosa* Roxb.

(a) T.S. of young pneumatothode (diagrammatic); (b) T.S. of mature pneumatothode (diagrammatic); (c) T.S. of horizontal root (diagrammatic); (d) T.S. of a portion of mature pneumatothode; (e) T.S. of a portion of lamina (diagrammatic); (f) T.S. of a portion of lamina; (g) View of epidermal cells with stomata of ventral surface; (h) View of epidermal cells with stomata of dorsal surface; (i) Stomata of ventral surface; (j) Stomata of dorsal surface.

parenchymatous with large cavities, compartmented by septa which consisting of sclerenchymatous cells and parenchymatous cells, *veins* - embedded in the mesophyll tissue surrounded by fibres, *petiole* - infrequent *stellate* hairs on peripheral surface; *epidermis* - one layered thick, *cortex*-multi-layered thick, compacted parenchymatous tissue, few sphaeraphides scattered in cortical cell; *endodermis* - one layered thick; *stele* - transverse section showing circular cylinder and accessory bundle in pith. *Pericycle* - multi-layered thick, sclerenchymatous tissue in ring, vascular bundles - closed collateral, phloem associated with sclereids and parenchyma ; xylem - mesarch traversed by parenchyma, *pith* - parenchymatous tissue having large mucilagenous cavities and accessory vascular bundles - closed collateral, phloem traversed by sclerenclymatous cells, while xylem intermingled with parenchyma.

Axis : Epidermal cell replaced by multi-layered thick well developed cork, *cortex*-multi-layered thick, .compacted, parenchymatous tissue having circular

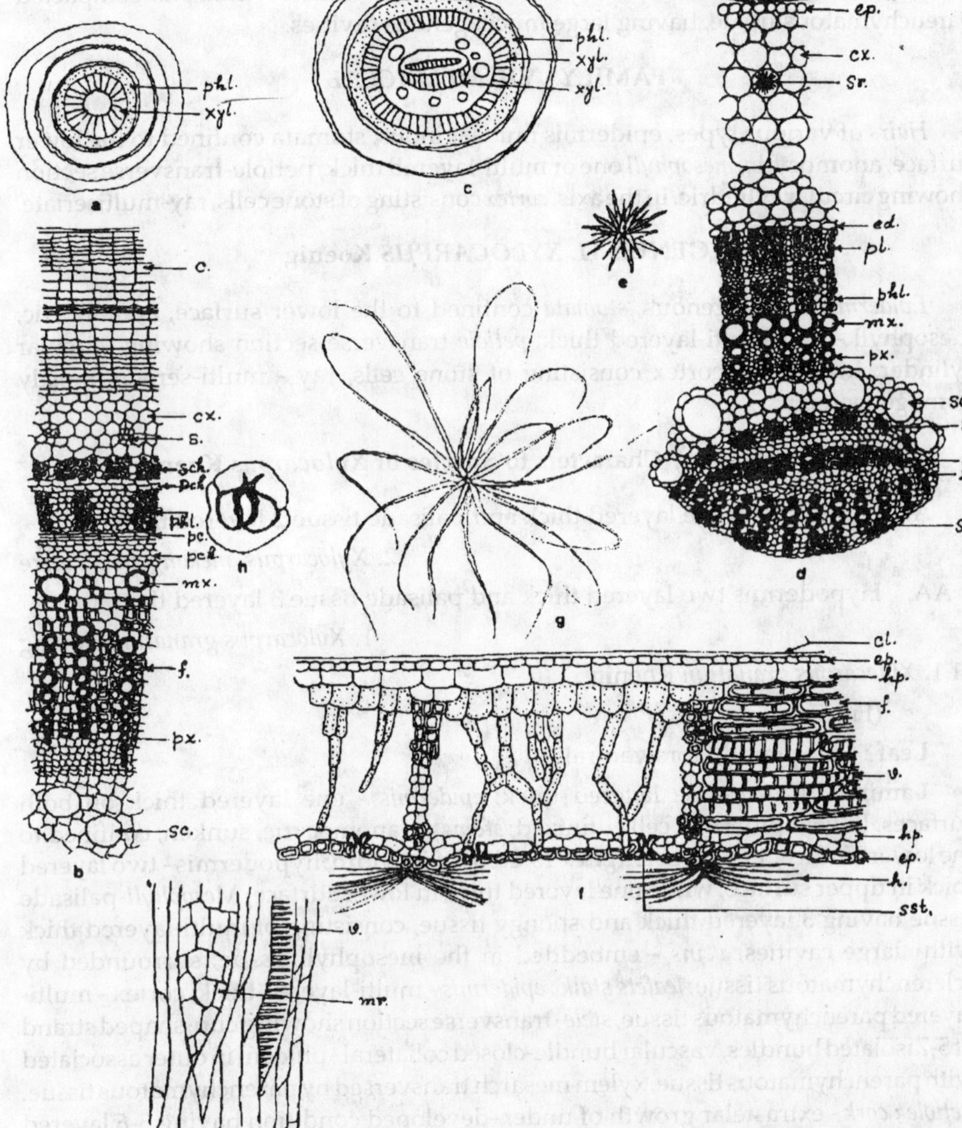

Fig. 75 : *Heritiera fomes* Buch. Ham.

(a) T.S. of stem (diagrammatic); (b) T.S. of a portion of stem.; (c) T.S. of petiole (diagrammatic); (d) T.S. of a portion of petiole; (e) Sphaeraphide; (f) T.S. of lamina; (g) Stellate hair; (h) Stomata; (i) L.S. of a portion of stem.

sclerenchyma patch. *Stele* - transverse section showing circular cylinder, *pericycle* - multi-layered thick sclerenchymatous patches interrupted by parenchyma cells, *vascular strand* open collateral, phloem-traversed by sclerenchyma bands and parenchymatous tissue, cambial ring in middle, xylem / wood-vessels medium sized, diffuse porous, scalariform perforation, vessels traversed by fibres and parenchymatous tissue, *parenchyma*- apotracheal, ray-multi-seriate, 3-6 celled width, *pith*-compacted parenchymatous tissue, having large mucilagenous cavities.

FAMILY - VII. *MELIACEAE*

Hairs of various types, epidermis mucilaginous, stomata confined to the lower surface, anomocytic, *mesophyll* one or multi-layered thick; petiole-transverse section showing circular cylindric. In the axis, *cortex* consisting of stone cells, ray-multiseriate.

GENUS-11. *XYLOCARPUS* Koenig

Epidermis mucilagenous, *stomata* confined to the lower surface, anomocytic, mesophyll - 1 or multi-layered thick, *petiole*-transverse section showing circular cylinder. In the axis-cortex consisting of stone cells, ray - multi-seriate, mostly homogeneous.

Anatomical key Characters to Species of *Xylocarpus* Koenig

A. Hypodermis one layered thick and palisade tissue 2 layered thick
2. *Xylocarpus mekongensis* Pierre

AA. Hypodermis two layered thick and palisade tissue 3 layered thick
1. *Xylocarpus granatum* Koenig

11.1. *Xylocarpus granatum* Koenig
(Fig. 76)

Leaf : Compound, dorsiventral.

Lamina: *Cuticle*–one layered thick; *epidermis* - one layered thick on both surfaces, lower epidermal cell cutinized; *stomata* - anomocytic, sunken, confined to the lower surface, 19 ± 1 μ in length x 15.2 ± 0.5 μ in width; hypodermis - two layered thick in upper surface, while one layered thick in lower surface; *Mesophyll*-palisade tissue having 3 layered thick and spongy tissue, consisting of multi-layered thick with large cavities, *veins* - embedded in the mesophyll tissue, surrounded by sclerenchymatous tissue; *leafets stalk : epidermis* - multi-layered thick, cortex - multi-layered parenchymatous tissue, *stele*- transverse section showing cup-shaped strand of 5-7 isolated bundles, vascular bundle-closed collateral - phloem in outer associated with parenchymatous tissue, xylem mesarch transversed by parenchymatous tissue. *Petiole* : *cork* - extrastelar growth of under–developed condition having 3-6 layered thick. *Cortex* - multi-layered thick parenchymatous tissue and patches of sclerenchymatous tissue in ring, *endodermis* - one layered thick, *stele - transverse* section showing circular cylinder, *vascular bundle* - closed collateral, phloem - traversed by parenchymatous tissue; xylem/wood-vessels medium sized, diffuse porous, traversed by sclerenchymatous tissue, parenchyma-paratracheal, *pith* - compacted parenchymatous tissue.

Axis : *Epidermis* - one layered thick in immature axis, epidermal layer replaced by cork in mature one, *cork* - multi-layered thick, *cortex* - multi-layered thick parenchymatous tissue and multi-layered thick dense sclerenchymatous ring below

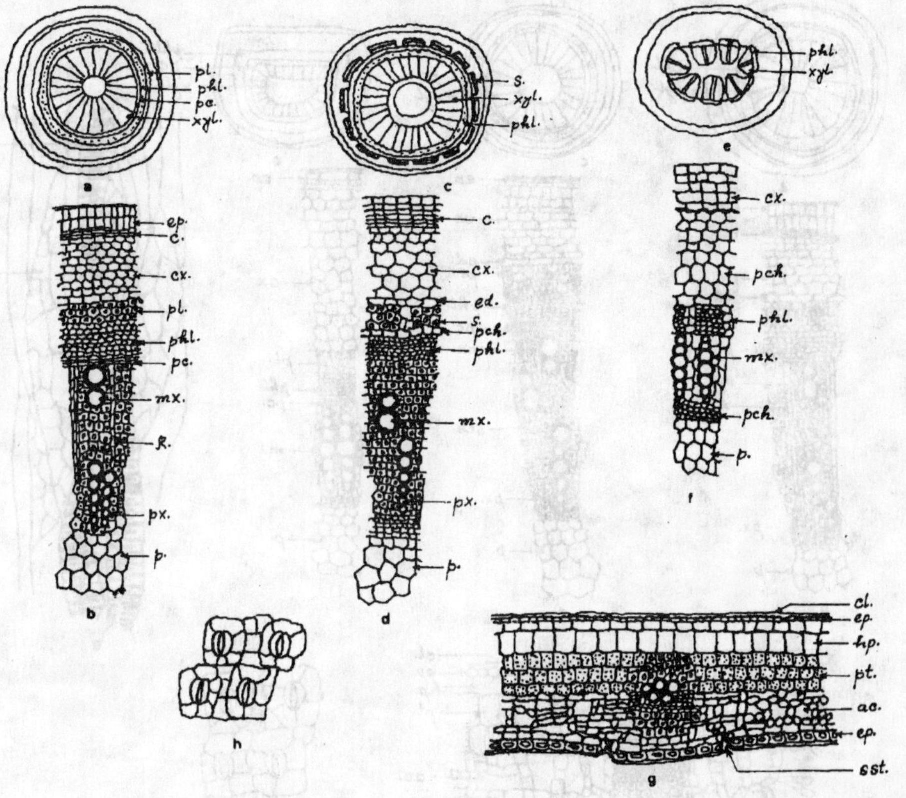

Fig. 76 : ***Xylocarpus granatum*** **Koen.**

(a) T.S. of stem. (diagrammatic); (b) T.S. of a portion of stem.; (c) T.S. of leaf axis (diagrammatic); (d) T.S. of a portion of leaf axis; (e) T.S. of stalk of leaflet (diagrammatic); (f) T.S. of a portion of leaf-let stalk; (g) T.S. of leaflet; (h) Stomata.

it, *stele*-transverse section showing circular cylindric, *vascular strand*–open collateral, phloem traversed by parenchymatous tissue, cambial ring in middle, xylem/ wood-vessels medium sized, diffuse porous, simple perforation, parenchyma - paratracheal, ray - multi-seriate, 2-4 celled width, homogeneous, *pith*-compacted parenchymatous tissues.

11.2. *Xylocarpus mekongensis* Pierre

(Plate - 155 and Fig. 77)

Leaf : Compound, dorsiventral.

Lamina: *Cuticle*–one layered thick, epidermis - one layered thick in both surfaces, lower epidermal cell thickly cutinized; *stomata* - anomocytic, sunken, confined to the lower surface, 16.4 ± 1 μ in length x 13.3 ± 0.5 μ in width; *hypodermis* - one layered thick below the upper epidermis and lower epidermis; *mesophyll* - palisade tissue consisting of two layered thick and spongy tissue having multi-layered thick with large air cavities *veins* - embedded in the mesophyll tissue and

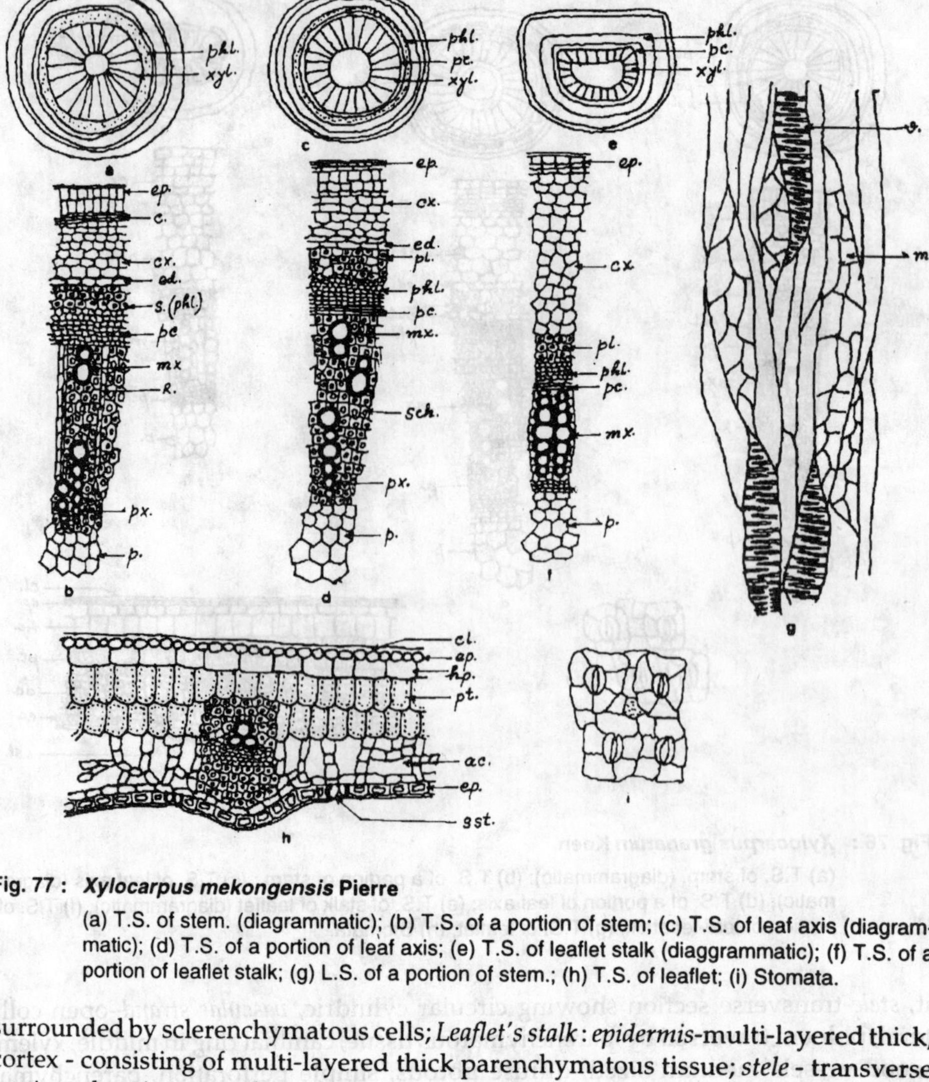

Fig. 77 : *Xylocarpus mekongensis* Pierre

(a) T.S. of stem. (diagrammatic); (b) T.S. of a portion of stem; (c) T.S. of leaf axis (diagrammatic); (d) T.S. of a portion of leaf axis; (e) T.S. of leaflet stalk (diaggrammatic); (f) T.S. of a portion of leaflet stalk; (g) L.S. of a portion of stem.; (h) T.S. of leaflet; (i) Stomata.

surrounded by sclerenchymatous cells; *Leaflet's stalk* : *epidermis*-multi-layered thick, cortex - consisting of multi-layered thick parenchymatous tissue; *stele* - transverse section showing cup-shaped strand, *pericycle* - consisting of sclerenchymatous tissue, *vascular bundle* - open collateral - phloem in outer traversed by parenchymatous tissue, cambial ring in middle, xylem in mesarch. *Petiole -epidermis* - two layered thick, *hypodermis* - 2 layered thick, vertically divided, *cortex* - multi-layered thick, compacted parenchymatous tissue; *stele* - transverse section showing circular cylinder, *pericycle* - multi-layered thick of sclerenchymatous tissue, *vascular bundle*- open collateral, phloem in outer associated with parenchymatous tissue, cambial ring in middle; xylem / wood-vessels medium sized, diffuse porous,traverse by sclereids, parenchyma-paratracheal, *pith* - compacted parenchymatous tissue.

Axis : *Epidermis* - one layered thick in immature axis, but epidermal layer replaced by cork in mature one. *Cork* - multi-layered thick, *cortex*-consisting of multi-layered thick parenchymatous tissue and multi-layered thick dense sclerenchymatous ring below it; *stele*-transverse section showing circular cylindric, *vascular bundle*-open collateral-phloem associated with parenchymatous tissue, cambial ring in middle; xylem/wood-vessels-medium sized, diffuse porous, simple perforation, *parenchyma* - paratracheal, ray - multi-seriate, 2-4 celled wide, homogeneous, *pith*-compacted parenchymatous tissue.

FAMILY - VIII. *EUPHORBIACEAE*

Hairs of different types, *epidermis* - consisting of one layered thick in both surfaces, *hypodermis* - consisting of single layered, *mesophyll*-consisting of multi-layered thick palisade tissue and spongy tissue, *veins*-traversed by sclerenchymatous tissue, ray-multiseriate, consisting of 2-4 cells wide.

GENUS - 12. *EXCOECARIA* L.

Two glands on the base of lamina, *epidermis* - consisting of one layered thick in both surfaces; *hypodermis* - one layered thick, mesophyll - palisade tissue consisting of multi-layered thick, *veins*-encircled by sclerenchymatous tissue, ray-multiseriate, consisting of 2-3 cells wide.

12.1. *Excoecaria agallocha* L.

(Fig. 78)

Leaf : Dorsiventral, petiolate.

Lamina : Two glands on the base of lamina, *cuticle* - one layered thick, *epidermis*-one layered thick on both surfaces; *hypodermis* - one layered thick; *stomata*-paracytic, confined to the lower surface, not sunken, 31 ± 1 μ in length x 13.3 ± 1 μ in width, mesophyll- palisade tissue-2 layered thick, upper layer having large cell, while lower layer having small cell, spongy tissue having multi-layered thick with large cavities, *veins* - embedded in the mesophyll tissue, traversed by sclerenchyma. *Petiole : epidermis*-one layered thick, *cortex* - consisting of multi-layered thick loosely arranged parenchymatous tissue; *endodermis* - one layered thick; *stele* - transverse section showing crescent shaped strand of 3 bundles; *pericycle* - one layered thick, vascular bundles - closed collateral, phloem associated with sclerenchyma parenchyma xylem mesearch in nature.

Axis : Epidermis-one layered thick in young axis, while epidermal layer replaced by cork, cork-multi-layered thick, *cortex*-consisting of multi-layered thick parenchymatous tissue, *stele*-transverse section showing circular cylinder, phloem-traversed by parenchyma layer, xylem/wood-vessels medium sized, diffuse porous, scalariform perforation, traversed by fibres, *parenchyma* - paratracheal, ray-multi-seriate, 2-3 celled width, homogeneous, *pith*-consisting of parenchymatous tissue.

Root : *Cork* - consisting of multi-layered thick, *cortex*-condensed by extrastelar and intrastelar growth, though cortex zone consisting of parenchymatous tissue with infrequent sclereids, *stele*-circular cylinder, vascular bundles-closed collateral, phloem-associated with sclerenchyma and parenchymatous tissue, xylem/wood-

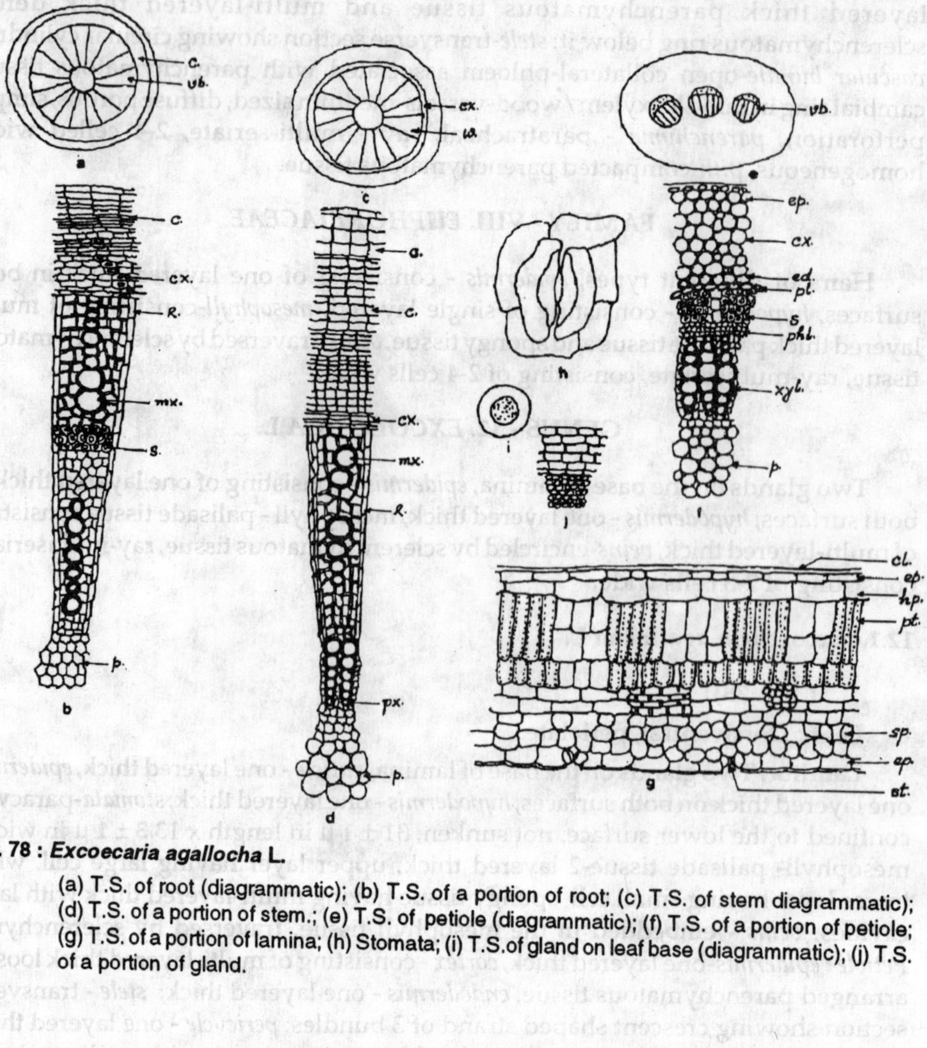

Fig. 78 : *Excoecaria agallocha* L.

(a) T.S. of root (diagrammatic); (b) T.S. of a portion of root; (c) T.S. of stem diagrammatic); (d) T.S. of a portion of stem.; (e) T.S. of petiole (diagrammatic); (f) T.S. of a portion of petiole; (g) T.S. of a portion of lamina; (h) Stomata; (i) T.S.of gland on leaf base (diagrammatic); (j) T.S. of a portion of gland.

medium sized, traversed by fibres and few sclereids, *parenchyma*-apotracheal, ray-uniseriate-homogeneous, *pith*-consisting of parenchymatous tissue.

FAMILY - IX. *MYRSINACEAE*

Hairs are glandular form found in few species, *mesophyll* having sclerenchymatous cells, *cork* arising superficially.

GENUS - 13. *AEGICERAS* Gaertner

Multi-cellular gland occur frequently; Mesophyll having sclerenchymatous cells, *cork*-arising superficially, ray—multiseriate, vessels-simple, perforation, spiral thickening.

13.1. ***Aegiceras corniculatum*** **(L.) Blanco**

(Plate - 156, Fig. 79)

Leaf : Dorsiventral, petiolate.

Lamina: *Cuticle*-one layered thick, *gland* : 14-16 celled gland surrounded by one celled layer jacket in ventral side, while numerous cell gland surrounded by one cells layer jacket in dorsal surface, epidermis -one layered thick in both surfaces, *Stomata*-anomocytic, 27.2 ± 0.5 μ in length x 15.8 μ in width, sunken, confined to the lower surface, hypodermis - 2 layered thick below the upper epidermis, mesophyll - palisade tissue 2 layered thick having small or large various shaped solitary sclereids, spongy tissue - loose parenchymatous tissue having circular, yellow or reddish brown colour cavities, small sclereides in groups, scattered in the spongy tissue, veins-embedded in the mesophyll tissue traversed by fibres; *Petiole* : cuticle-thick layered, epidermis - one layered thick, cortex - multi-layered thick, compacted parenchymatous tissue showing circular yellow or reddish brown cavities and brachysclereides, endodermis - one layered thick, stele-transverse section showing circular or semi-circular strand consisting of 5-6 bundles, pericycle, multi-layered thick sclerenchymatous tissue and multi-layered thick parenchymatous tissue in below, vascular bundle-closed collateral-phloem traversed by parenchyma, xylem- mesearch traversed by parenchymatous tissue, pith-consisting of sclerenchymatous cells.

Axis : *Epidermis*-replaced by cork, *cork* under developed, arising suprficially from the epidermis, *cortex*-multi-layered thick, compacted parenchymatous tissue, few solitary sclereids scattered in cortex, *stele*-transverse section showing circular cylinder, pericycle-multi-layered thick-sclerenchymatous tissue in ring, vascular *bundles*-open collateral, phloem associated with parenchyma tissue, cambial ring in middle, xylem/wood-vessels small to medium, diffuse porous, spiral thickening, simple perforation, *parenchyma*-paratracheal and apotracheal, fibres associated with vessels, ray-uniseriate, homogeneous, procumbent cells mostly, *pith*-consisting of parenchymatous tissue - few circular yellow or reddish brown cavities scattered in pith, besides these cavities, solitary sclereids and few secretory cells also occurred in pith.

Root : *Epidermis* replaced by cork, *cork*-multi-layered thick, rectangular cells placed horizontally, intrastelar growth occupied rest portion, ray-uniseriate, vertically placed, vessels-intermingled by parenchymatous tissue.

FAMILY — X. *RUBIACEAE*

Hairs of various types, *stomata* generally paracytic, mostly in the lower surface, petiolar stele in different form, *cork* arising from the cortex, ray-multi-seriate, fibres simple pitted.

Anatomical Key Characters to Genera of Rubiaceae

A. Leaf fleshy but stomata confined to the lower surface only

1. *Scyphiphora* Gaertn. f.

AA. Leaf isobilateral, stomata confined to the both surfaces

2. *Hydrophylax* L.

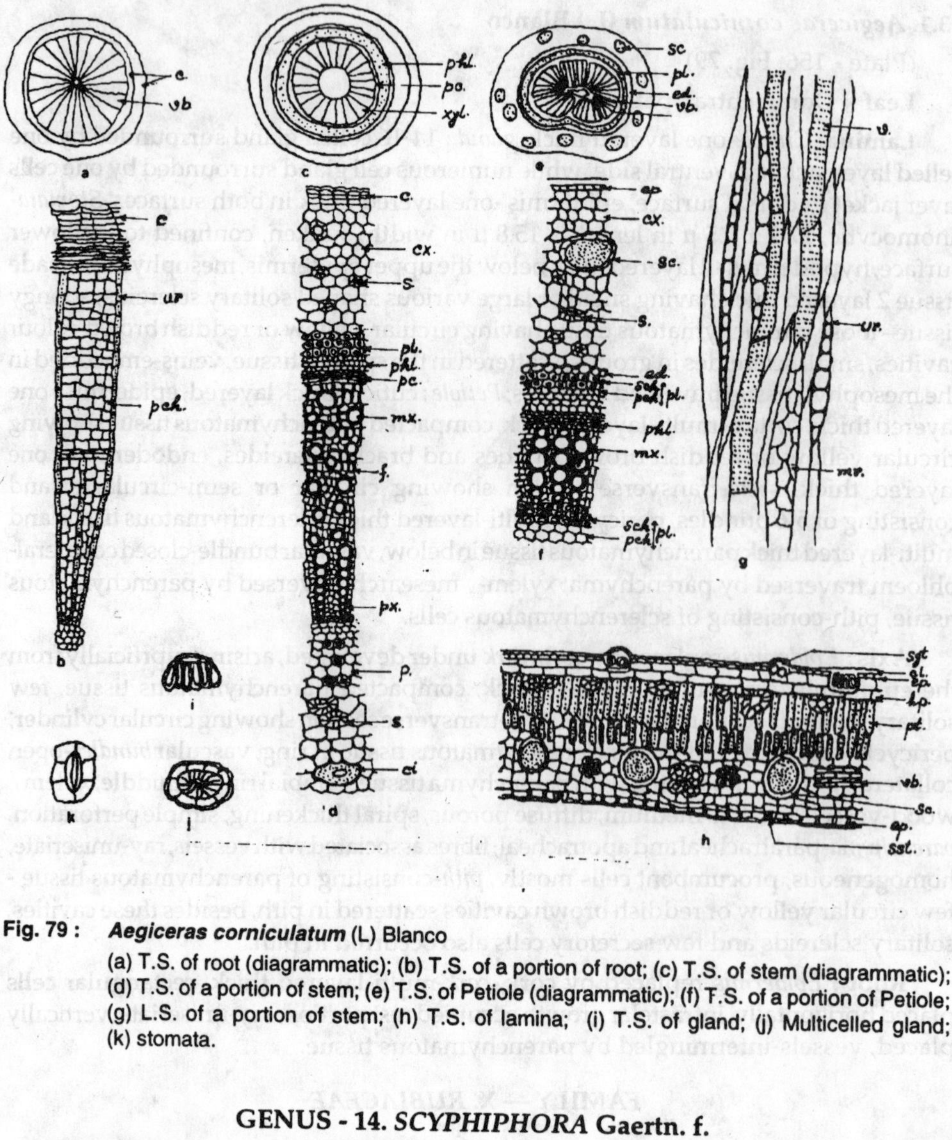

Fig. 79 : ***Aegiceras corniculatum*** **(L) Blanco**

(a) T.S. of root (diagrammatic); (b) T.S. of a portion of root; (c) T.S. of stem (diagrammatic); (d) T.S. of a portion of stem; (e) T.S. of Petiole (diagrammatic); (f) T.S. of a portion of Petiole; (g) L.S. of a portion of stem; (h) T.S. of lamina; (i) T.S. of gland; (j) Multicelled gland; (k) stomata.

GENUS - 14. *SCYPHIPHORA* Gaertn. f.

Stomata confined to the lower surfaces, paracytic, petiolar stele mostly in semi-lunar or crescent shaped, cork arising from the cortex, ray-multiseriate, fibres simple pitted.

14.1. *Scyphiphora hydrophyllacea* Gaertn. f.

(Fig. 80)

Leaf : Dorsiventral, petiolate.

Lamina: Cuticle–one layered thick; *epidermis* - one layered thick on both surfaces, upper epidermal layer thickly cuticularized *stomata* - paracytic, confined to the

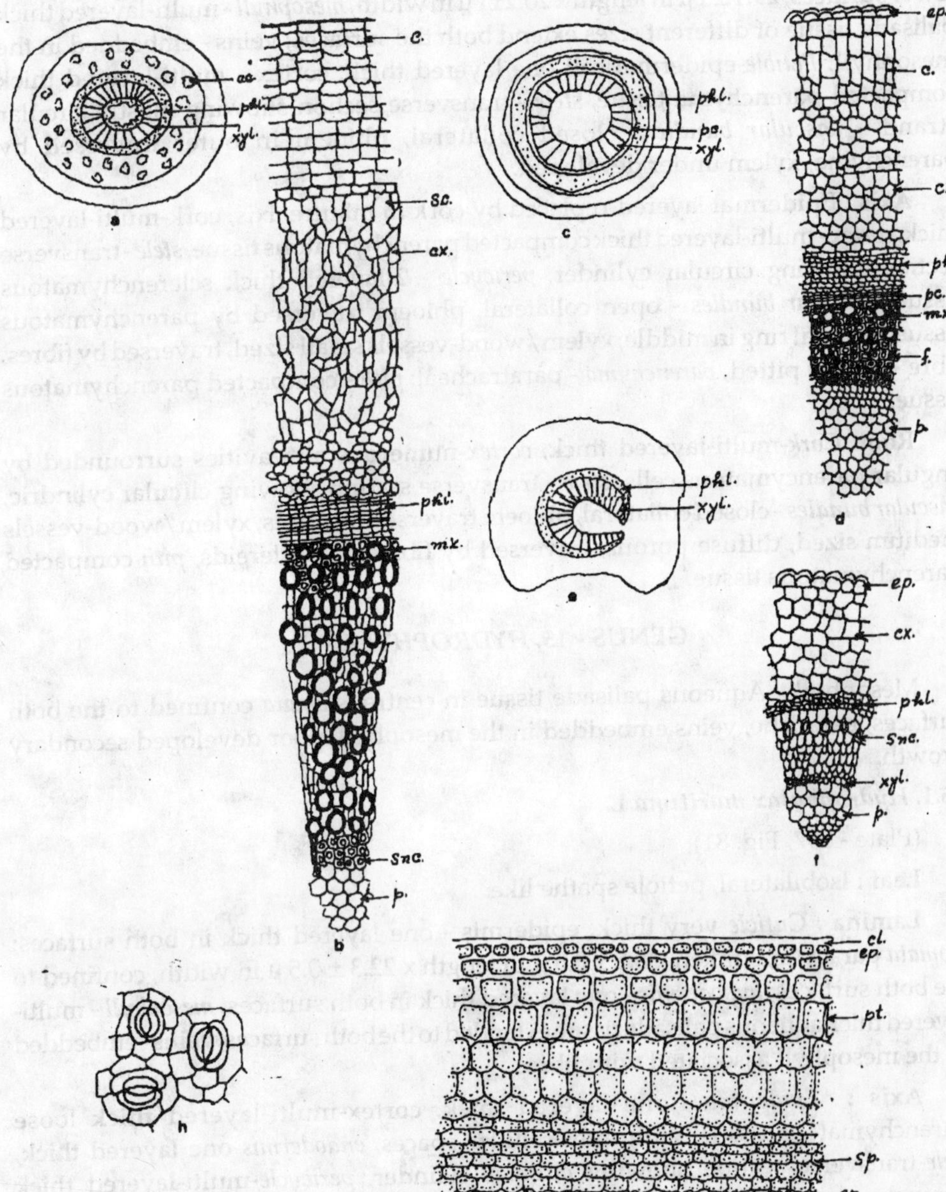

Fig. 80 : *Scyphiphora hydrophyllacea* Gaertn f.

(a) T.S. of root (diagrammatic); (b) T.S. of a portion of root; (c) T.S. of a stem. (diagrammatic); (d) T.S. of a portion of stem; (e) T.S. of petiole (diagrammatic); (f) T.S. of a portion of petiole; (g) T.S. of a portion of lamina;(h) Stomata.

lower surfaces, 29.1 ± 1 μ in length x 20.2±1 μ in width, *mesophyll* - multi-layered thick palisade tissue of different sizes extend both the surfaces; veins - embedded in the mesophyll; *Petiole*-epidermal cell one layered thick, cortex - multi-layered thick compacted parenchyma tissue, *stele* - transverse section showing mostly circular strand - *vascular bundle* - closed collateral, phloem in outer traversed by parenchyma, xylem under developed.

Axis : Epidermal layered replaced by cork in mature axis, cork-multi-layered thick, *cortex* - multi-layered thick compacted parenchymatous tissue, *stele* - transverse section showing circular cylinder, *pericycle* - 2-layered thick sclerenchymatous tissue, *vascular bundles* - open collateral, phloem traversed by parenchymatous tissue, cambial ring in middle; xylem / wood-vessels small sized, traversed by fibres, fibre - simple pitted, *parenchyma* - paratracheal; *pith* - compacted parenchymatous tissue.

Root: *Cork*-multi-layered thick, *cortex*-numerous air cavities surrounded by angular parencymatous cells, *stele* - transverse section showing circular cylindric, *vascular bundles* - closed collateral, phloem traversed by fibres, xylem / wood-vessels medium sized, diffuse porous, traversed by fibres and sclereids, *pith* compacted parenchymatous tissue.

GENUS - 15. *HYDROPHYLAX* L.

Mesophyll - Aqueous palisade tissue in centre, *stomata* confined to the both surfaces, paracytic, veins embedded in the mesophyll, poor developed secondary growth.

15.1. *Hydrophylax maritima* L.

(Plate - 157, Fig. 81)

Leaf : Isobilateral, petiole spathe like.

Lamina : *Cuticle* very thick, epidermis - one layered thick in both surfaces; *stomata* paracytic, sunken, 36.5 ± 0.5 μ in length x 22.3 ± 0.5 μ in width, confined to the both surfaces; *hypodermis* - one layered thick in both surfaces, *mesophyll* - multi-layered thick aqueous palisade tissue extended to the both surfaces, veins - embedded in the mesophyll associated with fibres.

Axis : *Epidermis* - one layered thick, cortex-multi-layered thick loose parenchymatous tissue, with inter-cellular spaces, *endodermis*-one layered thick, *stele*-transverse section showing circular cylinder, *pericycle*-multi-layered thick parenchyma tissue in ring; *Vascular bundle* - closed collateral, phloem traversed by fibres, xylem - small vessels, associated with fibres; *pith* - multi-layered thick compacted parenchymatous tissue.

FAMILY - XI. *PLUMBAGINACEAE*

Glands frequently occurred, few secretory cells found in this family, presence of medullary rays, parenchyma paratracheal, fibres - simple pitted.

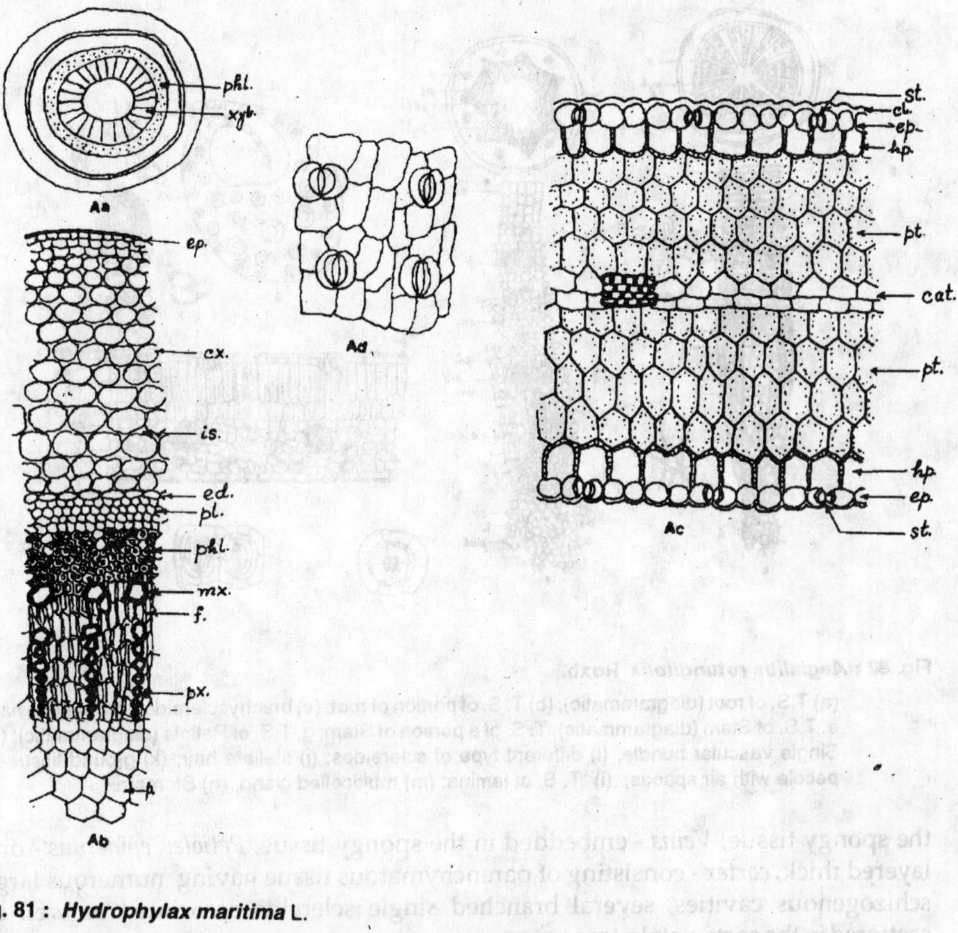

Fig. 81 : ***Hydrophylax maritima*** **L.**

(Aa) T. S. of stem (diagrammatic); (Ab) T. S. of a portion of stem; (Ac) T. s. of a portion of lamina; (Ad) Stomata.

GENUS - 16. *AEGIALITIS* R. Brown

Glands occurred frequently, anomalous secondary growth in cortex, *parenchyma* - paratracheal, fibres - simple pitted.

16.1. *Aegialitis rotundifolia* Roxb.

(Plate -158 and Fig. 82)

Leaf : Dorsiventral, petiolate.

Lamina : *Cuticle* - one layered thick, *glands* - 4/8 celled structured surrounded by one celled layered jacket in both the surfaces; *epidermis* - one layered thick in the both surfaces; *stomata*-anomocytic, sunken, confined to the lower surfaces, 28.5 ± 0.5 µ in length x 17.4 ± 0.5 µ in width, *mesophyll* - 2 layered thick palisade tissue rod shaped sclereids within the palisade tissue, spongy tissue consist of loose parenchymatous tissue, frequent branched sclereids and unbranched circular sclereids scattered in

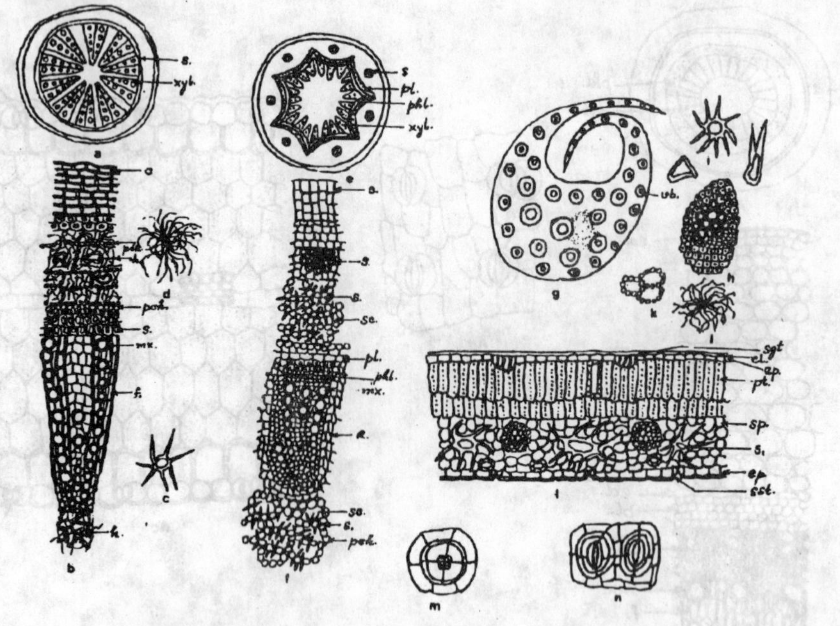

Fig. 82 : ***Aegialitis rotundifolia*** **Roxb.**

(a) T.S. of root (diagrammatic); (b) T. S. of portion of root; (c) brachysclereides; (d) stellate hair, e. T.S. of Stem (diagrammatic); T. S. of a portion of Stem; g. T.S. of Petiole (diagrammatic); (h) Single vascular bundle; (i) different type of sclereides; (j) stellate hair; (k) ground tissue of petiole with air spaces; .(l) T. S. of lamina; (m) multicelled gland; (n) Stomata.

the spongy tissue; *Veins* - embedded in the spongy tissue; *Petiole: epidermis* - one layered thick, *cortex* - consisting of parenchymatous tissue having numerous large schizogenous cavities, several branched single sclereids associated as well as scattered in the cortex, stele-transverse section showing numerous scattered vascular bundles in the cortex; *vascular bundle* - closed collateral, surrounded by sclerenchymatous tissue, phloem traversed by parenchymatous tissue, xylem mesarch.

Axis : *Epidermis* - one layered thick in the young axis, while epidermal layered replaced by cork - multi-layered thick; *cortex* - multi-layered thick parenchymatous tissue having numerous schizogenous cavities, branched star shaped sclereids scattered in the cortex, anomalous secondary thickening appeared as patch in cortex; *stele* - transverse section showing star shaped circular cylindric, *pericycle* - multi-layered thick sclerenchymatous tissue, *vascular strand*-closed collateral, phloem associated with parenchymatous tissue, xylem/wood-vessels small, semi-ring porous, traversed by parenchyma, parenchyma-paratracheal, ray - 2-3 celled width, *pith* - consisting of parenchymatous tissues including scattered star shaped single sclereids and schizogenous cavities.

Root: Epidermal layer replaced by multi-layered cork in mature axis, below the cork layer - multi-layered thick sclerenchymatous tissue, *cortex*-consisting of numerous large schizogenous cavities, which associated with dense sclerenchymatous hairs and circular unbranched sclereids, *stele* - transverse section showing circular cylinder of multi-seriate row bundles; *vessels* - medium sized, traversed by fibres in row, phloem associated with parenchymatous tissue within the two vessels row; *pith* - numerous schizogenous cavities with dense aggregating net like sclerenchymatous hairs.

FAMILY - XII. *PTERIDACEAE*

Stomata in ventral side only, xylem mesarch, bundles scattered in the cortex in case of frond.

GENUS - 17. *ACROSTICHUM* L.

Stomata confined to the ventral surface; *hypodermis* - one layered thick only in outer surfaces, bundles scattered in cortex, xylem mesarch.

17.1 *Acrostichum aureum* L.

(Plate -159, Fig. 83)

Frond axis : Epidermis - one layered thick, stomata -confined to the lower surfaces, 32.3±1 μ in length x 26.6±1 μ in width, *hypodermis* - one layered thick below the upper surfaces, *mesophyll* - multi-layered thick palisade tissue and multi-layered thick spongy parenchymatous tissue, veins - embedded in the mesophyll.

Frond axis : Epidermis - one layered thick, *cortex* - multi-layered thick parenchymatous tissue, *stele* - numerous bundle scattered in the cortex, *endodermis* - one layered thick encircle bundle, *pericycle* - one layered thick placed below the endodermal layer, *vascular bundles* - centric, xylem mesarch surrounded by scanty phloem traversed by parenchyma.

FAMILY - XIII. *ACANTHACEAE*

Spines occurred frequently, presence of various type of cystoliths in leaf, anomalous secondary growth include intra-xylary or inter-xylary phloem, vessels small to medium, rays-uniseriate or multiseriate.

GENUS - 18. *ACANTHUS* L.

Spines occurred frequently, glandular hair recorded frequently, cystolith not found frequently, *stomata* - diacytic, anomalaus secondary growth occurred in underdeveloped condition, ray uni-seriate mostly, homogeneous.

Anatomical Key Characters to the Species of Acanthus L.

1. Petiolar stele showing semilunar shaped vascular strand and two accessoy strands in both wings. *1. ilicifolius* L.

2. Petiolar stele showing crescent shaped vascular vascular strand. *2. volubilis* Wall.

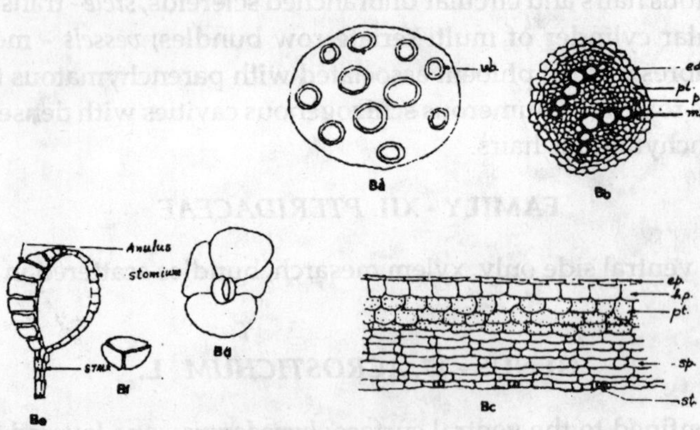

Fig. 83 : *Acrostichum anreum* L.

(Ba) T.S. of frond; (Bb) One vascular bundle of Frond; (Bc) T.S. of a portion of lamina; (Bd) Stomata; (Be) Sporangium; (Bf) Spore.

18.1. *Acanthus ilicifolius* L.

(Fig. 84)

Leaf : Dorsiventral, petiolate.

Lamina : *Cuticle*–one layered thick, *epidermis* - one celled thick both in upper epidermis and lower epidermis; lower epidermal cell thickly cutinized, *glands*–one celled structure surrounded by 5-6 jacket cells present in both the surfaces of the lamina, *stomata* - diacytic, not sunken, confined to the lower surface, 13.7 ± 1 μ in length x 10.4 ± 0.5 μ in width; *hypodermis* - two layered thick below the upper epidermis; *mesophyll* - palisade-two layered thick, spongy tissue having loose parenchymatous tissue with intercellular spaces, veins embedded in the mesophyll surrounded by sclerenchyma tissue; *Petiole - epidermis* - one layered thick, *cortex* - multi-layered thick loose parenchymatous tissue; endodermis - one layered thick; *stele* - transverse section showing semilunar strand and two accessory strands in both wings; *pericycle* - one layered thick, *vascular bundle* - closed collateral, phloem in outer and xylem in inner, mesarch.

Axis : *Epidermis* - two layered thick, *cortex* - multi-layered thick compacted parenchymatous tissue; stele - transverse section showing vascular strand; *pericycle* - two layered thick fibres, *vascular bundle* - open collateral, phloem in outer traversed by parenchyma and having some frequent sclereides patches, cambial ring middle, xylem - vessels small, traversed by fibres, indistinct intra-xylary phloem occurred, ray - uniseriate, homogeneous; *pith* - compacted parenchymatous tissue.

Aerial root (Stilt Root)

Root : Epidermal cell replaced by cork, cork multi-layered thick, *cortex*-parenchymatous tissue having numerous large schizogenous cavities, *endodermis*

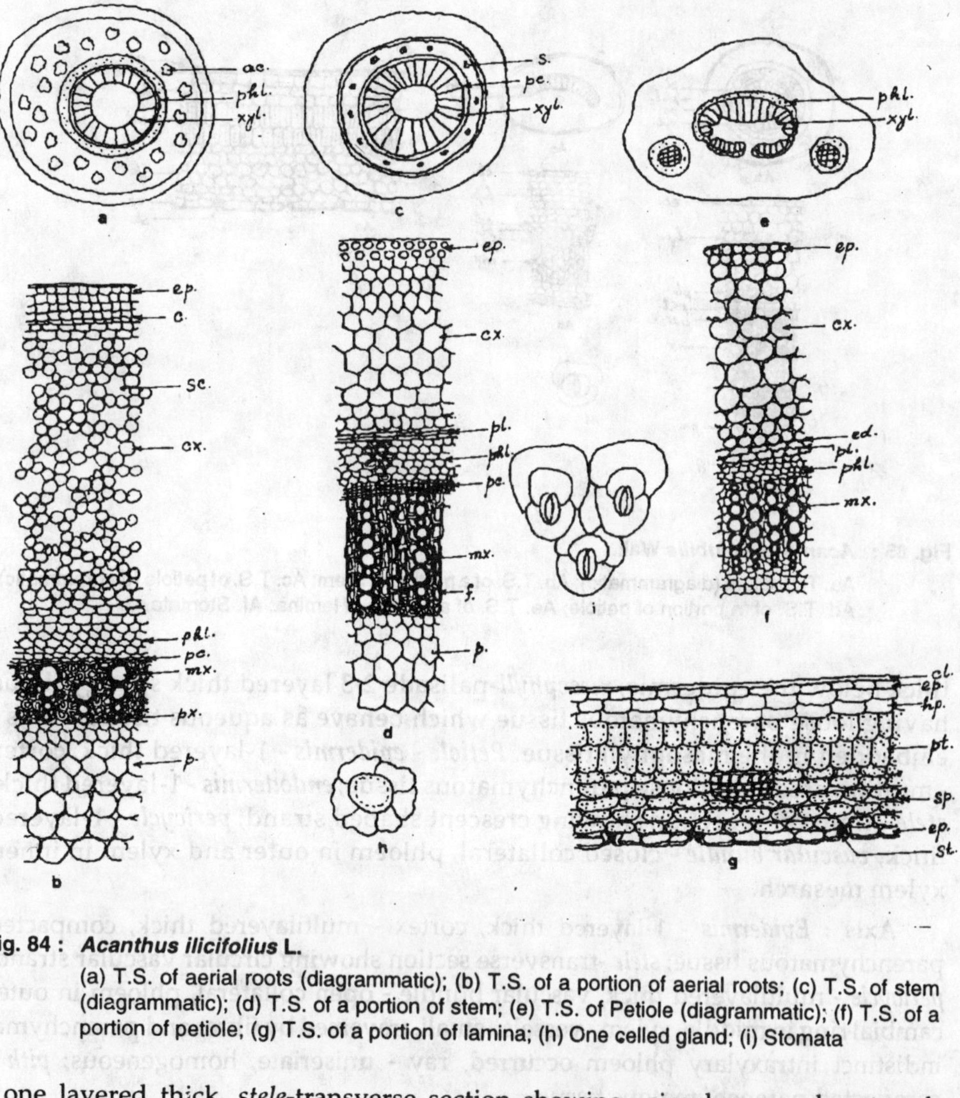

Fig. 84 : *Acanthus ilicifolius* L.
(a) T.S. of aerial roots (diagrammatic); (b) T.S. of a portion of aerial roots; (c) T.S. of stem (diagrammatic); (d) T.S. of a portion of stem; (e) T.S. of Petiole (diagrammatic); (f) T.S. of a portion of petiole; (g) T.S. of a portion of lamina; (h) One celled gland; (i) Stomata

- one layered thick, *stele*-transverse section showing circular vascular strand, *pericycle* - one layered thick, *vascular strand* - open collateral, phloem associated with parenchyma, cambial ring in middle, xylem - vessels simple, traversed by fibres, *pith* - compacted parenchymatous tissues.

18.2. *Acanthus volubilis* Wall.

(Fig. 85)

Leaf : Dorsiventral, petiolate.

Lamina : *Cuticle*-1-layered thick; *epidermis* - 1-celled thick both in upper epidermis and lower epidermis. *Stomata*-diacytic, not sunken, confined to the lower surface, 16 ± 1 μ in length x 12 ± 1 μ in width, hypodermis - 3-4 layered

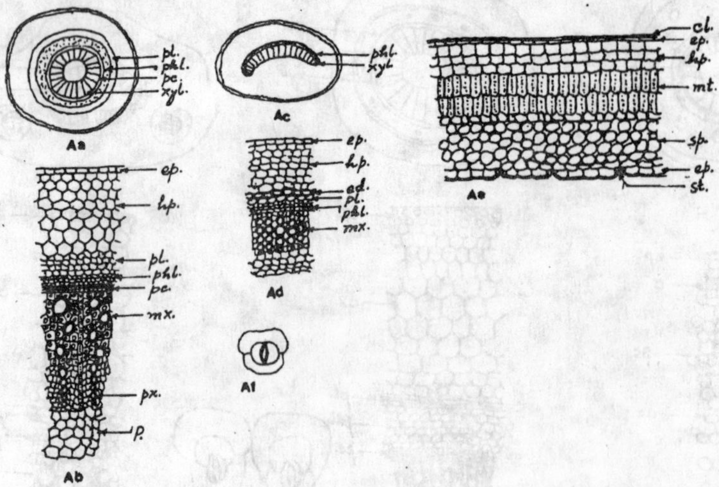

Fig. 85 : ***Acanthus volubilis*** **Wall.**

Aa. T. S. Stem (diagrammatic); Ab. T.S. of a portion of Stem; Ac. T.S. of petiole (diagrammatic); Ad. T.S. of a portion of petiole; Ae. T.S. of a portion of lamina; Af. Stomata.

thick below the epidermis, *mesophyll*-palisade 2-3 layered thick spongy tissue having loose parenchymatous tissue which behave as aqueous tissue; veins - embedded in the mesophyll tissue. *Petiole - epidermis* - 1-layered thick, cortex - multilayered thick loose parenchymatous tissue; *endodermis* - 1-layered thick, *stele* - transverse section showing crescent shaped strand; *pericycle* - 1-layered thick, *vascular bundle* - closed collateral, phloem in outer and xylem in inner, xylem mesarch.

Axis : *Epidermis* - 1-layered thick, cortex - multilayered thick, compacted parenchymatous tissue; *stele* -transverse section showing circular vascular strand; *pericycle* - multilayered thick, vascular bundle - open collateral, phloem in outer cambial ring in middle, xylem - vessels - small, traversed by fibres and parenchyma, indistinct intraxylary phloem occurred, ray - uniseriate, homogeneous; *pith* - compacted parenchymatous tissue.

FAMILY–XIV. *TILIACEAE*

Mucilagenous cells occurred mostly in the species, *stomata*-diacytic, *veins* embedded in the mesophyll, *petiole*-circular ring, *parenchyma*-apotracheal, ray-multiseriate, fibre-frequent.

Genus–19. *BROWNLOWIA* Roxb.

Mucilagenous cells occurred frequently, *stomata*–diacytic, *veins* embedded in the mesophyll, *petiole*–circular ring, *parenchyma*–apotracheal, vessels-medium sized, simple perforation, ray–multiseriate, fibre-frequent.

19.1. *Brownlowia lanceolate* (L.) Kosterm.

(Plate–160 and Fig. 86)

Leaf : Dorsiventral, petiolate.

Lamina : *Hairs* stellate, forming tufted condition, *cuticle*–one layered thick, *epidermis*-one layered thick in both surfaces, hypodermis-two layered thick below the upper epidermis; *stomata*–diacytic, confined to the lower surface, sunken, 9.5 ± 1 μ in length × 8.8 ± 1 μ in width, *mesophyll*-two layered thick palisade tissue, spongy tissue consisting of parenchymatous tissue with large frequent cavities; mesophyll tissue compartmented by horizontal large cells, *veins*-embedded in the mesophyll. *Petiole*–stellate hair on outer surface, *epidermis*-one layered thick, cortex-consisting of parenchymatous tissue; *endodermis*-one layered thick, stele-transverse section showing circular cylinder. *Pericycle*-multi-layered thick consisting of sclerenchymatous tissue, *vascular bundle*-closed collateral, phloem traversed by parenchymatous tissue and sclereids, xylem-mesarch, associated with fibres, ray-multiseriate, 2-3 celled width, concentric bundles occurred in pith.

Axis : *Epidermis*–one layered thick in young axis, while epidermal layer replaced by cork in mature one, *cork*-consisting of multi-layered thick, cortex-consisting of multi-layered thick compacted parenchymatous tissue, *stele*-transverse section showing circular cylinder, *pericycle*-consisting of multi-layered thick cells of sclerenchyma and parenchyma placed alternately in vertical row, *vascular bundle*-open collateral, phloem mixed with pericycle, multi-layered cambial ring in middle, xylem/wood-vessels medium sized, diffuse porous, simple perforation, traversed by sclerenchyma cells, *parenchyma*-apotracheal, ray-multi-seriate, 2-4 celled thick, homogeneous, *pith*-consisting of loosed parenchymatous tissue.

FAMILY–XV. *MALVACEAE*

Hairs mainly *stellate* and some species having peltate type of hairs. Mucilagenous cells are very common in the parenchymatous tissue. The petiolar stele showing 4-8 bundles of continuous ring or circular cylinder of continuous ring, the pith having parenchymatous tissue of thin walled. In the axis, vessels small to medium sized. Ray–very distinct.

Anatomical Key Characters to the Genera of Malvaceae

A. Generally *stellate* hairs occur on the lower surface of the leaf, stomata mainly paracytic

1. *Hibiscus* L.

AA. Generally peltate hairs occur on the lower surface of the leaf, stomata mainly anomocytic

2. *Thespesia* Solander ex Correa

GENUS–20. *HIBISCUS* L.

Generally stellate hairs occur on the lower surface of the leaf, mucilagenous cells are very common in the parenchymatous tissues, xylem forming a continuous ring,

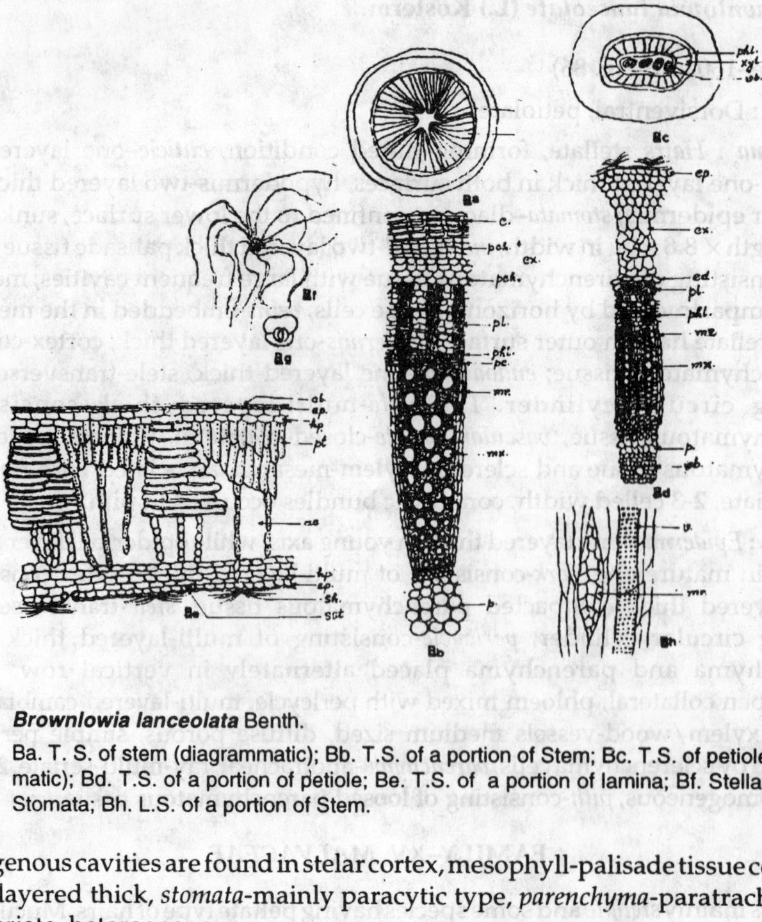

Fig. 86 : ***Brownlowia lanceolata*** **Benth.**

Ba. T. S. of stem (diagrammatic); Bb. T.S. of a portion of Stem; Bc. T.S. of petiole (diagrammatic); Bd. T.S. of a portion of petiole; Be. T.S. of a portion of lamina; Bf. Stellate hair, Bg. Stomata; Bh. L.S. of a portion of Stem.

schizogenous cavities are found in stelar cortex, mesophyll-palisade tissue consisting of one layered thick, *stomata*-mainly paracytic type, *parenchyma*-paratracheal, ray-multiseriate, heterogeneous.

Anatomical Key Characters to the Species of *Hibisus* L.

A. In the petiole, stele consisting of 7-8 bundles vascular strand of circular cylinder, several schizogenous cavities occur in the stelar cortex

Hibiscus tortuosus Roxb.

AA. In the petiole, stele consisting of continuous strand of circular cylinder, no schizogenous cavities in stelar cortex

Hibiscus tiliaceous L.

20.1. *Hibiscus tiliaceous* L.

(Fig. 87)

Leaf : Dorsiventral, petiolate.

Lamina : *Hairs*-Stellate hairs confined to the lower surface, *cuticle*-one layered thick on upper surface, epidermis-consisting of one layered thick, mucilagenous

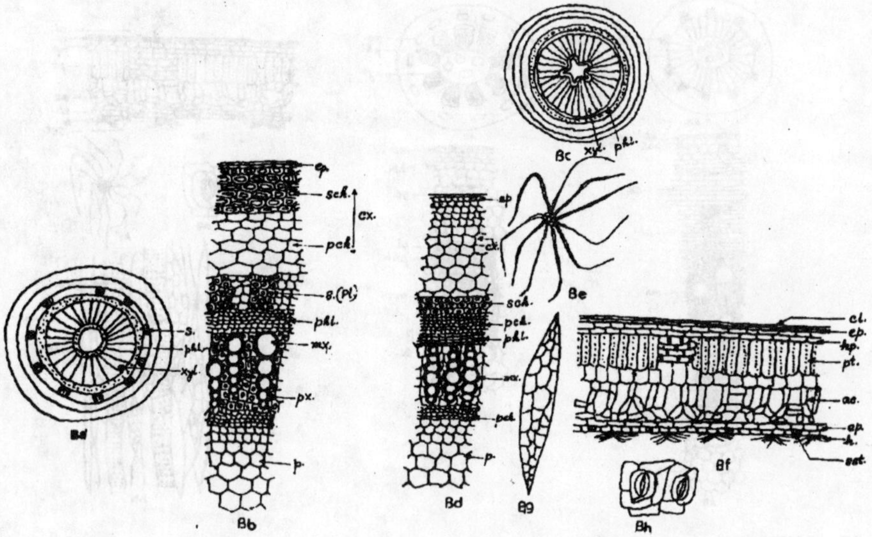

Fig. 87 : ***Hibiscus tiliaceous*** **L.**

(Ba) T.S. of stem (diagrammatic); (Bb) T.S. of a portion of stem.; (Bc) T.S. of petiole (diagrammatic); (Bd) T.S. of a portion of petiole; (Be) Stellate hair; (Bf) L.S. of a portion of lamina; (Bg) Ray; (Bh) Stomata.

cells below the epidermal layer; *stomata*-paracytic, confined to the lower surface, 15.8 ± 1 μ in length × 8.5 ± 1 μ in width, sunken, *mesophyll*-palisade tissue one layered thick, while spongy tissue multi-layerd thick of numerous air cavities, *veins*–embedded in the mesophyll tissue; *Petiole* : *epidermis*-one layered thick, *cortex*-consisting of parenchymatous tissue, *stele*-transverse section showing circular cylinder, *pericycle*-consisting of multi-layered thick sclerenchymatous cells; *vascular bundles*-open collateral-phloem in outer traversed by parenchymatous tissue, cambial ring in middle, xylem mesarch traversed by parenchymatous tissue, *pith*-consisting of parenchymatous tissue.

Axis : *Epidermis*–one layered thick, *cortex*-consisting of multi-layered thick of two types, upper layer-consisting of sclerenchymatous tissue, lower layer having parenchymatous tissue, *stele*-transverse section showing circular cylinder, *vascular bundles*-closed collateral, phloem in patches traversed by sclerenchymatous tissue, xylem/wood-vessels small to medium sized, simple perforation, diffuse porous traversed by firbres, *parenchyma*–paratracheal, ray-multiseriate, 2-3 celled thick, heterogeneous, *pith*-consisting of parenchymatous tissue.

20.2. *Hibiscus tortuosus* Roxb.

(Fig. 88)

Leaf : Dorsiventral, petiolate.

Lamina : *Hairs-stellate* hairs confined to the lower surface, *cuticle*-one layered thick on upper surface, *epidermis*-consisting of the one layered thick on both surfaces,

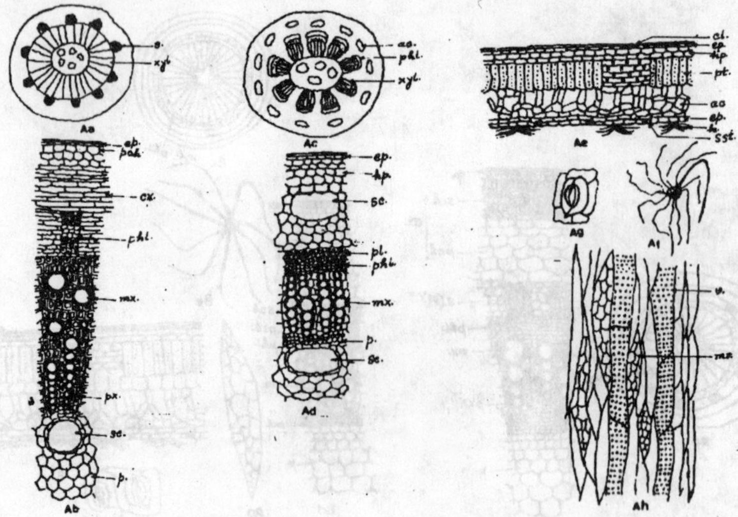

Fig. 88 : ***Hibiscus tortuosus*** **Roxb.**

(Aa) T.S. of stem (diagrammatic); (Ab) T.S. of a portion of stem.; (Ac) T.S. of petiole (diagrammatic); (Ad) T.S. of a portion of petiole; (Ae) T.S. of a portion of lamina; (Af) Stellate hair; (Ag) Stomata; (Ah) L.S. of a portion of stem.

mucilagenous cells below the epidermal layer, *stomata*-paracytic, confined to the lower surface, 17.1 ± 1 μ in length × 8.5 ± 1 μ in width, sunken, *mesophyll*-palisade tissue one layered thick, while spongy tissue multi-layered thick of numerous air cavities, *veins*–embedded in the mesophyll tissue; *Petiole*–epidermis-one layered thick, *cortex*-consisting of parenchymatous tissue; numerous schizogenous cavities occur in circular ring of cortex and frequently occur in pith, *stele*-transverse section showing 7-8 bundles of circular strand, *pericycle*-consisting of multi-layered thick sclerenchymatous tissue, *vascular bundle*-closed collateral, phloem in outer traversed by parenchymatous tissue, xylem-mesarch, traversed by parenchymatous tissue, *pith*-consisting of parenchymatous tissue.

Axis : *Epidermis*–one layered thick, *cortex*-consisting of multi-layered thick of two types-upper layer hexagonal while lower layer rectangular, *Stele*-transverse section showing circular cylinder, *vascular bundle*-closed collateral-phloem in patches traversed by sclerenchymatous tissue and interrupted by cortical extended parenchyma tissue, xylem/wood-vessels small to medium sized, simple perforation, diffuse porous, traversed by fibres, *parenchyma*-paratracheal, ray-multi-serite–2-3 celled width, heterogeneous, *pith*-consisting of parenchymtous tissue with schizogenous cavities.

GENUS–21. *THESPESIA* Sol. ex Corr.

Hairs–peltate hairs mostly, *epidermis*-one layered thick, mucilagenous cells below the epidermal layer, *mesophyll*-palisade tissue one layered thick, petiolar stele-having 3-5 vascular bundles of circular cylinder. In the axis, *pericycle*-consisting of sclerenchymatous tissue, ray-multiseriate, mostly homogeneous, few schizogenous cavities occur in pith or cortex.

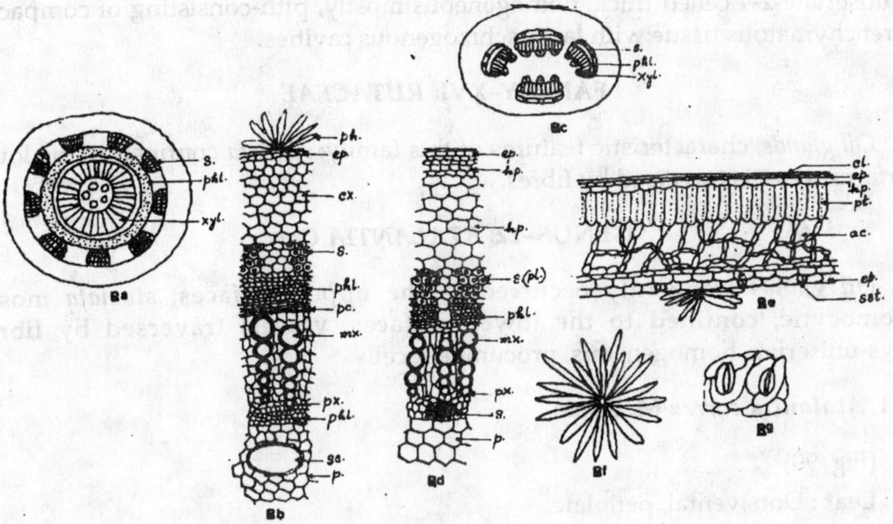

Fig. 89 : ***Thespesia populnea*** **(L.) Soland ex Corr.**

(Ba) T.S. of stem (diagrammatic); (Bb) T.S. of a portion of stem; (Bc) T.S. of petiole (diagrammatic); (Bd) T.S. of a portion of petiole; (Be) T.S. of a portion of lamina; (Bf) Peltate hair; (Bg) Stomata.

21.1. *Thespesia populnea* (L.) Sol. ex Corr.

(Fig. 89)

Leaf : Dorsiventral, petiolate.

Lamina : Hairs–peltate hairs confined to the lower surface, cuticle-one layered thick on upper epidermis, epidermis-one layered thick on the upper surface, mucilagenous cells below the upper epidermis; *stomata*-anomocytic, confined to the lower surface, sunken, 25.6 ± 1 μ in length × 19 ± 1 μ in width, mesophyll–palisade tissue one layered thick, spongy tissue consisting of multi-layered thick with inter-cellular spaces, veins-embedded in the mesophyll; *Petiole* : hair-peltate hairs sometime occur in the upper surface, epidermis-one layered thick, cortex-consisting of multi-layered thick, upper layer having small cells, while lower layer having large cells, frequent tannin cells in cortex; stele-transverse section showing 4 vascular bundles of different structures in circular ring, pericycle-consisting of multilayered thick sclerenchymatous patches, vascular bundle closed collateral; phloem in outer traversed by parenchymatous cells, xylem mesarch traversed by parenchymatous cells, pith-consisting of parenchymatous tissue having sclereids patch.

Axis : Hairs-peltate hairs recorded in the outer surface, epidermis-one layered thick, cortex-consisting of multilayered thick, compacted parenchymatous tissue, stele-transverse section showing circular cylinder, pericycle-consisting of sclerenchymatous patch, vascular bundle-open collateral, phloem in outer associated with parenchymatous tissue, multi-layered cambial ring in middle, xylem / wood-vessels medium sized, diffused porous in simple perforation, traversed by parenchymatous tissue, parenchyma-paratracheal, but few apotracheal, ray-multiseriate-2-4 celled thick, homogeneous mostly, pith-consisting of compacted parenchymatous tissue with large schizogenous cavities.

FAMILY–XVI. *RUTACEAE*

Oil glands, characteristic features of this family; *stomata* confined to the lower surfaces, vessels traversed by fibres.

GENUS–22. *ATALANTIA* Corr.

Oil glands frequently occurred in the upper surfaces; *stomata* mostly anomocytic, confined to the lower surfaces, vessels traversed by fibres. Rays-uniseriate homogeneous, procumbent cells.

22.1. *Atalantia correa* M. Roem

(Fig. 90)

Leaf : Dorsivental, petiolate.

Lamina : *Cuticle*-one layered thick, epidermis-one layered thick in both surfaces; *stomata*-anomocytic, slightly sunken confined to the lower surfaces, 19 ± 1 μ in length × 15.2 ± 1 μ in width, *glands*-oil glands one whole, surrounded by one celled jacket layer placed below the upper epidermis and within the mesophyll tissue, *mesophyll*-palisade multi-layered thick, spongy-multi-layered thick, veins-embedded in the mesophyll traversed by fibres. *Petiole*: *epidermis*–one layered thick, cortex-consisting of multi-layered thick, compacted parenchymatous tissue; stele-transverse section showing semi-circular strand, vascular bundle-closed collateral, phloem in outer associated with fibres and xylem in inner.

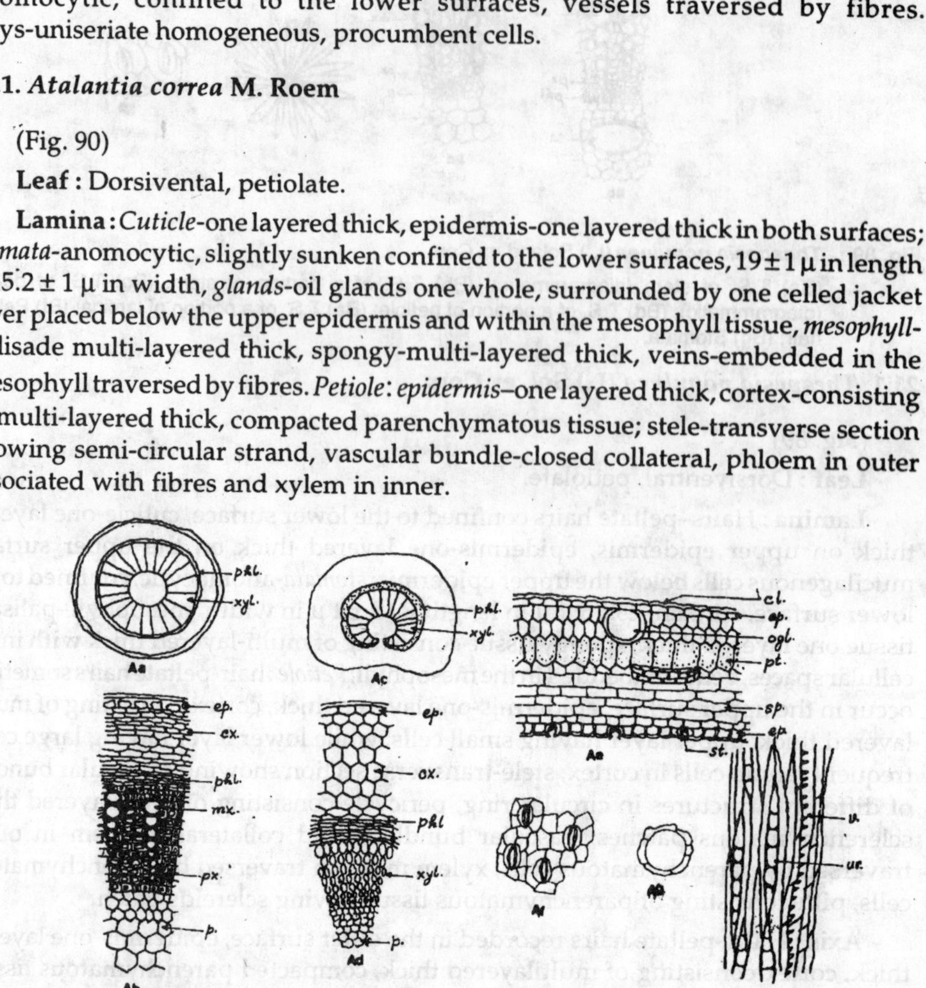

Fig. 90 : ***Atalantia correa*** M. Roeme

(Aa) T.S. of Stem (diag.); (Ab) T.S. of a portion of stem; (Ac) T.S. of Petiole (diag.); (Ad) T.S. of a portion of petiole, (Ae) T.S. of a portion of lamina; (Af) Stomata; (Ag) Oil gland; (Ah) L.S. of a portion of stem.

Axis : *Epidermis*–one layered thick, cuticularized, *cortex* multi-layered thick compacted parenchymatous tissue; *stele*-transverse section showing circular strand, vascular bundle–closed collateral-phloem outer traversed by parenchyma, xylem/wood-vessels small, simple pitted, traversed by fibres, *parenchyma* paratracheal, ray uniseriate, homogeneous, procumbent cells; *pith*-consisting of compacted parenchymatous tissue.

FAMILY–XVII. *FABACEAE*

Hairs of various types, *i.e.*, glandular and non-glandular; *epidermis* of the leaf may be papillose or mucilagenous. *Stomata* are generally paracytic type. *Petiole*-having continuous ring of vascular bundles. Woods are generally solitary; parenchyma paratracheal, rays multiseriate, phloem in outer and xylem in inner forming continuous ring.

Anatomical Key to the Genera of Fabaceae

A. In Rachis vascular bundles may be continuous ring or bundles of circular cylinder, two accessory bundles occur in two wings
B

AA. In Rachis vascular bundle are generally continuous ring of circular cylinder, but having no accessory bundles in the wings
C

B. Mesophyll tissue are vertically compartmented
3. *Derris* Lour.

BB. Several sclereids are longitudinally placed in mesophyll tissue, beneath the epidermal layer
1. *Cynometra* L.

C. Hypodermal layer present below the upper surface of the epidermis
4. *Dalbergia* L.

CC. Hypodermal layer absent
2. *Caesalpinia* L.

GENUS–23. *CYNOMETRA* L.

Lower epidermal cell sub-papillose; *stomata* paracytic, palisade-one layered thick, while spongy tissue multi-layered thick, *veins*-embedded in the mesophyll tissue, intermingled to sclereids, petiolar stele-circular cylinder. In axis, vascular bundles closed collateral, ray multiseriate, mostly homogeneous.

23.1. ***Cynometra ramiflora*** **L.**

(Fig. 91)

Leaf : Compound leaf, dorsiventral.

Lamina : *Epidermis* one layered thick in both the surfaces, lower epidermal cell sub-papillose, *stomata*-paracytic, 14.5 ± 1 μ in length × 7.6 ± 1 μ in width, confined to

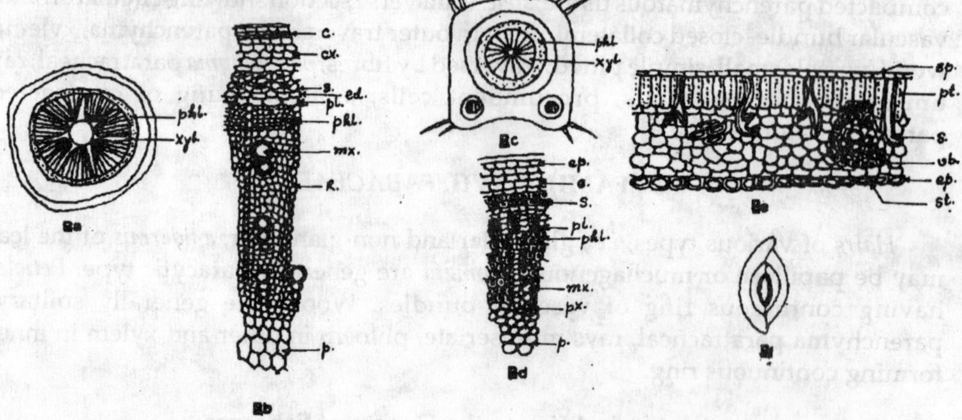

Fig. 91 : ***Cynometra ramiflora* L.**
(Ba) T.S. of stem (diagrammatic); (Bb) T.S. of a portion of stem; (Bc) T.S. of petiole (diagrammatic); (Bd) T.S. of a portion of petiole, (Be) T.S. of a portion of lamina; (Bf) Stomata.

the lower surface, not sunken, *mesophyll* one layered thick palisade tissue, while multi-layered thick spongy tissue with small intercellular spaces, branched sclereids placed vertically below the epidermal layer, *veins* embedded in the mesophyll tissue, traversed by sclereids; *Petiole–hairs* unicellular, *cuticle*-very thick, epidermis-one layered thick, *cortex*-multi-layered thick, sclerenchymatous tissue-multi-layered thick below the cortical layer, stele-transverse section showing circular cylindric, two accessory bundles in wing, pericycle-multi-layered thick consisting of parenchymatous tissue, *vascular bundle*-closed collateral, phloem in outer traversed by parenchymatous tissue, xylem in inner encircled of fibres, ray-uniseriate, homo-geneous, pith consisting of compacted parenchymatous cells.

Axis : *Epidermis*-one layered thick in young axis, while epidermal cell replaced by cork in mature axis, *cork*-multi-layered thick, *cortex*-multi-layered thick of parenchymatous tissue, traversed by below sclereids layers, *endodermis*-one layered thick, *stele*-transverse section showing circular cylinder, *pericycle*-one layered thick, *vascular bundles*-closed collateral, phloem in outer traversed by parenchymatous tissue, xylem/wood-vessels small sized, diffuse porous, scalariform perforation, *parenchyma*-paratracheal, ray-uniseriate, homogeneous, *pith*-consisting of loose parenchymatous cells.

GENUS–24. *CAESALPINIA* L.

Cuticle thin layer, *epidermis*-one layered thick, *hairs*-one celled thick in the lower epidermis. *Stomata*-paracytic, confined to the lower surface. *Mesophyll*-one layered thick palisade tissue, spongy parenchyma multi-layered thick with intercellular spaces. In the young axis, *cork*-multi-layered thick. *Stele*-transverse section showing circular cylindric.

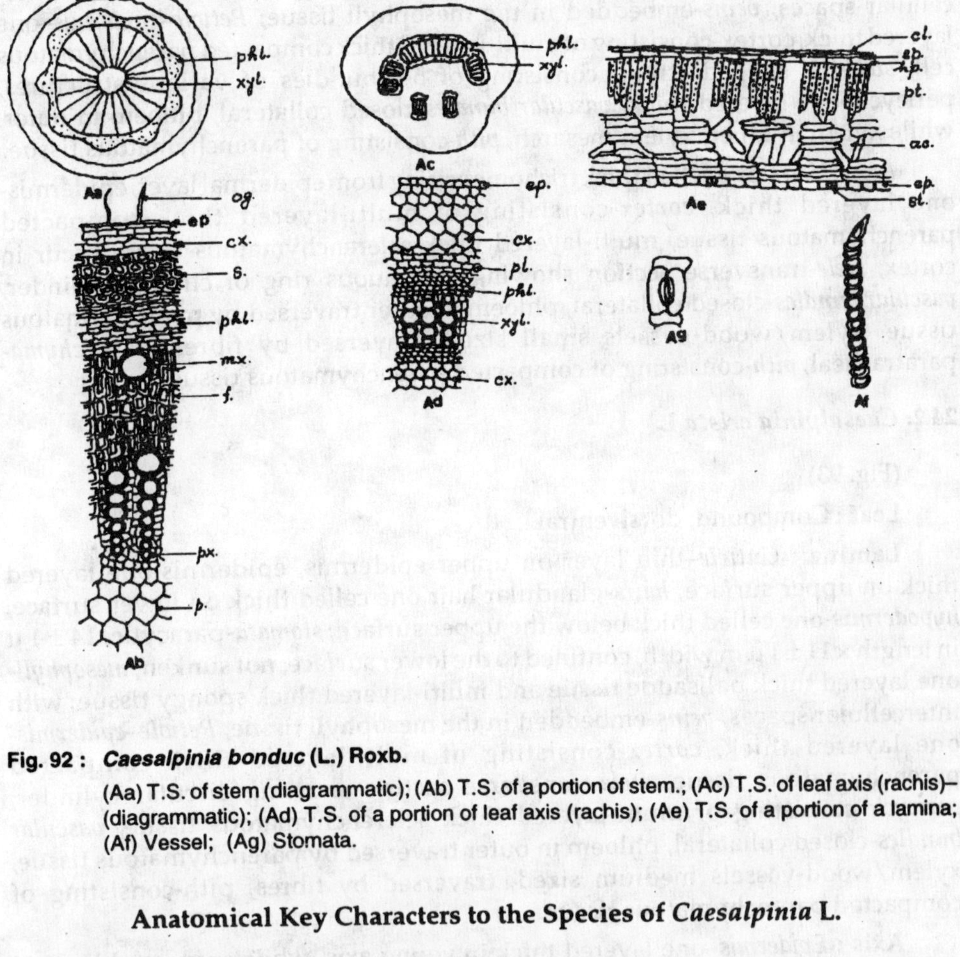

Fig. 92 : ***Caesalpinia bonduc*** **(L.) Roxb.**
(Aa) T.S. of stem (diagrammatic); (Ab) T.S. of a portion of stem.; (Ac) T.S. of leaf axis (rachis)–(diagrammatic); (Ad) T.S. of a portion of leaf axis (rachis); (Ae) T.S. of a portion of a lamina; (Af) Vessel; (Ag) Stomata.

Anatomical Key Characters to the Species of *Caesalpinia* L.

A. Petiolar stele showing circular cylinder

2. *Caesalpinia crista* L.

AA. Petiolar stele showing 'U' shaped strand, consisting of 5-6 bundles of various structure

1. *Caesalpinia bonduc* (L.) Roxb.

24.1. *Caesalpinia bonduc* (L.) Roxb.

(Fig. 92)

Leaf : Compound, dorsiventral.

Lamina : *Cuticle*-one layered thick, *epidermis*-one layered thick on upper surface, *hairs*-glandular hair one celled thick on lower surface, *hypodermis*-one layered thick below the upper epidermis; *stomata*-paracytic, $19 \pm 1\ \mu$ in length $\times$ 13.5

± 1 μ in width, confined to the lower surface, not sunken, *mesophyll*-one layered thick palisade tissue, while multi-layered thick spongy tissue with large intercellular spaces, *veins*-embedded in the mesophyll tissue; *Petiole–epidermis*–one layered thick, cortex-consisting of multi-layered thick compacted parenchymatous cell, *stele*-'U' shaped strand consisting of 5-6 bundles of various structures, pericycle-one layered thick, *vascular bundles*-closed collateral, phloem in outer, while xylem in inner, xylem mesarch, *pith*-consisting of parenchymatous tissue.

Axis : *Hairs*-one celled thick trichome arising from epidermal layer, epidermis-one layered thick, *cortex*-consisting of multi-layered thick compacted parenchymatous tissue, multi-layered thick sclerenchymatous tissue occur in cortex, *stele*-transverse section showing continuous ring of circular cylinder, *vascular bundles*-closed collateral, phloem in outer traversed by parenchymatous tissue, xylem/wood-vessels small sized, traversed by fibres, *parenchyma*-paratracheal, *pith*-consisting of compacted parenchymatous tissue.

24.2. *Caesalpinia crista* L.

(Fig. 93)

Leaf : Compound, dorsiventral.

Lamina : *Cuticle*–thin layer on upper epidermis, epidermis-one layered thick on upper surface, *hairs*-glandular hair one celled thick on lower surface, *hypodermis*-one celled thick below the upper surface; *stomata*-paracytic, 14 ± 1 μ in length × 11 ± 1 μ in width, confined to the lower surface, not sunken, *mesophyll*-one layered thick palisadde tissue and multi-layered thick spongy tissue; with intercellular spaces, *veins*-embedded in the mesophyll tissue; *Petiole–epidermis*-one layered thick, *cortex*-consisting of multi-layered thick compacted parenchymatous tissue, *stele*-transverse section showing circular cylinder, *pericycle*-consisting of multi-layered thick sclerenchymatous tissue, *vascular bundles*-closed collateral, phloem in outer traversed by parenchymatous tissue, xylem/wood-vessels medium sized, traversed by fibres, pith-consisting of compacted parenchymatous tissue.

Axis : *Epidermis*–one layered thick in young axis, while epidermal layered replaced by cork in mature axis, *cork*-multi-layered thick of under developed, cortex-consisting of sclerenchymatous cells. *Stele*-transverse section showing circular cylinder, *pericycle*-one layered thick, *vascular bundles*-open collateral, phloem in outer traversed by parenchymatous cells, multi-layered thick cambial ring in middle, xylem/wood-vessels small, diffused porous, scalariform perforation, traversed by fibres, *parenchyma*-paratracheal, *pith*-consisting of compacted parenchymatous tissue.

GENUS–25. *DALBERGIA* L.

Lower epidermis sub-papillose, upper eipdermis mostly papillose, hypodermis present below the epidermal cell; *stomata* mostly paracytic type; *petiole*-stele circular cylindric. Ray-multiseriate, mostly homogeneous.

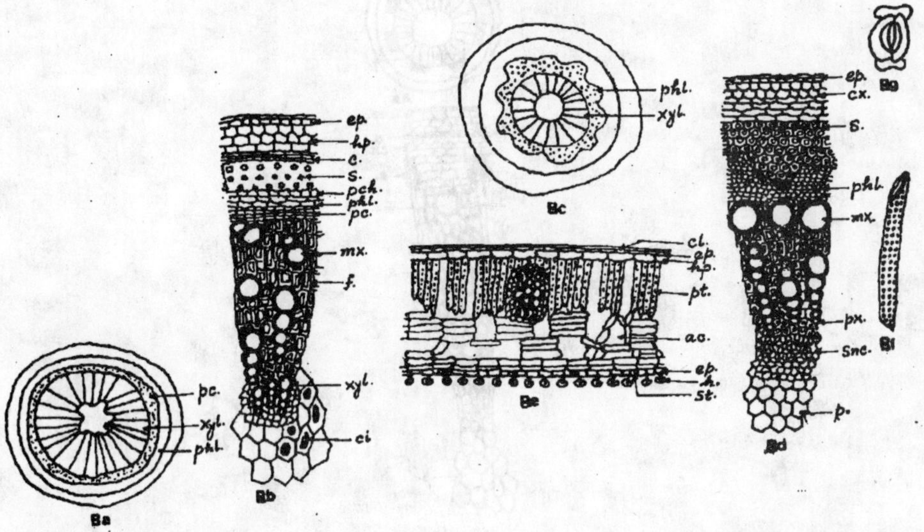

Fig. 93 : *Caesalpinia crista* L.

(Ba) T.S. of stem (diagrammatic); (Bb) T.S. of a portion of stem.; (Bc) T.S. of leaf axis (rachis) (diagrammatic); (Bd) T.S. of a portion of leaf axis (rachis) (Be) T.S. of a portion of a lamina; (Bf) Vessel; (Bg) Stomata.

25.1. *Dalbergia spinosa* Roxb.

(Fig. 94)

Leaf : Compound leaf, dorsiventral.

Lamina : *Cuticle*–thin layered on upper epidermis, *epidermis*-one layered thick in both surfaces, lower epidermal layer sub-papillose, *hypodermis*-one layered thick below the epidermal layer; *stomata*-paracytic, 14 ± 1 μ in length × 10 ± 1 μ in width, confined to the lower surface, not sunken, *mesophyll*-palisade tissue–one layered thick, while spongy tissue having air cavities, mesophyll tissue compartmented, *veins* embedded in the mesophyll tissue; *Petiole*–epidermis-one layered thick, *cortex*-consisting of compacted parenchymatous tissue, *stele*-continuous ring of circular cylinder, *pericycle*-one layered thick, *vascular bundle*-closed collateral; phloem in outer and xylem in inner traversed by tracheids, *pith*-consisting of compacted parenchymatous tissue.

Axis : *Epidermis*-one layered thick in young axis, while epidermal layered replaced by cork in mature axis, *cork*-multi-layered thick, *cortex*-consisting of parenchymatous tissue. Stele-continuous circular cylinder, *pericycle*-traversed by phloem tissue, *vascular bundle*-closed collateral–phloem in outer associated with sclereids patches in circular ring, xylem/wood-vessels-medium sized, diffuse porous, scalariform perforation, traversed by fibres, ray-multiseriate, 2-3 celled thick, homo-geneous, pith-consisting of compacted parenchymatous tissue.

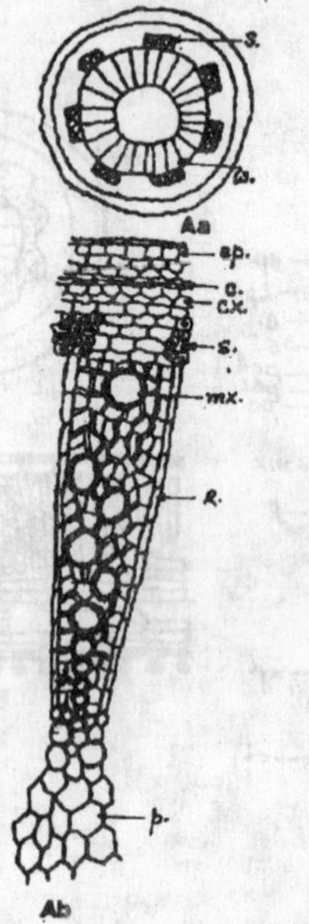

Fig. 94 : *Dalbergia spinosa* Roxb.
(Aa) T.S. of stem (diagrammatic); (Ab) T.S. of a portion of stem.

GENUS–26. *DERRIS* Lour.

Cuticle-thin layer on upper epidermis, *epidermis*-one layered thick in both surfaces, *stomata*-paracytic, *mesophyll*-palisade tissue may be one or two layers thick, *veins* embedded in the mesophyll tissue. In this axis, *epidermis* one layered thick, cork-multi-layered thick, *vascular bundle*-variable, *i.e.*, open collateral or closed collateral, *parenchyma*-paratracheal. Ray-multi-seriate thick.

Anatomical Key Characters to the Species of Derris Lour.

A. Palisade tissue 2 layered thick, compartmented by small parenchymatous cells

3. *Derris indica* Bennet

AA. Palisade tissue one layered thick, compartmented by large parenchymatous cells

1. *Derris scandans* Benth.

AAA. Palisade tissue two layered thick, upper layered small, while lower layer having large cells, compartmented

2. *Derris trifoliata* Lour.

26. 1. *Derris scandans* Benth.

(Fig. 95)

Leaf : Compound leaf, dorsiventral.

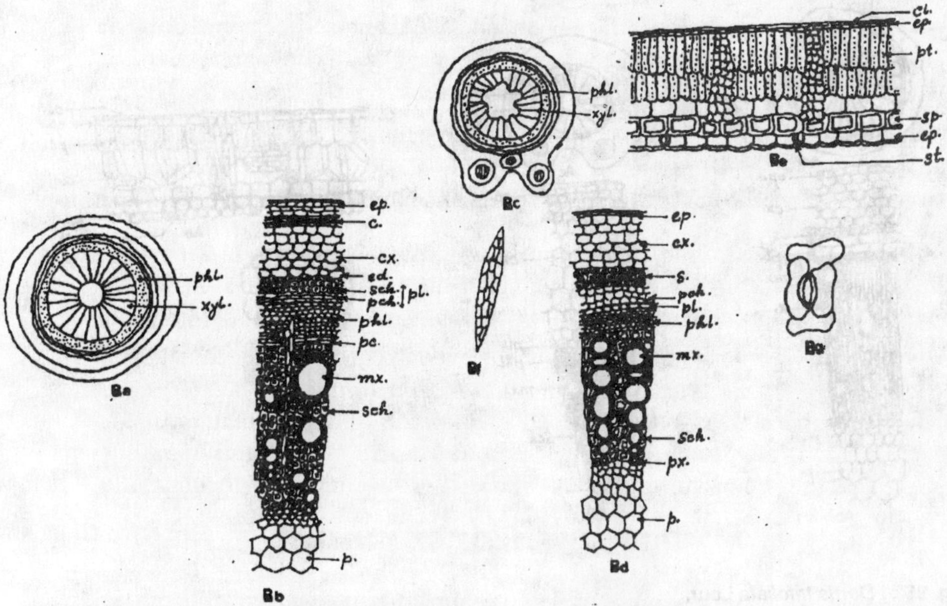

Fig. 95 : *Derris scandans* Benth.

(Ba) T.S. of stem (diagrammatic); (Bb) T.S. of a portion of stem; (Bc) T.S. of leaf axis (rachis) (diagrammatic); (Bd) T.S. of a portion of leaf axis (rachis); (Be) T.S. of a portion of lamina; (Bf) Ray cell; (Bg) Stomata.

Lamina : *Cuticle*–thin layered on upper epidermis, *epidermis*-one layered thick on both surfaces, in lower surface epidermal cell sub-papillose; *stomata*-paracytic, 22 ± 1 μ in length × 12 ± 1 μ in width, not sunken, confined to the lower surface, *mesophyll*-one layered thick-palisade tissue, spongy parenchyma having large schizogenous cavities, *mesophyll* tissue compartmented by longitudinal septa, consisting of parenchymtous cells, veins-embedded in the mesophyll tissue. *Petiole–epidermis*-one layered thick, *cortex*-consisting of parenchymtous cells arranged compactly, stele a continuous ring of circular cylinder, two accessory bundles in wing, *pericycle*-one layered thick, *vascular bundle*-closed collateral, phloem in outer associated with parenchymatous tissue, xylem in inner, intermingled to fibres, and traversed by tracheids, *pith*-consisting of compacted parenchymatous tissue.

Axis : *Epidermis*-one layered thick in young stem, while epidermal layered replaced by cork cell in mature axis, *cork*-multi-layered thick, *cortex*-consisting of compacted parenchymatous tissue, *endodermis*-one layered thick, stele-continuous ring of circular cylinder, *pericycle*-consisting of multi-layered thick, phloem in outer traversed by parenchymatous cells, xylem/wood-vessels medium sized, diffuse porous, scalariform perforation, parenchyma-mostly paratracheal, but sometimes apotracheal, ray multi-seriate, 2 layered thick, homogeneous, *pith*-consisting of large compacted parenchymatous cells.

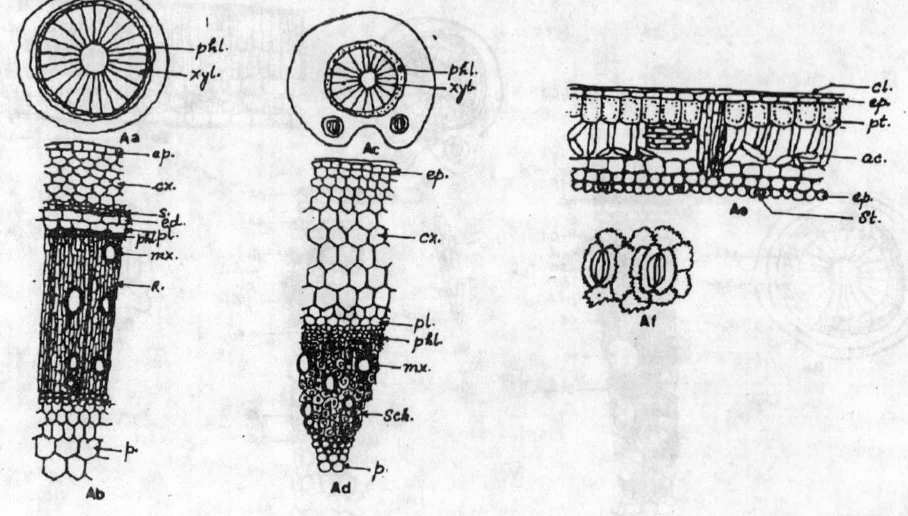

Fig. 96 : *Derris trifoliata* Lour.
(Aa) T.S. of stem (diagrammatic); (Ab) T.S. of a portion of stem; (Ac) T.S. of leaf axis (rachis) (diagrammatic); (Ad) T.S. of a portion of leaf axis (rachis); (Ae) T.S. of a portion of lamina; (Af) Stomata.

26.2. Derris trifoliata Lour.

(Fig. 96)

Leaf : Compound, dorsiventral.

Lamina : *Cuticle*–thin layered on upper epidermis, *epidermis*– one layered thick on both the surfaces; in lower surface, epidermal cell sub-papillose; *stomata*-paracytic, 21 ± 1 μ in length × 15.2 ± 1 μ in width, not sunken, confined to the lower surface; *mesophyll*-one layered thick palisade tissue, spongy tissue having parenchymatous cells, mesophyll tissue compartmented by longitudinal septa consisting of parenchymatous cells, *veins*-embedded in the mesophyll tissue. *Petiole-epidermis* one layered thick, *cortex* consisting of parenchymatous cells arranged compactly; *stele*–5 + 2 bundles of circular cylinder, two accessory bundles in wings. *Pericycle*-multi-layered thick consisting of sclerenchymatous cells, *vascular bundles* open-collateral; phloem in outer traversed by parenchymatous tissue, multi-layered thick cambial rings in middle, xylem-in inner associated with parenchymatous tissue, vessels medium sized, *pith*-consisting of parenchymatous tissue.

Axis : *Epidermis*–one layered thick in young stem, while epidermal layer replaced by cork cell in mature axis, *cork*-multi-layered thick, *cortex*-consisting of compacted parenchymtous tissue, *endodermis*-one layered thick, *stele*-continuous ring of vascular cylinder, pericycle-consisting of sclerenchymatous tissue, phloem in outer traversed by fibres, xylem/wood-vessels medium sized, diffuse porous, scalariform perforation, *parenchyma* paratracheal, ray-multiseriate, 3-5 layered thick, homogeneous, *pith*-consisting of large compacted parenchymatous cells.

26.3. *Derris indica* Bennet

(Fig. 97)

Leaf : Compound, dorsiventral.

Lamina : *Cuticle*–thin layer on upper epidermis, epidermis-one layered thick in both the surfaces, small cell in upper surface, while large one sub-papillose in lower surface; *stomata*-paracytic, 14.5 ± 1 μ in length × 9.5 ± 1 μ in width, confined to the lower surface, not sunken, *mesophyll*–2 layered thick palisade tissue, long, 1-2 layered thick parenchymatous cells, compactly arranged, mesophyll tissue compartmented by septa, consisting of parenchymatous cells; veins-embedded in the mesophyll tissue; *Petiole* (rachis) *epidermis*-one layered thick consisting of small cells, cortex-parenchymatous and arranged compactly, sclerenchymatous layer placed below the cortical layer, *endodermis*-consisting of multi-layered thick parenchymatous cells, stele-a continuous ring of circular cylinder, 3 accessory bundles in wings, *pericycle*-one layered thick; *vascular bundle*-closed collateral, phloem in outer associated with parenchymatous cells, xylem in inner, mixed with fibres, *pith*-consisting of compacted parenchymatous cells.

Axis : *Epidermis*–one layered thick in young axis, while epidermal layer replaced by cork, cork-multi-layered thick; *cortex*-multi-layered thick, consisting of parenchymatous cells-arranged compactly, *endodermis*-one layered thick. *Stele*-circular cylinder, *pericycle*-3 layered thick consisting of sclerenchymatous cells, *vascular bundle*-open collateral, phloem-multi-layered thick associated with parenchymatous cells, multi-layered thick cambial ring in middle, xylem/wood-vessels small to medium sized, diffuse porous, scalariform perforation, *parenchyma*-paratracheal, ray-multiseriate, 2-4 layered thick, homogeneous, *pith*-consisting of parenchymatous cells, compactly arranged.

FAMILY–XVIII. *ASCLEPIADACEAE*

Laticiferous tubes occur in both leaf and stem, associated with intra–xylary phloem. Trichomes are in different types, *stomata* are generally paracytic type and generally confined to the lower surface. In axis, cork arising from upper surface, distinct pericycle layer, phloem in outer and xylem in inner form continuous cylinder.

Anatomical Key Characters to the Genera of Asclepiadaceae

A. Stomata confined to the ventral side only

1. *Sarcolobus* R. Brown

AA. Stomata occur on both side (ventral and dorsal)

2. *Pentatropis* W. & A.

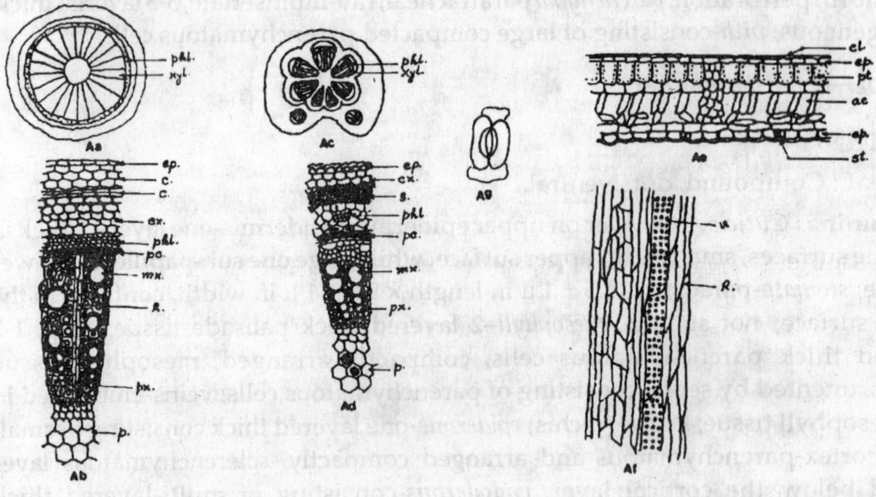

Fig. 97 : ***Derris indica*** **Bennett**

(Aa) T.S. of stem (diagrammatic); (Ab) T.S. of a portion of stem; (Ac) T.S. of leaf axis (rachis) (diagrammatic); (Ad) T.S. of a portion of leaf axis, (rachis); (Ae) T.S. of a portion of lamina; (Af) L.S. of a portion of stem; (Ag) Stomata.

GENUS–27. *SARCOLOBUS* R. Br.

Leaf : Paracytic type of stomata confined to the ventral side; not sunken, two layered thick palisade tissue; *petiole*-stele showing crescent shaped vascular strand (closed bicollateral). In axis, *cork* arising superficially, cortex consisting of 3 distinct horizons thick, *vascular strand* closed collateral forming continuous cylinder.

Anatomical Key Characters to the Species of *Sarcolobus* R. Br.

A. Length and width of the stomata are 27 ± 2 μ and 14 ± 1 μ respectively

2. *S. globosus* Wall.

AA. Length and width of the stomata are 21 ± 2 μ and 12.6 ± 1 μ respectively

1. *S. carinatus* Wall.

27.1. *Sarcolobus globosus* Wall.

(Fig. 98)

Leaf : Dorsiventral, petiolate.

Lamina : *Cuticle*-one layered thick, *epidermis*-one layered thick in both upper and lower surface, *hypodermis*-one layered thick; *stomata*-paracytic, 27.2 ± 1 μ in length × 14 ± 1 μ in width, confined to the lower surface, slightly sunken, *mesophyll* 2-layered thick palisade tissue, upper layers are rectangular in shape and lower layer cell square in shape, while in spongy tissue 6-7 layered spongy tissue thick, parenchymatous tissue along with large air spaces, veins embedded within the spongy parenchymatous tissue; *Petiole* : *epidermis*-one layered thick, *cortex*-multi-

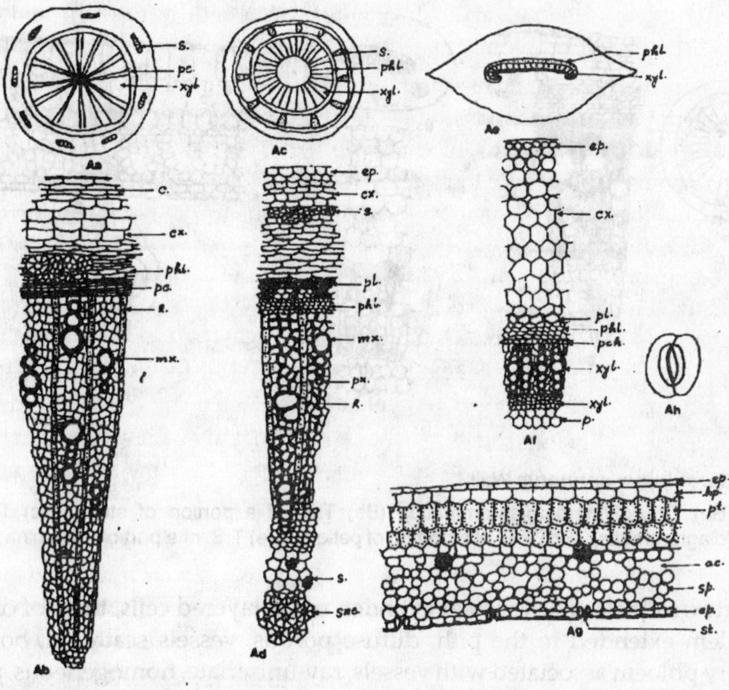

Fig. 98 : ***Sarcolobus globosus*** **Wall.**

(Aa) T.S. of aerial root (diagrammatic); (Ab) T.S. of a portion of aerial root; (Ac) T.S. of stem (diagrammatic); (Ad) T.S. of a portion of stem.; (Ae) T.S. of petiole (diagrammatic); (Af) T.S. of a portion of petiole; (Ag) T.S. of a portion of lamina; (Ah) Stomata.

layered thick, parenchymatous tissue, compactly arranged, *endodermis*-not distinctly separated. *Stele*-transverse section in the distal end showing crescent shaped vascular strand, *pericycle*-one layered thick, *vascular strand*-closed bicollateral, phloem in outer and inner and xylem in middle, xylem mesarch.

Axis : *Cork* usually in the epidermis, *epidermis*-one layered thick, cortex-having 3 horizons thick, 1st horizon consisting of 5 layered thick collenchymatous cell, 2nd horizon of 2 layered thick sclerenchymatous cells and 3rd horizon of multi-layered thick parenchymatous cells, *endodermis*-not distinctly separated, stele-transverse section exhibits circular vascular cylinder; *pericycle*-consisting of sclerenchymatous stone cells in patches interrupted by extended cortical parenchymatous cells; *vascular strand*-closed collateral, section of phloem multi-layered thick, xylem in the form of continuous cylinder; wood-vessel medium sized, diffuse porous, scalariform perforation, vessels associated with fibres, rays-uniseriate ray, homogeneous, *pith*-parenchymatous cells, compactly arranged, groups of sclereids scattered in the pith, idioblasts cells in pith.

Root : Root exposed due to eroded soil. In young root, epidermis–one layered thick, while in mature root, epidermis replaced by cork; *cork*-consisting of multi-layered thick; phellem cells in upper and multi-layered thick phelloderm cells in below, *cortex*-multi-layered thick parenchymatous cells, compactly arranged, groups of sclereids scattered

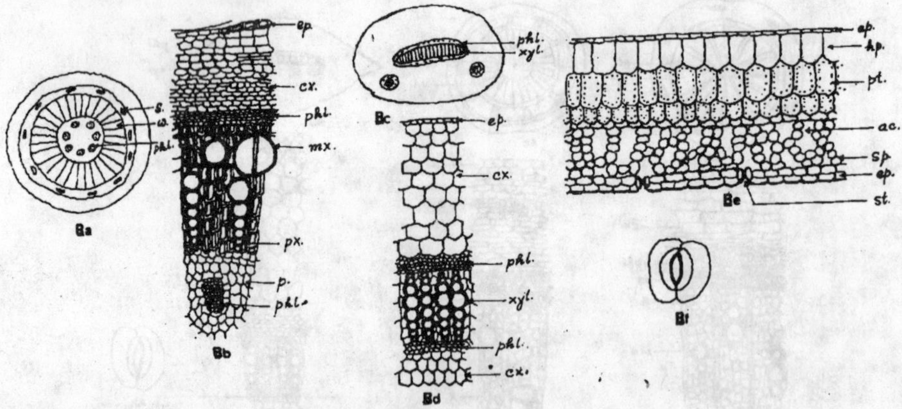

Fig. 99 : ***Sarcolobus carinatus*** **Wall.**

(Ba) T.S. of stem (diagrammatic); (Bb) T.S. of a portion of stem; (Bc) T.S. of petiole (diagrammatic); (Bd) T.S. of a portion of petiole; (Be) T.S. of a portion of lamina; (Bf) Stomata.

in the cortex; *stele*-circular vascular cylinder, multi-layered cells, thick of cambial ring, wood-xylem extended to the pith, diffuse porous, vessels scattered, border pitted, intraxylary phloem associated with vessels, ray-uniseriate, homogeneous, *parenchyma*-apotracheal. *Pith*-replaced by xylem cylinder in mature root.

27.2. *Sarcolobus carinatus* Wall.

(Fig. 99)

Leaf : Dorsiventral, petiolate.

Lamina : *Cuticle*–one layered thick, epidermis-one layered thick in both upper and lower surface; *hypodermis*–one layered thick on upper surface; *stomata*-paracytic, 21 ± 1 μ length ×12.6 ± 1 μ width, confined to the lower surface, slightly sunken, *mesophyll*–2-layered thick, palisade tissues, upper layered cell large, rectangular in shape and lower layered cell square in shape, while 5-6 layered thick parenchymatous tissue along with large air spaces; *Petiole* : epidermis-one layered thick, cortex-multi-layered thick parenchymatous tissue, compactly arranged, *endodermis*-not distinctly separated, stele-transverse section in distal end showing crescent shaped vascular strand, two isolated small strands occur in two wings of abaxial side, pericycle-one layered thick, *vascular strand*-closed bi-collateral; *phloem* in outer and inner and xylem in middle, xylem mesarch.

Axis : *Cork* usually in the epidermis, *epidermis*-one layered thick, lenticel formed by rupture of epidermal layer, cortex-having 3 horizons ; 1st horizon consisting of 3 layers thick collenchymatous cell; 2nd horizon of parenchymatous cells, compactly arranged; 3rd horizon of sclerenchymatous patch interrupted by extended cortical parenchyma, endodermis-not distinctly separated; stele-transverse section exhibits circular vascular strand, *pericycle*-consisting of 3 layered thick collenchymatous cells, *vascular strand*-closed collateral, secondary phloem-multi-layered thick,

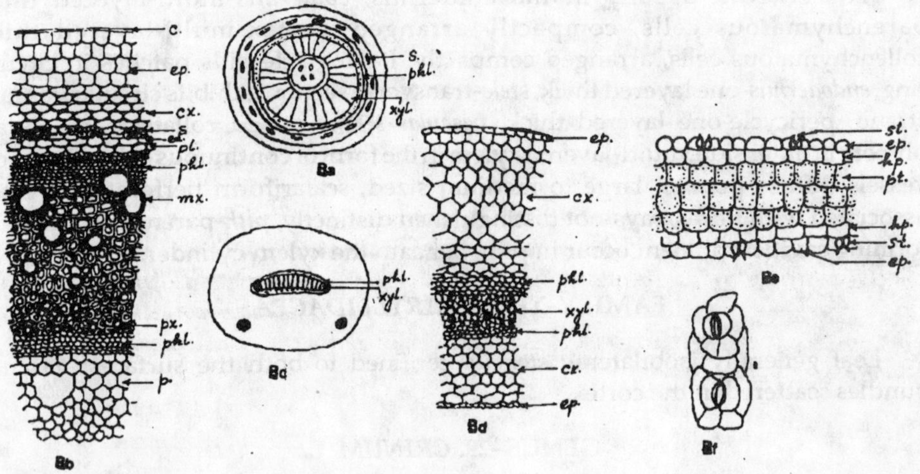

Fig. 100 : *Pentatropis capensis* Bull.

(Ba) T.S. of stem (diagrammatic); (Bb) T.S. of a portion of stem; (Bc) T.S. of petiole (diagrammatic); (Bd) T.S. of a portion of petiole; (Be) T.S. of a portion of lamina; (Bf) Stomata.

xylem in the form of continuous cylinder, wood-vessels medium sized, diffuse porous, scalariform perforation; xylem associated with fibres. *Pith*-parenchymatous cells, compactly arranged, some phloem patches scattered in pith.

GENUS–28. *PENTATROPIS* W. & A.

Leaf : *Stomata* on both sides, paracytic type, fleshy, multi-layered thick mesophyll tissue, associated with aqueous tissue due to fleshy nature; *Petiole*-crescent shaped vascular strand, closed bi-collateral. In axis-*cork* arising superficially in mature axis, *epidermis*-multi-layered thick, cortex of multi-layered thick, sclerenchymatous tissue, *stele*-vascular strand closed collateral, forming continuous cylinder.

28.1. *Pentatropis capensis* (L.f.) Bullock

(Fig. 100)

Leaf : Isobilateral, petiolate.

Lamina : *Cuticle*–one layered thick, *epidermis*-one layered thick in both upper and lower surfaces; *stomata*-paracytic, 20.2 ± 1 μ in length and 11.4 ± 0.5 μ in width, slightly sunken, confined to both upper and lower surfaces; *mesophyll* 3-layered thick, mesophyll tissue in middle, compactly arranged, square to ovoid shaped, one layered aqueous tissue on both sides of the mesophyll tissue, veins-embedded within the mesophyll tissue. *Petiole*-epidermis-one layered thick, cortex-multi-layered thick, parenchymatous tissue compactly arranged, *endodermis* not distinctly separated; stele-transverse section in distal end showing crescent shaped vascular strands, two isolated small strands occur on abaxial surface; *pericycle*-one layered thick, *vascular strand*-closed bi-collateral, xylem in middle and phloem in both sides of xylem, xylem mesarch.

Axis : Cork usually in the epidermis, *epidermis*-multi-layered thick, parenchymatous cells, compactly arranged, cortex-multi-layered thick, collenchymatous cells, arranged compactly, having sclereids patches in circular ring, *endodermis*-one layered thick, *stele*-transverse section exhibits circular vascular strand, pericycle-one layered thick, *vascular strand*-closed collateral, secondary phloem in outer side, multi-layered xylem in the form of continuous cylinder, wood-vessels diffuse porous, large to medium sized, scalariform perforation, vessels associated with fibres, rays-not distinguished distinctly, *pith*-parenchymatous cell forming cavities; phloem occur in pith, beneath the xylem cylinder.

FAMILY–XIX. *AMARYLLIDACEAE*

Leaf generally isobilateral; *stomata* confined to both the surfaces; vascular bundles scattered in the cortex.

GENUS–29. *CRINUM* L.

Stomata paracytic, confined to the both surfaces. *Mesophyll*-palisade as water storage tissue, vascular bundles scattered in the cortex.

29.1. *Crinum defixum* Ker-Gawler

(Fig. 101)

Leaf : Isobilateral.

Lamina : Cuticle–very thin, *epidermis*-one layered thick in both surfaces; *stomata*-paracytic, confined to both the surfaces, not sunken, $22.8 \pm 1\ \mu$ in length $\times 15.2 \pm 1\ \mu$ in width. *Mesophyll* multi-layered thick palisade as water storage tissue extend to the both epidermal layers, loosely arranged.

Rhizome : *Epidermis*-one layered thick, *cortex*-large celled, multi-layered thick, compacted parenchymatous tissue, mucilagenous contents occurred in few cortical cell; *stele* numerous bundles, gathered in the centre, *vascular bundle*-centric, surrounded by pericycle layer.

Root : *Epidermis*-one layered thick, cortex multi-layered thick, compacted parenchymatous tissue, *endodermis*-one layered thick; stele-transverse section showing circular ring, pericycle-one layered thick encircling, *vascular bundles*-radial, phloem arranged in periphery, traversed by fibres in small patch, xylem exarch.

FAMILY–XX. *CHENOPODIACEAE*

Hairs various types, *epidermis* having thick walled; *stomata* confined to both the surfaces generally, paracytic. In the axis, anomalous secondary thickening occurred.

GENUS–30. *SUAEDA* Forssk.

Epidermis having thick walled; *stomata* confined to both the surfaces, paracytic. *Mesophyll*–palisade tissue consisting of water storage tissue. In the axis, anomalous secondary thickening occurred.

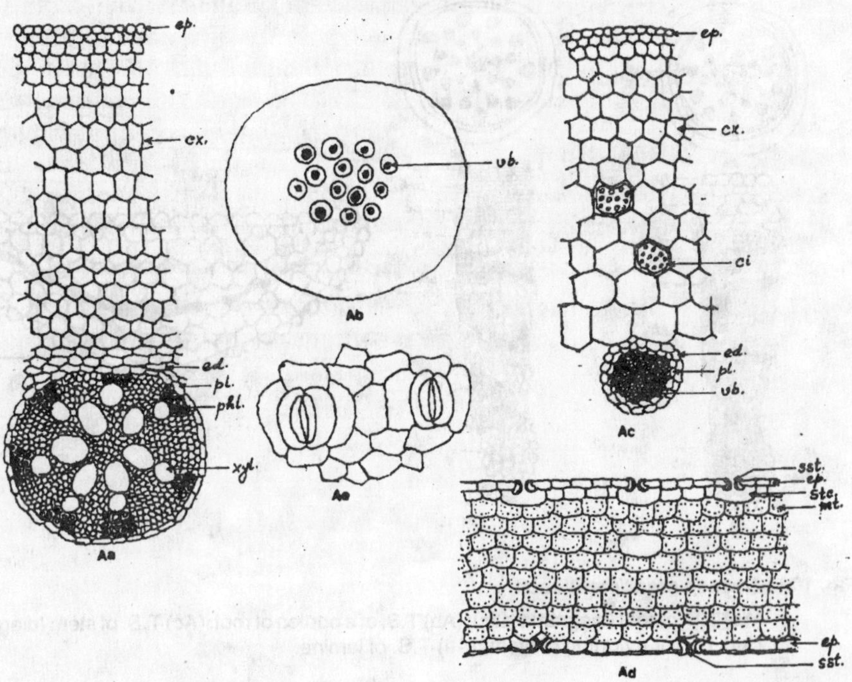

Fig. 101 : *Crinum defixum* Ker-Gawler

(Aa) T.S. of portion of root; (Ab) T.S. of Rhizome (diagrammatic); (Ac) T.S. of a portion of Rhizome; (Ad) T.S. of a portion of lamina; (Ae) Stomata.

30.1. *Suaeda maritima* Dumort.

(Fig. 102)

Leaf : Isobilateral, sessile.

Lamina : *Epidermis*-one layered thick, consisting of triangular thick walled cells; *stomata*-paracytic on both surfaces, *hypodermis*-one layered thick, *mesophyll*-consisting of water storage tissue. *Veins*-embedded in the mesophyll tissue.

Axis : *Epidermis*–one layered thick, *cortex*-consisting of multi-layered thick with few schizogenous cavities, *cortex*-consisting of parenchymatous tissue with numerous vascular bundles and secondary thickening occurred; vessels–small to medium sized, semi-ring porous, spiral thickening, traversed by parenchymatous tissue and sclerenchymatous tissue, *pith*-consisting of parenchymatous tissue.

Root : *Epidermis* (Epiblema)–cells replaced by cork, *cork*-multi-layered thick, *cortex*-consisting of multi-layered thick with schizogenous cavities, numerous vascular bundles in cortex and secondary thickening also occurred, vessels small to medium sized, semi-ring porous, spiral thickening, transversed by parenchymatous tissue and sclerenchymatous tissue.

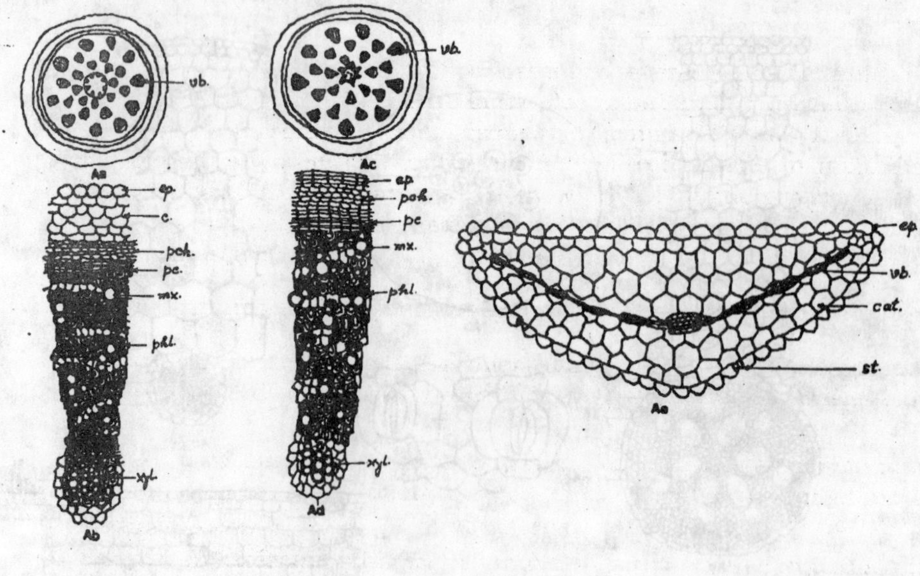

Fig. 102 : *Suaeda maritima* Dumort.

(Aa) T.S. of root (diagrammatic); (Ab) T.S. of a portion of root; (Ac) T.S. of stem (diagrammatic); (Ad) T.S. of a portion of stem; (Ae) T.S. of lamina.

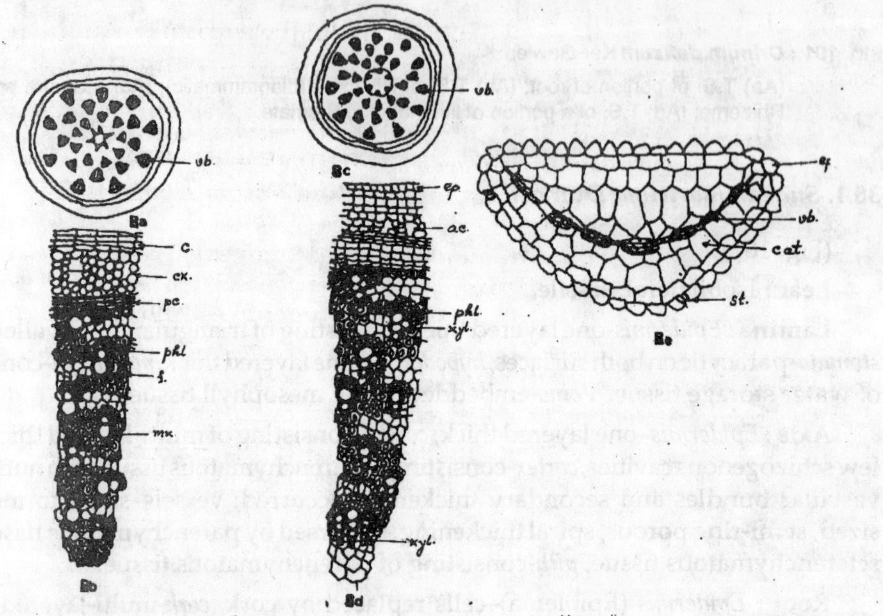

Fig. 103 : *Suaeda nudiflora* Roxb.

(Ba) T.S. of root (diagrammatic); (Bb) T.S. of a portion of root; (Bc) T.S. of stem (diagrammatic); (Bd) T.S. of a portion of stem; (Be) T.S. of lamina.

30.2. *Suaeda nudiflora* Roxb.

(Fig. 103)

Leaf : Isobilateral, sessile.

Lamina : *Epidermis*-one layered thick, consisting of triangular thick walled cell; stomata-paracytic, on both the surfaces, hypodermis-one layered thick, mesophyll-consisting of water storage tissue, *veins*-embedded in the mesophyll tissue.

Axis : *Epidermis*-one layered thick, *cortex*-consisting of multi-layered thick, *cortex*-consisting of parenchymatous tissue with numerous vascular bundles and secondary thickening occurred, vessels small to medium sized, semi-ring porous, spiral thickening, traversed by parenchymatous tissue and sclerenchymatous cells, pith-consisting of parenchymatous tissue.

Root : Epidermis (Epiblema)–cells replaced by cork, *cork*-multi-layered thick, *cortex*-consisting of multi-layered thick, numerous vascular bundles in cortex and secondary thickening also occurred, vessels small to medium sized, semi-ring porous, spiral thickening, traversed by parenchymatous tissue and sclerenchymatous tissue.

FAMILY–XXI. *BORAGINACEAE*

Hairs generally deciduous by touching, *petiole* have generally several bundles, anomalous secondary structure occurred infrequently.

GENUS–31. *HELIOTROPIUM* L.

Leaf isobilateral structure, *epidermis*-consisting of cells of thin walled, *stomata*-anomocytic, confind to both the surfaces; *mesophyll*-consisting of water storage tissue, *petiole* generally bi-collateral.

31.1. *Heliotropium curassavicum* L.

(Fig. 104)

Leaf : Isobilateral, petiolate.

Lamina : *Epidermis*-one layered thick; *stomata* anomocytic, confined to both the surfaces, not sunken, 22.8 ± 1 μ in length × 15.2 ± 1 μ in width, *mesophyll* consisting of water storage tissue, *veins*-embedded in the mesophyll tissue. *Petiole*-epidermis-one layered thick, *cortex*-consisting of parenchymatous tissue, *stele*-transverse section showing crescent shaped strand, *pericycle*-multi-layered thick, *vascular bundle*-open bi-collateral, xylem in middle and phloem in both sides of cambial ring.

Axis : *Epidermis*-one layered thick walled cells, *hypodermis*-one layered thick, cortex-consisting of multi-layered thick parenchymatous cells, *endodermis*-one layered thick, stele-transverse section showing circular cylinder, pericycle-one layered thick, *vascular bundle*-open collateral, phloem associated with parenchymatous tissue, cambial ring in middle, xylem / wood-vessels small sized, simple perforation, traversed by parenchyma, parenchyma-paratracheal, ray-2 celled thick, *pith*-consisting of large parenchymatous tissue.

Root : *Epidermal* cells replaced by cork in mature root, *cork* arising in middle portion of cortex, cortex multi-layered thick, stele-circular cylindric, xylem / wood-

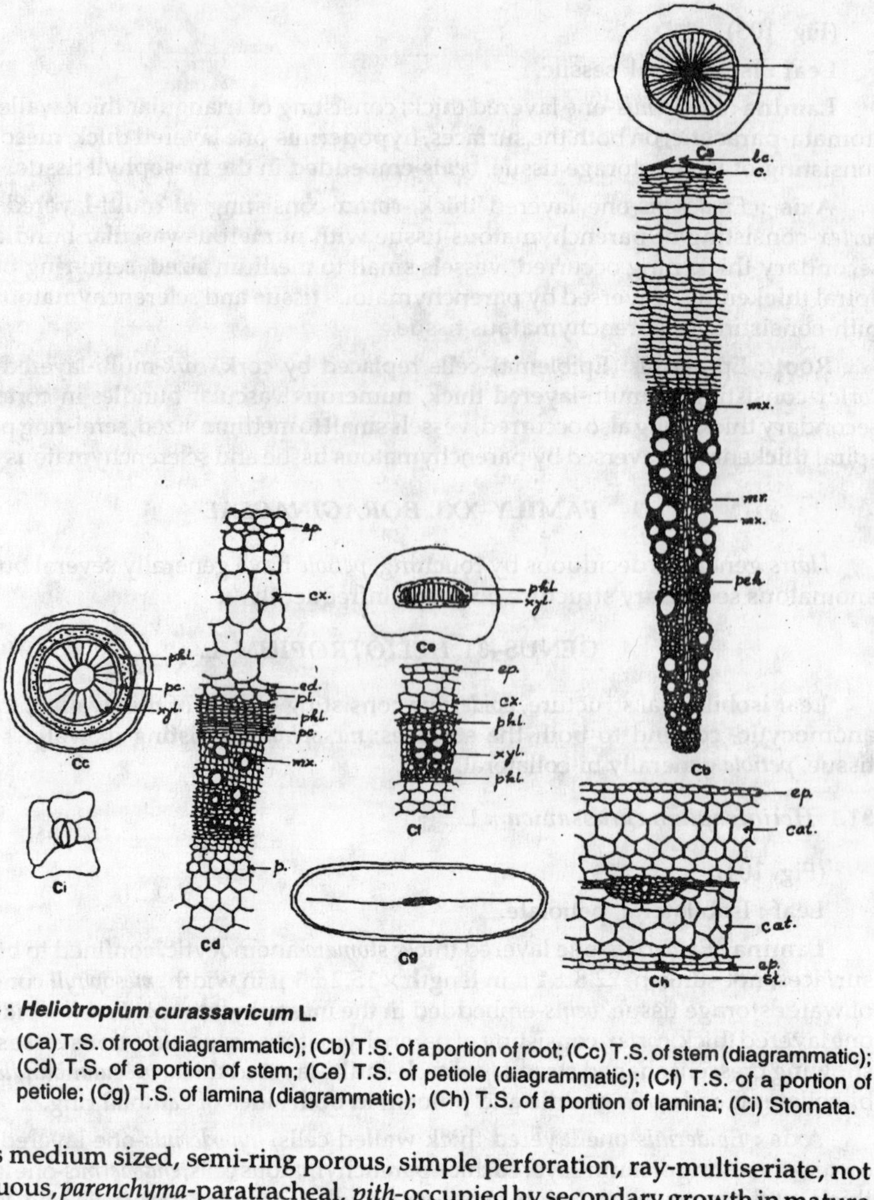

Fig. 104 : *Heliotropium curassavicum* L.

(Ca) T.S. of root (diagrammatic); (Cb) T.S. of a portion of root; (Cc) T.S. of stem (diagrammatic); (Cd) T.S. of a portion of stem; (Ce) T.S. of petiole (diagrammatic); (Cf) T.S. of a portion of petiole; (Cg) T.S. of lamina (diagrammatic); (Ch) T.S. of a portion of lamina; (Ci) Stomata.

vessels medium sized, semi-ring porous, simple perforation, ray-multiseriate, not continuous, *parenchyma*-paratracheal, *pith*-occupied by secondary growth in mature one.

FAMILY–XXII. *CONVOLVULACEAE*

Leaf-hair may have various types; presence of latex canal, *cuticle*-usually thin; *stomata* generally paracytic, *mesophyll*-palisade and spongy scattered, *veins*-embedded in the mesophyll, *petiole*-crescent shaped, *cork* not distinctly occurred. Vessels-variable in sized.

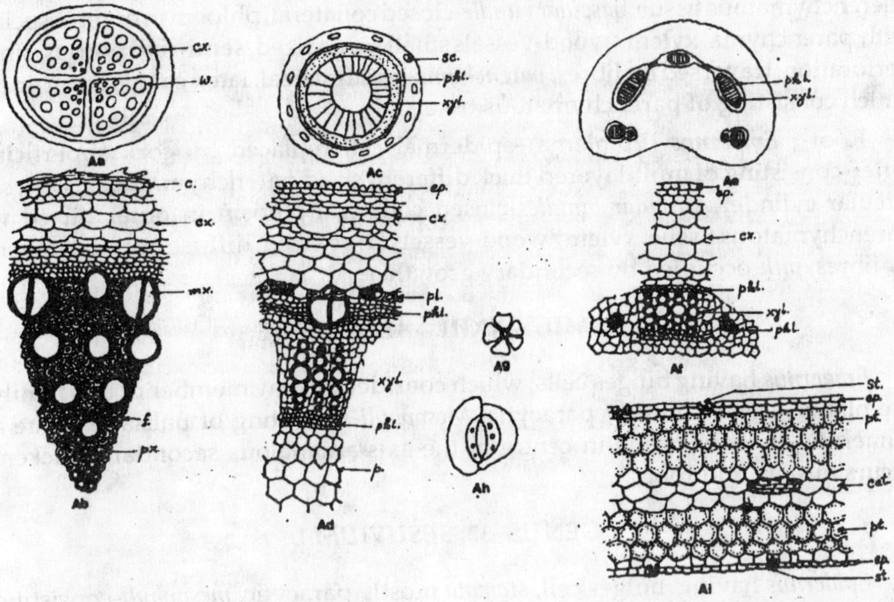

Fig.105 : *Ipomoea pes-caprae* (L.) Sweet

(Aa) T.S. of root (diagrammatic); (Ab) T.S. of a portion of root; (Ac) T.S. of stem. (diagrammatic); (Ad) T.S. of a portion of stem.; (Ae) T.S. of petiole (diagrammatic); (Af) T.S. of a portion of petiole; (Ag) Gland; (Ah) Stomata; (Ai) T.S. of a portion of lamina.

GENUS–32. *IPOMOEA* L.

Hairs of different types; glandular or non-glandular, *mesophyll*-palisade and spongy, inter-mixed; *petiole*-crescent shaped arc, presence of latex canal, veins-embedded in the mesophyll, *cork*-superficially occurred, vessels variable in size.

32.1. *Ipomoea pes-caprae* (L.) Sweet

(Fig. 105)

Leaf : Isobilateral, petiolate.

Lamina : *Gland*-one celled structure, bounded by 5-6- jacket cells, occurred on both the surfaces; cuticle usually thin, *epidermis*-one layered thick on both surfces; *hypodermis*-one layered thick in both surfaces; *stomata*-paracytic, confined to the both surfaces, slightly sunken, 29.7 ± 1 μ in length × 21 ± 1 μ in width, mesophyll-palisade tissue occupied major portion while in centre, presence of water storage tissue, veins-embedded in the mesophyll tissue. *Petiole*–epidermis-one layered thick, cortex-consisting of multi-layered thick parenchymatous tissue, *stele*-transverse section showing crescent shaped arc, *vascular bundle*-bi-collateral of 3 bundles and two accessory bundles in wings, xylem-mesarch in middle of which both sides covered by phloem, latex canal present in petiole.

Axis : *Epidermis*-one layered thick, *cork* found in tracing condition, cortex-multi-layered thick parenchymatous tissue with large schizogenous cavities, stele-transverse section showing circular cylinder, *pericycle*-one layered thick sclerenchymatous tissue, *vascular bundle*-closed collateral, phloem in outer associated with parenchyma, xylem / wood-vessels small-large sized, semi-ring porous, simple perforation, traversed by fibres, *parenchyma*-paratracheal, latex canals occur in *pith*- which consisting of parenchymatous tissue.

Root : *Epidermis* (Epiblema)–epidermal cell replaced by cork superficially, *cortex*-consisting of multilayered thick different sized parenchymatous tissue, *stele*-circular cylinder, *vascular bundles*-closed collateral-phloem in outer mixed with parenchymatous tissue, xylem / wood-vessels large sized, diffuse porous, traversed by fibres, *pith*-occupied by secondary growth.

FAMILY–XXIII. *AIZOACEAE*

Epidermis having bulges cells, which consider the few member of this family as 'ice plants', *stomata*-mostly paracytic. *Mesophyll*-consisting of palisade tissue and numerous aqueous tissue in centre in the axis-anomalous secondary thickening occurs in stem and root.

GENUS–33. *SESUVIUM* L.

Epidermis having bulges cell, *stomata* mostly paracytic, *mesophyll*-consisting of palisade tissue and aqueous tissue in centre, secondary thickening not much developed.

33.1. *Sesuvium portulacastrum* L.

(Fig. 106)

Leaf : Isobilateral, petiolate.

Lamina : Epidermis-one layered thick in both surfaces, cells globose in shape, cell wall thickly cutinized; *stomata*-confined to both the surfaces, paracytic, $29 \pm 1\ \mu$ in length $\times$ $25.3 \pm 1\ \mu$ in width in both surfaces, mesophyll-palisade in both surfaces while aquous tissue occur in centre, veins embedded in mesophyll. *Petiole*: epidermis-one layered thick, cortex-consisting of multi-layered thick loosed parenchymatous tissue, stele-transverse section showing 3 bundles of crescent shaped strand, vascular bundles-centric, *i.e.*, xylem in middle surrounded by phloem.

Axis : Epidermis-one layered thick, outer layered thickly cutinized, cortex-consisting of parenchymatous tissue loosely arranged, stele-circular cylinder, pericycle-one layered thick consisting of sclereids cells, vascular bundles-closed collateral, phloem in outer, xylem in inner traversed by fibres, vessels small, spiral thickening, parenchyma-paratracheal, pith-consisting of loosed parenchymatous tissue.

FAMILY–XXIV. *VERBENACEAE*

Leaf hair of various types, *hypodermis* of several layers; *stomata* confined to the lower side, palisade consisting of several layers mostly; *petiole-stele* showing various shaped strands. In the axis, *hypodermis* consisting of collenchymatous cells, *cork* originating superficially.

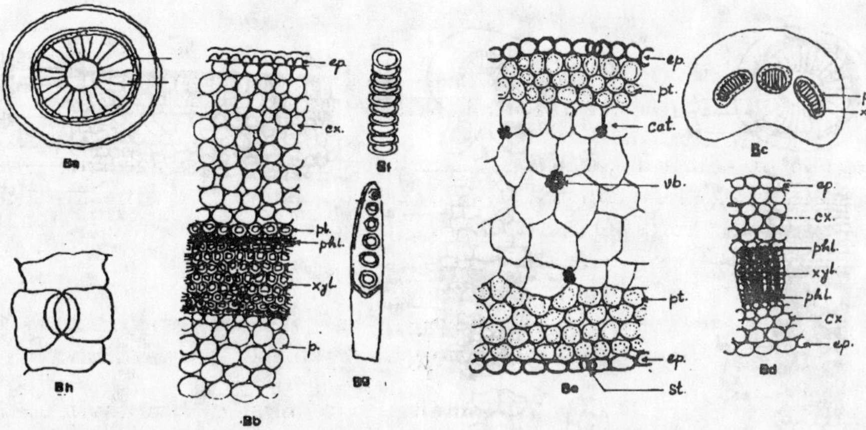

Fig. 106 : *Sesuvium portulacastrum* L.

(Ba) T.S. of stem (diagrammatic); (Bb) T.S. of a portion of stem.; (Bc) T.S. of petiole (diagrammatic); (Bd) T.S. of a portion of petiole; (Be) T.S. of a portion of lamina; (Bf) portion of vessel; (Bg) of T.S. portion of vessel; (Bh) Stomata.

GENUS–34. *CLERODENDRON* L.

Leaf hairs of various types, extra-floral nectaries occur on the lower surface of the leaf, *hypodermis* of several layer; *stomata* confined to the lower surface, anomocytic, *petiole*-accessory bundles in the wings, in the axis, *hypodermis* consisting of collenchymatous tissue, *pericycle* consisting of isolated bundles of sclerenchymatous cells, crystals in the pericycle.

34.1. *Clerodendron inerme* Gaertn.

(Fig. 107)

Leaf : Dorsiventral, petiolate.

Lamina : *Cuticle* one layered thick, hair of multi-cellular peltate, extra-floral nectaries on the lower surface, each gland surrounded by polygonal jacket cells, *epidermis*-one layered thick; *stomata*-anomocytic, 24 ± 1 μ in length and 14 ± 1 μ in width confined to the lower surface, not sunken, *mesophyll*-palisade tissue consisting of multilayered thick, spongy tissue-consisting of parenchymatous tissue, compactly arranged, *veins*-embedded in the mesophyll, *petiole-epidermis*-one layered thick, *cortex*-consisting of parenchymatous cells, compactly arranged, *stele*-transverse section showing semilunar strand, two solitary bundles occur in both wings, *pericycle*-one layered thick, *vascular-strand*-closed collateral, phloem in outer, xylem in inner, xylem mesarch.

Axis : *Hairs*–uniseriate, having 2-3 celled linear row, *epidermis*-one layered thick, *hypodermis*-consisting of collenchymatous tissues with lignified cells, *cortex*-consisting of parenchymatous cells, compactly arranged, scattered sclereids in cortex, *stele*-transverse section showing circular strand, *pericycle*-one layered thick, collenchymatous cells, phloem in outer traversed by parenchymatous cells, xylem/

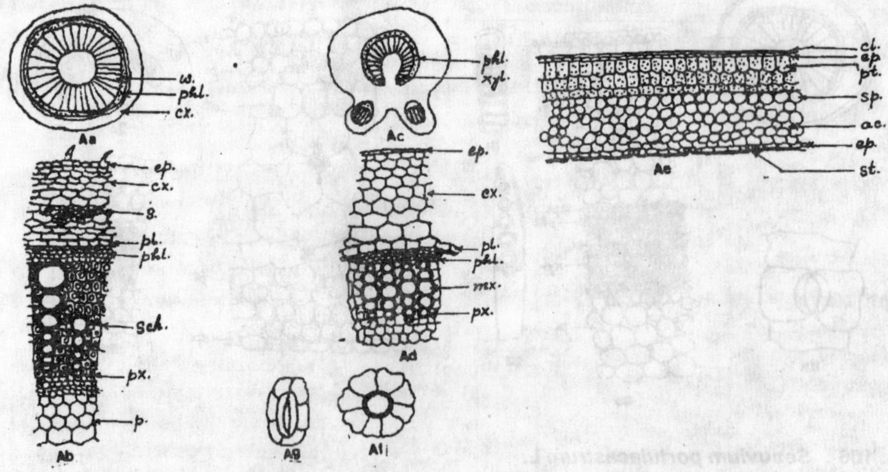

Fig. 107 : ***Clerodendron inerme*** **(L.) Gaertn.**

(Aa) T.S. of stem (diagrammatic); (Ab) T.S. of a portion of stem.; (Ac) T.S. of petiole (diagrammatic); (Ad) T.S. of a portion of petiole; (Ae) T.S. of a portion of lamina; (Af) One-celled gland; (Ag) Stomata.

wood-vessels medium sized, semi-ring porous, intravascular pitting very small, xylem traversed by fibres, ray-multi-seriate, 4-5 celled, fibre-septate, *pith*-consisting of parenchymatous tissue, compactly arranged.

FAMILY–XXV. *LORANTHACEAE*

The leaf hairs-complicated types. *Stomata* occur on both surfaces in isobilateral leaf and occur only to the lower surface of dorsiventral leaf. *Pericycle*-containing isolated strands of fibres, in the axis inter-xylary phloem occurred. Vessels very small, ray-multiseriate.

Key to the Genera of *Loranthaceae*

A. Pericycle consisting of sclerenchymatous cells of continuous ring

1. *Viscum* Tourn.

AA. Pericycle consisting of parenchymatous cells interrupted by sclereids patches

2. *Dendrophthoe* Mart.

GENUS–35. *VISCUM* Tourn. ex Linn.

Epidermal layers thickly cuticularized, *stomata* confined to the both surfaces, paracytic, *mesophyll*-associated with vascular bundles, *petiole*-stele showing closely placed of isolated vascular bundles. In the axis, xylem traversed by plenty of fibres and stone cells, ray-multiseriate-5-10 celled width.

35.1. *Viscum orientale* Willd.

(Plate–161 and Fig. 108)

Leaf : Isobilateral, petiolate.

Lamina : *Cuticle*-one layered thick, *epidermis*-one layered thick in both upper and lower surfaces, hypodermis-one layered thick below the upper surface; stomata-confined to the both surfaces, paracytic, 46.8 ± 1 μ in length and 15.8 ± 1 μ in width in both upper and lower surfaces, *mesophyll*-consisting of 5 celled thick, extended to the both surfaces, hexagonal cells, compactly arranged, *veins*-embedded in the mesophyll tissue, *petiole-epidermis*-consisting of one layered thick, cells more or less triangular shaped, thickly cuticularized, *cortex*-consisting of multi-layered thick, parenchymatous cells, compactly arranged, sphaeraphides scattered in cortex, *stele*-transverse section showing closely placed of 3 solitary bundles in a linear row, *vascular strand*-open collateral, phloem in outer associated with sclereids patches, traversed by parenchymatous cells, 2-layered cambial ring in middle, xylem-vessels medium sized, associated with parenchyma, ray-multiseriate, 4-6 celled width placed inter-bundles position, sclerenchyma-sclereids patch beneath the xylem strand.

Axis : *Epidermis*-one layered thick, covered by thick cuticle layer, *cortex*-consisting of multi-layered thick, compactly arranged, sclereids patches by regular intervals in a circular ring, *stele*-transverse section showing circular cylinder, *pericycle*-consisting of sclereids cells of 3 celled thick, *vascular strand*-closed collateral, phloem in outer associated with stone cells, xylem-traversed by fibres and frequent stone cells, ray-multiseriate, 2-3 celled width, upright cells, *pith*-consisting of parenchymatous cells, compactly arranged, in frequent brachysclereids and stones cells scattered in the pith.

GENUS–36. *DENDROPHTHOE* Mart.

Stomata confined to both the surfaces, paracytic, *mesophyll* extended to both the surfaces. In the axis-scattered sclereids in the cortex, *pericycle* not distinctly occurred, ray-multi-seriate, xylem traversed by fibres, brachysclereids, scattered in the pith.

36.1. *Dendrophthoe falcata (L. f.) Etting.*

(Fig. 109)

Leaf : Isobilateral, petiolate.

Lamina : *Cuticle*-one layered thick, *epidermis*-one layered thick in both upper and lower surfaces, *hypodermis*-one layered thick in both surfaces; *stomata*-paracytic, 33.5 ± 1 μ in length and 21 ± 1 μ in width, confined to both the surfaces. *Mesophyll*-consisting of 6-layered thick of which central 2-layers having comparatively small cell, *veins*-embedded in the mesophyll. *Petiole-epidermis*-one layered thick, cells are thickly cuticularized, cortex-consisting of multi-layered thick, parenchymatous cells compactly arranged. *Stele*-transverse section showing crescent shaped vascular strands, *vascular strand*-closed, collateral, phloem in outer and xylem in inner, xylem mesarch.

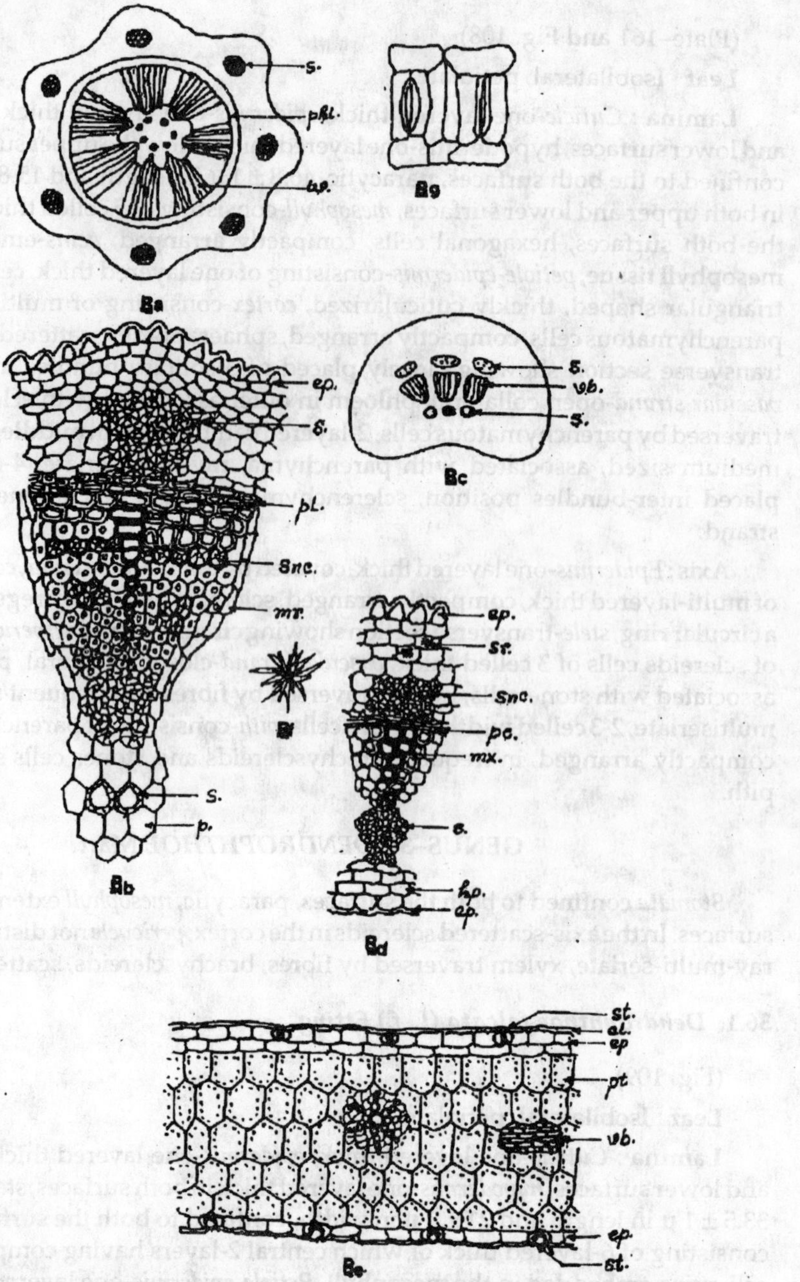

Fig. 108 : ***Viscum orientale*** **Willd.**

(Ba) T.S. of stem (diagrammatic); (Bb) T.S. of a portion of stem; (Bc) T.S. of petiole (diagrammatic); (Bd) T.S. of a portion of petiole; (Be) T.S. of a portion of lamina; (Bf) Sphaeraphide; (Bg) Stomata.

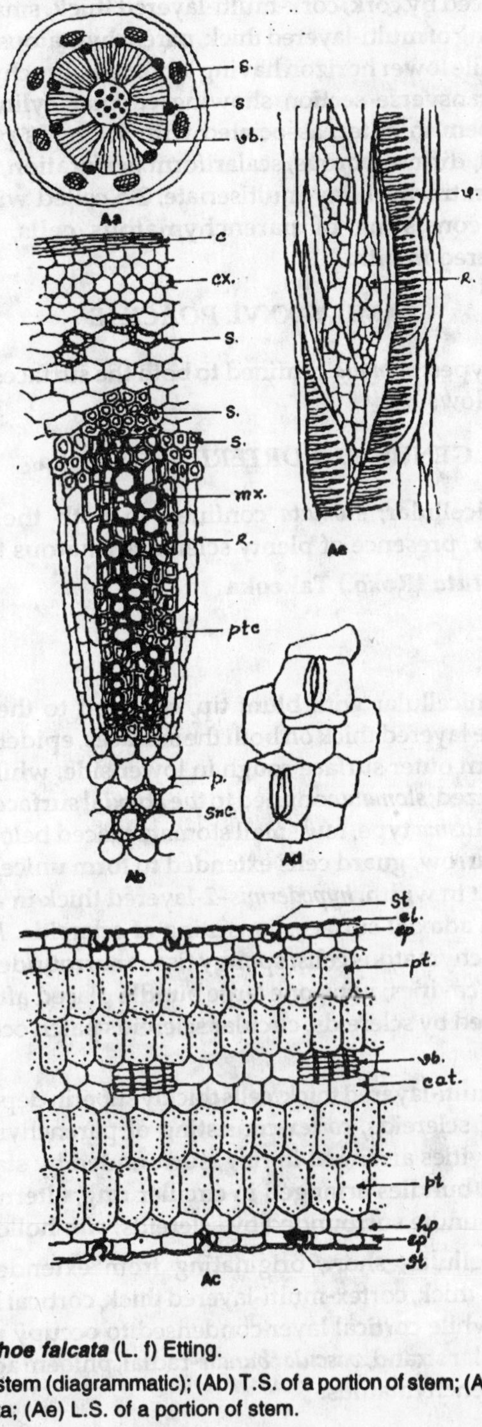

Fig. 109 : *Dendrophthoe falcata* (L. f) Etting.

(Aa) T.S. of stem (diagrammatic); (Ab) T.S. of a portion of stem; (Ac) T.S. of a portion of lamina; (Ad) Stomata; (Ae) L.S. of a portion of stem.

Axis : *Epidermis*-one layered thick in immature axis, while in mature one, epidermal layer replaced by cork, cork-multi-layered thick, small, cell, horizontally placed, *cortex*-consisting of multi-layered thick, parenchymatous cells, upper horizon having small cells, while lower horizon having large cells, brachysclereids, scattered in the cortex. *Stele*-transverse section showing circular cylinder, *vascular strand* closed collateral-phloem in outer, associated with sclereids patch, xylem/wood-vessels medium sized, diffuse porous, scalariform perforation, xylem traversed by fibres, *parenchyma*-paratracheal, ray-multiseriate, 2-5 celled width, homogeneous, upright cells, pith consisting of parenchymatous cells, loosely arranged, brachysclereids scattered in pith.

FAMILY–XXVI. *POACEAE*

Hairs of various types; *stomata* confined to both the surfaces, bundles scattered in the cortex, pith hollow.

GENUS–37. *PORTERESIA* Takeoka

Hairs mostly unicellular; *stomata* confined to both the surfaces, bundles scattered in the cortex, presence of plenty sclerenchymatous tissue, *pith*-hollow.

37.1. *Porteresia coarctata* (Roxb.) Takeoka

(Fig. 110)

Leaf : Isobilateral.

Lamina : *Hairs*-unicellular with blunt tip, confined to the furrow of adaxial surfaces, *epidermis*-one layered thick on both the surfaces, epidermal surface thickly silica deposited to form outer surface rough in lower side, while upper epidermal layer thickly cuticularized; *stomata* confined to the abaxial surface, 23.4 ± 1 μ in length × 8.5 ± 0.5 μ in width. *Alisma* type, few small stomata placed below the outer surface of the lateral side of furrow, guard cells extended to form unicellular hair, 7.2 ± 0.5 μ in length × 3.8 ± 0.5 μ in width. *hypodermis*–2-layered thick in abaxial surface and multi-layered thick in adaxial surface, consisting of sclereids. *Mesophyll*-palisade, associated with parenchymatous cells, spongy tissue surrounded the lateral side of the furrow, with large cavities; *veins*-one large bundle placed after 3 small bundles, each bundle surrounded by sclereids, circular sclereids patch occur in each ridge of adaxial side.

Stem : *Epidermis*-multi-layered thick, cells thickly suberin deposited. *Hypodermis*-multi-layered thick of sclereids, *cortex*-consisting of parenchymatous tissue, few large schizogenous cavities arranged in ring, surrounded by star shaped sclereids ring, *stele*-club shaped bundles arranged in circular ring, alternately with circular sclereids patch, each bundle surrounded by sclereids, *pith*-hollow cavity.

Root : Hair–unicellular, short, originating from extended epidermal cell, epidermis-one layered thick, cortex-multi-layered thick, cortical layer occupy much space in young root, while cortical layer condensed to occupy minimum space in mature root. *Stele*-circular strand, *vascular bundle*-radial, phloem adjacent to sclereids, xylem exarch, *pith* parenchymatous.

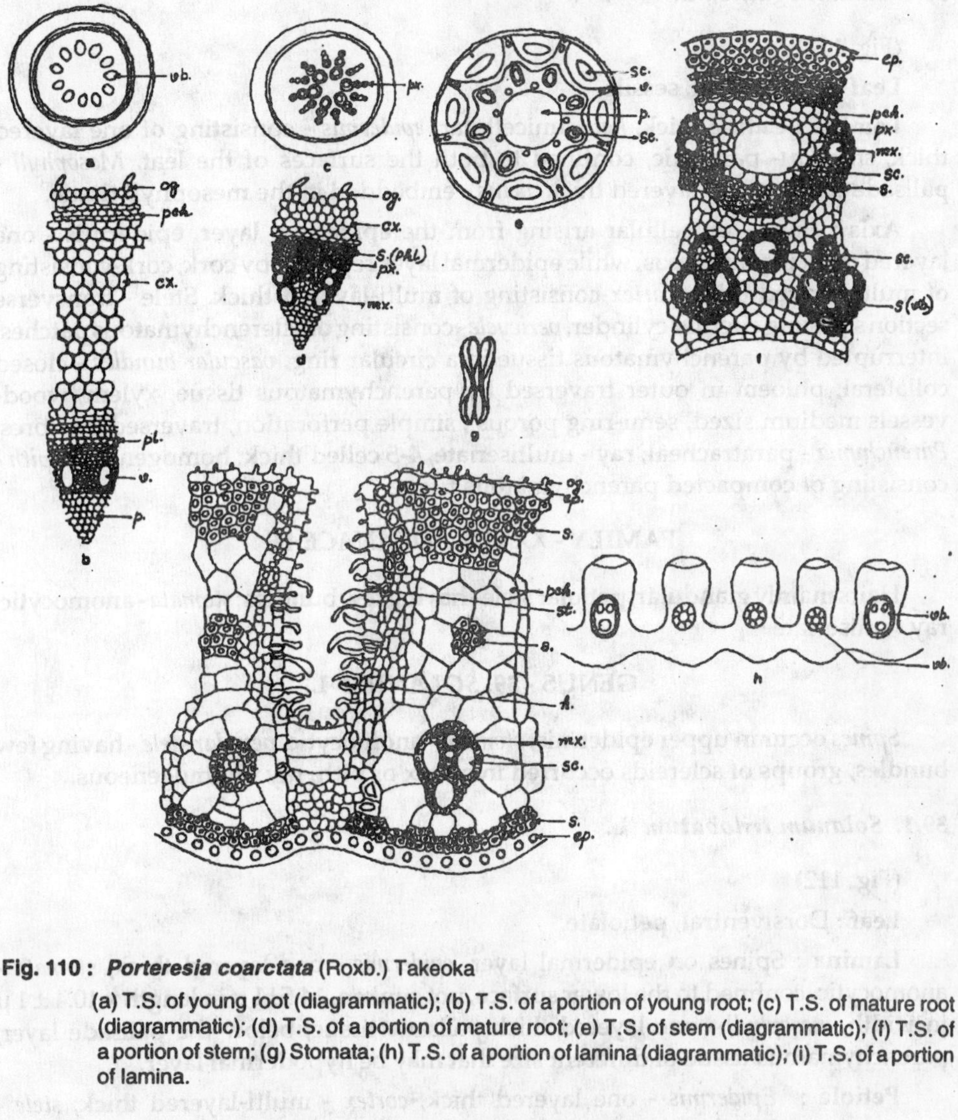

Fig. 110 : ***Porteresia coarctata*** **(Roxb.) Takeoka**

(a) T.S. of young root (diagrammatic); (b) T.S. of a portion of young root; (c) T.S. of mature root (diagrammatic); (d) T.S. of a portion of mature root; (e) T.S. of stem (diagrammatic); (f) T.S. of a portion of stem; (g) Stomata; (h) T.S. of a portion of lamina (diagrammatic); (i) T.S. of a portion of lamina.

FAMILY - XXVII. *TAMARICACEAE*

Mesophyll consisting of water storage cells; *stomata* mostly paracytic, confined to both the surfaces of the leaf.

GENUS -38. *TAMARIX* L.

Mesophyll consisting of water storage cells; *stomata* mostly paracytic, confined to both the surfaces of the leaf. Vessels small to medium sized, *parenchyma*-paratracheal. Rays-multiseriate.

38.1. *Tamarix gallica* L.

(Fig. 111)

Leaf : Isobilateral, sessile.

Lamina : *Cuticle* thick, *hair* -unicellular, *epidermis* - consisting of one layered thick; *stomata* - paracytic, confined to both the surfaces of the leaf. *Mesophyll* - palisade tissue multi-layered thick, *veins* - embedded in the mesophyll tissue.

Axis : *Hairs* - unicellular arising from the epidermal layer, epidermis - one layered thick in young axis, while epidermal layer replaced by cork, cork-consisting of multi-layered thick, *cortex*-consisting of multi-layered thick. Stele - transverse section showing circular cylinder, *pericycle* - consisting of sclerenchymatous patches, interrupted by parenchymatous tissue in a circular ring, *vascular bundles* - closed collateral, phloem in outer traversed by parenchymatous tissue, xylem/wood-vessels medium sized, semi-ring porous , simple perforation, traversed by fibres. *Parenchyma* - paratracheal, ray - multiseriate, 4-5 celled thick, homogeneous, *pith* - consisting of compacted parenchymatous tissue.

FAMILY - XXVIII. *SOLANACEAE*

Hairs mainly glandular, petiolar stele - having few bundles; *stomata* - anomocytic, ray - uniseriate.

GENUS - 39. *SOLANUM* L.

Spines occur in upper epidermis, *stomata* - anomocytic, *petiolar stele* - having few bundles, groups of sclereids occurred in cortex or pith, ray - homogeneous.

39.1. *Solanum trilobatum* L.

(Fig. 112)

Leaf : Dorsiventral, petiolate.

Lamina : Spines on epidermal layer, *epidermis* - one layered thick; **stomata** - anomocytic, confined to the lower surface, not sunken, 14.5±1 μ in length x 10.4 ± 1 μ in width, *mesophyll* - one layered thick palisade tissue, below the palisade layer, parenchymatous tissue of different size that may be hypodermal layer.

Petiole : *Epidermis* - one layered thick, *cortex* - multi-layered thick; *stele* - transverse section showing crescent shaped strand of 3 bundles; vascular *bundles-closed* bi-collateral - xylem in middle of which both sides phloem placed.

Axis : *Epidermis* - one layered thick, *cortex* - consisting of thick walled collenchymatous tissue, *endodermis* 2-layered thick : *stele* - transverse section showing circular cylinder, *pericycle*-one layered thick adjacent to sclereids, vascular *bundles* - closed collateral, phloem in outer associated with parenchyma, xylem/wood-vessels variable in shape, small to large, simple perforation, traversed by fibres, ray - uniseriate, homogeneous, *pith* - consisting of parenchymatous cells having groups of sclereids.

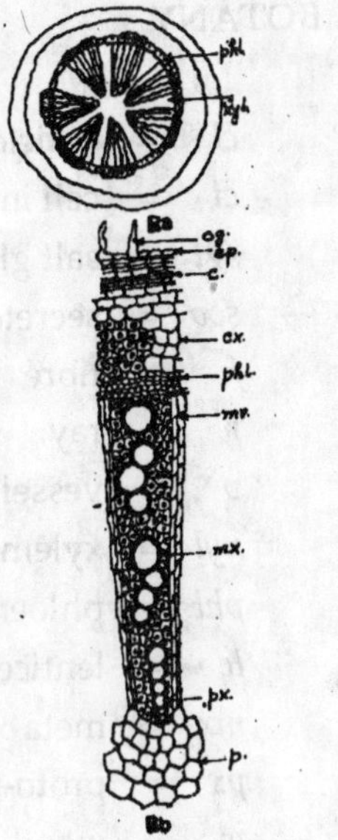

Fig. 111 : *Tamarix gallicha* L.

(Ba) T.S. of stem (diagrammatically); (Bb) T.S. of a portion of stem.

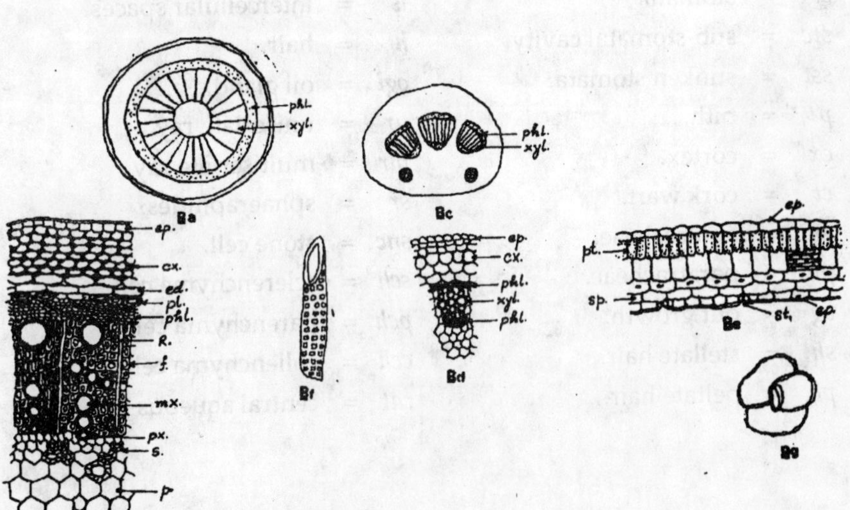

Fig. 112 : *Solanum trilobatum* L.

(Ba) T.S. of stem (diagrammatic); (Bb) T.S. of a portion of stem; (Bc) T.S. of petiole (diagrammatic); (Bd) T.S. of a portion of petiole; (Be) T.S. of a portion of lamina; (Bf) Vessel; (Bg) Stomata.

THE FOLLOWINGS ARE ABBREVIATION OF ANATOMICAL TERM IN BOTANY

c = cork.
ep = epidermis.
hp = hypodermis.
ac = air cavities.
sc = schizogenous cavities.
s = sclereids.
pl = pericycle.
ed = endodermis.
vb = vascular bundle.
cr = concentric ring.
mt = mesophyll tissue.
pt = palisade tissue.
sp = spongy parenchyma.
pc = primary-cambium.
sg = secondary growth.
st = stomata.
stc = sub-stomatal cavity.
sst = sunken stomata.
p = pith.
cx = cortex.
ct = cork wart.
ap = apotracheae.
ptc = paratracheae.
og = out growth.
sh = stellate hair.
ph = peltate hair.

cl = cuticle layer.
ci = cell inclusion
sgt = salt gland.
scv = secretory cavity.
f = fibre.
r = ray.
v = vessels.
xyl = xylem.
phl = phloem.
lc = lenticel
mx = meta-xylem.
px = proto-xylem.
w = wood.
sphl = secondary phloem.
hat = hypodermal aqueous tissue.
is = intercellular spaces.
h = hair.
ogl = oil gland.
ur = uniseriate ray.
mr = multiseriate ray.
sr = sphaeraphides.
snc = stone cell.
sch = sclerenchyma cell.
pch = parenchyma cell.
cch = collenchyma cell.
cat = central aqueous tissue.

22

CRITICAL ESTIMATION ON THE ANATOMICAL FEATURES OF MANGROVES & THEIR HALOPHYTIC ADAPTATIONS IN THE SALINE ZONES

MORPHO-ANATOMICAL FEATURES OF MANGROVES AND THEIR HALOPHYTIC ADAPTATIONS

Certain members of any plant community mostly live together, as because they may have the mutual physiological balance among themselves; and this physiological balance is developed by the modification of morphological and anatomical characters, which help these plants to survive even in hostile nature (Metcalfe, 1931). These unique modifications of morphological and anatomical features of any group of plants provide them continuous existence *via* modified reproduction means, followed by the evolutionary processes and protect them from the selective features of environmental stressed conditions (Metcalfe, 1931).

Specht (1981) has interpreted based on the review of fossil evidences and the modern distributions of mangrove species that the mangrove flora was originated in South-West and Northern Australia and was also distributed to New Guinea. Saenger (1982) has also concluded that the mangrove swamp is indeed a harsh and hostile environmental condition, in which mangrove flora exhibit the ranges of adaptations as physiological, morphological, anatomical and reproductive. The following halophytic mangrove families in the mangrove swamps of Sundarbans tidal forests have been analysed on the basis of their unique characters in respect to their halophytic adaptations.

FAMILY - I. *RHIZOPHORACEAE*

In the Indian, Sundarbans mangrove swamps, the Rhizophoraceae is the largest true mangrove family. The family-Rhizophoraceae, has 9 species under the 4 genera viz., *Rhizophora, Bruguiera, Ceriops* and *Kandelia.* All these members, generally grow on the river slopes and sometimes on flat ground, ridge forests in the river beds, creeks and canals sides or on the side of zig-zag river tributaries of the Sundarbans tidal swamps. As the coastal environment play very dynamic role in the swamp habitat, these mangrove flora have been provided with suitable adaptation and succession mechanism by the modifications of morphological, anatomical and reproductive structures. To survive in this typical intertidal saline environment, these mangrove flora must have adaptations of both the mechanical and the physiological processes, *viz.*, modification of vegetative parts and reproductive organs. The different modified parts of vegetative organs, *viz.*, leaf, bud, stem, root and canopy must help these plants to survive in this hostile environment; while, the modification of reproductive organs, *i.e.*, flowers, fruits, seeds and seed germination or the seedling development etc. largely help to protect their progenies in this typical tidal and saline environment. In the present observations, the modifications of vegetative and reproductive organs in relation to their halophytic intertidal adaptations on this harsh environmental conditions have been studied and discussed.

A. Vegetative Structures and Modifications among the members of Rhizophoraceae

(a) *Leaves* : Stace (1966) mentioned that mangroves show the xerophytic feature in response to water storing function, though his view has not been accepted by Uphof (1941).

(b) *Cuticle and Epidermis* : All these members of mangrove-Rhizophoraceae mostly have thick walled epidermis on upper surface and this epidermal layer is covered by thick, waxy cuticle (Artz, 1936; Stace, 1966 & Sidhu, 1975a). Furthermore, these thick waxy cuticle on upper surface, mostly cover cork wart and may reduce water loss on bright sunlight. Generally, single layered epidermis, on the upper surface are observed in the members of *Bruguiera* and *Ceriops* spp. : while, 2-3 layered thick epidermis are identified from the upper surface in the members of *Rhizophora* and *Kandelia* spp.

(c) *Hypodermis* : Bowman (1921) during his observation on several species of the family - Rhizophoraceae noted that the number of hypodermal layers are regulated by salt concentration in water media, where they grow; the plants growing on high salt concentration in water have much hypodermal layer in comparison to the plants growing on low salt concentration in water. Besides this statement, Bowman (1921) has also added that thicker leaves are found in salt concentrated sea water than that of the diluted less slaine sea water. According to Mullan (1931b) succulence is common character in mangrove leaves and this succulence depends on the nature of hypodermal aqueous tissue. Among these two members of the genus *Rhizophora, R. apiculata,* generally grow on higher salt concentration in soil and water phases in comparison to *R. mucronata.* However, *R. apiculata* shows 2-layered thick hypodermal tissues, while *R. mucronata* shows 3-layered thick hypodermal tissues; but, *R. apiculata* has large sized hypodermal aqueous tissue and occuppying large volume area than 3-layered thick hypodermal aqueous tissues of *R. mucronata*. Four members of the genus - *Bruguiera* exhibit 1-2 layered thick hypodermal aqueous tissue. *B. gymnorrhiza* and *B. cylindrica* generally grow on inner- or mid-estuary, at low salt concentration (25 ppt) in soil and water phases and shows 1-layered thick hypodermal aqueous tissue; while, *B. sexangula* and *B. parviflora* found on outer estuary, on high salt concentration (30-35 ppt) in the water and soil phases and they have 2-layered thick hypodermal aqueous tissues and seemed to be succulent leaves in comparison to the other members, *viz.*, *B. gymnorrhiza* and *B. cylindrica*. The genera, *Ceriops* have two members; *C. decandra* is found in diverse salt concentrated habitats, showing 2- layers of hypodermal aqueous tissue ; while, *C.tagal* distributed in inner- to mid-estuary of low salt concentration (15-25 ppt) in water and soil and exhibit two layers of hypodermal aqueous tissue, occupying smaller area. *Kandelia candel* also has 1-layered hypodermal aqueous tissue, occupying larger area, though *Kandelia candel* generally grow on low to medium salt concentration (20 ppt) in water and soil phases of mid estuary.

(d) *Stomata* : Mullan (1932) and Stace (1966) have mentioned that most of these mangrove genera have similar type of stomata, which are sunken in nature and placed on the lower surface of the leaf. Sidhu (1975a) did not trace any ecological significance of these stomata in relation to halophytic adaptation. All thess genera of the family - Rhizophoraceae have ***'anomocytic type'*** of stomata, which sunk beneath the level of epidermis, showing sub-stomatal cavity. Among these 9 members, *R. apiculata* shows largest size of stomata, while *B. cylindrica* has smallest size of stomata. *R. apiculata* exhibits lowest stomatal frequency (104.0/mm²) and stomatal index is 1.8 ± 0.5, while *K. candel* has largest stomatal frequency (260/mm²) and stomatal index is 10.13 ± 0.5μ. But it is uncertain to trace any concrete link on the stomatal frequency and stomatal index to the effect of salt concentration in water, though *R. apiculata* grow in higher saline zone in comparison to *K. candel*, grow on lesser saline habitats in the mid-and inner estuary.

(e) *Mesophyll Tissue* : The large spongy mesophyll tissue and strong palisade tissue are generally found in mangrove leaves (Baylis, 1940 & Stace, 1968). The edges of the vascular bundles are surrounded by tracheids (Baker, 1915; Watson, 1928 ; Rao & Sharma, 1968 and Breen & Jones, 1969). Sclereids are reported from *Rhizophora* spp. (Rao, 1957 and Shah & Sundarraj, 1965); these above mentioned sclereids and stone cells provide toughness and rigidity to the leaf, reduce damage from wilting (Malaviya, 1963). *R. mucronata* has 'H' shaped sclereids and other species show tracheids, scelereids which provide mechanical strength to the leaf from wilting and save from the natural calamities, like cyclone, sea surges etc.

Plant physiologists, like Stace (1966) and Sidhu (1975a) interpreted that presence of thick cuticle, waxy coating, sunken stomata and the distribution of cutinised and sclerenchymatous cells throughout the leaf may be responsible for the physiological dryness of the mangrove habitats and environment.

(f) *Salt Regulation* : Saenger (1982) mentioned that mangroves, growing in saline environment, absorb sodium and chloride ion. Three types of salt regulating mechanism, *viz.*, salt-exclusion, salt-extrusion and salt-accumulation have been suggested by different authors. Rains & Epstain (1967) and Scholander (1968) have suggested that all these members of the family - Rhizophoraceae possess salt-exclusion mechanism, *i.e.*, selective absorption mechanism. Among these 4 genera, only genus, *Rhizophora* exhibits cork wart, probably it excrete salt through this apertures, inspite of having selective absorption mechanism.

(g) *Wood Anatomy* : Janssonius (1950b) compared the wood of mangroves and non-mangrove species of the same family-Rhizophoraceae and he observed that total cross sectional area of vessels are generally greater in mangrove species; while, in mangroves individual vessel area is smaller in diam. This types of observation was previously drawn by Panshin (1932). The actual advantages of these modification have not been traced, but it may assume that it regulate water movement in the conducting tissues and also protect these vessels from rupturing. According to MaCro (1935) and Carlquist (1975) woods of mangrove - Rhizophoraceae have some distinctive features, *viz.*, (i) heavily barred, (ii) exclusively scalariform perforated plates, (iii) scalariform inter-vascular pitting, (iv) little vascicentric parenchyma, (v) numerous fine celled multiseriate rays, (vi) very few uniseriate rays, (vii) libriform fibres within conspicuous pits and (viii) unilaterally or bilaterally compound pitting between rays and vessels. MaCro (1935) has mentioned the presence of these above peculiarities, but he did not trace to signify these peculiarities in respect to their halophytic adaptation; whereas, Carlquist (1975) has concluded that the presence of above peculiarities are due to resist from collapse these vessels under tension. It has been observed during the present studies that *R. apiculata, R. mucronata, Kandelia candel* show more vessels per sq. mm and occupied large cross sectional area than that of the other 6 species, under 2 genera as *Bruguiera* spp. and *Ceriops* spp.

(h) *Root* : Several renowned mangrove workers, like Percival & Womersley (1975) , Gill & Tomlinson (1975-77) have listed various types of aerial roots, which are found in mangrove plants. Only 4 types of aerial roots occur in the members of the family Rhizophoraceae, as follows:

(1) *Stilt Roots* : Positively geotropic, arching (bow like) in the members of *Rhizophora* and straight in the members of *Ceriops, Kandelia candel;* aerial roots arise from trunk and moving downward to the substrate.

(2) *Knee Roots* : Negatively geotropic, arise from the horizontal positively geotropic cable roots or knee like formed by the modification of cable roots, occurred in the members of the genus *Bruguiera.*

(3) *Buttress Roots* : Flattened blade like structure, arising from the trunk base of the plants, occurred occasionally in *B. gymnorrhiza* and *K. candel.*

(4) *Aerial Roots* : Positively geotropic, arising from the trunk or branches but remain hanging and not touch the substrates, this is found in the members of the *Rhizophora.*

The atmospheric aeration functions of stilt roots has been remarked by Canoy (1975) and he observed that the increase in the number of stilt roots per sq. metre has direct relation to the increased temperature and simultaneously reduced oxygen concentration in air. These mangroves, growing at lower tidal level and slope land, have exhibited much array of the above ground roots. Presence of lenticels and aerenchymatous tissues in the above ground roots have reported by Brenner (1902), Chapman (1944), Gill & Tomlinson (1983) and they have been also interpreted the aeration mechanism of the above ground roots, which provide physiological service to the plants. Hosokawa, *et al.* (1977) observed that *Kandelia* sp. grow at flowing water exhibits buttress roots, but *K. candel* shows stilt roots when grow in normal water and no aerial roots form away from tidal inundation and surges. Generally, mangrove plants show poor underground roots; as such, the array of above ground roots provide mechanical support to the plants and protect them from different calamities, like land erosion, sea surges, cyclones, sea waves etc. It has been noted that density of stilt roots and number of branching of stilt roots increases on that portion of the plant which face to the river slopes or sides and is affected by various calamities, like land erosion, sea surges, stream flow etc.

B. Reproductive Structures and Modifications Among the Members of Rhizophoraceae

Reproductive Organs : Several Botanists like, Jones (1971), Grahm, *et al.*, (1975), Specht, *et al.* (1977), Byrness, *et al.* (1977), Naskar & Guha Bakshi, (1987) and Naskar (1993) have reported and observed that most of these mangroves species blossom during late winter to summer months. All these members of the family-Rhizophoraceae may blossom and flowers and fruits may present throughout the year, but most dominant flowering season is the summer months. From the field observations, it has been noted that the dominant flowering seasons exhibit during summer months and these have an environmental background. In the Sundarbans mangrove swamps, the salinity in water reaches higher during the winter to summer months and these high salt concentration become diluted by the freshwater during monsoon due to heavy rainfall and up-stream fresh-water supply. Juvenile mangrove plants mostly need freshwater or less saline water for their initial or early growth and development. During summer months most of these mangrove species have generally blossom, which produce mature seeds and seedling at the pre-monsoon or monsoon period. These Rhizophoraceae species produce good number of viviparous germinated seedlings without taking any dormant period of seed germination and these well grown seedlings/hypocotyles have fallen down in muddy soil with low concentrated salt. Otherwise, these viviparous seedlings or reproductive propagules may have to face physiological dryness or harsh climatic condition for viability in the presence of higher salinity in soil and water (Naskar, 1986).

According to Clifford & Specht (1979) the pollination mechanism of most mangrove species have been controlled or performed by winds, insects and birds. The two members of the *Rhizophora* have been pollinated by wind but the members of *Ceriops* are pollinated by moths. The large flowers of *Bruguiera* spp. (*B. gymnorrhiza* and *B. sexangula)* are pollinated by birds and the small flowers of *Bruguiera* spp. (*B. parviflora* and *B. cylindrica)* are also pollinated by butter-flies

(Tomlinson, *et al.*, 1978) ; the similar pollination mechanism have also been noted in the Sundarbans mangrove swamps. *Kandelia candel* is pollinated with the help of butter-flies.

The most distinctive, as well as remarkable features of the Rhizophoraceae members are vivipary or viviparous germination, *i.e.,* seed germination occurred within the fruits, while still hanging or attached to the mother trees. Hypocotyle of the viviparous germinated seedling enlarge through embryo development, without taking any dormancy period, which play a significant adaptation role in the mangrove environment. Several mangrove workers have put their opinion in respect of vivipary of Rhizophoraceae in the mangrove environment. According to MacNae (1968a) vivipary involves to rapid rooting. Joshi (1933 & 1972) have concluded that salt regulation mechanism and ionic balance are connected with this vivipary. Gill (1975) has put his opinion on the development of buoyancy and lastly Pannier & Pannier (1975) have interpreted the nutritional parasitism of vivipary. Saenger (1982) has also remarked that tidal buffering and wave borne objects provide threaten to survive these small propagules, while large propagules generally becomes out of threats for its vigour size and easily root development. All these above mentioned adaptation mechanisms have been studied time to time and have discussed along with the morphological discussion of each and every individual species.

C. Seeds and Seedlings Dispersal Mechanisms Among the Members of Rhizophoraceae

Propagules Dispersal and Establishment : The propagules of all these mangroves can float (Tomlinson, 1971; Gill, 1975; Steinve, 1975; Robinowitz, 1975 & 1978) and these propagules may also dispersed by water. According to Banus & Kolehmainen (1975), the buoyancy in the members of Rhizophoraceae are performed by radicles (hypocotyles). In the anatomical observation, the hypocotyles have shown circular array of the large air cavities below the epidermal layers. These cavities act as aeration and provide buoyancy mechanism. All these propagules mostly green in colour for having chloroplast and may take part in photosynthesis, until it becomes well established and produces the foliages. According to Robinowitz (1975) the large sized propagules can not establish its root in deep if water is insufficient. While, small propagules can not spread roots except shallow water. In our observation, the large propagules with sufficient water become erect and establish root quickly and show remarkable vegetative growth in the nutrient rich condition; while, large propagules in shallow water grow horizontally to the ground and vegetative growth and rooting pattern seems not to be good and show less performance.

FAMILY - II. *AVICENNIACEAE*

The family - Avicenniaceae is monotypic and the second largest one among these true mangrove members. This family has 3 distinct members in Indian Sundarbans under the single genus - *Avicennia* L. and single hybrid / variety has also been collected during field studies. These three members are *Avicennia alba* Blume, *Avicennia marina* Vierh. and *Avicennia officinalis* L. All these three members generally grow on the flat, intertidal lands of newly silted up soil zones. These members are considered as the pioneer plants of the Sundarbans mangals, which mostly establish its vegetative growth beyond the *Porteresia coarctata* meadow. These species grow in the physiological dry soil, on high concentrated saline soil and water phases and these lands inundate twice in a day with tidal sea water. As such, they exhibit halophytic adaptation by the modification of morphological, anatomical and reproductive structures. Their modification of both the morphological and anatomical structures in vegetative and reproductive organs, provide them smooth physiological balance to survive in this saline mangrove environment. From the literature review and field investigations it has been noted that these *Avicennia* spp. can grow or survive in varied ecological nitches, *i.e.*low saline to high saline, inter-tidal to above tidal zones and throughout the tropical, sub-tropical and occasionally in temperate zones of both old and new world. Their adaptation mechanism and different organs have been reviewed and stated below :

A. Vegetative Structure and Modifications Among the Members of Avicenniaceae

(a) *Leaves* : These *Avicennia* spp. also have shown the range of xerophytic features in respond to water storing function (Stace, 1966) and have thick walled epidermis, covered by waxy cuticle (Artz, 1936; Stace, 1966 & Sidhu, 1975a) to reduce water loss by reflecting bright sunlight in this physiological dry zones.

(b) *Epidermis* : All these three members of *Avicennia* L. have one layered, small celled thick epidermal layer, on both these surfaces of leaves, but their outer epidermal layer is covered by thick waxy cuticle.

(C) *Hairs* : Baker (1915) and Fahn & Shimony (1977) have mentioned that all these species of *Avicennia* L. show variously shaped hairs to cover salt glands and stomata for reducing water loss through these stomatal pores. These members of *Avicennia* L. also exhibit uniseriate tricellular hairs, including pear shaped terminal cells, arranged gregariously on the lower side and form velvet like appearance.

(d) *Stomata* : According to Mullan (1932) and Stace (1966) *Avicennia* spp. have also the sunken stomata with sub-stomatal cavity in the lower surface of the leaf. Clough (1979) has reported on the basis of photosynthetic study that stomata of *Avicennia* spp. remain open throughout the day in response to high temperature and probably induce the transpiration rates (Lewis & Naidoo, 1970 and Lugo, *et al.*, 1975). Among these three species of *Avicennia*, *A. marina* Vierh shows largest stomatal size (18.3 μ x 17 μ), while *A. officinalis* exhibits smallest stomatal size (13.7 μ x 10.45 μ) and *A. alba* Blume shows lowest stomatal frequency (327/mm^2), while *A. officinalis* L. shows highest stomatal frequency (483/mm^2). All these *diacytic stomata* are placed sunken in the lower surface with substomatal cavity, which also gregariously covered by hairs. But it has been unable to trace any ecological significance in relation to halophytic adaptation, though it is assumed that small aperture of sunken stomata transpire scanty amount of water by storing water in substomatal chamber and increase water holding capacity of leaves in these mangrove environments.

(e) *Hypodermis* : It has been discussed previously that succulence is common features in mangroves genera (Mullan, 1931b). Baylis (1940) has reported thal all mangrove species of *Avicennia* L. have multilayered thick hypodermal aqueous tissues. These species of *Avicennia* L. have 4-5 layered thick hypodermal tissue, placed below the upper epidermal layer, but exhibiting no hypodermal tissue in lower surface. Walter & Steiner (1936) have remarked that these succulence of leaves are the response to presence of chloride.

(f) *Mesophyll* : These members of *Avicennia* spp. show 2-3 layered thick large sized palisade tissues, which have similar behaviour to the hypodermal aqueous tissue as performing water storing function. Spongy mesophyll in these species of *Avicennia* L. occupy small area, consisting of compacted cells and interrupted by frequent stone cells. Presence of stone cells in spongy tissue probably provide toughness to the leaf. Several scientists like Baker (1915), Watson (1928), Rao & Sharma (1968), Breen & Jones (1969) have remarked that vascular bundles surrounded by tracheids; spongy mesophylls have stone cells (Bowman, 1921), undoubtedly provide toughness and rigidity to the leaf from wilting and other natural calamities; simultaneously, reduce transpiration rates by storing water (Malaviya, 1963).

(g) *Salt Regulation* : Three types of salt regulation mechanisms described by different plant physiologists, which have been classified by Scholander, *et al.* (1962). Of these 3 types, *i.e.*, salt exclusion, extrusion and accumulation; only two types of salt regulation *i.e*, salt exclusion and salt extrusion have been noted in these species of *Avicennia* L. (Rains & Epstain, 1967; Scholander, 1968 and Baylis, 1940).

Water is taken in and salt is largely filtered out by the process of salt exclusion and salt excreting through the glands is called salt extrusion. Baylis (1940) has also observed the salt extrusion mechanism in *A. marina*. Two types of salt glands have been found during this anatomical studies of the leaves, *viz.*, disc shaped single cell gland, surrounded by 5-6 jacket cells and multi-cellular gland, surrounded by 6-7 jacked cells. These two types of salt glands have not been found in both the surfaces of lamina and it is still unknown the specific function of these two types of glands. Mullan (1931a) and MacNae (1968a) mentioned that *Avicennia* L. have salt glands only in saline condition. But it has been observed that all these three species of *Avicennia* L., growing in pot and supplied by freshwater exhibit regular salt glands, as they have shown in saline condition. Generally, the *Avicennia* spp. grow on halophytic habitat, forming salt glands to control extra salt in physiological processes; which is obviously regulated by the genetic control. In halophytic condition, appearance of salt glands perform as salt regulation mechinary by excreting salt and help the plants to regulate the physiological processes.

(h) *Petiole* : Baylis (1940) interpreted that the presence of aerenchyma in the petiole has adapted for the aeration purpose. Members of *Avicennia* L.

exhibit air cavities towards endodermis, presumably to serve aeration function.

(i) *Stem* : According to Zamski (1978) anomalus secondary structure occurred in the members of *Avicennia* L., but he do not trace any ecological significance of this anomalous structure in response to halophytic adaptation. Zamski (1978) has mentioned that *Avicennia* L. show three successive cambial ring, consisting of alternative xylem and phloem which are derived from cortical parenchyma. It has been observed during these anatomical studies that immature shoot show one successive cambial ring, while two or three successive cambial rings formed through the gradual maturation. This formation of multi-successive cambial ring probably provide longevity to the stem and also uplift plenty of water to the upper most leaves. Baker (1915), Pansin (1932), Shedholme & Philipson (1966) and Gill (1971b) have concluded that anomalous wood structures and alternative concentric ring of xylem and phloem give unusual strength to protect conductive tissue and protect from the split of the trunk etc.

(j) *Root* : Saenger (1982) has concluded that to overcome these anaerobic soils, saline situation and devastating natural calamities, magroves form or develop its rooting system differently, which also provide both mechanical support and physiological function. Gill & Tomlinson (1977) have cited various type of underground rooting patterns in *Avicennia* spp., as (i) laterally spreading cable root, (ii) vertically descending anchor root and (iii) fine nutritive roots. All these members of *Avicennia* L. in Sundarbans swamps possess following types of rooting system:

1. *Pneumatophore* : The negatively geotropic unbranched roots in *A. alba* and *A. marina*, while unbranched but branched after injury or wounded has been found in *A. officinalis*.
2. *Aerial Roots* : Positively geotropic branched or unbranched roots, arising from the trunk or lower branches in *A. marina* and A. *officinalis*, but *A. alba* has not found any such aerial roots.
3. *Horizontally Spreading Cable Root* : Long spreading cable roots, parallel to the ground are present and support the trunk, absence of well developed strong anchor root in *Avicennia* spp.
4. *Anchor Root* : Anchor root is poorly developed, not found in deep soil.
5. *Nutritious Roots* : Hyphae like nutritious roots have spread in the upper horizon of the soil, for obtaining nutrients and essential elements from the soil.

Presence of aerenchymatous tissue and lenticels for the aeration purposes are common features of the above ground roots of *Avicennia* spp. (Brenner, 1902; Chapman, 1944 & 1947 and Gill & Tomlinson, 1971). These features are found in those plants which grow in the lower tidal levels and inundated frequently. These negatively geotropic roots perform gaseous exchange in the anaerobic soil condition in *Avicennia* (Scholander, *et al.*, 1955). According to Gessner (1967), mangroves show this air intake mechanism in that situation where poor soil aeration occurs. Saenger

(1982) has concluded that under continuous natural calamities or stressed situation, like soil erosion, sea surges, cyclones etc., these various above ground roots and cable roots are developed much to survive. Thom, *et al.* (1975) described that successful colonisation of *Avicennia* spp. occur at the harsh environmental condition in the Joseph Banafarte Gulf.

Baker (1915) has observed that lenticels in the cortex of pneumatophores come out through the apertures and appeared as black spots over the surface. Baker (1915) has also concluded that *A. marina* by developing spreading dense canopy maintain shade condition for its pneumatophores, which act more functional in that shade situation.

B. Reproductive Structures and Modifications Among the Members of Avicenniaceae

Reproductive Organs : Most mangroves commence their flowering on summer months and the background of the flowering commencement at the summer months have also been cited. Black & Roff (1972), Chanda (1977) and Clifford & Specht (1979) have observed that the pollination of *Avicennia* spp. is accelerated by bees. Perceval (1974) has observed the pollination mechanism of *A. germinans* by the bee, *Apis mellifera*. Chapman (1944) has interpreted that seed germination and enlargement of *Avicennia spp.* take place within the fruit, when it remain attached to the mother plant. But, their hypocotyles do not come out from the fruit like the members of Rhizophoraceae; this type of seed germination is known as *Cryptovivipary* or *incipient vivipary*. Tomlinson (1971), Gill (1975), Steinke (1975) and Rabinowitz (1975-78) have remarked that all these propagules of mangroves can float and are dispersed by water current. Clarke & Hannon (1969) have observed that dispersal of these propagules of *Avicennia* spp. mostly occurred in low tides, while Steinke (1975) has concluded that the pericarp of *Avicennia* spp. fruits provide buoyancy; Steinke (1975) has also interpreted that without pericarp these fruits lose its buoyancy and shedding of pericarp increase in medium salinity, as well as in high temperature; while shedding of pericarp become reduced in high and low salinity. According to Rabinowitz (1978) these propagules of *Avicennia* spp. are capable of growing roots, both in deep and shallow water. Among these three species of *Avicennia*, *A. officinalis* commence its flowering at late winter or pre-summer months, *A. marina* starts its flowring at pre-monsoon months while *A. alba* starts its flowering at monsoon to post-monsoon months. Simultaneously, fruiting period commence after 2-3 months latter, respectively. These minor variation of season to different species of the genera *Avicennia* L. have not been traced clearly, but it has been assumed that variable salinity concentration during different months may be responded for this variable flowering season. This view indicate that all these three species may not tolerate the similar salinity. The propagules of *A. marina* may tolerate higher saline condition than the propogules of *A. officinilis*, which may survive in medium saline condition and propagules of *A. alba* may like less saline condition for seedling establishment, during monsoon months and the substrate become less saline.

FAMILY - III. *SONNERATIACEAE*

Among the true mangrove members, the family-Sonneratiaceae has occupied the dominant position by exhibiting some specific morpho-anatomical features, adapted to these halophytic mangrove ecosystem. In Sundarban's mangrove swamps, three species, *viz.*, *Sonneratia apetala*, *Sonneratia griffithii* and *Sonneratia caseolaris* are available in different estuarine belt. The habitats of these three species of *Sonneratia* Linn.f. are distinct from one another, *viz.*, *S. caseolaris* is restricted in the inner estuary, *S. griffithii* found generally in mid- and outer-estuary and *S. apetala* showing its diverse range in the Sundarbans estuarine belts. In the Sundarbans esturaine belt, *Sonneratia apetala* grow on the flat land or river slopes or flats of newly silted up soil, beyond the *Porteresia coarctata* meadow like similar habitats of the members of *Avicennia* L. *Sonneretia caseolaris* grow, on river flats or ridge lands of inner or back estuary, generally in river side, reclaimed paddy fields or in brackishwater fish farms of the Sundarbans. *Sonneratia griffithii* is found in the mid- and outer-estuary, which can tolerate high salinity. Like other true mangrove members, the *Sonneratia* spp. survive in anaerobic soil of the saline zones and inundated twice per day. These *Sonneratia* spp. adapted best in these mangrove habitats by modification of morphological and anatomical structures. The modified vegetative and reproductive structures are adapted in these halophytic habitats for giving physiological balance to these species in harsh environmental condition.

A. Vegetative Structures and Modification Among the Members of Sonneraticeae

(a) *Leaves* : Artz (1936), Stace (1966) and Sidhu (1975a) have concluded that all members of the mangroves have thick walled epidermis covered by thick cuticle. As such, the *Sonneratia* spp. have also the typical leaves anatomy, like other mangroves, though these leaves partially isobilateral in nature.

(b) *Epidermis and Cuticle* : One layered thick epidermis covered by thick cuticle has been observed during anatomical studies in all these three members of *Sonneratia* Linn.f. studied from the Sundarbans.

(c) *Stomata* : As these three members of *Sonneratia* Linn.f. show isobilateral leaves, stomata are confined to both these surfaces and sunk beneath the epidermal layer, with substomatal cavity. *Anomocytic* type of stomata are found in all these three species of *Sonneratia* Linn.f. Among these three members, *S. griffithii* exhibits largest size of stomata (29.6 μ x 22.8 μ), while *S. caseolaris* show smallest size of stomata (19 μ x 10.1 μ) and *S. apetala* has medium sized stomata (25 μ x 13.3 μ); simultaneously, *S. griffithii* has highest stomatal frequency (233/mm^2). *S. caseolaris* show lowest stomatal frequency (128/mm^2) and *S.apetala* has medium stomatal frequency (183/mm^2). It has been discussed previously that *S. griffithii* grow in outer estuary and can tolerate high salinity. *S. apetala* grow on diverse habitat conditions, as it can tolerate diverse range of salinity, while *S. caseolaris* is restricted in inner estuarine belt of low saline zone. The presence of large sized stomata and highest stomatal frequency in *S. griffithii* has been probably interpreted that as *S. griffithii* grow in high saline zone it has to accumulate as well as excrete maximum salt through these stomatal aperture; but simulateneously a question may arise how do this species protect high rate of transpiration losses? For growing in low saline zone, *S. caseolaris* exhibits lowest stomatal frequency and smallest sized of stomata, while *S. apetala* growing in diverse saline belt show similar or medium sized of stomata like *S. griffithii* and medium stomatal frequency.

(d) *Mesophyll* : Walter & Steiner (1936) have concluded that succulence is common features in mangroves leaves, which may respond to the presence of chloride. According to Saenger (1982) *Sonneratia* spp. have isobilateral leaves, exhibits undifferentiated mesophyll tissue, which form central aqueous tissue. Among these three species *S. griffithii* show large number of central aqueous tissues, as it tolerate high salinity in water and probably deposit extra salt in aqueous tissue. While, *S. caseolaris* have central aqueous tissue, less than *S. griffithii*. Though, *S. apetala* shows 3-layers of central aqueous tissue, but individual cell have vigorous sizes and occupying maximum volume of the mesophyll layer in transverse section, and it tolerate high salt concentration as *S. griffithii*. According to Bowman (1921), Watson (1928) and Malaviya (1963) presence of stone cells and sclereids in leaves give rigidity and toughness to the leaves, reduce damage from wilting. *S. griffithii* have shown frequent sclereids in its leaves, while *S. caseolaris* and *S. apetal* show infrequent sclereids in leaves. A conclusion may be drawn in favour of frequent sclereids in the leaves of *S. griffithii* for rigidity and toughness as it grow in outer estuary, facing maximum natural calamities like sea surges, cyclones, wave tides etc.

(e). *Salt Regulation* : Three types of salt regulations mechanism were discussed by Scholander, *et al.* (1962). All these salt regulation processes were observed by different authors, time to time in all these members of *Sonneratia* Linn. f. Rain & Epstain (1967) and Scholander (1968) have mentioned salt exclusion mechanism found in *Sonneratia* spp., while Walter & Steiner (1936) observed salt extrusion mechanism through salt glands in *S. alba* (this species has been collected from Goa Coast). Joshi, *et al.* (1975b) have shown the salt accumulating mechanism in the old leaves of *Sonneratia* spp. Linn.f. Presence of succulent leaves, appearance of glands in dorsal sides in all these three members of *Sonneratia* Linn.f. indicating the salt extrusion and salt accumulation mechanism, respectively. But, salt exclusion mechanism is selective absorption mechanism, performed by roots and it has been reported in *Sonneratia* spp. (Scholander, *et al.*, 1968).

(f) *Petiole* : Presence of aerenchymatous tissue and appearance of sclereids presumably provide aeration process and toughness, respectively to their petioles.

(g) *Stem* : Janssonius (1950) has postulated that like other mangroves, *Sonneratia* spp. show more vessels per square millimetre, occupying larger total cross sectional area ; Venkateswarlu & Rao (1964) have supported Janssonius's statement after their observations on mangrove species. Three species, *viz.*, *S. caseolaris*, *S. apetala* and *S. griffithii* are available in Sundarban swamps show mostly similar features as more vessels per square millimetre, occupying larger total cross sectional area but individual vessel pores are smaller in size. Janssonius (1950) has remarked that this features of mangrove species may possibly increase the resistance to water movement in the conducting tissues and Reinders Gouwentak (1950) has added that it reduce rupturing the water column in narrow vessels.

(h) *Root* : According to Saenger (1982), as mangrove grow on anaerobic soil, mangrove possesses different types of rooting system. Gill & Tomlinson (1975-1977) have observed the different types of rooting system of mangroves. The family - Sonneratiaceae bear negatively geotropic branched or unbranched roots, the pneumatophores, arising from the horizontal cable root system. According to Saenger (1982) like other mangrove members, the family - Sonneratiaceae also possess the greatest array of above ground roots at lower tidal levels. Gill (1975) has concluded that pneumatophores are the distinct adaptive part of mangrove species. It has been observed, during the anatomical studies that the pneumatophores of *Sonneratia* Linn. f. showing following features :

(i) much lenticels in above surface,
(ii) in immature root, larger cortical zone with frequent air cavities,
(iii) in mature root, cortical zone reduce in area by extensive secondary growth,
(iv) cortical zone show excessive and frequent unbranched sclereids,
(v) secondary growth showing much vessels per sq millimetre area in cross section.

It may conclude from the above features of the *Sonneratia* spp. that presence of lenticels denote aeration requirements in anaerobic condition as immature root contains much lenticels, while mature roots having secondary growth and provide mechanical support for its vigorous growth in case of continued inundated areas. Presence of excessive sclereids give extra strength and much vessels increase water movement in harsh environmental situation. Besides, these physiological and mechanical characteristics, pneumatophores perform photo-synthesis activities as the cells of outer surface of immature one contain chloroplast.

B. Reproductive Structures and Modifications Among the Members of Sonnera tiaceae

Reproductive Organs : According to Jones (1971), Graham, *et al.* (1975), Specht, *et al.* (1977), Byrnes *et al.* (1977) most of these mangrove species exhibit their flowering commencement at summer month. Muller (1969), Faegri & Vander Pijt (1971) have described the nature of pollen grain and pollination mechanism of the family - Sonneratiaceae; pollination of *Sonneratia* spp. mostly performed by bats and moths.

In the inner estuary of Sundarbans, *S. caseolaries* has blossomed almost throughout the year, though the dominant flowering and fruiting season is summer.

Previously discussed that all mangroves propagules are capable of floating (Tomlinson, 1971 ; Gill, 1975; Steinke, 1975 and Robinowitz, 1975-1978). According to Saenger (1982) several seeded fruits in *Sonneratia* spp. may disperse as the reproductive unit. No significant and specific development procedures have been observed in the reproductive unit of *Sonneratia* spp., though multiseeded fruits have been noted with hypogeal germination, after short dormant period.

FAMILY - IV. *COMBRETACEAE*

The tropical family - Combretaceae has several genera, of which only two genera, *i.e., Laguncularia* and *Lumnitzera* are available in the mangrove swamps. *Lumnitzera* L., a major elements of mangrove flora (Tomlinson, 1986) is found in the frequnent tidal inundated mangrove swamps of the Sundarbans. Like other mangrove members, *Lumnitzera* sp. modify its vegetative and reproductive parts to adapt in anaerobic tidal influenced soil.

A. Vegetative Structures and Modifications Among the Members of Combretaceae

(a) *Leaf* : Like other mangroves its leaves show a range of xeromorphy in response to water conserving functions (Stace, 1966 and Millar, *et al.*, 1975).

(b) *Epidermis and Cuticle*: *Lumnitzera* sp. has thick walled epidermis, covered by thick cuticle (Artz, 1936; Stace, 1966 and Sidhu, 1975a). It has been observed that one layered thick epidermis covered by thick cuticle, which may respond to the physiological dryness on the mangrove environment and this observation has also been made by Saenger (1982).

(c) *Mesophyll* : Succulence is a common feature in *Lumnitzera* sp. like other mangroves leaves (Mullan, 1931b) and that succulence is responsible due to presence of chloride salt (Saenger, 1982). According to Saenger (1982) *Lumnitzera* sp. shows partially isobilateral leaves, as such transverse section of the leaves show undifferentiated mesophyll cells, which form central aqueous tissues. It has been observed that mesophyll tissue is placed beneath the epidermal layer of both the surfaces, while aqueous tissue are found in the centre.

(d) *Stomata* : According to Mullan (1932) and Stace (1966) most of the mangrove genera have similar type of stomata, sunken in nature. In *Lumnitzera* sp. stomata is *anomocytic*, sunken, 23 μ in length and 16.6 μ in width and having the number 130/mm².

(e) *Salt Regulation* : Joshi, *et al.* (1975b) suggested that *Lumnitzera* sp. is the salt accumulating mangrove, which deposit sodium in the stems, pneumatophores, bark and in the older leaves. Joshi, *et al.* (1975b) have also remarked that the old senescent leaves deposit sodium and chloride ; simultaneously, potassium and phosphorus withdrawn by leaf fall and by this means it excrete excess salt.

(f) *Stem* : According to Janssonius (1950) mangrove wood show more vessels per sq millimetre, while individual pore is small but total cross sectional area is larger. It has been observed during this anatomical studies that *Lumnitzera* sp. wood shows more vessels per sq millimetre, while individual pore is small. This modification of anatomical studies may uplift more water through the conducting tissues (Janssonius, 1950). The appearance of large stone cells in the pith is probably help to keep intact the pith from any shrinkage.

B. Reproductive Structures and Modifications Among the Members of Combretaceae

Reproductive Organ : According to different workers, most of the mangroves commence its flowering during the summer month. Tomlinson, *et al.* (1986) observed that pollination in *Lumnitzera racemosa* has been performed by insects. There is not much information about the reproductive organ of *Lumnitzera* sp., though it has been observed during the field investigations that seeds are without vivipary and exhibit hypogeal germination. This species is commonly restricted in the eastern coastal area of the mangrove swamps of the Sundarbans and shows very poor species distribution.

FAMILY -V. *ARECACEAE*

The monocotyledon family - Arecaceae has several tropical genera, of which two genera, *viz.*, *Phoenix* and *Nypa* are present in the mangrove inundated swamps of the Sundarbans. The genus *Phoenix* has one species, *P. paludosa* in the tidal inundated swamps, while monogeneric flora, *N. fruticans* is also found in coastal sea facing areas of tidal inundated mangrove swamps in the Sundarbans. *N. fruticans* is treated as major elements of mangrove flora (Tomlinson, 1986), but Tomlinson (1986) did not refer or describe this *Phoenix* sp. as mangrove; though this species show most of these features like major mangrove members. These two species adapt to the mangrove swamps by modifying its vegetative parts.

A. Vegetative Structures and Modifications Among the Members of Arecaceae

(a) *Leaf* : Likewise the mangroves it has thick walled epidermis, covered by thick cuticle.

(b) *Epidermis and Cuticle* : One layered thick epidermis covered by thick cuticle.

(c) *Hypodermis* : Succulence is not common features in their leaves and these common palms respond to the presence of chloride (Walter & Steiner, 1936). In *Nypa* sp. and *Phoenix* sp. hypodermis consist of one layered thick aqueous tissue.

(d) *Stomata* : Stomata is *Pandanus* type, sunken in nature, 30.1 μ in length x 20 μ in width, in *Nypa* sp. and 20 μ in length x 12 μ in width in *Phoenix* sp., 233 stomata/mm² in *Nypa* sp. and 383 stomata/mm² in *Phoenix* sp. The sunken type of stomatal behaviour respond to the transpiration rate (Lewis & Naidoo, 1970 and Lugo. *et al.*, 1975). Stomata is found in leaf axis also in *Nypa* sp., probably perform to aeration purpose in anaerobic soil condition.

(e) *Mesophyll* : Mesophyll in both *Nypa* sp. and *Phoenix* sp. consisting of palisade tissue and spongy tissues are associated with sclereids and stone cells. Besides mesophyll tissue, vascular bundles, surrounded by stone cells provide toughness and rigidity to the leaf, reduce damage from wilting and involve in conserving water (Malaviya, 1963).

(f) *Salt Regulation* : There is not much information about the salt regulation mechanism in *Nypa* sp. and *Pohenix paludosa*. It is noted from the anatomical features, both these species have no salt gland, absence of succulent leaf; both the species regulate salt probably by effective selective absorption mechanism, *i.e.*, exclusion, water is taken up and salt is filtered out.

(g) *Root* : *Nypa* sp. having no specialised rooting system, but fibrous root exhibit large air cavities in cortical zone. This air cavities probably perform aeration purpose in the anaerobic soil condition. *Phoenix* sp. sometimes exhibits specialised rooting system, *i.e.*, *Pneumatothodes* in compacted anaerobic soil condition and that is also supported by Jost, 1887. Simultaneously, no specialised aerial rooting systerm is found in the loose soil condition. These negatively geotropic aerial roots, *i.e.*, Pneumatothodes have large air cavities in cortical region; however, mature pneumatothodes show more cavities than young pneumatothods. These air cavities are associated with sclereids bands, which probably save the cortical zone from any shrinkage. These pneumatothodes perform aeration purpose (Jost, 1887). It has also been observed during the field investigations that density of pneumatothodes are proportionate to the much compacted anaerobic soil condition.

(B) Reproductive Structures and Modifications Among the Members of Arecaceae

Reproductive Organ : Most species of mangroves commence flowering in the summer months (Jones, 1971) these palms also blossom during summer. In *Nypa* sp. the pollination is taken part by insects (Uhl, 1972) and particularly by Drosophilid flies (Issig, 1973). In *Phoenix* sp. pollination is taken part by wind. According to Tomlinson (1971) *Nypa* sp. shows viviparous fruit, which is dispersed by water. Vivipary fruit has some adaptive significances as rapid rooting (MacNae, 1968), salt regulation (Joshi, 1973), ionic balance (Joshi, *et al.*, 1972), development of buoyancy (Gill, 1975) etc., while *Phoenix* sp. has small fruit, dispersed by water and germination pattern is hypogeal.

FAMILY - VI. *STERCULIACEAE*

The tropical and sub-tropical family - Sterculiaceae has several genera of which only one genus *Heritiera* Ait. is found in mangrove swamps of the Sundarbans. The genus *Heritiera* Ait. have three species (Banerjee, *et al.*, 1989), which are distributed in the mangrove areas, while one species, *H. fomes* Buch. Ham. has been collected and studied from Sundarbans. *H. fomes* Buch. Ham. has been treated as 'minor elements of mangrove flora' (Tomlinson, 1986). Like other major and minor elements of mangrove flora, *H. fomes* Buch. Ham. modify its vegetative and reproductive parts to adapt in the halophytic environment.

A. Vegetative Structures and Modifications Among the Members of Sterculiaceae

(a) *Leaf* : Leaves of *H. fomes* shows a range of thick and coriaceous in respect to water conserving function (Miller, *et al.*, 1975).

(b) *Epidermis and Cuticle* : *H. fomes* has thick walled epidermis covered by thick cuticle and 'stellate scales' (Jones, 1971). Saenger (1982) has suggested that the presence of stellate hairs, which covers salt glands and stomata and is probably reduce water loss via these stomatal apertures.

(c) *Hypodermis* : 2-several layers hypodermal aqueous tissue found in *Heritiera* Ait. (Stace, 1966, Gressner, 1967), but it has been observed during the anatomical studies that *Heritiera fomes* shows only 1-layer thick hypodermal aqueous tissue. It is previously discussed that hypodermal aqueous tissue respond to the presence of chloride.

(d) *Stomata* : According to Mullan (1932) and Stace (1960) most of the mangrove genera having similar type of stomata, sunken in nature. In *Heritiera fomes* Buch. Ham. stomata is *diacytic* type, sunken and 12.6 μ in length and 14.5 μ in width and have the numbers, ranges between 866-1000/mm².

(e) *Mesophyll Tissue* : 1-layered thick palisade tissue and spongy parenchymatous zone become compartmented by sclereids and sometimes by horizontally placed vessels; which form large cavities. Presence of sclereids and horizontally placed vessels provide toughness and rigidity to the leaves and is probably to conserve water (Malaviya, 1963).

(f) *Salt Regulation* : No evidences have been cited about the salt excretory mechanism in *Heritiera* Ait.; as such, no references have been discussed here though it has been observed during the field study and *in vivo* culture of *Heritiera fomes* seedling that seedling needs fresh water for its good germination and developmental stages. Even it has been observed that the mature trees of 3 years old shows standard growth in freshwater condition than in saline water.

(g) *Petiole* : In *Heritiera fomes*, petiole shows accessory vascular bundle patches in pith and secretory like cavities. Beside these, petiole shows circular thick sclerenchymatous ring, is probably to give strength the petiole from wilting and any other damages.

(h) *Stem* : According to Janssonius (1950) mangrove wood shows more vessels per sq millimetre, while individual pore is small but total cross sectional area is larger. But this features, in case of *Heritiera littoralis*, are not found (Reinders Gouwantak, 1953). It has been observed during anatomical studies that vessels are arranged vertically as semi-ring porous pattern. Few schizogenous cavities are occurred in pith.

(i) *Root* : According to Saenger (1982), *Heritiera* spp. shows upward thickening of roots, i.e., root buttresses. Beside the root buttresses, it has been observed during field study, *Heritiera fomes* exhibiting flattened negatively geotropic aerial roots, arising from long cable horizontal roots. The presence of root buttresses and flattened aerial roots provide mechanical support to the plant and aerial roots presumably perform aeration mechanism.

(B) Reproductive Structures and Modifications Among the Members of Sterculiaceae

Reproductive Organ : Like other mangrove members, *Heritiera fomes* commence its flowering during summer months. Pollination in *H. fomes* is performed by honey bees and flies. According to Sidhu (1975b), seeds in *Heritiera* sp. germinate after a short resting period.It has been observed that dry fruit coat, *i.e.*, pericarp help the buoyancy in *Heritiera fomes*. It has also been studied that the germination pattern is hypogeal, and seeds prefer freshwater more than saline water during its germination. Even early developmental stage of seedling exhibit vigorous growth in freshwater condition in comparison to their growth in saline water.

FAMILY - VII. *MELIACEAE*

The family - Meliaceae has diverse genera, distributed in diverse environmental situation, *viz.*, mesophytic, halophytic both in tropical and temperate region. The family - Meliaceae have two halophytic genera, *viz.*, *Xylocarpus* Koen. and *Aglaia* Lour. and these are available in the Sundarbans. Only two species under the genus, *Xylocarpus, viz.*, *X. granatum* Koen. and *X. mekongensis* Pierre. are found in Sundarbans mangroves swamps, while one species *Aglaia cucullata* (Roxb.) Pellegrin has been reported in Sundarban mangrove biosphere, but it has not been collected during this field investigation. The genera *Xylocarpus* Koen. are described as true mangroves by different workers, as it exhibit various halophytic features, such as bearing pneumatophore, buttress, sunken stomata, hypodermal aqueous tissue etc. These various adaptive behaviours of the vegetative and reproductive parts are reviewed in below as follows :

A. Vegetative Structures and the Modifications Among the Members of Meliaceae

(a) *Leaves* : Their pinnately compound leaves have thick walled epidermis, covered by thick cuticle.

(b) *Epidermis and Cuticle* : 1-layered thick epidermal layer in both surfaces, lower epidermal layer more thick, covered by thick cuticle.

(c) *Stomata* : Like other mangrove members, *Xylocarpus* spp. have sunken stomata but having no sub-stomatal cavity (Sidhu, 1975a). Both the two species, *viz.*, *X. granatum* and *X. mekongensis* have anomocytic type of stomata and exhibiting sunken behaviour, but no sub-stomatal cavity. Among these two species of *Xylocarpus* Koen., *X. granatum* has large sized (19 μ × 15.2 μ) stomata and exhibiting lower stomatal frequency–516/mm^2, while *X. mekongensis* has small sized (16.4 μ ×13.3 μ) stomata and showing higher stomatal frequency-905/mm^2. Both these two species grow on mid- and outer-estuary and tolerate same saline condition.

(d) *Hypodermis* : Stace (1966) has observed that 1-layered thick hypodermal aqueous tissue found in *Xylocarpus* spp. but *X. granatum* shows two layers of asymmetrical hypodermal aqueous tissue in the upper surface and *X. mekongensis* shows one layer thick hypodermal aqueous tissue in upper surface. According to Saenger (1982) succulence in leaves depend on the presence of well developed water storing hypodermal aqueous tissue. Walter & Steiner (1936) have stated that succulence is due to the presence of chloride.

(e) *Mesophyll* : Both the members of *Xylocarpus* Koen. show multi-layered thick palisade tissues and multi-layered thick spongy parenchymatous tissue; spongy tissues contain well developed air cavities. Presence of stone cell provide toughness and rigidity to the leaf and damage from wilting and to conserve water (Malaviya, 1963). Stone cells surrounding the conducting elements are observed in the leaves of *Xylocarpus* (Watson, 1928).

(f) *Salt Regulation* : Joshi, *et al.* (1975b) has concluded that like other mangroves, *Xylocarpus* spp. deposit sodium and chloride in the stem, bark of pneumatophore and the older leaves, through the method of salt accumulation processes. Saenger (1982) has also observed the removal of excess salt deposited in older leaves and subsequent annual leaf fall in case of deciduous species like *Xylocarpus* Koen. In our present investigations the annual leaf fall has been noted during late winter months at the Sundarbans.

(g) *Stem* : No anomalous secondary structure are found, vessels are well decorated by xylem fibres and rays.

(h) *Root* : Percival & Womersley (1975) and Gill & Tomlinson (1975-1977) have concluded that to perform physiological function regularly in anaerobic soil, *Xylocarpus* spp. show unbranched negatively-geotropic roots, *i.e.*, pneumatophore. Saenger (1982) observed secondary thickening in the pnemuatophores of *Xylocarpus* spp. Troll & Dragendroff (1931), Metcalf (1931), Chapman (1944) and Gill (1975) have concluded the presence of pneumatophores indicate as obvious adaptive mechanism. In Sundarbans mangrove forest, both the species of *Xylocarpus* Koen. show unbranched and infrequent branched pneumatophores, which exhibit well developed secondary thickening. The development of secondary thickening provide mechanical support to the trunk in continued soil eroded land.

B. Reproductive Structures and Modifications Among the Members of Meliaceae

Reproductive Organs : Jones (1971), Graham, *et al.* (1975), Specht (1977), Byrnes, *et al.* (1977) have pointed out that most of the mangroves have the flowering season in summer months. Sidhu (1975b) observed the seed germination pattern of *Xylocarpus* spp. and concluded that in *Xylocaprus* spp. seeds germinate after a short dormant period. According to Saenger (1982) small seeds face large threat by sea wave, surges etc., while large seeds face minimum threat; so non-viviparous seeds, like the seeds of *Xylocarpus* spp. face minimum threat by water movement.

Propagules Dispersal and Establishment : According to several mangrove researchers, *viz.*, Tomlinson, 1971; Gill, 1975; Steinke, 1975; Robinowitz, 1975 & 1978; Naskar, 1993 and Naskar & Guha Bakshi, 1987 mangroves propagules are buoyant. Multi-seeded fruits of *Xylocarpus* spp. can float in sea water with the presence of endosperm (Saenger, 1982). In Sundarbans mangrove swamps, the multi-seeded fruit float a few days and seeds become detached one another after dissociation of fruit coat. Seeds germinate after a short dormant period and germination type is hypogeal.

FAMILY–VIII. *EUPHORBIACEAE*

The widely distributed family–Euphorbiaceae has several genera, of which the genus *Excoecaria* L. has one/two species in the tidal zone; among them only one species *E. agallocha* L. is treated as 'minor elements of mangrove flora' (Tomlinson, 1986). This species is widely distributed and it is also available in inundated mangrove swamps of the Sundarbans. To grow in halophytic environmental situation, *E. agallocha* L. modify its vegetative structure and reproductive structure very much similar to the other mangrove members.

A. Vegetative Structure and Modifications Among the Members of Euphorbiaceae

(a) *Leaves* : *Excoecaria agallocha* L. has thick walled and two layered epidermis, covered by thick cuticle and which may be responsible to conserve the water.

(b) *Hypodermis* : 1-layered thick hypodermal aqueous tissue is found in *Excoecaria agallocha* L., which respond to succulent behaviour of leaves (Saenger, 1982).

(c) *Stomata* : In *Excoecaria agallocha* the stomata is *paracytic* type, 31 μ in length and 13 μ in width and having the number-250/mm².

(d) *Mesophyll Tissue* : Palisade tissue consisting of 2-layered thick of asymmetrical cell layer and spongy tissue with frequent air spaces. Vascular bundles surrounded by fibres and few sclereids. The presence of air cavities associated with fibres and sclereids respond to store water and reduce damage from wilting, respectively (Malaviya, 1963).

(e) *Salt Regulation* : *Excoecaria agallocha* L. shows effective and selective absorption mechanism, *i.e.*, salt exclusion (Rains & Epstain, 1967; Scholander, 1968). In this mechanism, water is taken in and salt is selectively filtered out. Joshi, *et al.*, (1975b) have shown that *Excoecaria agallocha* deposit its salt through salt accumulating mechanism. Joshi, *et al.* (1975b) have also concluded that old scerescent leaves deposit sodium and chloride and simultaneously, withdraw potassium and phosphorus by leaf fall. Saenger (1982) has concluded that *Excoecaria agallocha* L. excretes excess salt through leaf fall as it is found in deciduous plants.

(f) *Stem* : Transverse section of the stem showing semi-ring porous vessels, which is traversed by continuous rays. Base of the trunk show several galls which may act as aeration purpose and also excrete excess salt.

(g) *Root* : Saenger (1982) has observed that *Excoecaria agallocha* L. has no specialised aerial root system, as such its cable root spreads on the surfaces areas for the assimilation of oxygen and is found to avoid waterlogged area. It has been observed that cable roots in *Excoecaria agallocha* L. shows multi-lenticels on the outer surface, probably act for the aeration purpose.

B. Reproductive Structures and Modifications Among the Members of Euphorbiaceae

Reproductive Organ : Catkin like male inflorescence with 2-celled pollen grain found in *Excoecaria* (Venkateswarlu & Rao, 1975) and it is wind pollinated (Clifford

& Specht, 1979). It has been observed that the germination pattern in *Excoecaria* sp. is epigeal.

FAMILY–IX. *MYRSINACEAE*

The tropical family-Myrsinaceae has several genera, of which only one genus, *Aegiceras* Gaertn. f. is found in mangrove area of the Sundarbans. The genus *Aegiceras* Gaertn. f. has also two mangrove associated species, of these *Aegiceras corniculatum* (L.) Blanco is available in the Sundarbans mangrove swamps and teated as 'minor element of mangrove species' (Tomlinson, 1986). As it grows in the halophytic environment, *Aegiceras corniculatum* (L.) Blanco modify its vegetative and reproductive parts like other mangrove members.

A. Vĕgetative Structures and Modifications Among the Members of Myrsinaceae

(a) *Leaves* : *A. corniculatum* has thick walled epidermis, covered by thick cuticle, like other mangrove species of the inter tidal zone.

(b) *Hypodermis* : Stace (1966) concluded that hypodermal aqueous tissue present in *Aegiceras* sp. It has also been observed that in *Aegiceras corniculatum* 2-layered thick hypodermal tissue occurred beneath the epidermal layers. It has also been cited previously that the appearance of hypodermal aqueous tissue respond to the presence of chloride (Walter & Steiner, 1936).

(c) *Stomata* : In *Aegiceras corniculatum*, the stomata is *anomocytic* type, sunken in nature; 27.2 μ in length and 15.8 μ in width and having the number–228/mm^2.

(d) *Mesophyll Tissue* : 2-layered thick palisade tissues and multi-layered thick compacted spongy tissue are found in *Aegiceras corniculatum* (L.) Blanco. Mesophyll tissue has frequent sclereids and stone cells. The presence of stone cells and sclereids provide toughness and rigidity to the leaf and reduce damage from wilting and conserving water.

(e) *Salt Regulation* : *Aegiceras* sp. regulate salt through the selective salt absorption mechanism, *i.e.*, exclusion (Rains & Epstein, 1967). Cardale & Field (1971) have concluded that salt excreted through the salt gland is known as extrusion, occurred in *Aegiceras* sp. Cardale & Field (1971); Bostrom & Field (1973) had done ultrastructural studies on salt glands of *Aegiceras* sp. and described that salt gland consisting of 24-40 celled compacted secretory cells. It has also been studied that salt gland in *Aegiceras* sp. appeared in both upper surface of lamina and mesophyll tissue and consisting of 16-32 cells, having a large basal cell. Cardale & Field (1971) have also concluded that *Aegiceras* sp. exhibits its salt gland in both saline and non-saline environment. It has also been observed that mature salt gland exhibit vacant cavity by dissolving its adjacent secretory cells. The vacant cavity is known as schizogenous secretory cavity.

(f) *Petiole* : In *Aegiceras corniculatum*, the petiole has frequent schizogenous secretory cavities and appearance of these cavities is probably to excrete salt.

(g) *Stem* : Like Rhizophoraceous members, *Aegiceras* sp. exhibits similar wood characters. The wood of *Aegiceras* sp. shows more vessels per square millimetre, while individual pores are smaller, but total cross sectional area are more than non-mangrove members. In immature stem, the pith shows frequent sclereids and large stone cells.

(h) *Root* : No aerial roots are found.

B. Reproductive Structure and Modifications Among the Members of Myrsinaceae

Reproductive Organs : *Aegiceras* sp. has scented flower, so its pollination is performed by honey bee, *Apil dorsata* in the Sundarbans.

The seeds germinate and embryo develop within the fruit, but not come out through the pericarp and this type of germination is known as *crypto-vivipary*, occurred in *Aegiceras* (Carey & Fraser, 1932). Crypto-vivipary exhibits similar type of adaptive mechanism, *i.e.*, seed and seedling developments like viviparous members. According to Saenger (1982) seedlings in *Aegiceras* sp. behave as propagules, which may disperse and establish in suitable condition. Clarke & Hannon (1969) have observed that propagules in *Aegiceras* sp. disperse with high tidal current. After few days of floating, the crypto-vivipary may settle on the silty soil, establish and develop.

FAMILY–X. *RUBIACEAE*

The widely distributed family-Rubiaceae has several genera, of which two genera such as *Scyphiphora* and *Hydrophylax* have been collected during studies from the coastal areas of the Sundarbans. *Scyphiphora* sp. is treated as 'minor elements of mangrove' flora (Tomlinson, 1986), while *Hydrophylax* sp. is known as herbaceous coastal flora. Both these genera are found in inundated mangrove swamps of the Sundarbans. *Scyphiphora* sp. has only one species, *S. hydrophyllacea* Garten.f. and *Hydrophylax* has one species, *H. maritima* L.; both these species are common in the eastern zone of mangrove swamps in Sundarbans. As these plant grow in inundated halophytic area, they exhibit modified vegetative parts for adaptations.

A. Vegetative Structures and Modifications Among the Members of Rubiaceae

(a) *Leaves* : Member of the mangrove, *Scyphiphora hydrophyllacea* Gaertn.f. has thick walled epidermis, covered by thick cuticle.

(b) *Epidermis and cuticle* : 1-layered thick epidermis, covered by thick cuticle.

(c) *Stomata* : *Scyphiphora hydrophyllacea* has *paracytic* stomata, sunken in nature with 29 μ in length × 20 μ in width, frequency 189 stomata/mm^2 and *Hydrophylax maritima* has 36.5 μ in length × 22.3 μ in width stomata with frequency of 100 stomata/mm^2.

(d) *Mesophyll Tissue* : Both the species exhibit mesophyll tissue, which form central aqueous tissue extended to the upper and lower sufraces. Rao & Wan (1969) have reported that stone cells present in the leaves of *Scyphiphora* sp. and the appearance of stone cells provide toughness to the leaf. Succulence is common feature in mangrove leaves (Mullan, 1931b) and this succulence behaviour in response to be presence of chloride. It has been observed that central aqueous tissue maintain the succulent behaviour of leaves in both the *Scyphiphora* sp. and *Hydrophylax* sp.

(e) *Salt Regulation* : Though sufficient informations are not available about salt regulation mechanism in *Scyphiphora* sp. and in *Hydrophylax* sp., yet it has been observed from anatomical features, *i.e.*, presence of central aqueous tissue, nature of stomata and it may be concluded that both these species regulate its salt by salt accumulating mechanism.

(f) *Root* : In *Scyphiphora hydrophyllacea* cortex showing frequent air cavities which probably act aeration function, as such *S. hydrophyllacea* exhibits above surface ground roots.

B. Reproductive Structures and Modifications Among the Members of Rubiaceae

Reproductive Organ : Like other mangrove members, *Scyphiphora* sp. and *Hydrophylax* sp. commence its flowering during the summer months and pollination in *Scyphiphora* sp. is taken part by insects. It has not been observed the seed germination and seedling developmental stage, but *Scyphiphora* sp. and *Hydrophyllax* sp. are non-vivi-parous species.

FAMILY–XI. *PLUMBAGINACEAE*

The family–Plumbaginaceae has several genera, of which only one genus, *Aegialitis* Br. is distributed in the mangrove swamps of Sundarbans. The genus *Aegialitis* Br. has few species of which only one species, *i.e., Aegialitis rotundifolia* Roxb. is common as 'minor elements in mangrove swamp' (Tomlinson, 1986). Like other mangrove members, *Aegialitis rotundifolia* Roxb. modify its vegetative and reproductive parts for adaptation in halophytic environment.

A. Vegetative Structures and Modifications Among Members of Plumbaginaceae

(a) *Leaves* : *Aegialitis rotundifolia* Roxb. has thick walled epidermis, covered by thick cuticle.

(b) *Epidermis and Cuticle* : 1-layered thick epidermal cell, covered by thick walled cuticle.

(c) *Hypodermis and Mesophyll* : According to Saenger (1982), the leaf of *Aegialitis* sp. do not exhibit any hypodermal layer and mesophyll tissue. But it has been observed through the anatomical studies that the leaf of *Aegialitis* sp. has no hypodermal layer, but show one layer thick vertically arranged palisade tissue and compacted spongy parenchymatous tissue. Mesophyll tissue exhibit frequent stone cells and branched sclereids, which provide toughness and rigidity to the leaf and reduce damage from wilting and conserving water.

(d) *Stomata* : In *Aegialitis rotundifolia* Roxb. the stomata is anomocytic type, sunken in nature and 28.5 μ in length and 17 μ in width and the number is 133/mm^2.

(e) *Salt Regulation* : *Aegialitis* sp. excretes its salt through the selective salt absorption mechanism, *i.e.*, exclusion (Pains & Epstain, 1967). Atkinson, *et al.* (1967) have concluded the salt excretes through the salt gland is known as extrusion, occurred in *Aegialitis* sp. It was also observed that salt glands in *Aegialitis* sp. counting 4 to 8 palisade like cells surrounded by 1-2 epidermal jacket layers. Salt glands in *Aegialitis* sp. occurred in both the surfaces of the lamina, below the epidermal layer; as such it is seemed to be depressed nature.

(f) *Petiole* : Baylis (1940) remarked that appearance of air cavities in petiole to serve aeration purpose, this is also one type of adaptation mechanism. In *Aegialitis* sp. petiole show numerous air cavities and scattered vascular bundles in cortex. Appearance of frequent sclereids in cortex is probably to provide strength the cortical zone and maintain shape from shrinkage.

(g) *Stem* : In, *Aegialitis rotundifolia* cortex have air cavities, which is associated with frequent numerous branched sclereids. Vascular bundles show anomalous secondary growth. Pith has also air cavities, associated with branched sclereids. But presence of air cavities in stem is probably to serve aeration purpose in anaerobic situation, though exact function of air cavities in stem is unknown.

(h) *Root* : Cortex shows air cavities, associated with sclereids and stellate hair like fibres. Secondary growth with distinct ray, found in root.

B. Reproductive Structure and Modifications Among the Members of Plumba ginacae

Reproductive Organ : In *Aegialitis rotundifolia*, pollination is performed by insects Crypto-vivipary are noted in *Aegialitis* sp., crypto-vivipary exhibits similar type of adaptive mechanism, *i.e.*, seed and seedling development like other viviparous members.

FAMILY–XII. *ACANTHACEAE*

The family–Acanthaceae has widely distributed genera, of which the genus, *Acanthus* L. has the common species, *A. ilicifolius* L. in the mangrove swamps of the Sundarbans. *Acanthus ilicifolius* L. is treated as mangrove associated flora (Tomlinson, 1986), exhibit few adaptive behaviours for growing in mangrove area. *A. volubilis* Wall. is another rare species with few halophytic characters, has recently been collected from the Western part of the Sundarbans.

A. Vegetative Structures and Modifications Among the Members of Acanthaceae

(a) *Leaf* : *Epidermis and Cuticles* : It has also been observed that 1-layered thick epidermis covered by thick cuticle for both these species of *Acanthus* L.

(b) *Hypodermis* : Two layered thick hypodermal aqueous tissue, which maintain succulence in the leaf of *Acanthus ilicifolius* L.

(c) *Stomata* : Stomata is *diacytic* type, 18.3 μ in length × 14 μ in width and frequency 439/mm^2, situated in the lower surface of the leaves.

(d) *Mesophyll* : 2-layered thick, palisade tissue and loosely arranged spongy parenchymatous with inter-cellular spaces, probably help aeration mechanism. Vascular bundles surrounded by fibres and presence of sclereids provide toughness and rigidity to the leaf.

(e) *Salt Regulation* : *Acanthus* spp. excretes excess salt through salt gland; the process is called extrusion. Mullan (1931b) also concluded that *Acanthus* sp. show no salt gland in freshwater condition. Joshi, *et al.* (1975b) remarked that *Acanthus* sp. is restricted its growh in less saline condition. It has been observed that *Acanthus* spp. shows its salt glands in both the leaf surfaces.

(f) *Root* : *Acanthus* spp. form its aerial stilt roots, arising from the base of the stem. These roots act both mechanical support and perform physiological function, as it has air cavities in cortical region. It has been observed *in vitro* that bending branch produce aerial roots, which move towards ground, but when this bending branch forcely fixed to upright for one month, stilt roots restrain the growth.

B. Reproductive Structures and Modifications : Among the Members of Acanthaceae

(g) *Reproductive Organ* : Like other mangroves, *Acanthus* spp. commence its flowering during summer and monsoon months. The flower is pollinated by bumble bee, butter-fly etc. Sidhu (1975b) remarked that *Acanthus* spp. show *crypto-vivipary, i.e.,* seed germination provide rapide rooting, salt tolerance, ionic balance and buoyancy mechanism have also highlighted by MacNae, *et al.* (1968a), Joshi, *et al.* (1933) and Gill (1975) in the mangrove environment.

FAMILY–XIII. *TILIACEAE*

The tropical family-Tiliaceae has widely distributed genera of which one genus, *i.e., Brownlowia* Roxb. has one species, *i.e., B. lanceolata* is distributed in the mangrove swamps of the Sundarbans, as 'mangrove associated flora' (Tomlinson, 1986). In this saline environmental condition of the Sundarbans, this species shows some adaptive behaviours.

A. Vegetative Structures and Modifications Among the Members of Tiliaceae

(a) *Leaf* : Stellate hairs occurred in the lower surface of the lamina. According to Saenger (1982) this type of tomentum covering stomata reduce the water loss. Presence of sunken stomata with sub-stomatal cavity check the water loss by preserving water in sub-stomatal cavity.

(b) *Stomata* : The stomata is *diacytic* type, sunken in nature, 9.5 μ in length × 8.5 μ in width, having number 839/mm^2.

(c) *Mesophyll Tissue* : Mesophyll zone is compartmented and exhibiting large air cavities, presence of stone cells presumably provide toughness and rigidity to the leaf.

(d) *Stem* : Like Rhizophoraceae members, wood showing more vessel per sq. millimetre area and total cross sectional areas of the vessel are larger, while individual vessel pore is small. This type of feature are related to uplift the water in plants.

B. Reproductive Structures and Modifications Among the Members of Tiliaceae

Reproductive Organ : No significant feature have been found in reproductive organs, in relation to halophytic adaptation.

FAMILY–XIV. *MALVACEAE*

The widely distributed family-Malvaceae has several genera, of which two are recognised as mangrove associated flora, *viz., Thespesia* Sol. ex Corr. and *Hibiscus* L. *Thespesia* has one common species found in mangrove area is *T. populnea*, while *Hibiscus* L. has two species available in the mangrove area are as *H. tiliaceus* L. and *H. tortuosus* Roxb. These three species show their adaptation features by modifying vegetative structures.

Vegetative Structures and Modifications Among the Members of Malvaceae

(a) *Leaf* : Lower surface of the lamina cover by stellate hairs in *Hibiscus* spp., while peltate hairs occur in *Thespesia* sp. These hairs probably reduce the water lose through the stomatal aperature. Presence of sunken stomata check the water lose by preserving water.

(b) *Mesophyll* : Presence of palisade tissue and spongy parenchyma with inter-cellular spaces; probably help aeration purpose; vascular bundle surrounded by sclereids, provide toughness and rigidity to the leaf.

(c) *Stem* : In young stem, secretory cavities found in pith region, but the significant role in relation to halophytic adaptation have not been studied.

FAMILY–XV. *ASCLEPIADACEAE*

The tropical family–Asclepiadaceae has several genera, of which the two genus, *viz., Sarcolobus* R. Br. and *Pentatropis* W. & A. are present in Sundarbans mangals. Two species, *viz., S. globosus* Wall. and *S. carinatus* Wall. and one species of *Pentatropis* W. & A. have been collected and studied from the mangrove swamps of the Sundarbans. These three species show their adaptation features for growing in mangrove environment.

Vegetative Structures and Modifications Among the Members of Asclepiadaceae

(a) *Leaf* : Epidermal layer thick walled and hypodermal layer placed between the epidermal layer. Stomata are of paracytic type.

(b) *Mesophyll Tissue* : Palisade tissue behave as aqueous tissue in *Sarcolobus* spp., while palisade tissue form central aqueous tissue in *Pentatropis* sp. The aqueous tissue is responsible to preservation of cloride (Mullan, 1931b). Spongy parenchyma has large air spaces, which probably respond to the aeration purpose.

(c) *Stem* : The wood showing semi-diffuse porous, but there is no significant role of these wood in relation to halophytic adaptation.

The reproductive structures have not been found specific to the halophytic adaptations.

FAMILY–XVI. *CHENOPODIACEAE*

The family–Chenopodiaceae has several genera, of which one genus *Suaeda* Forssk. has only two species, *viz., S. nudiflora* Roxb. and *S. maritima* Dumort. are available in the mangrove swamps of the Sundarbans. These plants are found to grow in the tidal flats or hedges of the Sundarbans and modified its vegetative structures.

Vegetative Structures and Modifications Among the Members of Chenopodiaceae

(a) *Leaf* : Isobilateral leaf has 1-layered thick epidermis and central aqueous tissue within the epidermal layers. The central aqueous tissue is probably responsible for presence of chloride. These flora accumulate salt in the leaf, which is succulent in nature. Vascular bundles are surrounded by fibres, which probably provide toughness to the leaf.

(b) *Stem*: Anomalous secondary growth is found in stem. Secondary thickening associated with inter-xylary phloem; these are scattered in the cortex. The presence of the anomalous structure throughout the cortex is probably due to the supply of water to the leaf. These secondary growth provide mechanical support to these herb in these tidal zones.

(c) *Root* : Like stem, root also exhibit anomalous secondary thickening. Inter-xylary phloem scattered throughout the cortex, probably to absorb, as well as uplift water to the leaf via stem.

FAMILY–XVII. *TAMARICACEAE*

The family–Tamaricaceae is well known for their xerophytic habit. This family has the genus–*Tamarix* L. of which one species *T. gallica* L. is collected from the Sundarbans mangrove swamps, as back mangal flora. The *Tamarix gallica* grow in mangrove forests by modifying its vegetative structures.

Vegetative Structures and Modifications Among the Members of Tamaricaceae

(a) *Leaf*: Leaf is short and triangular, this may be to reduce transpiration rate. Presence of sclereids in leaf provide toughness and rigidity to the leaf.

(b) *Stem* : Wood is diffuse porous which have more vessels per sq. millimeter, like xerophytes or other mangrove members. The presence of dense vessels may help to uplift adequate water to the leaf.

FAMILY–XVIII. *LORANTHACEAE*

Most of the members of this family are epiphytic, of which, *Viscum* L. and *Dendrophthoe* (L. f.) Etting. are common. *V. orientale* Willd. and D. falcata (L. f.) Etting. are collected and studied from the mangrove swamps of the Sundarbans forests. These epiphytic members are studied in relation to their adaptative mechanism in the halophytic situation.

Vegetative Structures and Modifications Among the Members of Loranthaceae

(a) *Leaf* : Isobilateral leaves have thick walled epidermis, covered by thick cuticle; stomata is sunken in nature, *paracytic* type, 46 μ in length × 15.8 μ in width in *Viscum* sp.; while, 33.5 μ in length × 21.0 μ in width in *Dendrophthoe* sp. The thick cuticle, sunken stomata may be responsible for conserving water.

(b) *Mesophyll* : Palisade tissue forming central aqueous tissue, which gives leaf as succulence. Succulence may be due to chloride. Vascular bundles surrounded by sclereids and fibres probably protect the leaf from wilting.

(c) *Salt Regulation* : No proper significance has been studied in favour of salt regulatory mechanism in *Viscum orientale* and *D. falcata.*

(d) *Petiole* : The vascular bundles are surrounded by gregarious stone cells and sclereids. The presence of stone cells and sclereids provide toughness and strengthen the vascular bundle, which supply water and minerals to the leaf without any shrinkage.

(e) *Stem* : As it is the epiphytic plant, stem is not so developed, vessels diffuse porous, group of stone cells and frequent sclereids are occurred densely throughout the stem, except epidermis. The presence of stone cells and sclereide probably provide strength to the stem as it remain intact from any natural calamities, like cyclones, sea surges etc.

FAMILY–XIX. *SOLANACEAE*

The family–Solanaceae has the genus *Solanum* L.; one species, *S. trilobatum* L. is collected in mangrove reclaimed thickets as back mangal. *S. trilobatum* L. is twining shrub, which exhibit its adaptation like other back mangals or xerophytic flora.

Vegetative Structures and Modifications Among the Members of Solanaceae

(a) *Leaf* : Epidermis one layer thick, palisade tissue 1-layered thick and hypodermal aqueous tissue also 1-layered thick. Stomata is *anomocytic,* 14.5 μ in length × 10.4 μ in width, spongy parenchyma compacted.

(b) *Salt Regulation* : No significant role played by *S. trilobatum* L. in respect of salt regulation mechanism.

(c) *Stem* : Diffuse porous vessels in stem and frequent sclereids may provide toughness to the stem from any natural calamities.

FAMILY–XX. *AIZOACEAE*

The family-Aizoaceae has the genus *Sesuvium* L. and one species, *S. portulacastrum* L. is collected from the mangrove swamps of the Sundarbans. To grow in halophytic environment this flora exhibits few special adaptive features.

Vegetative Structures and Modifications Among the Member of Aizoaceae

(a) *Leaf* : 1-layered thick walled epidermis in both surfaces and mesophyll tissue placed below the epidermal layer. Aqueous tissue found in the centre, is responsible for the presence of chloride. The succulent nature of the leaf is presumably due to accumulate salt.

(b) *Stem* : Prostrate stem is fleshy and has large cortex, which is occupied by major portion of the total cross sectional area. The cortex consisting of parenchymatous aqueous tissue, which form the stem succulent. This succulence nature of stem is probably for response to the chloride.

FAMILY–XXI. *BORAGINACEAE*

The family-Boraginaceae has the genus, *Heliotropium* L., of which one species, *H. curassavicum* L. is collected and studied from the mangrove swamps of the Sundarbans. For growing in the saline habitat it exhibit adaptive features by modifying its vegetative structure.

Vegetative Structures and Modifications Among the Member of Boraginaceae

(a) *Leaf* : 1-layered thick walled epidermis in both the surfaces is found; within these epidermal layer, central aqueous tissue is identified, which conserve water in the leaf. It has been concluded that this succulent behaviour is probably due to accumulation of salt in the leaf.

(b) *Stem* : Fleshy prostrate stem in cross section, showing large area occupied by parenchymatous tissue, which supposed to preserve water and salt.

(c) *Root* : Secondary growth occurred in root showing more vessels per sq. millimetre and occupying large cross sectional area; which probably supply plenty of water to the leaf via stem.

FAMILY–XXII. *RUTACEAE*

The family–Rutaceae has the genus, *Atalantia* along its one species, *A. correa* M. Roem. is collected and studied from the mangrove swamps of the Sundarbans. This flora is treated as 'back mangals', adapted in the mangroves forest of Sundarbans.

Vegetative Structures and Modifications Among the Member of Rutaceae

(a) *Leaf* : 1-layered thick epidermis, covered by thick cuticle is probably to check water loss via the stomatal aperture. Presence of oil glands is probably to excrete salt with aromatic oils. Presence of sclereids surrounding vascular bundles provide rigidity and toughness to the leaf from wilting.

(b) *Stem* : Stem showing no significant role for adapting in halophytic situation.

FAMILY–XXIII. *VERBENACEAE*

The widely distributed family–Verbenaceae has the genus *Clerodendron* L. of which one species, *C. inerme* Gaertn. is available in Sundarbans forest. *Clerodendron inerme* is treated as 'back mangals', which adapt in the halophytic zone by modifying its vegetative structure.

Vegetative Structures and Modifications Among the Member of Verbenaceae

(a) *Leaf* : 1-layered thick, epidermis, covered by thick cuticle and provide the leaf to reduce transpiration rate. The leaf has hypodermal aqueous tissue, while growing in halophytic region, otherwise it exhibits normal hypodermal tissue. Presence of salt gland is probably to excrete salt by salt extrusion mechanism.

(b) *Stem* : No significant features have been found in the wood, though presence of intra-xylary phloem to act as substitute to the conductive tissue in halophytic region.

FAMILY–XXIV. *AMARYLLIDACEAE*

The monocotyledon family–Amaryllidaceae has mainly aquatic genera, of which one genus *Crinum* L. has one species; *C. defixum* Ker. Gawler, is found as mangrove associated flora in tidal inundated river bank of inner estuary. *Crinum defixum* shows few features for the adaptation in halophytic area.

Vegetative Structures and Modifications Among the Member of Amaryllidaceae

(a) *Leaf* : Isobilateral leaves, with succulent nature, exhibits sunken stomata with sub-stomatal cavities in both surfaces. Sunken stomata with substomatal cavities probably respond to reduce water loss by preserving water in the sub-stomatal cavities.

(b) *Mesophyll* : Mesophyll tissue forming central aqueous tissue, which is responsible to the succulent nature of the leaf, by preserving chloride.

(c) *Stem* : Rhizomatous stem having scattered vascular bundles and large area occupied by parenchymatous tissue, which probably preserve excess salt. By this way this species may adapt in halophytic area.

(d) *Root* : No significant role has been investigated in the anatomical structure of root to respond the halophytic adaptation.

FAMILY–XXV. *FABACEAE*

The tropical family–Leguminosae or Fabaceae has widely distributed genera of which four genera, *viz.*, *Derris* Lour., *Dalbergia* Linn. f., *Cynometra* L. and *Caesalpinia* L. have been collected and studied from the mangrove swamps of the Sundarbans. All these are treated as back mangal flora of the Sundarbans mangrove swamps. The genus *Derris* Lour. having three species *viz.*, *D. indica* Bennet, *D. trifoliata* (L.) Lour. and *D. scandens* Roxb.; the genus *Dalbergia* has single species *viz.*, *Dalbergia spinosa* Roxb., the genus *Cynometra* has single species, *C. ramiflora* L. and the genus *Caesalpinia* has two species as *C. bonduc* Roxb. and *C. crista* L.; all these species have been collected from the mangrove swamps of Sundarbans. These species have mesophytic features, as such they shows mesophytic adaptive characters.

A. Vegetative Structures and Modifications Among the Member of Fabaceae

(a) *Leaf*: Mesophyll with compacted palisade tissue, spongy tissue with distinct inter-cellular spaces and *paracytic* stomata are the characteristic features of these genera. All these characters are seemed to be adapted in mesophytic zone. So, all these species do not naturalised in halophytic zones or areas. Among these genera, *Cynometra* L. and *Dalbergia* Linn.f. sometimes distributed in inner estuary of mangrove area.

B. Reproductive Structures and Modifications Among the Member of Fabaceae

(b) *Reproductive Organ* : All these species commence its flowering during summer months. Pollination is taken part by insects. Fruits are dispersed by water and pericarp provide buoyancy mechanism to disperse the propagules. The germination pattern is hypogeal.

FAMILY–XXVI. *CONVOLVULACEAE*

The family-Convolvulaceae has the genus *Ipomoea* L. of which one species *I. pes-caprae* (L.) Sweet. is available in the mangrove swamps of the Sundarbans. *I. pes-caprae* (L.) Sweet. is treated as mangrove associates, but found in sandy beach of the Sundarbans coastal area.

Vegetative Structures and Modifications Among the Member of Convolvulaceae

(a) *Leaf* : Mesophyll tissue form central aqueous tissue, extending to both surfaces. Isobilateral leaf showing stomata, which are present in both the surfaces of the leaf. The aqueous tissue provide the leaf succulent nature, while salt gland in both the surfaces excrete salt. So by these two modifications, *viz.*, salt accumulation in case of succulence and salt extrusion in respect of salt gland, the plant probably remove its excess salt.

(b) *Stem* : No significant role in respect of halophytic adaptation have been observed in the stem, though large area of cortical region expected to conserve the water and accumulate salt.

(c) *Root* : Large vessels occurred in the root, probably translocate plenty of water to the leaf via stem and presence of gregarious sclereids in the vessels, provide toughness and rigidity to the leaf.

FAMILY–XXVII. *POACEAE*

The family–Poaceae has the genus *Porteresia,* of which one species *P. coarctata* (Roxb.) Takeoka is collected and studied from the tidal inundated silted up river flats and beds. As it grows in halophytic region of the mangrove forests, *P. coarctata* (Roxb.) Takeoka exhibits its adaptive features by modifying vegetative structure.

Vegetative Structures and Modifications Among the Member Poaceae

(a) *Leaf*: The leaves show regular groves in dorsal surfaces and wavy structures in ventral surface. *Epidermis* 1-layered thick, covered by thick silica layer; found in ventral surface of the leaf. Sclereids thickening is found below the epidermal layer of ventral surface. In dorsal surface, epidermal layer covered by short spines consisting of silica. Sclereids patches are situated below the epidermal layer of dorsal surface. All these characters provide toughness and rigidity to the leaf from wilting.

Mesophyll tissue consisting of large sized parenchymatous tissues and vascular bundles, surrounded by sclereids and presence of sunken stomata probably conserve water by reducing transpiration rate.

Presence of aerenchymatous tissue below the epidermal layer of the grooves is probably to act aeration mechanism.

(b) *Salt Regulation* : An interesting salt excreting mechanism is present in *P. coaractata.* According to Ball (1984) this grass flora grow in salty area, it deposit its salt within the one celled hairs, placed in the upper surface of the grooves of lamina and as a result salt is washed out by the falling of mature hairs. By this mechanism this flora excrete its excess salt.

(c) *Stem* : Epidermal layers are associated with silica, make the stem hardy and rigid, as such this flora stand erect in the inundated mangrove swamps. Presence of vascular rings and large cavities surrounded by sclereide patch may uplift plenty of water to the leaf. Mixed size of cavities seem to be act as buoyancy purpose.

(d) *Root* : Though the flora has fibrous roots, however mature root may penetrate into the deep horizon, while young root showing shallow in nature. Mature root exhibiting secondary growth, which probably act as anchoring mechanism and simultaneously translocate water and minerals. On the other hand young root exhibiting large area of cortical zone, consisting of parenchymatous tissue, which probably absorb nutrients. The deep rooted condition of this flora provides mechanical strength and protect the flora from land erosion, sea surges and sea waves etc.

FAMILY–XXVIII. *PTERIDACEAE*

The family–Pteridaceae has the genus *Acrostichum* L., of which one species. *A. aureum* is well known as mangrove fern in the Sundarbans. This flora adapting the mangrove swamps by modifying its vegetative and reproductive structure.

A. Vegetative Structures and Modifications Among the Member of Pteridaceae

(a) *Leaf* : Frond has 1-layered thick epidermal tissue and below it, 2-4 layered thick hypodermal aqueous tissue are placed; succulence is common features in mangrove leaves (Mullan, 1931b) and it respond to the presence of chloride (Saenger, 1982). *Acrostichum* sp. has 2 to several layer hypodermal aqueous tissue (Stace, 1966).

(b) *Salt Regulation* : *Acrostichum* sp. shows its selective absorption mechanism, *i.e.*, exclusion (Rains & Epstain, 1967 and Scholander, 1968) in this process roots absorb water selectively without salt.

B. Reproductive structures and Modifications Among the Member Pteridaceae

(c) *Reproductive Organ* : Reproductive organ, as spores are dispersed by water and after few days it germinate on the mud area. Spore having two walls outer extine and inner intine, provide buoyancy to spread the suitable places.

23

OVERVIEW ON THE MANGROVES ANATOMY & SUCCESSIONS

Saenger (1982) illustrated that it is difficult to distinguish these mangroves and non-mangroves flora, particularly on the basis of reproductive parts. It has been drawn a general conclusion on the basis of field and laboratory observations of the vegetative and reproductive parts of the mangroves, mangrove associates and back mangal flora. These features have been summarized in respect of the halophytic adaptation given in Table–63.

Table 63 : Morpho-Anatomical Features in Respect of Halophytic Adaptations

Veg. parts	*Morphology*	*Anatomy*	*Function in respect of halophytic adaptation*
Leaf	Generally fleshy, flattened, shiny, large sized, dark green	Cuticle (Waxy coating);	Reflect bright sunlight to check the transpiration rate.
		Tomentum (Hair);	Cover the stomata to reduce water loss.
		Salt; Gland	Excrete excess salt.
		Sunken stomata and associated with sub- stomatal cavity;	Conserve water to reduce transpiration rates.
		Hypodermal aqueous tissues/central aqueous tissue.	Maintain succulence by response to the presence of chlorine ion (cl⁻).
Stem	Solid, trunk base erect stout, umbrella like canopy	No. of vessels more per sq. milimetre occupy large cross sectional area, while individual vessel small,	To uplift plenty of water through these vessels and in case of any conductive column become damaged, another vessel act as substitute.
		Scalariform performation, vascicentric parenchyma, fine celled multiseriate ray, libriform fibre.	Protect the vessels from any collapse under high tension.

Veg. parts	*Morphology*	*Anatomy*	*Function in respect of halophytic adaptation*
Root	Aerial roots—Pneumatophores, Stilt roots, Root buttress and Above ground roots.	Lenticels, air cavity	Inhale and exhale air to regulate physiological process. Act for aeration purpose. Provide mechanical support, absorb nutrients, check land erosion etc.

Reproductive organ	*Morphology*	*Anatomy*	*Functions in respect of halophytic adaptation*
Seed germination Vivipary	Hypocotyle long enlarge outside ot the pericarp	Lenticels on the outer surface.	Seeds germinate and hypocotyle develop in the mother plant to avoid the high salinity in water.
2. Cryptovivipary	Hypocotyle small, hidden within the pericarp	no lenticels observed	

Table 64 : Classification of Different Categories of Halophytic Flora

Categories	*Sl. No.*	*Scientific Names of the flora*	*Halophytic character adapted by flora*				
			Leaf	*Stem*	*Root*	*Reproductive parts*	*Remarks*
T	1.	*Rhizophora mucronata* Lamk.	+++	+++	+++	+++	These true mangrove flora obviously grow in anaerobic soil, inundated twice per day; can tolerate high salinity in water and seemed to be found only in halophytic environmental condition or mangal habitats in the tropical and the sub-tropical coasts of the old World.
R	2.	*R. apiculata* Blume	+++	+++	+++	+++	
U	3.	*Bruguiera gymnorrhiza* Lamk.	+++	+++	+++	+++	
E	4.	*B. sexangula* Poir.	+++	+++	++	+++	
	5.	*B. parviflora* W. & A.	+++	+++	++	+++	
	6.	*B. cylindrica* (L.) Blume	+++	+++	++	+++	
M	7.	*Ceriops decandra* (Griff.) Ding Hou	+++	++	+++	+++	
A	8.	*C. tagal* (Perr.) Robinson	+++	++	+++	+++	
N	9.	*Kandelia candel* (L.) Druce	+++	+++	++	+++	
G	10.	*Avicennia alba* Blume	+++	+++	+++	++	
R	11.	*A. marina* Vierh.	+++	+++	+++	++	
O	12.	*A. officinalis* L.	+++	+++	+++	++	
V	13.	*Sonneratia caseolaris* Eng.	+++	+++	+++	–	
E	14.	*S. apetala* Buch.–Ham.	+++	+++	+++	–	
S	15.	*S. griffithii* Kurz.	+++	+++	+++	–	
	16.	*Xylocarpus granatum* Koen.	++	++	+++		
	17.	*X. mekongensis* Pierre	++	++	+++	–	
F	18.	*Heritiera fomes* Buch.–Ham.	+++	++	++	–	
L	19.	*Aegiceras corniculatum* (L.) Blanco	+++	++	–	++	
O	20.	*Aegialitis rotundifolia* Roxb.	+++	+++	+	++	
R	21.	*Lumnitzera racemosa* Willd.	+++	++	+	–	
A	22.	*Scyphiphora hydrophyllacea* Garten.f.	+++	++	+	–	
	23.	*Excoecaria agallocha* L.	++	++	+	–	

(Contd.)

Categories	Sl. No.	Scientific Names of the flora	Halophytic characters adapted by flora: Leaf	Stem	Root	Reproductive parts	Remarks
	24.	*Nypa fruticans* (Thumb) Wurmb.	++	++	+	++	
	25.	*Phoenix paludosa* Roxb.	+	++	++	–	
	26.	*Porteresia coarctata* (Roxb.) Takeoka	+++	++	+	–	
MANGROVE ASSOCIATE FLORA	27.	*Acrostichum aureum* L.	++	–	–	+	These mangrove associated flora grow in the fringe area of mangrove swamps; inundated 1-5 times per month during spring tide and also found in mesophytic area.
	28.	*Brown lowia lanceolata* Benth.	+++	+	–	–	
	29.	*Acanthus ilicifolius* L.	++	–	+	–	
	30.	*A. volubilis* wall.	++	–	+	–	
	31.	*Clerodendron inerme* Gaertn.	++	+	–	–	
	32.	*Atatanlia correa* M. Roem	++	–	+	–	
	33.	*Viscum orientale* Willd.	+++	+	–	–	
	34.	*Dendrophthoe falcata* (L.f.) Etting.	+++	+	–	–	
	35.	*Ipomoea pes-caprae* (L.) Sweet	++	–	+	–	
	36.	*Suaeda maritima* Dumort.	+	+	+	–	
	37.	*S. nudiflora* Roxb.	+	+	+	–	
	38.	*Hibiscus tiliaceous* L.	++	+	–	–	
	39.	*H. tortuosus* Roxb.	++	+	–	–	
	40.	*Thespesia populnea* Soland ex. Corr.	++	+	+	–	
	41.	*Sarcolobus globosus* Wall.	+	–	+	–	
	42.	*S. carinatus* Wall.	+	–	–	–	
	43.	*Hydrophylax maritima* L.	+	–	–	–	
	44.	*Crinum defixum* Ker Gawl.	+	–	+	–	
	45.	*Cynometra ramiflora* L.	+	+	–	–	
	46.	*Pentatropis capensis* Bull.	+	–	–	–	
	47.	*Heliotropium curassavicum* L.	+	–	+	–	
	48.	*Sesuvium portulacastrum* L.	+	–	–	–	
	49.	*Tamarix gallicha* L.	+	+	–	–	
BACK MANGAL FLORA	50.	*Dalbergia spinosa* Roxb.	+	+	–	–	These back mangal flora generally grow in mangrove adjacent area and infrequently or very rarely inundated one time per month during spring tide.
	51.	*Derris scandans* Benth	+	+	–	–	
	52.	*D. indica* Bennet	+	+	–	–	
	53.	*D. trifoliata* Lour.	+	+	–	–	
	54.	*Caesalpinia bonduc* (L.) Roxb.	+	+	–	–	
	55.	*C. crista* L.	+	+	–	–	
	56.	*Solanum trilobatum* L.	+	+	–	–	

Those flora which exhibit more features given in Table 64 seemed to be mangrove flora, however all the flora in Sundarban swamps do not show all features mentioned in Table 63. But it should be illulstrated that the flora having maximum characters (mentioned in Table 64) which are suitable for succession in halophytic environmental condition, mentioned generally as true mangroves, mangrove associates and others are back mangals. These flora have been listed (Table 64) as true mangroves, mangrove associates and back mangal on the basis of they possessing halophytic characters or lacking of halophytic characters given below :

'+++'	denote maximum mangrove characters present
' ++'	denote medium mangrove characters present
' + '	denote minimum mangrove characters present
' – '	denote absence of mangrove characters

24

PALYNOLOGY OF MANGROVES : ITS IMPACT ON THE PAST AND PRESENT EVOLUTIONAL, ECOLOGICAL AND BIODIVERSITY ESTIMATION

The unique mangrove ecosystem adapt best in the shallow sea coasts, shelter estuaries and silted up deltaic zones of the tropics and sub-tropics; which also well established and spread with characteristic environmental condition, *viz.*, within the temperature ranges between 20°C-30°C, moderate to high monsoon precipitation, *i.e.*, 1000-3000 mm / annum, high humidity but with coastal aridity. The step by step natural and successive development of these mangroves and mangrove ecosystem cannot be artificially replaced with afforesting other plants or by the artificial plantation of mangroves; as their sustainable productivity, coastal zone stability and other ecological functions cannot be maintained with the **'man-made mangroves'**. The gradual, slow or quickly changed ecological or geomorphological conditions on the sea-land interphase in global mangrove ecosystem alongwith the recent human thrust or pressure and other increasing biotic factors and / or interaction are inevitable; these cause day to day change of the mangroves biodiversity and other ecological factors and also degrade the ecosystem. As such, mangroves are now very much dynamic and this ecosystem is mostly seem very much fragile throughout the Globe.

The palynological studies of living flora in mangroves and the knowledge of living mangrove pollen keys help to identify the palaeo-palynological sediment samples of mangroves from the geological past of the different palaeogeographical zones of different strata of the earth.

This palynological studies also help to ascertain the palaeo-ecology and the palaeo-climate of the geological past and from different palaeo-geographical zones. The line of ancestor, evolution or changing features of the land and sea may also be drawn based on these palaeo-palynological studies and for determination of the geological age by carbon dating techniques. This palynological studies and pollen key can also help to identify these mangrove specimens for taxonomic needs and/ or to estimate the biodiversity and the palaeo-geographical and present distribution of the mangroves in the Globe. Digging out of the sedimentary soil, rocks and the palaeo-sediments, deposited pollens and spores samples and identification of these palaeopollens following the palaeo-palynological methods and comparing these with the living mangrove pollens and spores will surely help to assess the past mangrove species diversity and palaeo-ecology. This fossil mangrove pollens and spores must help to draw palaeo-geographical distribution of mangroves and this may consider the history and the evolutionary trend, rate of sedimentation, accression of coastal zone or the past geological landmarks and palaeo-mangroves forms in place to place.

The present days' knowledge and studies on mangrove pollens and spores of different core-zones or back-mangal regions and knowledge of their ecological features and adaptation natures must be helpful if that knowledge can be applied and compared with the fossil records or sedimentary deposited pollens and spores of mangroves.

With the valuable studies made by the eminant world renowned palynologist, Kiener (1973) it was possible to draw the world mangrove zones more critically from both the Old and New World Tropics (Map-43)

Studies of the recent mangrove pollens and spores deposited sediments have been conducted by the world renowned palynologists of the different parts of the Globe and based on these studies and reports it was possible to highlight and compare the past and present world mangrove distribution and their evolutionary trends (Thanikaimoni, 1987).

Mangrove vegetation mapping during the Quartenary and Tertiary age were possible to draw and there also emphasized the world distribution of mangroves and the estimation of these origin of mangroves during these geological age were possible. Though the line of research of Tertiary sediments and Quarternary sediments are different; but based on these fossil evidences and palaeo-palynological studies, assessment of these origin and distribution of mangroves in the palaeo-geographical areas were only possible.

This pollens studies may also adequately highlight the palaeoclimate, sea level rise and the coastal zone formation and evolution trend of the mangroves during the Quartenary and Tertiary period.

In view of this above, the pollen morphology and pollen structures of these mangrove species of Indian Sundarbans are mentioned below. For these valuable studies on mangrove palynology the description and pollen morphology of mangroves were taken from the publication of Thanikaimoni (1987).

PALYNOLOGY OF THE INDIAN MAJOR MANGROVES

1. RHIZOPHORACEAE (Tribe-Rhizophorae):

(i) Rhizophora apiculata Blume— Fig. 113 (Sketches 1 & 2)

(ii) Rhizophora mucronata Lamk.— Fig. 113 (Sketches 3 & 4)

(iii) ***Bruguiera gymnorrhiza*** (L.) Lamk— Fig. 113 (Sketches 5 & 6)

(iv) ***Bruguiera sexangula*** (Lour.) Poir— Fig. 113 (Sketches 7 & 8)

(v) ***Bruguiera cilindrica*** (L.) Blume— Fig. 113 (Sketches 9 & 10)

(vi) ***Bruguiera parviflora*** W. & A. ex Griff—Fig. 113 (Sketches 11 & 12)

(vii) ***Ceriops decandra*** (Griff.) Ding Hou—Fig. 113 (Sketches 13 & 14)

(viii) ***Ceriops tagal*** (Perr.) Robin—Fig. 113 (Sketches 15 & 16)

(ix) ***Kandelia candel*** (L.) Druce—Fig. 113 (Sketches 17 & 18)

Morphology of Pollen Grains—

Isopolar pollen; tricolporate; without pseudocolpi; ectoaperture long (longicolporate), generally free, pollen longiaxial to equiaxial, sometime breviaxial, endoperture perpendicular to the ectoaperture; aperture 2-partite (ectoaperture, endoaperture), costae colpi present, costae pori discontinuous at the equator; fastigiate, ectoaperture constricted at the equator, endoaperture perpendicular to the ectoaperture; infratectum faintly columellate, exine scabrous to finely reticulate; costae promient; costae prominent; endoaperture wide, with uneven lateral ends; pollen sometimes tetracolporate or dicolporate.

2. AVICENNIACEAE : *Avicennia* spp.

(x) Avicennia alba Blume— Fig. 113 (Sketches 19 & 20)

(xi) Avicennia officinalis L.— Fig. 113 (Sketches 21 & 22)

(xii) Avicennia marina (Forsk.) Vierh.—Fig. 113 (Sketches 24 & 25)

Morphology of Pollen Grains

Isopolar pollen; tricolporate; without pseudocolpi; ectoaperture long (longicolporate), generally free, pollen longiaxial to equiaxial, sometime breviaxial, endoaperture perpendicular to the ectoaperture; aperture 2-partite (ectoaperture, endoaperture), costae colpi present, costae pori discontinuous at the equator, non fastigiate, ectoapertute wide at the equator, endoaperture circular or elliptic (parallel to the actoaperture); reticulate, infratectum columellate; costae absent; sometimes discolporate with ectoapertures free or fusing at poles or 6-pericolporate or in aperture without content.

The structures of pollens distinguished this *Avicennia* spp. from Verbenaceae (Mukherjee & Chanda, 1973).

3. SONNERATIACEAE : *Sonneratia* spp.

(xiii) Sonneratia griffithii Kurz.—Fig. 113 (Sketches 25 & 26)

— Colpi and pseudocolpi more or less readily distinguishable; exine smooth to perforate at polar region, verrucate-rugulate at the equatorial region.

(xiv) Sonneratia caseolaris (L.) Engler—Fig. 113 (Sketches 27 & 28)

— Colpi and pseudocolpi rarely distinguishable, aperture almost reduced to pores (triporate); exine smooth-perforate at polar region, verrucate-rugulate at equitorial region.

(xv) Sonneratia apetala Buch. Ham—Fig. 113 (Sketches 29 & 30)

— Colpi and pseudocolpi rarely distinguishable, aperture almost reduced to pores (triporate); exine uniformly rugulate.

Morphology of Pollen Grains :

Isopolar pollen, triporate, with 3 pseudocolpi; 4 exine verrucate-rugulate at the equatorial region, smooth or perforate or verrucate-rugulate at polar region, colpi and pseudocolpi sometimes poorly or not differentiate; abnormal pollen 4 or 6 apertures.

4. COMBRETACEAE—*Lumnitzera* sp.

(xvi) *Lumnitzera racemosa* Willd.—Fig. 114 (Sketches 31 & 32)

Morphology of Pollen Grains

Isopolar pollen; tricolporate; with 3 pseudocolpi; 6-lobbed in polar view, more or less elliptic in meridional view, rugulate-reticulate, not distinctly collumellate, endoaperture permendicular to the ectoaperture; polar axis 28-37 um, dom shaped in polar optical section; exine reticulate.

5. ARECACEAE—*Nypa fruticans* (Thunb.) Wurmb & *Phoenix* sp.

(xvii) *Nypa fruticans* (Thunb.) Wurmb.— Fig. 114 (Sketches 33 & 34).

(xviii) *Phoenix paludosa* Roxb.—Fig. 114 (Sketches 35 & 36)

Morphology of Pollen Grains

Heteropolar pollen, monosulcate, reticulate.

6. MELIACEAE—*Xylocarpus* spp.

(xix) *Xylocarpus granatum* Koeing.— Fig. 114 (Sketches 37 & 38)

(xx) *Xylocarpus mekongensis* Pierre—Fig. 114 (Sketches 39 & 40)

Morphology of Pollen Grains

Isopolar pollen; tricolporate; without pseudocolpi; ectoapertures short (brevicolporate) free, pollen breviaxial to equiaxial; tetracolporate-pentacolporate, rarely tricolporate, finely reticulate scabrous, fastigiate, costae colpi and costae pori present.

7. MYRSINACEAE : *Aegiceras* spp.

(xxi) *Aegiceras corniculatum* (L.) Blanco— Fig. 114 (Sketches 41 & 42)

Morphology of Pollen Grains

Isopolar pollen; tricolporate; without pseudocolpi; ectoaperture long (longicolporate), generally free, pollen longiaxial to equiaxial, sometime breviaxial, endoaperture perpendicular to the ectoaperture; aperture 2-partite (ectoaperture, endoaperture), costae colpi present, costae pori discontinuous at the equator; fastigiate, ectoaperture constricted at the equator, endoperture perpendicular to the ectoaperture; infratectum faintly columellate, exine scabrous to finely reticulate; costae prominent; endoaperture papiliform; rarely tetra-(peri) syncolporate.

8. EUPHORBIACEAE—*Excoecaria* sp.

(xxii) Excoecaria agallocha L.—Fig. 114 (Sketches 43 & 44)

Morphology of Pollen Grains

Isopolar pollen; tricolporate; without pseudocolpi; ectoaperture long (longicolporate), generally free, pollen longiaxial to equiaxial, sometime breviaxial, endoaperture perpendicular to the ectoaperture; aperture 2-partite (ectoaperture, endoaperture), costae colpi present, costae pori discontinuous at the equator; 6. fastigiate, ectoaperture constricted at the equator, endoaperture perpendicular to the ectoaperture; infratectum distinctly columellate, exine reticulate, endoaperture narrowly elliptic.

9. AEGIALITIDACEAE—*Aegialitis* sp.

(xxiii) Aegialitis rotundifolia Roxb.—Fig. 114 (Sketches 45 & 46)

Morphology of Pollen Grains

Isopolar pollen; tricolpate; densely pilate, nexine, honey comblike, aperture membrane operculoid; pollen rarely 2-syncolpate or 6-pericolpate.

10. STERCULIACEAE : *Heritiera* spp.

(xxiv) Heritiera fomes Buch.-Ham.—Fig. 114 (Sketches 47 & 48).

Morphology of Pollen Grains

Isopolar pollen; tricolporate; without pseudocolpi; ectoaperture long (longicolporate), generally free, pollen longiaxial to equiaxial, sometime breviaxial, endoaperture perpendicular to the ectoaperture; aperture 3-partite (ectoaperture, mesoaperture and endoaperture), nexine endosculpture, sexine reticulate; pollen rarely disyncolporate.

11. Acanthaceae : *Acanthus* spp.

(xxvi) Acanthus ilicifolius L.—Fig. 114 (Sketches 51 & 52)

Morphology of Pollen Grains

Isopolar pollen; tricolpate, reticulate with a free columellar element in the lumina, 3 lobed in polar view.

12. TILIACEAE : *Brownlowia sp.*

(xxvii) Brownlowia lanceolata Benth. — Fig. 114 (Sketches 53 & 55)

Morphology of Pollen Grains

Isopolar pollen; tricolporate; without pseudocolpi; ectoapertures short (brevicolporate), free pollen breviaxial to equiaxial; tricolporate; reticulate-fossulate, costae discontinuous at the equator, endoaperture parallel to the ectoaperture.

13. PTERIDACEAE : *Acrostichum* spp.

(xxviii) *Acrostichum aureum* L.—Fig. 114 (Sketches 57 & 58)

Morphology of Spores

Heteropolar spore with triradiate tetrad mark.

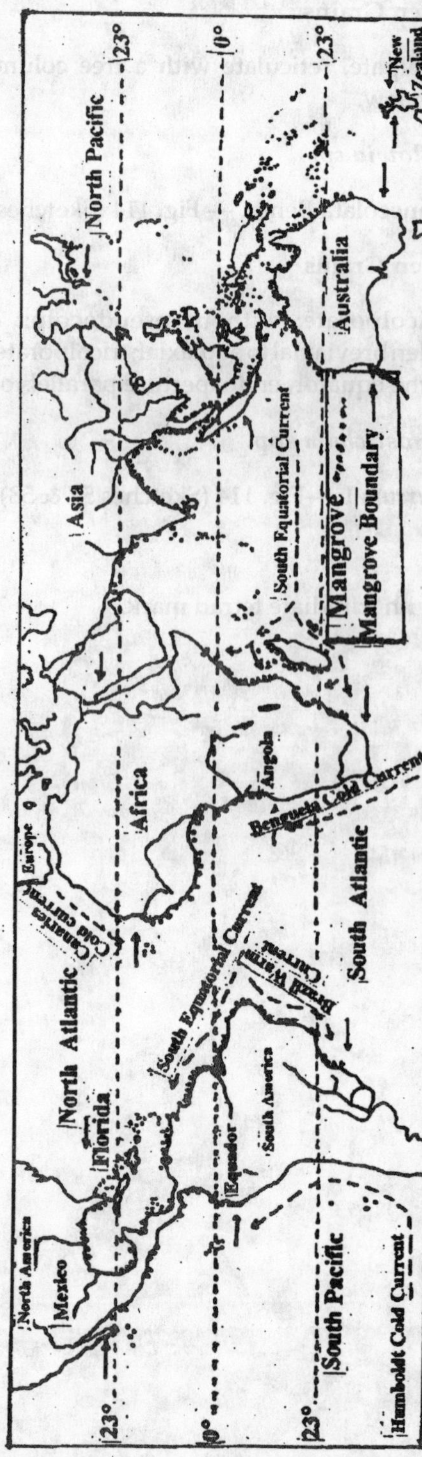

Map 43 : Recent World Distribution of Mangroves, Based on Palynological Studies (Modified after Kiener, 1973)

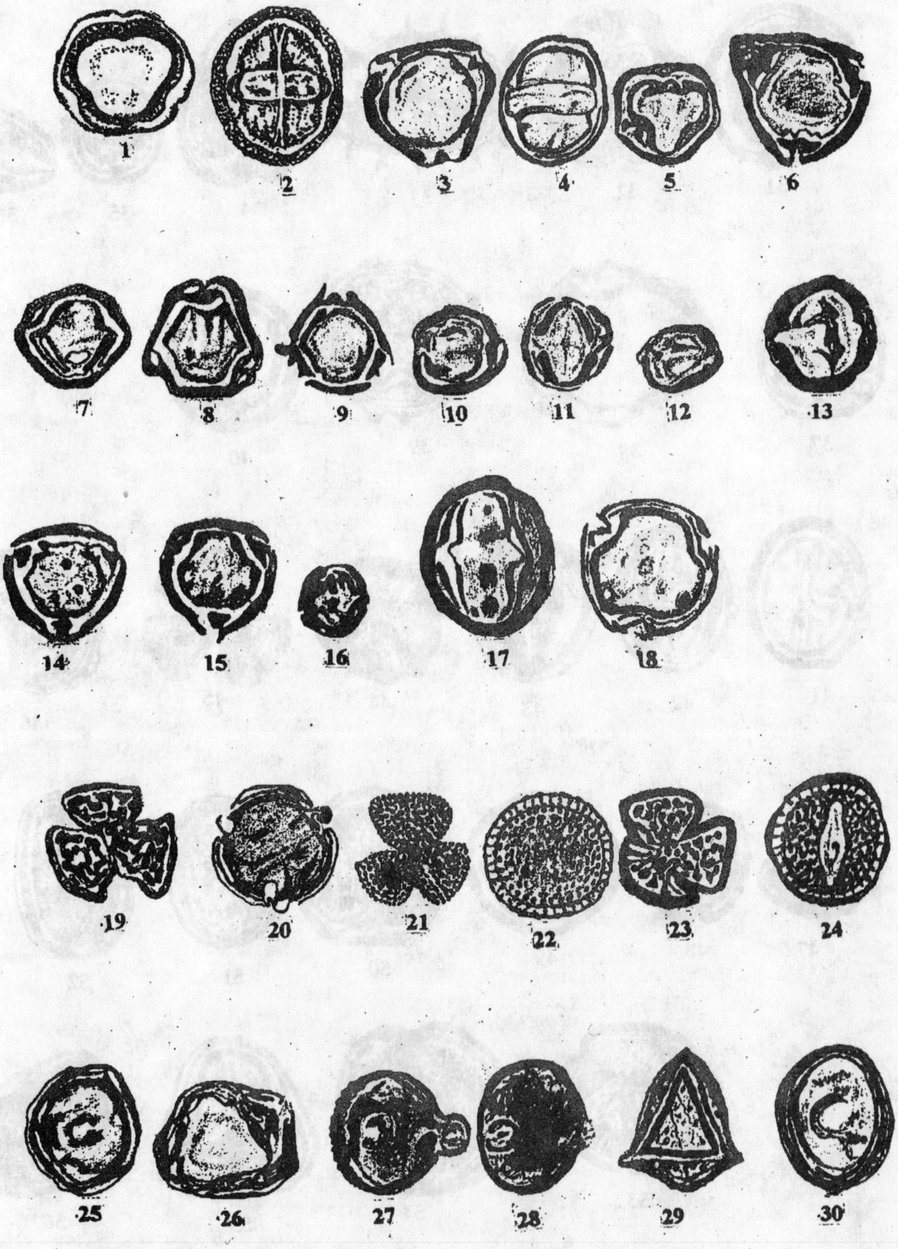

Fig. 113 : Pollen Grains of the Manjor Mangrove Species of Indian Sundarbans

1 & 2 -Rhizophora apiculata; 3 & 4-R. mucronata; 5 & 6-Bruguiera gymnorrhiza; 7 & 8-B. sexangula; 9 & 10-B. cylindrica; 11 & 12-B. parviflora; 13 & 14-Ceriops decandra; 15 & 16-C. tagal; 17 & 18-Kandelia candel; 19 & 20-Avicennia alba; 21 & 22-A. officinalis; 23 * 24-A. marian; 25 & 26-Sonneratia griffithii; 27 & 28-Sonneratia caseolaris; 29 & 30-Sonneratia apetala. (Source; Thanikaimoni, G., 1987)

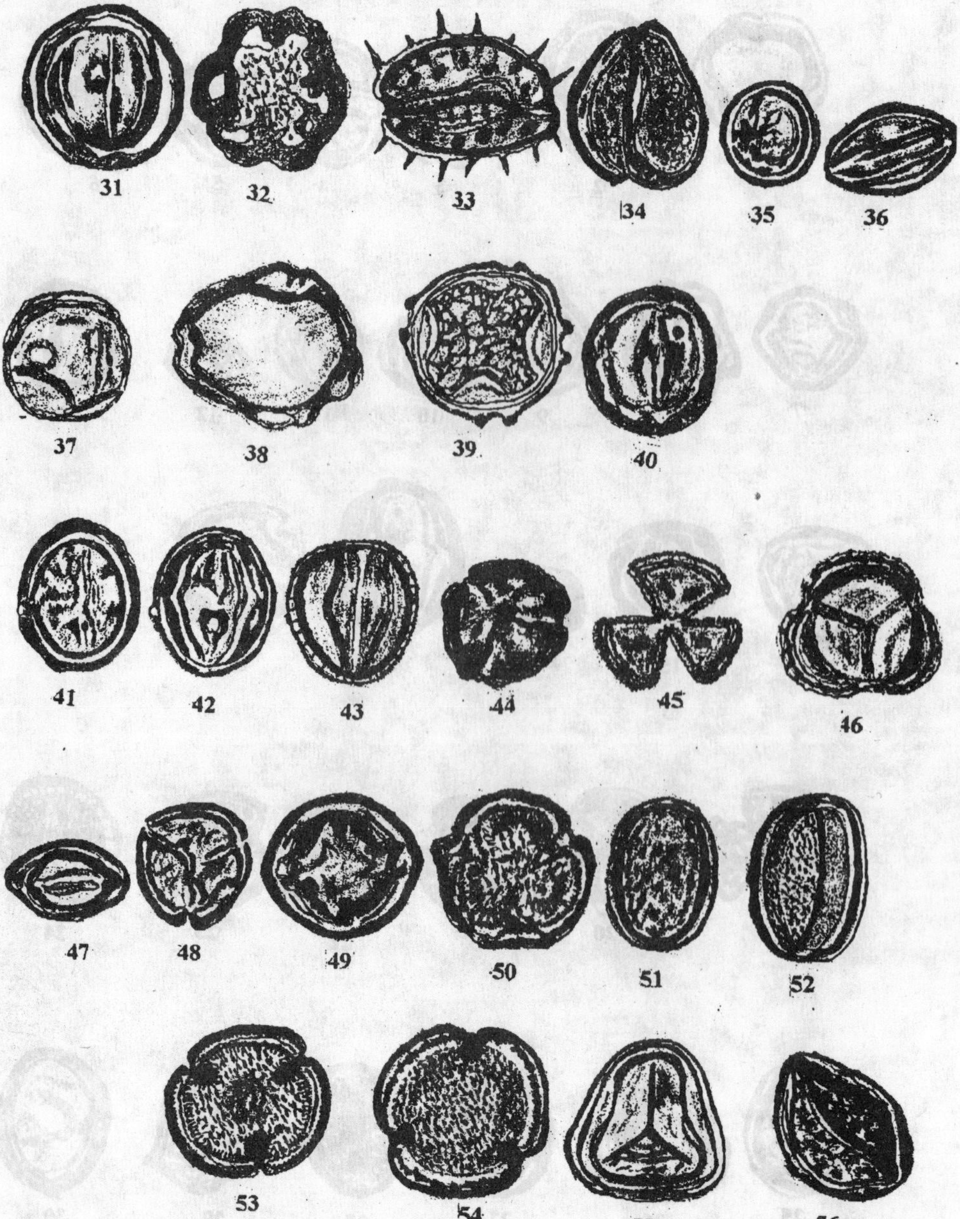

Fig. 114 : Pollens & Spores of the Mangroves & Associates of Indian Sundarbans

31 & 32-Lumnitzera racemosa; 33 & 34-Nypa fruiticans; 35 & 36-Phoenix paludosa; 37 & 38-Xylocarpus granatum; 39 & 40-X. mekongensis; 41 & 42-Aegiceras corniculatum; 43 & 44-excoecaria agallocha; 45 & 46-Aegialitis rotundafolia; 47 & 48-Heritiera fomes; 49 & 50-scyphiphra hydrophyllacea; 51 & 52-Acanthus ilicifolius; 53 & 54-Brownlowia lanceolata; 55 & 56-Acrostichum aureum; (Source : Thanikaimoni, G. 1987)

25

FILAMENTOUS ALGAL FLORA OF THE SUNDARBANS-ATTACHED WITH THE SOIL, MANGROVE ROOTS AND STEM OR TRUNK BASES

The algal flora in this saline habitats are interesting and some what distinct from the freshwater habitat algae, i.e., these groups of algae can withstand saline water and survive well in the saline habitats; their physiological functions are so set up that they can grow and perform the physiological activities in the saline water ranges between 10-40 ppt salinity, where the fresh water algae do not grow or survive. For their direct exposure to the saline habitats these algae are very much limited in terms of species composition and distribution. Here, in the Sundarbans intertidal mangrove habitats the salt tolerant algae are found from three distinct habitats, i.e., free floating planktonic form, benthic form-which grow on mature soil substratum rich with organic matter and further enrich the soil; besides these, in the less consolidated soil the cyanophyta and the chlorophyta grow as the pioneer photosynthetic life forms and by this process the soil turns suitable for the growth of higher plants, like the mangroves; the third distinct form is the periphytic algae attached on the periphery of aerial roots, like pneumatophores, stilt roots, prop roots, root buttresses, flank roots and also on the trunk bases of regularly tidal inundated branches of these mangroves. Frequently, and periodical tidal submergence and simultaneous exposure in air and light in between two high tidal inundation phases these algae fit well by receiving the moisture and nutrients from the tidal waters and tidal water carried silt or sediment and they can also get the opportunity to take part in the process of photosynthesis during the exposure of these algal habitats at low tides. Besides these, this periphytic algal flora are also largely utilised by the mangrove dwelling or mangrove habitat mollusca, crustaceans or shell fish and fin fish species and the large number of herbivorous estuarine fauna during tidal inundation of these mangrove habitat zones. As such, few periphytic and benthic algal flora have been collected from such habitats and identified these algae (Plate-162).

1. DIVISION : CYANOPHYTA

Family–*OSCILLATORIACEAE*

A. Genus : *Phormidium* Kuetz.

1. *Phormidium fragile* (Meneghini) Gomont (Fig. G, No. 14).

Sheath diffluent, thallus mucilaginous, trichomes, slightly curved, cells more or less square shaped, constriction present at the cross-walls, end cell attenuated, cell 1.5 μ–2.5 μ in width × 1.3 μ–2.3 μ in length.

Locality : Canning, attached in the periphery of pneumatophores of *Avicennia* sp.

B. Genus : *Lyngbya* Ag.

2. *Lyngbya lutea* (Ag.) Gom. (Fig. F, No. 2)

Sheath having no distinct colour, filaments more or less coiled, sheat 1.5 μ–2.5 μ in width, trichomes 2.5 μ–5 μ in width × 1 μ–4 μ in length, calyptra present.

Locality : Netidhopani, attached in the periphery of trunk bases and pneumatophores of *Avicennia* spp.

3. *Lyngbya hieronymusii* Lemm. (Fig. F, No. 8)

Sheath firm, homogeneous, filament single and straight, cells 11 μ–14 μ in width × 2.5 μ–5 μ in length.

Locality : Canning, attached with the pneumatophores of *Avicennia* spp.

4. *Lyngbya confervoides* C. Ag. ex Gomont (Fig. F, No. 13)

Sheath colourless, slightly firm, filament straight, cells 10 μ–18 μ in width × 2.5 μ–4 μ in length, end cell curved.

Locality : Bakkhali, attached with trunk bases of *Excoecaria* sp. and *Avicennia* sp.

5. *Lyngbya martensiana* Menegh. ex Gomont (Fig. F, No. 9 and 10)

Sheath thick, 2 μ in width, filament loosen, trichome straight, cells 5 μ–10 μ in width × 2 μ–4 μ in length, end cells covered.

Locality : Durgadwani, attached with the pneumatophores of *Avicennia* sp.

6. *Lyngbya semiplena* (C. Ag.) J. Ag. ex Gomont. (Fig. F, No. 6 and 7)

Sheath colourless, filament caespitose, trichome most constricted at the cross walls, end slightly attenuated, curved trichome, cell 8 μ–12 μ in width × 2 μ–3.5 μ in length.

Locality : Datta forest, attached with the trunk bases of *Excoecaria* sp.

C. Genus : *Microcoleus* Desmazieres

7. *Microcoleus chthonoplastes* Turet ex Gomont (Fig. F, No. 14-16)

Trichomes are spirally inter-woven, covered by thick gelatenous sheath, end cell acute, individual cells 3 μ–6 μ in width × 3μ–8 μ in length.

Locality : Basanti, attached with the pneumatophores of *Avicennia* sp.

D. Genus : *Hydrocoleum* Kutz.

8. *Hydrocoleum lyngbyaceum* Kuetz. ex Gomont (Fig. F, No. 11 and 12)

Sheath thick, not transparent, filament firm, cells 2.5 μ–6 μ in width × 1 μ–2 μ in length, not granulated.

Locality : Bakkhali, attached with the trunk bases of *Excoecaria* sp.

E. Genus : *Schizothriz* Kuetz.

9. *Schizothrix lamyi* Gomont (Fig. G, No. 17)

Sheath uneven, lamellated, yellowish, both ends tapering, trichome curved, slightly constricted at cross wall, cells 2.5 μ–3.5 μ in width × 2.5 μ–7 μ in length.

Locality : Sajnekhali, attached in the periphery of pneumatophores of *Avicennia* spp.

F. Genus : *Oscillatoria* Vaucher

10. *Oscillatoria princeps* Vaucher ex Gomont (Fig. G, No. 19)

Trichome straight, end cell curved, spirally coiled, cells 18 μ–44 μ in width × 4 μ–7 μ in length.

Locality : Frezergaung, attached in the trunk bases of intertidal mangroves.

11. *Oscillatoria jasorvensis* Vouk. (Fig. F, No. 1)

Trichoma 2 μ–3 μ in width × 0.5 μ–1.5 μ in length, straight end cell curved.

Locality : Jharkhali, attached in the periphery of submerged grasses.

12. *Oscillatoria subbrevis* Schemidle (Fig. F, No. 5)

Trichome slightly coiled, cells not constricted, 7 μ–10 μ in width × 2 μ–3 μ in length, end cell curved, calyptra not found.

Locality : Sajnekhali, attrached with the pneumatophores of *Avicennia* spp.

13. *Oscillatoria tenuis* Ag. (Fig. F, No. 3 and 4)

Trichome straight, end cell ovoid shaped with thick membrane, cells 7μ–8 μ in width × 2.5 μ–3.5 μ in length.

Locality : Sudhanyakhali, attached with the trunk bases of *Avicennia* spp.

FAMILY–*HYELLACEAE*

G. Genus : *Xenococcus* Thuret

14. *Xenococcus chaetomophae* Setchell et Gardner (Fig. G, No. 15)

Cells arranged differently, having various shape and size, cells–5 μ–8 μ in width.

Locality : Basanti and Gosaba Island; attached with the trunk bases of *Aegialitis*

15. *Xenococcus cladophorae* (Tilden) Setchell et Gardner (Fig. G, No. 16)

Cells arranged regularly, having various shaped and sized, colony formed, cells 2.5 μ–5.5 μ in width × 3 μ–8 μ in length.

Locality : Sajnekhali; attached with the trunk bases of Mangrove trees.

FAMILY–*NOSTOCACEAE*

H. Genus : *Anabaena* Bory

16. *Anabaena variabilis* Kg. (Fig. G. No. 18)

Thrichome having no sheath, flexible, cells 4 μ–5 μ in width × 2.5 μ–5 μ in length, constricted at the cross wall.

Locality : Datta forest, attached on the soil in intertidal zones.

17. *Anabaena oryzae* Fritsch (Fig. G, No. 20)

Trichome straight, heterocyst barrel shaped, terminal and inter-calary, cells- 2.5 μ—3.5 μ in width.

Locality : Drugadwuani, attached on the soil substratum in intertidal zones.

DIVISION : CHRYSOPHYTA

FAMILY–*VAUCHERIACEAE*

I. Genus : *Vaucheria Christensen*

18. *Vaucheria erythrospora* Christensen (Fig. G, No. 4 and 5)

Plant monoecious, antheridia and oogonia placed side by side, filament 35 μ–40 μ in diameter, antheridia 10 μ–16 μ in diameter, oogonia 50 μ in diameter, oogonia upwardly placed, beak slightly bent.

Locality : Bagna forest, attached on the hard soil substrate in the intertidal zones

19. *Vaucheria prescotti* Islam (Fig. G. No. 10 and 11)

Plant dioecious, antheredium elliptic, stalked or sessile, 50 μ–60 μ in diameter × 80 μ–120 μ in length, open through apical pore, rhizoids are found, act as anchore mechanism, oogonium single, elliptical, 100 μ–120 μ in diameter × 180 μ–250 μ in length, filament 40 μ–70 μ in diameter.

Locality : Netidhopani, attached on the soil substratum in intertidal zones.

20. *Vaucheria mayyanadensis* Erady (Fig. G, No. 8 and 9)

Plant dioecious, antheridium single, lateral, scattered, elliptic, open through pore, 20 μ–30 μ in diameter × 90 μ–110 μ in length, oogonium-80 μ–95 μ in diameter × 110 μ –125 μ in length, solitary, sessile, filament erect, anchored to the soil by Rhizoids; 20 μ–40 μ diameter, branched.

Locality : Sudhanyakhali, attached on the trunk of mangroves and soil substratum.

21. *Vaucheria pronosperma* Islam (Fig. G, No. 6 and 7)

Plant monoecious, antheridium near to oogonium, 25 μ in diameter, oogonium elliptic, beak curved towards the filament, filament 30 μ–50 μ in diameter, oospore fill up the oogonium cavity.

Locality : Netidhopani attached on the hard soil in intertidal zone.

DIVISION :— CHLOROPHYTA

FAMILY–*ULVACEAE*

J. Genus : *Enteromorpha* Link

22. *Enteromorpha clathrata* (Roth) Grev. (Fig. F, No. 18)

Plants primarily attached to the substratum; then floating freely, greenish 20-100 mm long, cylindrical, slender, cells rectangular, arranged side by side, vegetative cells 15 μ–25 μ width × 20 μ–35 μ in length.

Locality : Basanti Island

23. *Enteromorpha intestinalis* (L.) Link (Fig. F, No. 17)

Plants flattened, unbranched, intestine like, contorted, tubular, cells more or less square, 14 μ–20 μ in width × 12 μ–18 μ in length, chloroplasts seemed to be distinct.

Locality : Minakhan, Kulti area, attached on soil substrate in fisheries.

24. *Enteromorpha prolifera* (Mueller) J. Ag. (Fig. G, No. 2 and 3)

Plants branched, hyphae like branches arising from the main axis; 1.5 mm in width, cells variable shaped, 12 μ–16 μ in width × 14 μ–26 μ in length, pyrenoids prominent.

Locality : Sajnekhali, attached in the soil substratum, in the intertidal zones.

K. Genus : *Ulva* L.

25. *Ulva patengensis* L. (Fig. E, No. 5-8)

Thallus flattened; more or less square shaped, distinct chloroplast light green, pyrenoids present, 20 μ–40 μ in width × 30 μ–50 μ in length.

Locality : Jharkhali Island, attached at the trunk bases of *Avicennia* spp.

26. *Ulva lactuca* L. (Fig. E, No. 16)
Plant cells broader than long, ovate to elliptic, dark green, thin, margins distinct, frond irregularly split, short, attched to the substratum.

Locality : Frezergaung

FAMILY–*CLADOPHORACEAE*

L. Genus : *Lola* Hamel

27. *Lola implexa* (Harv.) Hamel (Fig. E, No. 3 and 4)
Absence of rhizoid, cells 20 μ–30 μ in diameter, 30 μ–60 μ in length.

Locality : Sajnekhali, attached in the soil within the intertidal zones.

28. *Lola capillaris* (Kg.) Hamel (Fig. E, No. 1 and 2)
Filament slightly bent, vegetative cells 40 μ–70 μ in diameter, 80 μ–200 μ in length, cells cylindrical, rectangular, septed, rhizoid present.

Locality : Sudhanyakhali, attached on the consolidated soil.

29. *Lola tortuosa* (Dillw.) Chapman (Fig. G, No. 1)
Filament contorted and tortuose, cells 15 μ–20 μ in diameter 24 μ–60 μ in length, rhizoid absent.

Locality : Sajnekhali, attached on mangrove trunk bases.

M. Genus : *Cladophora* Kuetz.

30. *Cladophora echinus* (Bias.) Kuetz. (Fig. H, No. 1)
Thallus stout, 3 mm to 25 mm tall, branched, branching dichotomoustly arranged, geen, cell wall thick, cell 20 μ–120 μ in diameter × 200 μ–700 μ in length.

Locality : Bagna forest, attached with *Avicennia* trunk bases.

31. *Cladophora patentiramea* (Mont.) Kuetz. (Fig. H, No. 2 and 3)
Thallus filamentous, green coloured, branched, branching arising from one side, septa is distinct and from that junction places branching developed, vegetative cells 80 μ–110 μ in diameter × 500 μ–950 μ in length, holdfast 25 μ–50 μ in diameter × 200 μ–240 μ in length.

Locality : Harinbari forest, attached on the trunk bases of *Avicennia* sp.

32. *Cladophora sundarbanensis* Islam. (Fig. H, No. 8 and 9)
Thallus form, erect, green, branched, thalloid like creeping, rhizoids 15 μ–30 μ

in diameter, vegetative cell cylindrical 25 μ–50 μ in diameter × 200 μ–230 μ in length, cell wall 2.5 μ in thick.

Locality : Harinbari forest, attached on the mature soil in intertidal zones.

N. Genus : *Rhizoclonium* Kuetz.

33. *Rhizoclonium riparium* (Ruth) Harvey (Fig. E, No. 11-15)

Filament stout, vegetative cell 22 μ–27 μ in diameter × 40 μ–60 μ in length, wall delicate, rhizoids distinct.

Locality : Sajnekhali, attached on the pneumatophores of *Sonneratia* spp.

34. *Rhizoclonium hookeri* Kuetz. (Fig. E, No. 10)

Filaments having branches which formed in right angle to the main axis, cell wall 210 μ in thick, rhizoids frequent, vegetative cell cylindrical, sometimes swollen, 60 μ–100 μ in diameter × 75 μ–110 μ in length.

Locality : Sajnekhali, attached on the pneumatophores of *Sonneratia* spp.

O. Genus : *Chaetomorpha*

35. *Chaetomorpha linum* (Muell.) Kuetz. (Fig. E, No. 9)

Filaments yellowish green, cells rectangular, cylindrical, wall slightly thick, vegetative cell 75 μ–110 μ, in diameter × 100 μ–200 μ in length.

Locality : Sajnekhali, attached on the pneumatophores of *Sonneratia* spp.

36. *Chaetomorpha area* (Dillw.) Kg. (Fig. E, No. 19)

Filament obscure, cell wall thick, rhizoids present, disc shaped, filament 45 μ–55 μ in diameter × 45 μ–75 μ in length.

Locality : Sudhanyakhali, attached on the pneumatophores of *Sonneratia* spp. and *Xylocarpus* sp.

37. *Chaetomorpha gracilis* Kg. (Fig. E, No. 18)

Filament curved, cells cylindrical, variable shaped, cells 50 μ–110 μ in diameter × 60 μ–200 μ in length, cell wall thin, base rhizoid like.

Locality : Sudhanyakhali

38. *Chaetomorpha brachygona* Harvey (Fig. E, No. 17)

Filaments unbranched, cell wall thick, rhizoid like base, basal cell too much long, vegetative cell 60 μ–200 μ in diameter × 100 μ–350 μ in length.

Locality : Sajnekhali, attached on the trunk bases of *Excoecaria* sp.

FAMILY–*CODIACEAE*

P. Genus : *Boodleopsis* Gepp & Gepp

39. *Boodleopsis sundarbanensis* Islam (Fig. G, No. 12 and 13)

Thallus dichotomously branched, green, rhizoids, main axis 100 μ–140 μ in diameter, lateral branches gradually furcated, cell wall thick 12 μ–15 μ in diameter.

Locality : Sajnekhali, attached on the intertidal zone and on brick.

FAMILY–*ZYGNEMATACEAE*

Q. Genus : *Spirogyra*

40. *Spirogyra* sp. (Fig. E, No. 20)

Filament form silky like mat in habitat, vegetative cell septed, cells rectangular, chloroplasts distinctly found, pyrenoids prominent, cells–150 μ–200 μ in diameter × 300 μ–550 μ in length.

Locality : Ramgopalpore, floating on shallow stagnant water.

DIVISION— RHODOPHYTA

FAMILY–*RHABODONIACEAE*

R. Genus : *Catenella* Greville

41. *Catenella impudica* (Mont.) Ag. J. (Fig. F. No. 19)

Thallus 150 mm or more height, fibrous holdfast, rhizomatous axis, branches stout, erect, flattened, slender, 3 cm-5 cm long.

Locality : Netidhopani, attached on the buried tree trunk.

42. *Catenella repens* (Light foot) Batters. (Fig. H, No. 4)

Thallus dark pink-violet, creeping, branches divided dorsiventrally, segments elliptic or ovate, 3 mm width and 9 mm–12 mm long of single segment, haptera rhizoidal.

Locality : Netidhopani, attached on the trunk bases of *Excoecaria* sp.

43. *Catenella nipae* Zanardini (Fig. F, No. 20)

Thallus pink-violet, prostrate, gregarious, segmented, segments upto 8 mm-10 mm long, 3 m-4 m broad, branching dichotomously forked, segments stout, oblong, swollen.

Locality : Sajnekhali, attached on the trunk bases.

FAMILY–*RHODOMELACEAE*

S. Genus : *Pterosiphonia* Roth

44. *Pterosiphonia pennata* (Roth) Falkenberg (Fig. F, No. 21 and 22)

Thallus stout, 3 cm–5 cm tall, pinnately arranged main axis having two rows consisting of large cells in mid region, while several cells arranged serially both sides of the mid region.

Locality : Sajnekhali, attached on the pneumatophores of *Xylocarpus* spp.

T. Genus : *Bostrychia* Montagne

45. *Bostrychia radicans* (Mont.) Montagne

Thallus purple in colour, prostrate, pinnately branched, branching sometimes aggregated, plant anchored to the substratum by haptera, curved, plants slightly siphonous.

Locality : Harinbari forest, attached on the trunk bases of *Xylocarpus* sp.

46. *Bostrychia tenella* (Vahl.) Ag. J.

Thallus stout, reddish-brownish, erect, pinnately arranged, tip erect and straight, main axis with two rows consisting of large cells in mid region, branches spreading, plant siphonous, division corticate.

Locality : Bagna forest, attached on the pneumatophores of *Sonneratia* sp.

U. Genus : *Acanthophora* Lamk.

47. *Acanthophora specifera* (Vahl.) Boerg. (Fig. H, No. 6 and 7)

Thallus erect, bushy branching, 30 cm tall, branching exhibiting spine like appearance, main axis 3mm–4 mm in diameter, branches and branchlets aggregate to form bushy appearance, branches become bent at the tip.

Locality : Bagna forest attached on the truck bases on the *Sonneratia* sp.

V. Genus : *Polysiphonia* Greville

48. *Polysiphonia denudata* (Dillwyn) Kuetz.

Thallus 12 cm–20 cm tall, upper branches irregularly arranged, while lower branches strongly arranged dichotomously, branches erect in apical portion, main axis 0.5 mm–1.5 mm in diameter, central portion consisting of 4-6 cells.

Locality : Ramgopalpore, attached on the trunk bases of *Excoecaria* sp.

49. *Polysiphonia mollis* Hooker & Harvey (Fig. H, No. 5)

Thallus erect, epiphyte, 10 mm tali, light purple colour, anchored by holdfast, sometimes free floating, reproductive body formed within the mother axis, reproductive organ elliptic shaped, main axis consisting of 4-8 celled thick, branching irregularly formed.

Locality : Ramgopalpore attached on the trunk bases of *Excoecaria* sp.

FAMILY–*DELESSERIACEAE*

W. Genus : *Caloglossa* J. Ag.

50. *Caloglossa leprieurii* (Montagne) J. Ag. (Fig. F, No. 23 and 24)

Thallus having bisurfaces, branches spreading and erect reddish-violet, branching blade 2 mm–5 mm width, secondary blades formed proliferously and irregularly arranged, nodes like structure are found at the junction of branching.

Locality : Sajnekhali, attached on the trunk bases and pneumatophores of *Sonneratia* spp.

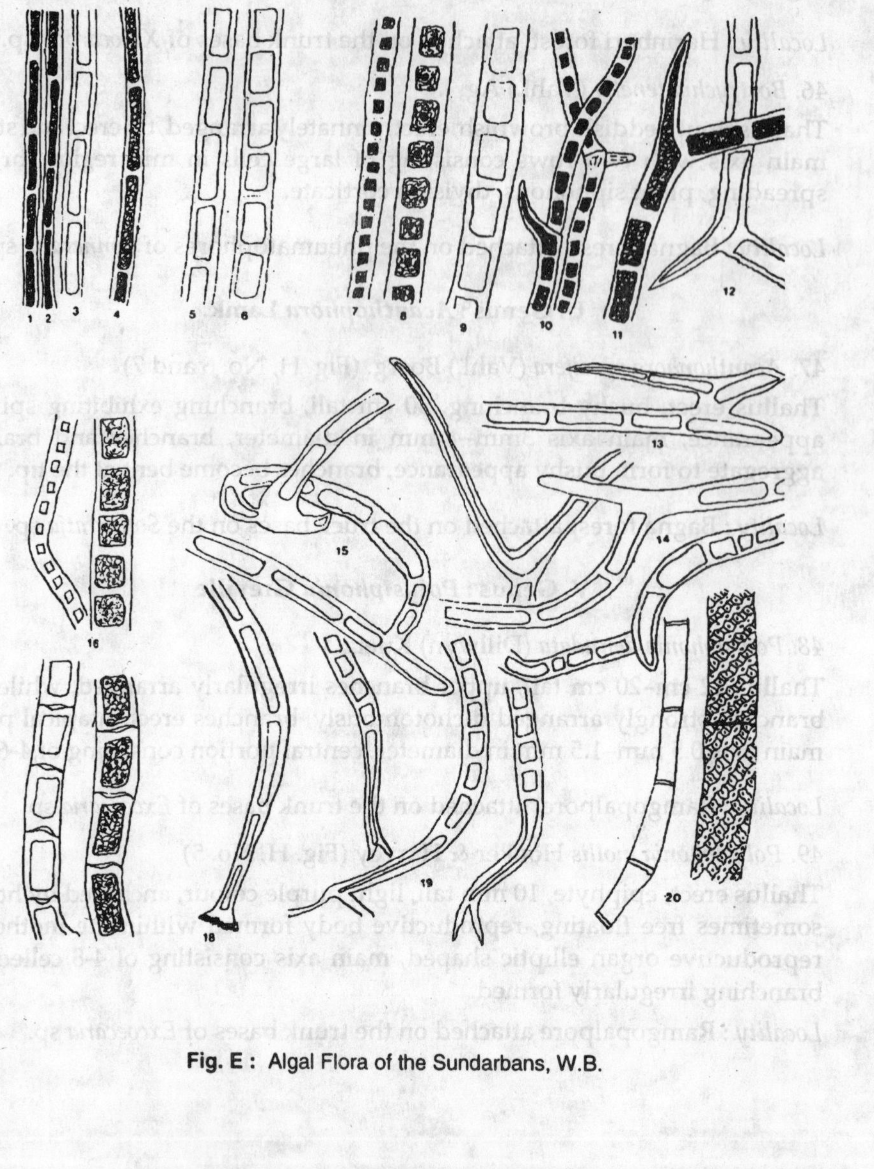

Fig. E : Algal Flora of the Sundarbans, W.B.

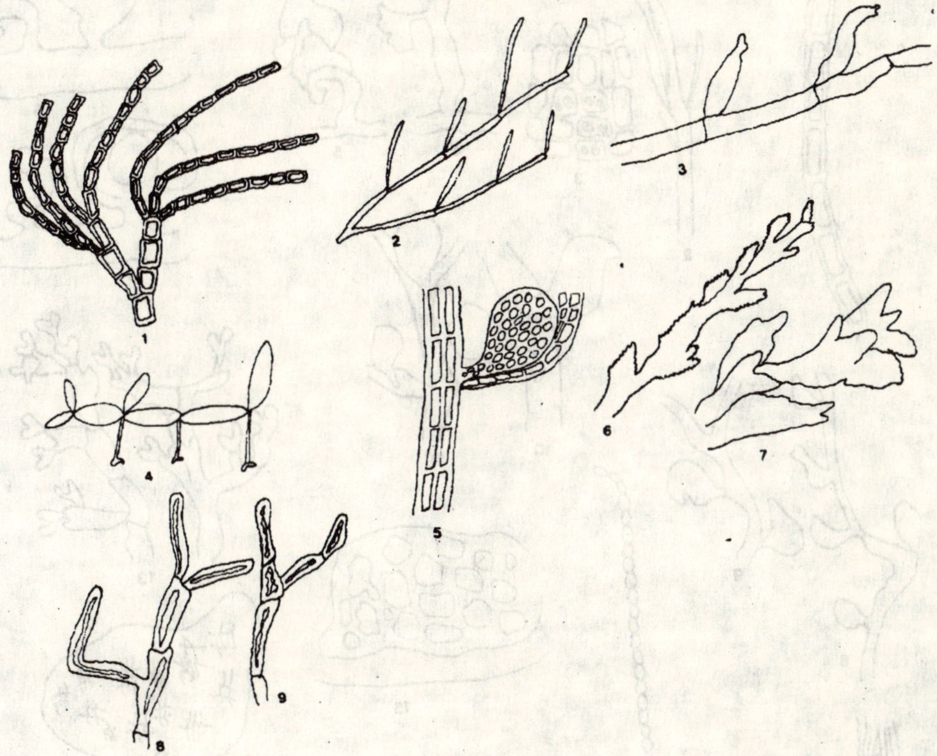

Fig. F : Algal Flora of the Sundarbans, W.B.

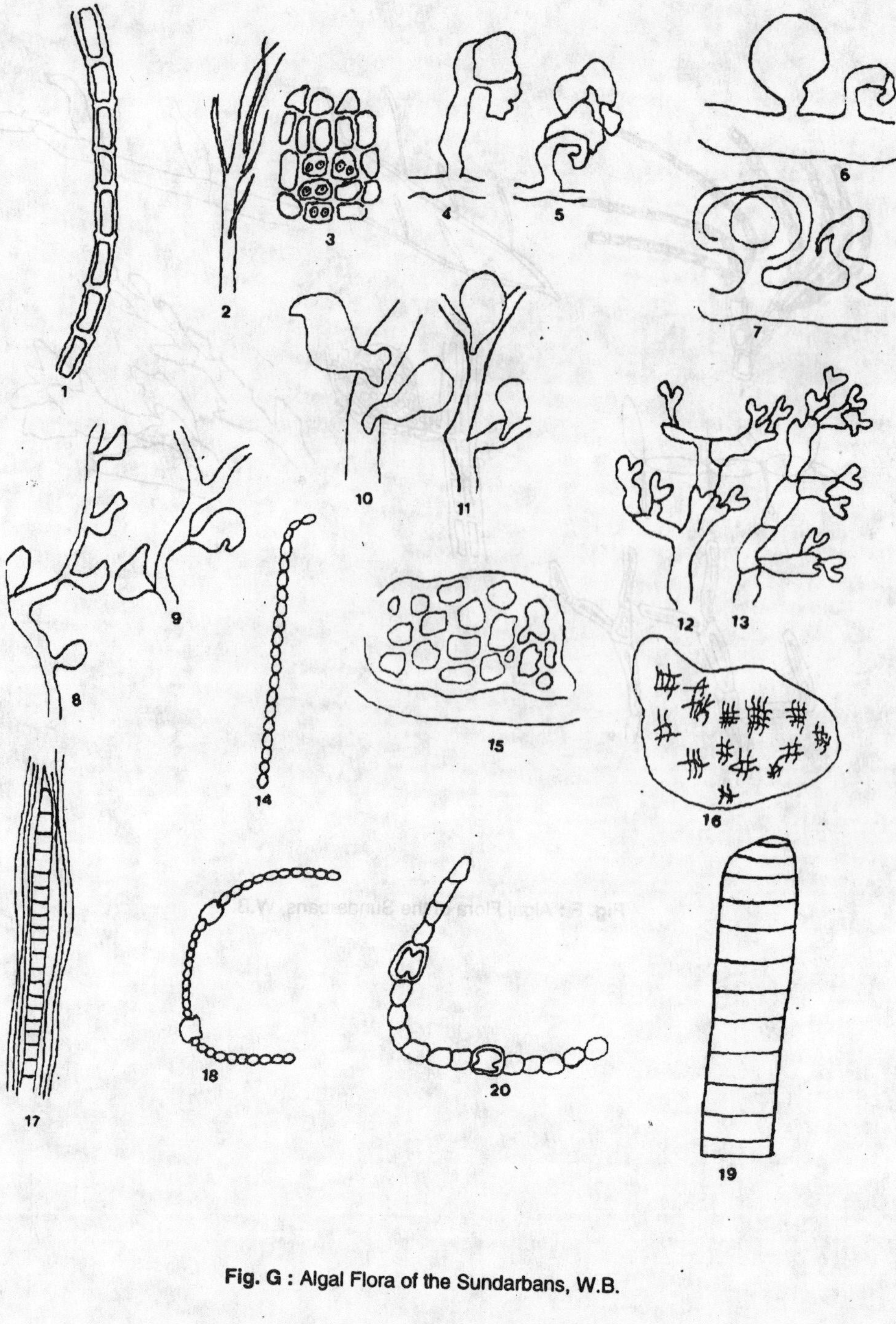

Fig. G : Algal Flora of the Sundarbans, W.B.

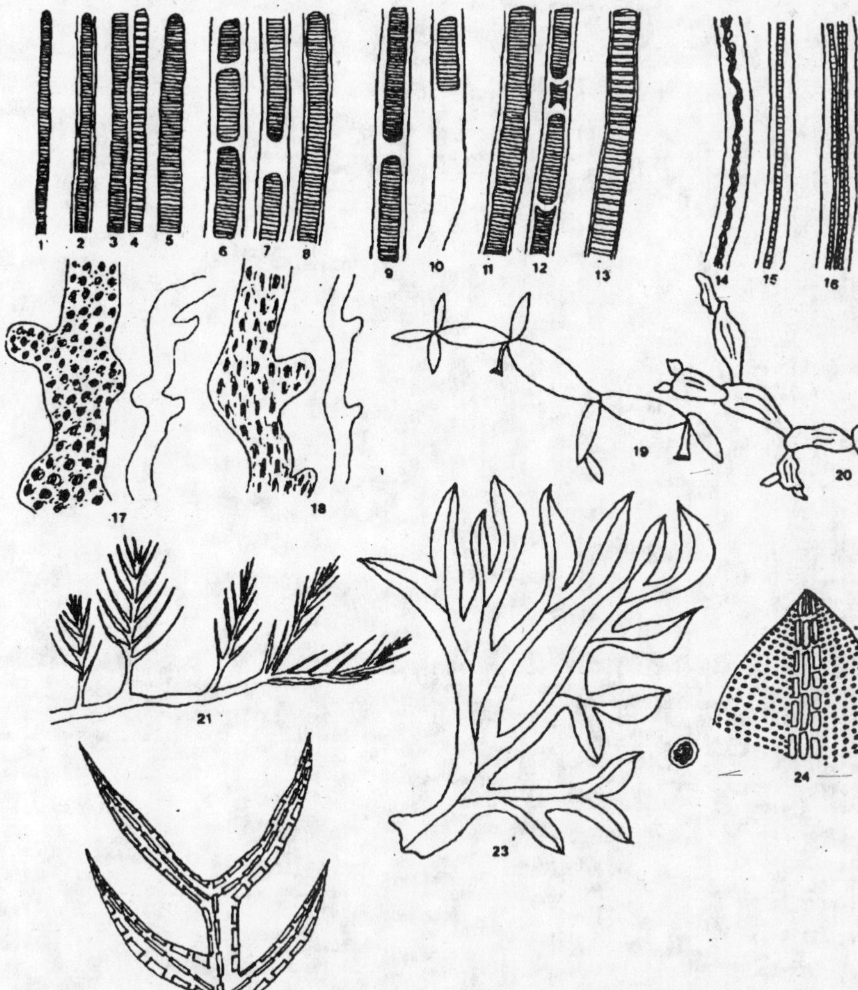

Fig. H : Algal Flora of the Sundarbans, W.B.

26

SELECTED BIBLIOGRAPHY ON THE MANGROVES

Aboul–Abbas-en-Nebaty (1230). *In : Ibn el Beitar Kitab al dschami al-Kibar*. Cited by Leclerc, 1877, Traite'des simples notices, extraits des manuscrits de la bibliotheque nationale, Paris, 23 : 1, 25 : 1, 26 : 1.

Adrim, M., A. Djamali & A.V. Toro (1984). Komunitas ikan di dacrab mangrove gugus pulau Pari. *In* : E. Soemodihardjo (ed), *Prosiding Seminar II Ekosistem Mangrove;* LIPI.

Agate, A.D. (1984). *Final Report : Microbiology of Mangrove Ecosystems of Southern Thailand.* UNDP/ UNESCO Regional Project on Research and Training Pilot Programme on Mangrove Ecosystems in Asia and the Pacific. UNESCO, Paris.

Agate, A.D., V.S. Humnabalkar, A. Chalermpongse & S. Aksornkoae (1985). Isolation of iron oxidizing bacteria from the mangrove in Southern Thailand ; *Biovigyanam J. Life Sci.*

Ahmad, N. (1966). Fish and fisheries of the Sundarbans. *In : UNESCO, Scientific problems of the humid tropical zone deltas and their implications : In Proceedings of the Dhaka Symposium,* 24 February–2 March, 1964, Paris, pp. 271-279.

Ahmed, E. (1972). *In : Coastal Geomorphology of India*. Chapter–II : 126–131. Orient Longman Limited, New Delhi.

Aksornkoae, Sanit (1987). Country Reports : Thailand : *In : Mangroves of Asia and the Pacific : Status and Management*. Technical Report of UNDP/UNESCO Research and Training Pilot Programme on Mangrove Ecosystem in Asia and the Pacific (RAS/79/002). pp. 231-261.

Aksornkoae, S., G. Wattayakorn & W. Kaitpraneet (1978). *Final Report : Physical and Chemical Properties of Soil and Water in Mangrove Forest at Amphoe Khlung, Changwat Chantaburi, Thailand,* UNESCO.

Albrigat, I.J. (1976). *In situ* degradation of mangrove tissue, *New Zealand Journal of Marine and Freshwater Research 10* : 385-389.

Alim, A. (1984). Soil and plant interaction in mangrove ecosystem. *UNDP/UNESCO Training Seminar on Geology, Sedimentology, Erosion and Accretion in Mangrove Areas,* Jan. 14–Feb. 5, Dhaka. 11 pp.

Amarasinghe, M.D. & W.K.T. Perera (1984). A preliminary survey of peripheral vegetation communities of Puttalam lagoon and Dutch Bay. *Proc 40th Ann. Sess. SLAAS.*

Amarasinghe, M.D. & S. Balasubramaniam (1992a). Structural properties of two types of mangrove stands on the north-western coast of Sri Lanka. *Hydrobiologia 247* : 17-27.

Amarasinghe, M.D. & S. Balasubramaniam (1992b). Net primary productivity of two mangrove forest stands on the north western coast of Sri Lanka. *Hydrobiologia 247* : 37-47.

Amjad, M. & Nadir Khan (1983). *The State of Forestry in Pakistan*, Pakistan Forest Institute, Peshawar.

Anderson, R. (1862). *Catalogue of Plants Indigenous in the Neighbourhood of Calcutta with Directions for Examination and Preservation of Plants*, Calcutta, India.

Andrews, T.J., B.F. Clough & G.J. Muller (1983). Photosynthetic gas exhchange properties and carbon isotope ratios of some mangroves in North Queensland, *Proc. 2nd Intern. Symp. Biol. Manage, Mangr.* W. Junk Publ.

Anonymous (1948). *First working plan for 24-Parganas Forest Division*, Government of West Bengal, Calcutta.

Anonymous (1948-1976). Wealth of India (Raw Materials). Vols. I-XII, CSIR, New Delhi.

Anonymous (1957). *Proceeding Mangrove Symposium*, Faridabad (1959).

Anonymous (1961). International Code of Nomenclature for Cultivated Plants, *Regnum Vegetalis*, Vol. 22. Utrecht-Netherlands.

Anonymous (1965). *West Bengal Forests : Centenary Commemoration Volume*, Calcutta.

Anonymous (1965). *Atlas of the Laccadive, Minicoy and Amindivi Island.* Prepared under the Direction of Shri Murkot Ramunny, Administrator, Union Territory of Laccadive.

Anonymous (1976). *Galley Reach–Mangrove Vegetation : Lighting Strikes,* I : 25,000 Office of Forests, Port Moresby, Papua New Guinea.

Anonymous (1977). *Forest Types : Gulf of Papua*, Map No. 2687, I : 150,000, Office of Forests, Port Moresby, Papua New Guinea.

Anonymous (1978). *Collected Reports from the 3rd Workshop on 'All India Co-ordinated Research Project on Brackishwater Prawn and Fish Farming'*, ICAR, New Delhi, India.

Anonymous (1979). *State-of-the Art. Report on Mangrove Ecosystems.* Govt. of India, DST, New Delhi, pp. 1-42.

Anonymous (1980). *State of Art : Mangrove Resources of the Philippines, Bakawan* **1** (1) : 3-6.

Anonymous (1980). Network of research on mangrove. *Bakawan 1* (1) : 7.

Anonymous (1983). *1983 Statistical Year Book of Bangladesh.* Bangladesh Bureau of Statistics, Govt. of Bangladesh.

Anonymous (1983). *Global Status of Mangrove Ecosystems* (Commission on Ecology Paper-3), IUCN-Switzerland. pp. 1-85.

Anonymous (1984). *Project Report on Mangrove Ecosystem of Sundarbans in Virgin and Reclaimed Areas with special reference to productivity*, Dept. of Science & Technology, New Delhi.

Anonymous (1984). A profile of the Indian Mangrove. *Bakawan Newsl.* 3 (3) 10.

Anonymous (1984). *Annual Fisheries Statistics*, Fisheries Division, Ministry of Agriculture, Malaysia.

Anonymous (1985). *Mangrove Management in Thailand, Malaysia and Indonesia*, FAO Environment. Paper No. 4, Rome.

Anonymous (1987). *Country Report : Fiji. In : Mangroves of Asia and The Pacific : Status and Management.* Technical Report of UNDP/UNESCO Research and Training Pilot Programme on Mangrove Ecosystems in Asia and the Pacific (RAS/79/002). pp. 281-298.

Anonymous (1987). *Country Report* : **Philippines**. *In : Mangroves of Asia and The Pacific : Status and Management.* Technical Report of UNDP/UNESCO Research and Training Pilot Programme on Mangrove Ecosystems in Asia and the Pacific (RAS/79/002). pp. 175-210.

Anonymous (1987). ***Mangrove in India.*** *Status Report.* Government of India, Ministry of Environment & Forests, New Delhi. December, 1987, pp. 1-150.

Anonymous (1989). *Integrated Resource Development of the Sundarbans Reserve Forest.* Project Document BGD/84/056/A/01/12/UNDP. 55 pp.

Anonymous (1989). *Conservation of Mangroves in India.* Ministry of Environment & Forest, New Delhi. pp. 1-72.

Anonymous (1990). Declaration of Management Strategies. *In : International Symposium on 'The Ecology of Mangroves and Related Ecosystems'*, Department of Zoology, University of Nairobi, Mombasa, Kenya. 20-30 September, 1990.

Anonymous (1992). *Caostal Environment.* Remote Sensing Application Mission, RSAM/SAC/COM/SN/11/92. ISRO, Ahmedabad-380053. pp. 1-114.

Anonymous (1995). Hugli Malta Estuary, West Bengal. *Estuarine Ecosystem*, ZSI Series, Part-2. pp. 1-542.

Ansari, T.A. (1985). *Management plan for Mangrove of Pakistan.* Government of Sind, Forest Department Project.

Ansari, T.A. (1987). ***Pakistan.*** *Country Report.* ***In*** *: Mangroves of Asia and The Pacific : Status and Management* (Ed. Umali, M. Ricardo, *et al.*). Natural Resources Management Center and National Mangrove Committee, Ministry of Natural Resources, Philippines, Quenzon City, Metro Manila, November, 1986. pp. 151-173.

Anthony, M.K.D., V.N. Panakalaro & Sarkaranarayana (1974). Studies on the monsoonal flow of the Mandovi Estuary. *Mahasagar Bull. Natn. Inst. Oceanogr.* 7 : 69-72.

Arroyo, C.A. (1977a). Flora of Philippines Mangrove, ***In*** *: Proc.* of *Intern. Workshop on Mangr. Estuar. Area Dev.* Indo-Pacific Region, Manila, Philippines. pp. 137-143.

Arroyo, C.A. (1977b). Vegetation Structure of Mangrove Swamps. ***In*** *: National Symp./Workshop on Mangrove Research & Development*, Manila.

Arroyo, C.A. (1979). Flora of Philippine Mangrove. ***In*** : Mangr. & Estuar. Veg. Southeast Asia. *BIOTROP Spec. Publ. 10* : 33-34.

Artz, T. (1936). Die Kutikula einiger afrikanischer Mangrove Pflanzen, *Ber. d. bot. Gesell.* ***54*** : 247-60.

Aruchelvam, K. (1968). Mangroves of Ceylon. *For. 8* (3 & 4) : 59-62.

Assemien, P. (1969). Pollens fossiles de *Rhizophora* a Bogue (Basse Vallee du Senegal) Pollen Spores, Vol. XI. No. I. pp. 73-81.

Atkinson, M. R., G.P. Findlay, A.B. Hope, M.G. Pitman, H.D.W. Saddler & K.R. West (1967). Salt Regulation in the Mangrove, *Rhizophora mucronata* Lamk. and *Aegialitis annulata* R. Br. *Aust. J. Biol. Sci.* ***20*** : 589–599.

Attims, Y. & G. Cremer (1967). *Les radicelles capillaries des paletuviers dans une mangrove de cote d'Ivoire, Adansonia Series* ***2*** (7) : 547-51.

Aubreville, A. (1964). *Problems de la mangrove d'hier at d'anjourd-hui Addisouia* ***4*** : 19-23.

Backer, C.A. & C.G.G.J. Van Steenis (1951). Sonneratiaceae, *Flora Malesiana* I, ***4*** : 280-289.

Backer, C.A. & R.C. Backhuizen Van der Brink Jr. (1963 & 1965). "Flora of Java", V*ol. I* (1963), ***Vol. II*** (1965), Noorahoff, Groningen.

Bacon, P.R. (1975). Recovery of a Trinidadian Mangrove swamps from Attempted Reclamation. ***In*** : G.E. Walsh, S.C. Snedakar & H.J. Teas (eds.). *Proceeding of the International Symposium on Biology and Management of Mangroves.* Institute of Food & Agriculture Sciences, University of Florida, Gainesville, Florida, Vol. II. pp. 805-915.

Badiane, S. (1986). The Mangrove in Casamance. *The-Casamance-Estuary : Envrionment-Fisheries-Economics.* pp. 207-218.

Bagchi, S., A. Matilal, A.K. Shaw & B.B. Mukherjee (1988). Lipids and waxes in leaves of some mangrove plants of Sundarban, India. *Indian J. Mar. Sci.* ***17*** (2) : 150-152.

Baguli, T.K., A.K. Ghosh & K.R. Naskar (1993). Fishery potentialities of the estuarine water of the Sundarbans mangrove ecosystem in India. *ABST. : 194. Asia Pacific Symp.* on *Mangrove Ecosystem.* Research Centre & Dept. of Biology. The Hong Kong University of Science & Technology, Hong Kong.

Baguli, T.K., A.K. Ghosh & K.R. Naskar (1994). Availability and Adaptability of different species of Fish, Prawn and Crab in the Mangrove Ecosystem of the Lower Ganga Delta and their Impact in the Ecosystem. *ABST. 100. West Bengal State Science Congress.* West Bengal State Council of Science & Technology, Calcutta, 28th Feb.–1st March, 1994.

Baguli, T.K., A.K. Ghosh, D.C. Ghosh & K.R. Naskar (1995). Studies on the Fish and Prawn Parasitic Diseases of the Estuarine Ecosystems of the Sundarbans. *ABS. 52. National Symp. on Perspective in Biodiversity.* The Zoological Society, Calcutta. Indian Institute of Chem. Biology, Calcutta.

Bahuguna, Anjali, S. Nayak, K.N. Prasad, S.A. Kadri & C.R. Trivedi (1990). *Mapping of long term-Changes in the tidal wetland areas of the Gulf of Kachchh.* Scientific Note. Space Application Centre, Ahmedabad. IRS-UP/SAC/MCE/SN/29/91, 15 p.

Baharudin, J. (1984). The concept of sustained yield management in the mangrove ecosystem. *Proceedings of the Workshop of Productivity of the Mangrove Ecosystem : Management implication*. ***In*** : Ong, J.E. & Gong, W.K. (eds.), University Sains, Malaysia, Penang Malaysia, pp. 115-120.

Baines, G.B.K. (1979). *Mangrove for National Development : A Report on the Mangrove Resources of Fiji.*

Baines, G.B.K. (1979). Mangrove resource management in Pacific Island Nation. Fiji, pp. 728-739. ***In :*** *Proceedings Asian Symp. Mangr. Environment Res. Management*, University of Kualalumpur, Malaysia.

Baker, R.T. (1915). The Australian 'Grey Mangrove' (*Avicennia officinalis* L.). *Journal and proceedings of the Royal Society of New South Wales*, ***49*** : 257–288.

Bakhuizen, Van der Brink R.C. (1921). Revisio generis *Avicennia, Bull. Jard. bot. Buitenz. Ser.* ***3*** : 199-226.

Balachandra, L. (1988). A. comprehensive account of the mangrove vegetation of Andaman and Nicobar Islands. *Indian For.* ***114*** : 741-751.

Bal, A.R. & S.K. Dutt (1984a). Effect of soil salinity on growth of *Coix lachrymajobi* L. *Indian J. Plant Physiol.* **27** (4) : 398-400.

Bal, A.R. & S.K. Dutt (1984b). Studies on the occurrence and distribution of salt glands and stomata in some mangrove species. *J. Indian Soc. Coastal agric. Res.* **2** (1) : 5-9.

Balakrishnan, N.P. (1979). Andaman Islands–Vegetation and Floristics. ***In*** : C.J. Saldanha (ed.), *Andaman, Nicobar and Lakshadweep–An Environmental Impact Assessment*, Oxford and IBH Publishing Co. Pvt. Ltd., New Delhi, pp. 55-68.

Balasubramaniam, S., P. Rasiah & M.L. Ratnapala (1979). Mangrove vegetation of Vadamarachchi lagoon. *Phyta* **1** (4) : 21-23.

Balasubramanian, T. & V.K. Venugopalan (1984). Dissolved organic matter in Pitchavaran mangrove environment, Tamil Nadu, South India, pp. 496-513. *Proceeding Asian Symp. Mangr. Environ. Res. Management*, University of Malaya, Kualalumpur, Malayasia.

Ball, M.C. & G.D. Farguhar (1984a). Photosynthetic and stomatal responses of two mangrove species, *Aegiceras corniculatum* and *Avicennia marina* to long term salinity and humidity conditions. *Plant Physiol.* ***74*** : 1-6.

Ball, M.C. & G.D. Farguhar (1984b). Photosynthetic and stomatal responses of the grey mangrove, *Avicennia marina* to transient salinity condition. *Plant Physiol.* ***74*** : 7-11.

Bandyopadhyay, A.K. (1966). Forests of Sundarbans. *In : Centenary Commemoration Volume*, West Bengal Forest (1964) pp. 166-175, Published by the D.F.O. Planning and Statistical Cell, Writers' Building, Calcutta.

Bandyopadhyay, A.K. (1984). Soil and water characteristics of the mangrove forest of Sundarbans (India). **In** : *Report 2nd Introductory Training Course on Mangrove Ecosystems*. UNDP/UNESCO Regional Project RAS/79/002.

Bandyopadhyay, A.K. (1985a). Management Techniques for increased crop production in denuded mangal lands of Sundarbans. *Nat. Symp. on Biology, Utilization and Conservation of Mangroves*. 18-20 Nov., 1985, *Abstract No. 56*, P. 29. Shivaji University, Kolhapur.

Bandyopadhyay, A.K. (1985b). The Sundarbn mangrove experience on reclamation. **In** : *Workshop on the Conversion of Mangrove Areas to Paddy Cultivation. Los Banos, Laguna, Philippines*, 1-3 April (1985) *UNDP/UNESCO Research and Training Pilot Programme on Mangrove Ecosystem in Asia and the Pacific*. pp. 65-69.

Bandyopadhyay, A.K. (1986a). *Characterization and management of coastal salt affected soils. CSSRI, Annual Report*, pp. 184-188.

Bandyopadhyay, A.K. (1986b). Soil and water characteristics of the mangrove forests of Sundarbans (India). *Indian For.* ***112*** : 58-65.

Bandyopadhyay, A.K. (1986c). A note on the effect of liming on soil properties and crop growth in acid saline soils of Sundarbans. *J. Indian Soc. Coastal agric. Res.* ***4*** (1) : 73-75.

[illegible]dyopadhyay, A.K. (1987). A note of Effect of Reclamation through Marginal Embankments on Coastal Ecosystem of Sundarbans. *J. Indian Soc. Coastal Agric. Res.* **5** (1) : 39-41.

Bandyopadhyay, A.K. & Dipak Sarkar (1987). Occurrence of Acid Saline Soils in Coastal area of Sundarban Area of West Bengal. *J. Ind. Soc. Soil. Sci.* 35 : 542-54.

Bandyopadhyay, A.K. (1992). Role of Mangroves in Island Ecosystems with particular reference to the Bay Islands. **In** : *Proceedings of the Project Formulation Workshop for establishing a Global Network of Mangrove. Genetic Resource Centres for Adoption to Sea Level Rise,* January, 15-19, 1991 (eds.) S.V. Desmukh & R. Mahalingam Padroo.

Bandyopadhyay, B.K., A.K. Bandyopadhyay & H.S. Sen (1987). Physico-Chemical Characteristics of Coastal Saline Soils of India. *J. Indian Soc. Coastal Res.* 5 (1) : 1-14.

Banerjee, A.N. (1931). *History and Hydraulics of the the River System near Calcutta,* Calcutta.

Banerjee, J. (1959). Regeneration and Exploitation of Mangroves. *Proc. Mang. Symp.* Faridabad, India, pp. 124-128.

Banerjee, L.K. (1980). *On the Vegetation and Flora of Tidal Forest in the Regional Deltaic setting between the Dhamra and the Devi River of the Mahanadi Delta in Orissa.* Ph.D. Thesis, Kalyani University. (Unpub.)

Banerjee, L.K. (1984). Vegetation of the Bhitarkanika Sanctuary, Orissa State, *J. Econ. Tax. Botany* 5 : 1065-1079.

Banerjee, L.K. (1985). Comparative study on mangroves of the Sundarbans and that of the Mahanadi Delta in Eastern India. ***The Mangroves : In** : Proc. Nat. Symp. Biol. Util. Cons. Mangroves* : Nov. 1985 : 527-530, Kolhapur, India. L.J. Bhosale (ed.)

Banerjee, L.K. (1987). Ecological Studies on the Mangals of Mahanadi Estuarine Delta, Orissa, *Trop. Ecol.* 82 : 117-125.

Banerjee, L.K. & T.A. Rao (1985). Mangals of Mahanadi Delta, Cuttack District, Orissa State, India. *Marine Plants : Papers presented at the All India Symposium on Marine plants, their Biology, Chemistry and Utilization,* Dona Paula, Goa, Oct. 30-Nov. 1, 1983 Krishnamurthy, V. & Untawale, A.G. (eds.) 137-152.

Banerjee, L.K., A.R.K. Sastry & M.P. Nayar (1989). ***Mangrove in India, Identification Manual.*** Botanical Survey of India, Govt. of India, pp. 1-113.

Bangladesh Country Report (1978). *Presented at UNESCO Regional Seminar on Humen Uses of the Mangrove Environment and Management Implication,* 4-8 December, Dhaka, 36 pp.

Banus, M.D. & S.E. Kolehmairen (1975). Floating, rooting, growth of red mangrove (*Rhizophora mangle* L.) Seedlings : Effect on expansia of mangroves in South-Western Puerto Rico. **In** : G.E. Walsh, S.C. Snedakar & H.J. Teas (eds.). *Proceedings of the International Symposium on Biology and Management of Mangrove.* Institute of Food and Agricultural Science, University of Florida, Gainesville, Florida, Vol. *I*, pp. 370-384.

Barlrw, B.A. (1967). Parasitic flowering plants, *Australian Natural History Magazine,* 15 : 365-368.

Barth, H. (1982). The biogeography of mangroves. Chap. 3. ***In** : Tasks for Vegetation Science,* (eds.) D.N. Sen & K.S. Rajpurahit, Vol. 2. The Hague, Junk.

Barton, E.S. (1903). List of Marine Algae collected at Maldive and Laccadive Islands by J.S. Gardiner Expedition. *J. Linn. Soc. London Bot.* 35 : 475-482.

Bastardo, H. (1988). Preliminary study of the decomposition process in the mangrove forest of the Tacarigua lagoon. *Acta Cient. Veuez,* 39 (2) : 184–185.

Basu, K. (1965). Cultivation of Mangrove seedlings away from their natural habitat.

Basu, A.K. & B.B. Ghosh (1970). Observations on diurnal variations in some selected stretch of Hooghly estuary (India). *Schweie. Z. Hydrd.,* 32 (1) : 271-283.

Baylis, G.T.S. (1940). Leaf anatomy of the New Zealand mangrove. *Trans. R. Soc. N.Z.* 70 : 164-170.

Baylis, G.T.S. (1950). Root Systems of the New Zealand mangrove. *Trans. R. Soc. N.Z.* 78 : 509-514.

Beard, J.S. (1967). An inland occurrence of mangrove. *Western Australian Naturalist.* 10 : 112-115.

Beddome, R.H. (1883). *Handbook to the Ferns of British India,* Ceylon and Malay Peninsula, p. 440.

Beeckman, H., E. Gallin & E. Coppejans (1990). Indirect gradient analysis of the mangal formation of Gar Bay (Kenya), *Silva Gandavensis.* 54 : 57–72.

Beer, S.A., Shomer Ilan & Y. Waisel (1976). Salt Stimulated Phosphoenolpyruvate carbonylase in Cakile maritime, *Physiol. Plant.* **34** : 293-295.

Bell, L. A.J. (1985). Coastal Zone Management in Locoter Samoa. *Report of the Third South Pacific National Parks and Research Conference*, held in Apia, Western Samoa, 1985. (Collected Key Issue and Case Study Papers) 2 : 57-53.

Bentham, G. & J.D. Hooker (1862). *Genera Plantarum*, Vol. *I*. London : Reeve.

Berry, A.J. (1963). Faunal Zonation in Mangrove Swamps. *Bull. Nat. Muz. Sing* **32** : 90-98.

Berry, A.J (1972). The Natural History of W. Malaysian Mangrove Fauna. *Malaya. Nat. J.* **25** : 135–162.

Bharucha, F.R. & B.S. Navalkar (1942). Studies in the ecology of mangroves : The Chloride-content of sea water, soil solution and the leaf cell sap of mangrove. *J. Univ. Bombay* **10** : 97-106.

Bhaumik, S.M., G.N. Roy & A. Choudhury (1983). Bacteria from the Mangrove Swamps of Sundarbans Complex. *Proc. All India Symp. on Marine Plants*, NIO, Goa.

Bharucha, F.R. & V.S Shirke (1947). A study of important Catabolic changes in the seedlings of *Avicennia officinalis* L. *J. Univ. Bombay*, **15** : 1-14.

Bharucha, F.R. & P.V. Rangasekar (1957). Studies on free Amino acids and Organic acids of Halophytes of Bombay. *NATURWESE* **15** : 44 : 469.

Bhat, S.R. & A.G. Untawale (1987). *Bruguiera cylindrica* (L.) Bl. (Rhizophoraceae)–a new locality record from the West Coast of India. *Curr. Sci.* **56** (11) : 604-606.

Bhattacharya, A.K. & S. Chaudhuri (1985). The role of vegetation and moisture in dune formation : A study from the tidal island deposits on the deltaic Sundarbans, facing Bay of Bengal. *Nat. Symp. on Biology, Utilization and Conservation of Mangroves*, Kolhapur, 18-20 Nov. 1985, *Abst. No. 8*, pp. 4-5.

Bhattacharya, R.K. (1976). New salt tolerant rice varieties for coastal saline soils of Sundarbans (W.B.). *Sci. & Cult.* **42** : 122-123.

Bhattacharyya, Asok Kumar & Somnath Chaudhuary (1986). The role of vegetation and moisture in dune formation : A study from the tidal island deposits of the deltaic Sundarbans facing Bay of Bengal. *The Mangroves : Proc. Nat. Symp. Biol. Util. Cons. Mangroves.* Nov. 1985 : 170-174. (ed.) L.J. Bhosale, Shivajee University, Kolhapur.

Bhattacharya, S.K. & S.M. Chatterjee (1972). Brickfields and Pollution. *Proc. All India Conference on Abatement of Environmental Pollution.* pp. 134-146, Calcutta, Dec. 14-16, 1972.

Bhosale, L.J. (1978) *Ecophysiological studies on the Mangrove from Western Coast of India.* Shivaji University Publication, Kolhapur.

Bhosale, L.J. (1979). Distribution of Trace elements in the leaves of mangroves. *Ind. J. Marine Sci.* **8** : 58-59.

Bhosale, L.J. (1981). Photosynthetic carbon metabolism in a salt marsh plant, *Buguiera gymnorrhiza*, Lamk.–1. 14 CO_2 light fixation products, enzyme activities and aspartate utilisation. *Environ. exp. Bota*, **21** (2) : 163-170.

Bhosale, L.J. (1986). ***The Mangroves.*** *Proc. Nat. Symp. on Biol., Util. & Conserv. of Mangroves.* (ed.) Shivajee University, Kolhapur.

Bhosale, L.J. & B.A. Karadge (1975). Photosynthetic Efficieny and Productivity of Mangroves of Western India. *Bull. Dept. Mar Sci. Univ. Cochin*, **7** (1) : 205-211.

Bhosale, L.J. & L.S. Shinde (1983). Significance of Cryptovivipary in *Aegiceras corniculatum* (L.) Blanco. Chap. 14, in *Biology and Ecology of Mangroves*, (ed.) H.J. Teas. *Tasks for Vegetation Science* 8. The Hague : Junk.

Bhosale, L.J. & L.S. Sinde (1985). Nitrate reductase activity in Mangrove. Marine Plants paper presented at the *All India Symposium on Marine Plants, their Biology, Chemistry and Utilization*, Dona Paula, Goa, Oct. 30–Nov. 1 (1983), V. Krishnamurthy & A.G. Untawale (eds.) 1985. 299-306.

Bhosale, N.B., V.K. Dhargalkar, S.G.P. Matondkar & Sayeeda Bukhari (1976). Biochemical composition of mangrove leaves from Goa. *Ind. J. Marine Science* **5** (2) : 239-241.

Bhowmik, S.K., G.N. Roy & S. Barman (1985). A list of twenty-four strains of bacteria from the decomposed litter of mangrove areas of Sundarbans Complex. *Nat. Symp. on Biology, Utilization and Conservation of Mangroves, Abstract No. 72*, p. 38.

Bhovichitra, M. (1982). Mining undisturbed and disturbed mangrove estuarine ecosystems in relation to coastal fisheries. *Proc. NRCTJSPS Rattanakosin Bicem. Joint Seminar Sci. Mangr. Resources.* **1982** : 244-261.

Biebl, R. & H. Kinzel (1965). Blattabu und Salzhausalt von *Laguncularia racemosa* (L.) Gaertn. f. and anderer Mangrove baume auf Puerto Rico. *Ost. Bot. Zeit.* **112** : 56–93.

Bina, R.T., E.N. Jora, B.R. Lorenzo & Jr. de Jesus (1978). Mangrove Inventory of the Philippines using Landsat Multispectral Data and the IMAGE-100 System, NRMC Res. Monogr. 2 : 1-8.

Bird, E.C.F. (1971). Mangroves as land builders, *Victorian Naturalists, 88* , 189-197.

Bird, E.C.F. (1982). Management problems in tropical estuaries. ***In*** : Bird, C.F.F., S. Soegiarto and K. Soegiarto (eds.), *Proceedings of the Workshop on Coastal Resources Management in the Cilacap Region* : 9-16. LIPI-UNU, Jakarta-Tokyo.

Bird, E.C.F. & M.M. Barson (1977). Measurement of Physiographic changes on mangrove-fringed estuaries and coast lines, *Marine Research in Indonesia. 18*, 73-80.

Birowo, S. & H. Uktolseya (1982). The coastal estuarine waters of the Cilacap Region. pp. 27-29. ***In*** : C.F.F. Bird, S. Soegiarto & K. Soegiarto (eds.). *Proc. Workshop Caostal Resource Manage.* Cilacap Region. LIPI–UNU.

Birowo, S. & H. Uktolseya (1983). Oceanography of Sunda Straitp. *In* : C.E.F. Bird, *et al.* (eds.). *Proc. Workshop on Coastal Resources.* Manage. Krakatua and Sund Strait Region, Indonesia.

Biswas, K. (1927). Flora of the Salt Marshes. *J. Dept. Sci. Cal. Univ.* **8** : 1-47.

Biswas, K. (1934). A comparative study of Indian species of *Avicennia,* Notes. *Royal Bot. Gard.*, **18** : 159-166.

Biswas, A., S.K. Bhaumik, G.N. Roy & A. Choudhury (1983). Cellulolytic bacteria from the decomposed litter in mangrove patches of Sundarbans Complex. *Proc. All India Symp. on Marine Plants,* NIO. Goa.

Blaber, S.J.M. (1986). Feeding selectivity of a guild of Piocivorous fish in mangrove areas of North West Australia. *Aust. J. Mar. Freshwater Res.*, **37** (3) : 329-336.

Blasco, F. (1974). The mangroves of India, *Inst. Frac. Pondicherry Tran. Sect. Sci. Tech.*, **14** : 168.

Blasco, F. (1975). *The Mangroves in India* (translated by K. Thanikaimoni from LES MANGROVES DE L'INDE), *Institute Francais de Pondicherry, Inde, Sri Aurobinda Ashram,* India, pp. 1-175.

Blasco, F. (1977). Outlines of ecology, botany and forestry of the mangals of the Indian subcontinent. *Chap. 12. **In*** : V.J. Chapman (ed.). *Ecosystem of the World : Wet Coastal Ecosystems.* ***Vol. I,*** New York : Elseiver, pp. 241-260.

Blasco, F. (1982). *An inventory of the Gangetic mangroves from Landsat satellite data and SPOT stimulations.* Conference Leminaire franco-asidique 22-26 Nov. 1982.

Blasco, F. (1984). Climate factors and the biology of mangrove plants. ***In*** : *The Mangrove Ecosystem :* Research method (ed. S.C. Snedaker & J.G Snedaker) UNESCO, pp. 18-35).

Blasco, F. & C. Caritini (1974). Mangrove de Pichavaram : Phytogeographe et Palyanologie, *Travanax. ex. Documents de Geographie Tropicale,* No. 8, pp. 164-185.

Blasco, F., C. Caritini, S. Chandra & G. Thanikaimoni (1974). Main Characteristics of Indian Mangroves. *Symp. Int. Biol. Mgt.* Honolulu, **1** : 71-87.

Blasco, F., S. Chanda & G. Thanikaimoni (1975). Main characteristics of Indian Mangroves. ***In*** : G. Walsh, S.C. Snedeker & H.J. Teas (eds.). *Proc. Internal, Symp. on Biol. and Managt. of Mangroves. Inst. Food and Agril. Sci. Univ., Florida, **Vol. I.*** pp. 71-83.

Blasco, F., R. Kerrest & C. Marius (1985). Considerations of some ecological factors influencing the biology of Indian Mangroves. *The Mangroves : Proc. Nat. Symp. Biol. Util. Cons. Mangroves,* Nov. 1985 : 135-145. L.J. Bhosale (ed.), Kolhapur, Shivajee University.

Blatter, E. (1905). The Mangroves of Bombay Presidency and its Biloogy, *J. Bombay Nat. Hist. Soc.* **16** : 644-656.

Blatter, E. (1908). On the Flora of Kutch, Part-I. *J. Bombay Nat. Hist. Soc.*, **18** (4) : 756-777.

Blatter, E. (1909). On the Flora of Kutch, Part-II. *J. Bombay Nat. Hist. Soc.*, **19** : 157-176.

Blatter, E. (1928). The Flora of the Indus Delta. *J. Indian Bot. Soc.*, **20** : 22-43.

Bleeker, P. (1983). *Soils of Papua New Guinea.* Canberra : Austra. Nat. Univ. Press.

Bole, P.V., F.R. Bharucha (1954). Osmotic Relations of the leaves of *Avicennia alba* Bl. *J. Univ. Bombay*, **22** : 50-54.

Bole, P.V. & J.M. Pathak (1988). *Flora of Saurashtra.* B.S.I., Calcutta.

Bonnittee, W. (1978). The Rhizophorae. *Symp. on Managrove and Estuarine Vegetation in South East Asia.* Biotrop. Spl. Publication No. 10, pp. 23-31.

Boonruang, P. (1984). The rate of degradation of mangrove leaves. *Rhizophora apiculata* and *Avicennia marina* at Phuket Island, Western Peninsula of Thailand. ***In*** : Soepadma, E. *et al.* (eds.). *Proc. Asian Symp. Mangr. Env. Res. Mang.*, Univ. Malaya, Kualalumpur, Malayasia.

Boon, P.I. & S. Cain (1988). Nitrogen cycling in salt-marsh and mangrove sediments at Western Port, Victoria. *Aust. J. Mar. Freshwat. Res.* **39** (5) : 607-623.

Bordillon, T.F. (1980). *The Forests of Travancore.* Govt. Press, Trivandrum.

Bostrom, T.E. & C.D. Field (1973). Electrical potentials in the salt gland of *Aegiceras*. ***In*** : W.P. Anderson (ed.). *Ion Transport in Plants.* Academic Press, London, pp. 385-392.

Boto, K.G. (1982). Structure, function and management. ***In*** *: Mangrove Ecosystems in Australia.* B.F. Clough (ed.), pp. 239-257. ANU Press, Canberra.

Boto, K. G. & J.S. Bunt (1981). Tidal export of particulate organic matter from a northern Australian mangrove system. *Estuar. Coast. Shelf Sci.* **13** : 247-255.

Boto, K.G. & J.S. Bunt (1982). Carbon export from mangroves. *In : Cycling of carbon, nitrogen, sulpher and phosphorous in terrestrial and aquatic ecosystem* (eds.). J.R. Freney & I.E. Galbally, Springerverlag, pp. 105-110.

Boto, K.G., J.S. Wellington & J.T. Wellington (1984). Variations in mangrove forest productivity in northern Australia and Papua New Guinea, *Estuar. Coast. Shelf Sci.* **19** : 321-329.

Boughey, A.S. (1963). Diting of tropical coastal regression (based on roots of mangrove *Rhizophora mucronata*). *Nature. Lond.* **200** : 600.

Bowman, H.H.M. (1916). Physiological studies in *Rhizophora. Proc. Nat. Acad. Sci., U.S.A.* **2** : 685-688.

Bowman, H.H.M. (1917). Ecology and physiology of the red mangroves. *Proc. Amer. Philos. Soc.* **56** : 589-672.

Bowman, H.H.M. (1921). Histological variations in *Rhizophora mangle* L., *Papers of the Michigan Academy of Science, 22,* 129–134.

Branner, W. (1902). Uber die Luftwurzetn Von *Avicennia tomentosa. Berichte der Deutschen Botanischen Gesellschaft.* **20** : 175-187.

Brass, L.J. (1938). Botanical results of the Archbold Expeditions. XI. Notes on the vegetation of the Fly and Wassi Kussa Rivers, British New Guinea. *J. Arnold. Arb.* **19** : 175-190.

Breen, C.M. & B.J. Hill (1969). A mass mortality of mangroves in Rosi Estuary, *Transactions of the Royal Society of South Africa.* **38** : 285-303.

Breen, C.M. & I.D. Jones (1969). Observations on the anatomy of foliar nodes of young *Bruguiera gymnorrhiza, Jounral South African Botany,* **35** : 211-218.

Breteler, F.J. (1969). The Atlantic Species of *Rhizophora. Acta. Bot. Neerl.* **18** : 434-444.

Breteler, F.J. (1977). America's Pacific Species of *Rhizophora. Acta. Bot. Neerl.* **26** : 225-230.

Brown, W.W. & A.F. Fischer (1918). Philippines mangrove swamps. *Bull. Bur. For. Philipp. Is.* **22** : 9-215.

Brown, W. H. & A.F. Fischer (1920). Philippines mangrove swamps. *In : Minor products of Philippines forests,* (ed.) W.H. Brown, *Vol. I. Phil. Bur. Forest. Bull* **22** : 1-125.

Brown, W. H. &E.D. Merrill (1920). Philippine palms and palm products. *In : Minor Products of Philippine Forests,* (ed.) W.H. Brown, *Vol.I. Phil. Bur. Forest. Bull.* **22** : 127-248.

Brown, J.M.A., H. Outred & F.C. Hill (1969). Respiratory metabolism in mangrove seedlings. *Plant Physiol* **44** (2) : 287-294.

Browne, F.G. (1955). *Forest Trees of Sawarak and Brunei and their Products.* Sawarak Govt. Printer, Kuching.

Buckley, T.A. (1929). Mangrove bark as a tannin material. *Malay For. Rec.* **7** : 40 pp.

Budiman, A. & D. Darnaedi (1984). Struktur Komunitas moluska di hutan mangrove Morowali, Sulawesi Tengah, p. 175-182. *In* : S. Soemedihardjo (eds.). *Prosiding Seminar II Ekosistem Mangrove.* LIPI.

Budiman, A. & K. Kartawinata (1984). Pattern of settlement and uses in mangroves with special reference to Indonesia : *Paper for Workshop on human population; mangrove resources ; human induced stresses; human health,* Oct. 2-7, 1984. Bagor, Indonesia, 1984.

Bunt, J.S. & W.T. Williams (1980). Studies in the analysis of data from Australian tidal forests (Mangrove). I. Vegetational sequences and their graphic representation, *Aust. J. Ecol.* **5** : 385-390.

Bunt, J.S. & W.T. Williams (1981). Vegetational relationships in the mangroves of tropical Australia. *Mar Ecol. Prog. Ser.* **4** : 349-359.

Bunt, J.S., W.T. Williams & N.C. Duke (1982). Mangrove distributions in north east Australia, *J. Biogeog.* **9** : 111-120.

Bunt, J.S. (1987). *Australia, Country Reports* : Mangroves *of* Asia and the Pacific : Status and Management, Technical Report of the UNDP/UNESCO Research and Training Pilot Programmes on Mangrove Ecosystems in Asia and the Pacific (RAS/79/002), pp. 263-280.

Bunt, J.S., K.G. Boto & G. Boto (1979). A survey method for estimating potential levels of mangrove forest primary production, *Marine Biology,* **52** : 123-128.

Bunt, J.S. & W.T. Williams (1981). Vegetational relationships in the mangroves of tropical Australia. *Mar. Ecol.* **4** : 349-359.

Bunt, J.S. (1984). Mangrove waters in nutrient production, cycling and transport. In : *Lectures on Second introductory training course on mangrove ecosystem,* Goa, Nov. 1-25, 1984.

Burbridge, P.R. (1984). The management and planning of mangrove resource in Asia, pp. 27-42. Soepadmo, E., A.N. Rao & D.J. Macintosh (eds.) ***1984.*** *Proceedings Asian Symp. Mangr. Environ. Res. Management,* University Malaya, Kualalumpur, Malaysia.

Burbridge, P.R. & Koesoebiano (1980). Management of Mangrove Exploitation in Indonesia. Centre for Natural Resources Management and Environmental Studies, Bogor Agricultural University. PPSPL/Research Report/007.

Burbridge, P.R. & Koesoebiano (1984). Management of Mangrove Exploitation in Indonesia, pp. 740-760. Soepadmo, E., A.N. Rao & D.J. Macintosh (eds.). *Proceedings Asian Symp. Mangr. Environ. Res. Management,* University Malaya, Kualalumpur, Malaysia.

Burchett, M.D., C.D. Field & A. Pulknowanick (1984). Salinity, growth and root respiration in the grey mangrove, *Avicennia marina. Physiol. Plant,* **60** : 113-118.

Burchett, M.D., C.D. Field & A. Pulkownik (1985). Aspects of growth and maintenance in *Avicennia marina* seedlings in a range of salinities.

Burns, B.R. & Ogdenj (1985). The demography of the temperate mangrove, *Avicennia marina* (Forsk.) Vierh. at its Southern limit in New Zealand. *Aust. J. Ecol.,* **10** (2) : 125-133.

Burkill, I.H. (1935). *A dictionary of the economic products of the Malay Peninsula.* 2. Vols. Oxford University Press.

Byrnes, N.B., S. L. Evierist, S.T. Reynolds, A. Specht & R.L. Specht (1977). The vegetation of Lizard Island, North Queensland. *Proceedings of the Royal Society of Queensland,* **88** : 1-15.

Caldwell, G.S. (1986). Predation as a selective force on foraging herons : effects of plumage colors flocking. *Auk.* **103** (3) : 494-505.

Camacho, A.S. (1984). Some aspects of the management and utilization of Philippine mangroves in relation to fisheries development. *(Paper presented during the 19th convention of the biology Teacher Association of the Philippines BIOTA),* Iloilo City, Philippines.

Camacho, A.S. (1985). Saline rice-fish culture. A prospectus, pp. 105-111. ***In*** : *Workshop on the Conservation of Mangroves areas to Paddy Cultivation, held from 1-3 April 1985,* Los Banes, Laguna Philippines. UNDP/UNESCO. Regional Mangrove Project for Asia and the Pacific.

Camacho, Arsenio S. & Teodora Bagarinao (1986). Impact of fish pond development on the mangrove ecosystem in the Philippines. *Tech. Report on Mangrove,* UNDP/UNESCO. pp. 383-405.

Camilleni, J.C. & G. Ribi (1983). Leaf thickness of mangroves growing in different salinities, *Biotropica.* **15:** 139-141.

Camilleni, J.C. & G. Ribi (1986). Leaching of dissolved organic carbon (D.O.C.) from dead leaves, formation of flakes from D.O.C. and feeding on flakes by crustaceans in mangroves. *Mar. Biot.* ***91*** (3) : 337-344.

Candolie, A.C. de (1841). *Second memoire Surla famille des Myrsinaceae Ann. Sci. nat. Series* ***2***, 16 : 1-46.

Canoy, M.J. (1975). Diversity and stability in a Puerto Rican *Rhizophora mangle* L. forest. ***In*** : G.E. Walsh, S.C. Snedaker & H.J. Teas (eds.). *Proceedings of the International Symposium on Biology and Management of Mangroves. Inst. Food & Agri. Sci., Univ. Florida*, Gainesville, Florida, ***Vol.1***. pp. 344-356.

Caratini, C., F. Blasco & G. Thanikaimoni (1973). Relations between the pollen spectra and the vegetation of a south Indian mangrove. *Pollen et spores*, ***15*** : 281-292.

Caratini, C. & F. Blasco (1980). Les mangroves De L'Snde. *Trar. Doc. Geogr. Tropicale.* ***39*** : 91-106.

Caratini, C., G. Thanikaimoni & C. Tissot (1980). Mangroves of India : Palynological study and recent history of the vegetation. *IV. Inst. Palynol. Conf.*, Lucknow (1976-77). ***3*** : 49-59.

Caratini, C., S. Chanda & G. Thanikaimoni (1974). Main Characteristics of Indian Mangroves. *Proc. Inst. Symp. Biol. Manag. Mangrove.* ***1*** : 71.

Cardale, S. & C.D. Field (1971). The Structure of the Salt Gland of *Aegiceras corniculatum. Plants*, ***99*** : 183-191.

Cardale, S. & C.D. Field (1975). Ion transport in the salty lands of *Aegiceras. Proc. Mangroves Symposia* (Eds.). G. Walsh, S. Snedakar & H. Teas, University of Florida, Gaines Ville, Florida, ***2*** : 608-614.

Carlquist, S. (1975). *Ecological Strategies of Xylem Evolution.* University of California Press, Berkeley. 259 pp.

Carreno, I., S. Pujol & E. Zoppi-de-Rea (1989). *Metamysidopsis Insularis* Brattegard (Mysidacea) Field study of the larval cycle. *Crustaceans.* ***56*** (2) : 127–130.

Carey, G. (1934). Further investigations on the Embryology of viviparous seeds. *Proc. Linn. Soc. N.S.W.* ***59*** : 392-410.

Carey, G. & L. Fraser (1932). The Embryology and seedlings development of *Aegiceras majus* Gaebert. *Proc. Linn. Soc. N.S.W.* ***57*** : 341-360.

Carter, J. (1959). Mangrove succession and coastal changes in S.W. Malaya. *Trans. Inst. Brit. Geogr.* ***26*** : 79–88.

Cawkell, E.M. (1964). The utilization of mangroves by African birds. *Ibis* ***106*** : 251-253.

Chaffey, D.R., F.R. Millar & J.H. Sandom (1985). *A forest inventory of the Sundarbans, Bangladesh.* Main Report, Overseas Development Administration, England, 196 pp.

Chai, P.K. (1980). Mangrove forests of Sarawak. *Proc. Workshop Mangr. Estuar. Veg. Malaysia Univ.*, Partanian, Malaya.

Chai, P.K (1982). *Status and strategy for the management and utilization of Mangrove Forests in Sarawak*, Sarawak For. Dept. Malaysia (Unpublished Report).

Chai, P.K. (1982). *Ecological studies of mangrove forest in Sarawak.* Ph.D. Thesis. University of Malaysia, Kualalumpur. (unpub.)

Chai, P.K. & K.K. Lai (1984). Management and utilization of mangrove forest in Sarawak, pp. 705-795. *Proceedings Asian Symp. Mangr. Environ. Res. Management*, University of Malaya, Kualalumpur, Malaysia.

Chaiglom, D. (1982). Biological and Ecological Studies of insects in Mangrove Forests. *Proc.* NRCT-JSPS Rattanakosim Bicen. *Joint Seminar Sci. Mangr. Resources, 1982* . 227–229.

Chakrabarti, K. (1978-79). A preliminary study on the fishery resources of the mangrove swamps of Sundarbans, West Bengal. *Indian Fish. Assoc. 8-9* : 44-48.

Chakraborti, K. (1979). Sundarbans–a unique habitat for tiger—a study on its diversity. *Proc. Internatn. Symp. on Tiger*, New Delhi, 22-24 Feb. 1979.

Chakraborti, K. (1984). Dynamics of flora and fauna diversity in the mangroves of Sundarbans in comparison to laterite forests of South Bengal and North Bengal forests–A bio-ecological study. *Indian J. Forestry* 7 (8) : 220-232.

Chakraborto, K. (1985). Sundarbans mangroves biomass, productivity and resource utilization–An indepth study. *Nat. Symp. on Biology, Utilization and Conservation of Mangroves. Abst.* No. 31, p. 17.

Chakraborti, K. (1985). Sundarbans mangroves–An ecostatistical appraisal on diversity. *Nat. Symp. on Biology, Utilization and Conservation of Mangroves. Abst.* No. 30.

Chakraborti, K. (1985). The Sundarban forests where life began. *Sanctuary,* **5** (4) : 324-335 and 359.

Chakraborti, K. (1986). Plant and animal dynamics in Sundarban Mangrove–An ecological study. *Mahasagar,* **19** (2) : 147-151.

Chakrabarti, K. (1988). Genetic and species diversity of animal vegetation dynamics of Sundarbans mangrove South Bengal Laterite Research of West Bengal and North Bengal forests and ecological study. *Indian For.* **112** (5) : 407-416.

Chakraborty, K. (1991). *Man, Plant and Animal Interaction.* Calcutta : Darbari Prakashan. 130 pp.

Chakraborty, K. (1992). *Man-Eating Tiger.* Calcutta : Darbari Prakashan.

Chakraborti, K. & A.B. Chaudhuri (1972). Wild life Biology of the Sundarbans forests–Honey production and behaviour pattern of the Honey bee. *Science & Culture,* **38** : 28-31.

Chakraborti, K. & A.B. Chaudhuri (1979). An ecological reconnaissance of Tiger, *Panthera tigris* Linnaeus, habit and conservation of forests and Wildlife. *Proc. Internatn. Symp. on Tiger,* New Delhi, 22-24 Feb., 1979.

Chakraborti, K. & A.B. Chaudhuri (1979). Project Tiger–A premier and intricate conservation Projects. *Proc. Internatn. Symp. on Tiger,* New Delhi, Sundarbans, S.D.B., Govt. of West Bengal.

Chakrabarti, A.K., N.C. Baral & N. C. Chattapadhyay (1988). Ecological Studies of the Mangrove Flora of Bhagbatpur (Sundarbans). *J. Indian Soc. Coastal agric. Res.* **6** : (1) : 1-4.

Chakraborti, R.K., D.D. Halder, N.K. Das, S.K. Mandal & M.L. Bhowmick (1986). Growth of *Penaeus monodon* Fabricius under different environmental conditions, *Aquaculture* **51** : 189-194.

Chakrabarti, R.K., M.L. Bhowmik & Apurba Ghosh (1987). On a New Method for Nursery Rearing of *Penaeus monodon* (Fab.) at High Stocking Density. *J. Indian Soc. Caostal agric. Res.* **5** (2) : 427-429.

Chakraborti, N.M., P.K. Chakraborti & A. Ghosh (1987). Carnivorous Fishes and their Culture Possibilities in Coastal Areas. *J. Indian Soc. Coastal agric. Res.* **5** (1) : 307-311.

Chakraborty, P.K. & K.R. Naskar (1988). Role of mangroves in estuarine fisheries development. In : *Conservation and Resource Management of Inland Capture Fisheries Resource of India.* (Eds.) A.G. Jhingran & V.V. Sugunan, I.F.S.I., India, pp. 229-233.

Champion, H.A. (1936). A preliminary survey of the forest types of India and Burma. *Indian For. Rec.* (N.S.) **1** (1) : 286.

Champion, H.G., S.K. Seth & G.M. Khattak (1965). Forest types of Pakistan. pp. 87-95. Pakistan Forest Institue, Peshawar.

Champion, H.G., S.K. Seth & G.M. Khattak (1965). *Manual of Silviculture for Pakistan,* Pakistan Forest Institute, Peshawar, 543 p.

Champion, H.G. & S.K. Seth (1968). *A revised survey of the forest types of India.* Manager of Publication, Delhi, p. 404.

Chanda, S. (1977). An–ecofloristic survey of the mangrove of Sundarbans, West Bengal, India. *Trans. Bose Res. Inst.* **40** (1) : 5-14.

Chanda, S. & B.B. Mukherjee (1969). Radio Carbon Dating of two Microfossileferous Quarternary Deposits in and around Calcutta. *Sci. & Cult.* **35** : 275–276.

Chanda, S. & S.C. Datta (1986). Prospects and Problems of a Mangrove Ecosystem in Western Sundarbans (India). *Trans. Bose Res. Inst.* **49** (3) : 47-57.

Chandlor, M.E.J. (1957). Note on the occurrence of mangroves in the London Clay. *Proc. Geol. Assoc.* **62** : 271-272.

Chang, H.T. (1984). Coastal and river bank erosion of mangroves in Peninsular Malaysia. *Paper presented at the NARU Conference and Workshop on Coasts and Tidal Wetlands of the Australian Monsoon Region,* November, 1984, Darwin, Australia.

Chang, H.J., Razani Ujang & F.E. Putz (1982). Growth and natural regeneration of a stand of *Rhizophora apiculata* tree at the Matang Mangrove following thinning. *Paper presented at the Symposium on Mangrove Forest Ecosystem Productivity*, Bogor, Indonesia.

Changprai, C. (1984). Mangrove soils and their utilization in Thailand. *Proceeding Asian Symp. Mangr. Environ. Res. Management*, University of Malaya, Kualalumpur, Malaysia.

Chansang, H. (1984). Compatibility of alternative uses of mangroves and estuarine resources. *Proceeding Asian Symp. Mangr. Environ. Res. Management*, University of Malaya, Kualalumpur, Malaysia.

Chapman, V.J. (1940). The function of the pneumatophores of *Avicennia nitida* Jacq. *Proc. Linn. Soc., Lond. 152* (3) : 228-233.

Chapman, V.J. (1947). Secondary thickening and lenticels in *Avicennia nitida* Jacq. *Proc. Linn. Soc., Lond. 158* : 2-6.

Chapman, V.J. (1964).*Coastal Vegetation*, Pergamon Press, 233.

Chapman, V.J. (1962a). Respiration studies of mangrove seedlings. *I. Bull. Mar. Sci. Gulf. Carribb.* **12** **(1):** 137-167.

Chapman, V.J. (1962b). Respiration studies of mangrove seedlings. *II. Bull. Mar. Sci. Gulf. Carribb.* **12** **(2):** 245-263.

Chapman, V.J. (1969). Lagoons and mangrove vegetation. **In** *: Lagunas Costeras, Un Simposio, Mexico.* UNAM-UNESCO, pp. 504-514.

Chapman, V.J. (1970a). Mangrove phytosociology. *Trop. Ecol.* **2** : 1-19.

Chapman, V.J. (1970b). *Sea weeds and their uses.* Camelot Press, London.

Chapman, V.J. (1972). *Mangrove of the World*, Leonard Hill, London.

Chapman, V.J. (1975a). The salinity problem in general–its importance and distribution with special reference to natural halophytes. **In** : A. Poljakoff-Mayber and I. Gale (eds.). *Plants in saline Environments*, Berlin, Springer-Verlag.

Chapman, V.J. (1975b). Mangrove biogeography. **In** : *Proceedings of the International Symposium on biology and management of mangrove* (eds.) G.E. Walsh, S.C. Snedakar & H.J. Teas. ***Vol. I*** : 3-22.

Chapman, V.J. (1975c). *Mangrove Vegetation.* Cramer, Lehre, 425 p.

Chapman, V.J. (1976a). Mangrove bio-geography. *In : Wet Coastal Ecosystem* (Ed.) V.J. Chapman, Elesevier Scientific Publishing Company, Amsterdam, pp. 1-29.

Chapman, V.J. (1976b). *Mangrove Vegetation.* Valduz : J. Cramer, 447. p.

Chapman, V.J. (1977a). *Ecosystem of the World : Wet Coastal Eco-systems.* ***Vol. I,*** New York : Elsevier.

Chapman, V.J. (1977b). Introduction, pp. 3-15. *In : Ecosystems of the World. I, Wet Coastal Ecosystems.* (Ed.) V.J. Chapman, Elsevier Scientific Publishing Company.

Chapman, V.J. (1977c). *Wet Coastal Formations of Indo-Malesia and Papua New Guinea*, pp. 1-29. *In : Ecosystems of the World–I, Wet Coastal Ecosystems.* (Ed). V.J. Chapman, Elsevier Scientific Publishing Company.

Chapman, V.J. (1978). *Coastal Vegetation.* 2nd Edition. Pergamon Press, Oxford, 292 p.

Chapman, V.J. & J.W. Ronatdson (1958). The mangrove and salt marsh flats of the Auckland Isthumus : *New Zealand Dep. Sci. Ind. Res. Bull.* **125** : 1-79.

Chapman, V.J. & D.J. Chapman (1961). *The Algae*, Macmillon, London, p. 481.

Chapman, V.J. & D.J. Chapman (1973). *The Algae*, 2nd Edition, Macmillion.

Chappell, J. & J. Grindrod (1984). Chenier plain formation in Northern Australia, pp. 197-231. *In* : B.G. Thom (ed.). *Coastal Geomorphology in Australia.* Australia : Acad. Press.

Chattaway, M.M. (1932). The wood of Sterculiaceae. I. Specialisation of the vertical wood parenchyma within the sub family Sterculiaceae, *New Phytol.* **31** : 119-132.

Chattaway, M.M. (1938). The wood anatomy of the family Sterculiaceae. *Phil. Thans. R. Soc. B.* **228** : 313-365.

Chatterjee, D. (1939). Studies on the endemic flora of India and Burma. *J. Asist. Soc. Bengal*, **5** : 19-67.

Chatterjee, D. (1958). Symposium on mangrove vegetation. *Sci & Cult.* **23** : 329-335.

Chatterjee, D. (1959). The Nature and Distribution of Mangrove plants. *Proc. Mangr. Symp.*, Calcutta. pp. 132-135.

Chatterjee, D. (1962). Floristic patterns on Indian vegetation. *Proc. Sumn. School Bot.*, Darjeeling (1960), pp. 32-42, New Delhi.

Chattopadhya, G. N. & L. N. Mandal (1980). A study on the Physico-chemical Characteristics of some Brackishwater Fish Pond Soils of West Bengal. *Proc.* Coastal Aquaculture, CMFRI, Cochin, India, pp. 50-66.

Chattopadhyay, G. N. & R. K. Chakraborty (1980). A Comparative study on the Nature and Properties of Some Brackish Water Fish Pond Soil. *Proc. Symp. Coastal Aquaculture*, CMFRI, Cochin, India, pp. 12-16.

Chattopadhyay, G. N. & L. N. Mandal (1980). Distribution of Different Inorganic Forms of Phosphorus in Brackishwater Fish Pond Soils of West Bengal, *J. Inland Fish. Soc. India*,12 (1) : 25-29.

Chattopadhyay, G. N. & L. N. Mandal (1982). Influence of Manure and Fertilizers on Some Brackishwater Fish Pond Soils of West Bengal, *J. Inland Fish Soc. India* **29** : 191-200.

Chattopadhyay, G. N., A. Ghosh and P. K. Saha (1983). A Note on the Possibilities of Salinisation of Soils under Paddy-cum-Brachishwater Aquaculture in Coastal Saline Soils. *J. Ind. Soc. Coastal agric. Res.* **3 (1)** : 63-66.

Chattopadhyay, G.N., A.K. Roy & P.R. Das (1991). Determination of Available Phosphorus in Brackish-water Fish Pond Soils, *J. Indian Soc. Coastal agric. Res.* **9** (1 & 2) : 77-79.

Chattopadhyay, N., M. Dutta Gupta & S.K. Gupta (1986). Studies on Characteristics of some mangrove soils of the Sundarbans ecosystem. *J. Indian Soc. Coastal agric. Res.*, **4** (1) : 17-23.

Chaubal, P.D. & S.Y Kotmire (1983). Pollen morphology of some mangrove and their associates. *Marine Plants*, Papers presented at the All India Symposium on Marine Plants, their biology, chemistry and utilization, Dona Paula, Goa, Oct. 30–Nov. 1, 1983. V. Krishnamurty & A.G. Untawale (eds.). ***1985***, 311-314.

Chaubal, P.D. & S.Y. Kotmire (1985). Meliosopalynological Aspects of Mangroves. II : Pollen Spectrum. *The Mangroves : Proc. Nat. Symp. Biol. Util. Cons. Mangroves*, Nov. 1985 : 401-403. (ed.) L.J. Bhosale), Shibajee University, Kollapur.

Chaudhuri, A.B. (1978). Mangrove environment of Sundarbans, West Bengal, India, *Proc. Regional Seminar on Human uses of the mangrove environment and management implication*. Paper presented in UNESCO Symp., Dacca, Bangladesh.

Chaudhuri, A.B. (1983). Endangered aquatic and mangrove environment in India (Problem of the site identification and mapping). *Proc. Intn. Seminar in Environmental maps and attases and Thematic Mapping Organization*, Calcutta.

Chaudhuri, A.B. (1988). Biotic Communities in endangered and mangrove environment *Sci. & Cult.* **54** (6) : 182-187.

Chaudhury, A.B. & K. Chakraborty, (1973b). Wildlife biology of the Sundarbans forests–A study of the birds of Sundarbans with special reference to the breeding biology of open billed stork, little cormorant and large egrets. *Sci. & Cul.*, **39** : 8-16.

Chaudhuri, A.B. & K. Chakraborty (1973a). Wild Life Biology of the Sundarbans Forests. *Sci. & Cult.*, **40** : 210-212.

Chaudhuri, A.B. & K. Chakraborty (1989). *Şundarbans Mangroves* (Ecology & Wild Life). Jugal Kishore & Co. (Pub. Div.), Dehra Dun-248001, India, pp. 1-210.

Chaudhuri, A.B. & A. Choudhury (1994). *Mangrove of the Sundarbans*, Volume One : *India* . IUCN, Bangkok.

Chaudhuri, R.A. (1984). Management of Mangrove Ecosystem. *Paper presented in UNDP/UNESCO Training Seminar on Geology, Sedimentology, Erosion and Accretion in Mangrove Areas*, Dhaka. 7 p.

Chaudhury, A.M. (1960). *Working Plan for the forests of Sundarbans Division for the period from 1960-61 to 1979-80.*

Chauhan, S.K., V.K. Tyagi & M.L. Nagar (1980). Mycoflora of soil around Pneumatophores of *Sonneratia acida* L. in Andaman Islands. *J. Indian Bot. Soc.* **59** : 281-285.

Chavan, S.A. (1985). Status of mangrove ecosystem in Gulf of Kutchh. Paper presented at the Symposium on '*Endangered* Marine Animals and Marine Parks'; held from 12-16 January, 1985 at Cochin, India.

Chen, Ronghuai Ling Peng (1989). Absorption and removal of mercury by mangrove seedlings. *ACTA-SCI-CIRCUMSTANT-HUANJING-KEXUEXUEBAD. Vol. 9*, pp. 218-224.

Chengappa, B.S. (1944). Andaman Forests and their Regeneration. *Ind. Forester, 70* : 297-304.

Chia, L.S. (1982). *Singapore : Coastal zone resource and management*, pp. 241-251. *In* : C. Soysa, L.S. Chia & W.L. Collier (eds.). *Mand. Land and Sea, Agricultural Dev. Coun.*, Bangkok, Thailand.

Chirputkar, M.B. (1969). *Physiological Studies in Marine plants in Bombay.* (Ph.D. Thesis) Univ. Bombay.

Chong, V.C. (1980). Maturation and breeding ecology of the white prawn, *Penaeus merguiensis* De Man. *Proc. 5th Intern. Symp. Prop. Ecol.*, Kualalumpur, Malaysia.

Chou, L.M., S.H. Ho, H.W. Khoo, T.J. Lam, D.H. Murphy & W.H. Tan (1980). The present state of mangrove ecosystems in S.E. Asia and the impact of pollution. *Singapore, South China Sea Fish Dev. Coor. Prog., Manila.*

Choudhury, A. (1979). Mangrove environment of Sundarbans, West Bengal, India, *Invited Seminar paper Submitted in National Mangrove Workshop of Australia.* 18-22, April, 1979.

Choudhury, A., S. Misra, A. Ghosh & J. Datta (1983). Changes of leaf surface waxes in the adaptation of flooded plants of Sundarbans Mangove forest. *All India Symposium on Marine Plants*, N-10, Goa.

Choudhury, J.K. (1984). Land formation process along the coast of Bangladesh. *Paper presented in the UNDP/UNESCO Training Seminar on Geology, Sedimentology, Erosion and Accretion in Mangrove Areas in Dhaka.* 23 p.

Choudhuri, P.K.R. (1983). An approach note on aritficial regeneration of mangroves in Sundarban. *Marine Plants, Paper presented at the All India Symposium on Marine Plants their biology, chemistry and utilization*, Donapaula, Goa, Oct. 30–Nov. 1, 1983, V. Krishnamurty & A.G. Untawale (eds.). *1985*. 315-322.

Choudhury, M.K. & P. Sanyal (1985). Vegetation dynamics of mangroves of Indian Sundarbans–economic importance. *Marine Plants : Papers presented at the All India Symposium on Marine Plants, their biology, chemistry and utilization*, Donapoula, Goa, Oct. 30–Nov. 1, 1983. V. Krishnamurty & A.G. Untawale (eds.). 325-34.

Choudhury, R.N. (1978). An appraisal of the problems and prospects of the Sundarbans, West Bengal. *Nat. Geogr. J. India. 24* (1 & 2) : 86-89.

Choudhury, S.K., N. Mallick & J. Sen (1962). The history of the vegetation of the lower Gangetic plains. *Proc. Indian Sci. Congress III*, pp. 259-260.

Christensen, B. (1978). Biomass and primary production of *Rhizophora apiculata* Bl. in a mangrove in Southern Thailand. *Aquat. Bot. 4* : 43-52.

Christensen, B. (1983). Mangroves–what are they worth ? *Unasylva, 35* : 2-15.

Christensen, B. & S. Anderson Wium (1977). Seasonal growth of mangrove trees in Southern Thailand, *I.* The Phenology of *Rhizophora apiculata* Bl. *Aquat. Bot. 3* : 281-286.

Chuang, S.H. (1961). *On Malayan Shores.* Singapore : Muwu Shosa.

Chunkan, K., N. Tangtham & W. Niyom (1982). Estaurine Hydrobiological Characteristics of Mangrove Forests in Ranong, Southern Thailand. *Proc. NRCT-JSPS. Rattanakosin bicen. Joint Seminar Sci. Mangr. Resources. 1982* : 134-143.

Cintron, G. & Y. Novelli Schaeffer (1985). Characteristics and structural development of the mangroves of North and South America. *Cienc. Interam.* 25 (1–4) : 4-15.

Clark, J. (1974). *Coastal Ecosystems–Ecological consideration for management of the Coastal Zone.* The Conservation Foundation, Washington D.C.

Clarke, C.B. (1884). In : J.D. Hooker, *Fl. Brit. India. 4* : 481.

Clarke, C.B. (1896). Presidential Address to the Linnean Society on the Sundarbans of Bengal. *Proc. Linn. Soc., London. 32* : 14-29.

Clarke, L.D. & N.J. Hamon (1967). The mangrove swamp and salt marsh, communities of the Sydney district. *I.* Vegetation, soil and climate. *J. Ecol. 55* : 753-771.

Clarke, L.D. & N.J. Hannon (1969). The mangrove swamp and salt marsh communities of the Sydney District. II. The holocoenotic complex with particular reference to physiography. *Journal of Ecology, 57* : 213-234.

Clarke, R.L. & J.C. Guppy (1988). A transition from mangrove forest to freshwater wetlands in the monsoon tropics of Australia. *J. Biogeogr.* **15** (4) : 665-684.

Clifford, H.T. & R.L. Specht (1979). *The vegetation of North Strandbroke Island, Queensland* (with notes on the fauna of mangrove and marine meadow ecosystems by M.M. Specht). University of Queensland Press, St. Lucida. 141 p.

Clough, B.F. (1982a). Growth and Salt balance of the mangroves, *Avicennia marina* (Forsk.) Vierh. and *Rhizophora stylosa* Griff. in relation to salinity. *Aust. J. Plant. Physiol.* **11** : 419-430.

Clough, B.F. (1982b). *Mangrove Ecosystems in Australia* : Structure, Function and Management. Austr. Canberra : Austr. Nat. Univ. Press.

Clough, B.F. (1985). Photosynthetic Mangroves. *The Mangroves : Proc. Nat. Symp. Biol. Util. Cons. Mangroves.* Nov. 1985 : 80-88. (ed.) L.J. Bhosale, Kolhapur, Shivajee University.

Clough, B.F. & P.M Attiwill (1975). Nutrient cycling in a community of *Avicennia marina* in a temperate region of Australia. *In* : *Proceeding of the International Symposium on Biology and Management of Mangroves.* (eds.) G.E. Walsh, S.C. Snedaker & H.J. Teas. **1** : 137-146.

Clough, B.F., T.J. Andrews & I.R. Cowan (1982). *In* : *Mangrove Ecosystems in Australia. Structure, Function and Management,* B.F. Clough (ed.), pp. 193-210, ANU Press, Canberra.

Clough, B.F. & P.M. Attiwill (1982). Primary Productivity of Mangroves. PP. 213-222. *In* : B.F. Clough (ed.). *Mangrove Ecosystem in Australia : Structure, Function and Management.* Australian Institute of Marine Science.

Clough, B.F., K.G. Boto & P.M. Attiwill (1983). *Mangrove and sewage : A re-evaluation,* pp. 151-161. *In* : H.J. Teas (ed.). *Biology and Ecology of Mangroves.* The Hague : Dr. W. Junk Publ.

Cluesener, M. & S.W. Breckle (1987). Reasons for the limitation of mangrove along the west coast of northern Peru. *Vegetatio, 68* (3) : 173-177.

Coaldrake, J.E. (1961). Ecosystem of the coastal low lands ("Wallum") of Southern Queensland CSIRO *Aust. Bull. No. 283,* 138 p.

Cohen, A.D. & T.D. Davies (1989). Petrographic/botanical composition and significance of the peat of Florida Bay. *Mar. Sci.* **44** (1) : 515-516.

Collette, B.B. (1983). Mangrove fishes of Papua New Guinea. *In* : H.J. Teas (ed.). *Biology and Ecology of Mangroves.* The Hague : Dr. W. Junk Publ.

Collins, M.S. (1921). On the mangrove and salt marsh-vegetation near Sydney with special reference to cabbage tree creek, Port Hacking, *Proc. Linn. Soc. N.S.W.* **46** : 376-397.

Cook. M.T. (1907). The embryology of *Rhizophora mangle. Bull. Torrey Bot. Club.* **34** : 271-277.

Cook. T. (1908). *The flora of the Presidency of Bombay.* (Ed.) Taylz and Francis, London.

Cooper, R.E. & S.A. Pasha (1935). The osmotic and suction pressure of some species of the mangrove vegetation. *J. Indian Bot. Soc.* **14** : 109-120.

Coorv, R. (1977). Wild nursery of the mangroves. *Nat. Geol.* **151** : 669-689.

Corlett, R.T. (1986). The mangrove under story : Some additional observations. *J. Trop. Ecol.* **2** (1) : 93-94.

Corlett, Richard (1987). Country Reports on Mangroves—Singapore. Tech. Reports on Mangroves—Singapore. (RAS/79/002). pp. 211-218

Corner, E.J.H. (1978). The freshwater Swamp-forest of South Johore and Singapore, *Gard. Bull. Singapore, Suppl. I,* 1-266.

Cottrell, D.J. (1989). Holocene evolution of the north eastern coast and islands of Florida Bay. *Mar. Sci.* **44** (1) : 516.

Cown, J.M. (1928). The flora of Chakaria Sundarbans. *Rec. Bot. Surv. India.* **11** (2) : 197-225.

Cragg, M. Simon (1986). *Country Reports : Mangroves–Papua New Guinea.* Tech. Reports on Mangroves, UNDP/UNESCO (RAS/79/002), pp. 299-309.

Cragg, S. (1983). The mangrove ecosystem of the Purari Delta. pp. 295-324. *In* : Petr. (ed.). *The Purari : Tropical Environment of a high Rainfall River Basin*, The Hague, Dr. W. Junk Publ.

Crowley, G.M., P. Anderson, A.P. Kershaw & J. Grindrod (1990). Palynology of a Holocene marine transgressive sequence, lower Malgravl River Valley, north-east Queensland. *Aust. J. Ecol.* **15** 231-240.

Crownwell, R.B. (1937). Working Plan for Godavari Lower Division. Forest Dept. Andhra Pradesh.

Cundell, A.M., M.S. Brown, R. Stanford & R. Mitchell (1979). Microbial degradation of *Rhizophora mangle* leaves immersed in the sea. *Estuarine Coastal and Shelf Science*, **9** : 281-286.

Curtis, S.J. (1933). *Working plans for the forests of the Sundarbans Divisions*, Bengal, India, for the period from 1931 to 1951. Govt. of W. Bengal, Calcutta.

Dagar, J.C. (1982). Some ecological aspects of mangrove vegetation of the Andaman and Nicobar Islands in India. *Sylvatrop Phillip. For. Res. J.* 7 : 177-216.

Dagar, J.C. & H.S. Dagar (1986). Mangroves and some coastal plants in Ethnobotany of the tribals of Andaman and Nicobar Islands. *J. Andaman Sci. Association.* **2** (2) : 33-46.

Dagar, J.C. (1987). Mangrove vegetation, its structure, ecology, management and importance with special reference to Andaman and Nicobar Islands. *In : Proc. Symp. Management of Coastal Ecosystems and Oceanic Resource of the Andamans*, 17-18 July, 1987, Andaman Sci. Association (CARI), Port Blair. pp. 8-23.

Dagar, J.C. & A.K. Sharma (1988). Some growth parameters of seedlings of *Rhizophora apiculata* Bl.–a prominent mangrove of the Andamans. *J. Andaman Sci. Assoc.* 4 : 47-50.

Dagar, J.C. & A.K. Sharma (1989). Multiple viviparity in mangroves. *J. Andaman Sci. Assoc.* **5** (1) : 72-73.

Dagar, J.C., A.D. Mongia & A.K. Bandyopadhyay (1991). *Mangrove of Andaman and Nicobar Islands* : Oxford & IBH Publishing Co. Pvt. Ltd., New Delhi. pp. 1-166.

Dahlgren, R.M.T. (1980). A revised system of classification of the Angiosperms. *Bot. J. Linn Socl.* **80** : 91-124.

Daiber, F.C. (1960). Mangroves : the tidal marshes of the tropics, *Estuar. Bull.* **5** : 10-15.

Dale, I.R. (1938). *Kenya mangroves. Z. Weltfortstw.* **5** : 413-421.

Damodaran, R. (1972). Miobenthos of the mud banks of the Kerala Coast. *Bull. Dept. Mar. Sci., Uni. Cochin,* **6** : 1-126.

Damodara, Naidu W. (1986). Tintinnid swarms of Porto Novo waters. *Mahasagar;* **19** (1) : 23-27.

Darnaedi, D. & A. Budiman (1984). Analisis vegetasi hutan mangrove Morowali, Sulawesi Tengah, pp. 162-172. *In* : Soemodihardjo (ed.). *Prosiding Seminar II Ecosistem Mangrove.* LIPI.

Darnell, R.M (1990). Mapping of the biological resources of the continental shelf. *AM. Zool. 1990.* **30** (1) : 15-21.

Darsidi, A. (1984). Pengeloloaan hutan mangrove di Indonesia. pp. 19-28. *In* : Soemodihardjo (ed.). *Prosiding Seminar II. Ecosistem Mangrove.* LIPI.

Das, B.M. (1921). Mangrove swamps of the Sundarbans forest Division, a valuable source of transtuffs. *J. Indian Ind. Labour.* **1** : 487-499.

Das, B.M. (1950). Transtuff resources of the Sundarbans forest Division. *Tanmer,* **4** (7) : 17-18.

Das, D.K. (1960). Mangrove of East Pakistan, their ecology and utilization, *Proc. 4th Congress, PIOSA Karachi,* Section D., 6, p.

Das, D.K. (1984). Properties of woods in relation to various uses. *Bulletin-7,* Wood Anatomy Serico. Forest Research Institute, Chittagong, 19 pp.

Das, G.K. & A.K. Bhattacharyya (1991). Pattern of Sedimentation with Reference to Erosional and Depositional Areas along Meander Bends in the Upper Estuarine Stretch of the Thakuran River, West Bengal. *J. Indian Soc. Coastal agric. Res.,* **9** (1 & 2) : 37-42.

Das, H.P. (1977). Food of the Grey Mullet, *Muzil cephalus* (L.) From the Goa Region, *Mahasagar Bull. NIO.* **10** (1 & 2) : 34-43.

Das, N. (1987). Biological Observation on the Burrowing Ghost Shrimp, *Thalassina anomala* (Herbst) from Sundarbans, West Bengal, India. *J.Indian Soc. Coastal agric. Res.*, *5* (1) : 333-338.

Das, P.R. (1969). A preliminary note on the toxicity of the plant *Derris trifoliata* Lour. on Fishes. *J. Bombay Pharm. Manuf.* 7 (4) : 197-200.

Das, P.R. (1972). Some observation on the macro-vegetation in and around 'Bheris' of Sundarbans, W. Bengal, *J.Inland Fish Soc. India. 4* : 210-211.

Das, S. (1971). The Vegetation and the ecology of the coastal belt of Bangladesh. *Forestdale News.* **4** (3) : 36-44 pp.

Das, S. (1980). Mangrove forest of Bangladesh (Management, exploitation and economics). *Paper presented at the combined 4th and 5th Annual Bangladesh Science Conference at Rajshahi University.* 9 p.

Das, S. & N.A. Siddiqi (1985). Effect of time and method of storage on the survivality of uprooted Keora (*Sonneratia apetala*) seedlings (in Print) Bano Biggyan Patrika.

Das, Sauren & Monoranjan Ghose (1993). Morphology of Stomata and Leaf Hairs of some Halophytes from Sundarbans, West Bengal. *Phytomorphology,* **43** (1 & 2) : 59-70.

Das, T.H., D. Sarkar, S. Das & D.N. Ghoshal (1991). Characterization of organic Matter in Some Soils of Coastal Sundarban, West Bengal. *J. Indian Soc. Coastal agric. Res.* *9* (1 & 2) : 27-31.

Datta, S. C. & S. Chanda (1978). Prospects and Problems of mangrove ecosystem in western Sundarbans (Indian). *Proc. International Congress in Ecology,* Israel.

Davey, J.E. (1975). Notes on the mechanism of pollen release in *Bruguiera gymnorrhiza. J.S. Afr. Bot.* **41** : 269-72.

David, A. (1959). Effect of Calcutta sewage upon the fisheries of the Kulti Estuary and the connected cultivable fisheries. *J. Asiat. Soc. Bengal* (N.S.), *1* (4) : 339-363.

Davidson, J. (1984). Assistance to the forestry sector of Bangladesh. Field Document No. 4. FAO/UNDP Project BGD/79/017. 182 p.

Davidson, J. (1985). Field visit to Bogi Mangrove Silviculture Station and Dhangmari Seed Centre, BFRI and the Sundarbans Littoral Forest. UNDP/FAO Project BGD/79/017.

Davies, J.F. (1972). Geographical variation in Coastal Development (ed. K.M. Clayton), Oliver & Boyd, Edinburg, 204 p.

Davis, J.H. (Jr.) (1940). The Ecology and Geologic role of mangroves in Florida. *Carnegic Inst. Wash. Publ. 517* : 303-412.

Day, J.W. (Jr.), W.H. Conner, F. Ley Lou, R.H Day & A.M. Navaro (1987). The productivity and composition of mangrove forests, Laguna de Treminos Mexico. *Aquat. bot 27* (3) : 267-284.

De, Rathindranath (1990). *The Sundarbanas,* Oxford University Press, India.

De, R.N. & P.N. Sreedharan (1991). Promising Rice Culture for Coastal Saline Area. *J. Indian Soc. Coastal agric. Res., 9* (1) : 273-277.

Dehadrai, P.V. (1970). Changes in the environmental features of the Zuari and Mandovi estuaries in relation to tides. *Proc. Indian Acad. Sci.* **72** : 68-80.

Den Hartog, C. (1970). *The sea grasses of the World.* Amsterdam, North Holland.

Desikachari, T.V. (1959). *Cyanophyta,* I.C.A.R., New Delhi.

De Silva, M.W.R.N. (1985). Report in Studies on Mangroves of Sri Lanka, University of Peradeniya, Sri Lanka.

Dev, S.C. (1956). Paleoclimatology and Geophysics of the Ganga Delta, Geographical Review of India, *28* : 11-18.

Dey, A.C. (1959). Utilization of Mangrove bark. *Proc. Mangr. Symp.* 129-133.

Dicks, B. (1986). Oil and the black mangrove, *Avicennia marina* in the northern Red Sea. *Mar. Pollut. Bull 17* (1) : 500-503.

Ding, Hou (1957). A conspectus of the genus *Bruguiera* (Rhizophoraceae). *Nova, Guinea. n.s. 8* (1) : 163-171.

Ding, Hou (1958). *Rhizophoraceae, Flora Malesiana* **1** (5) : 429-493.

Ding, Hou (1960). A review of the genus *Rhizophora* with special reference to the Pacific species. *Blumea, 10* : 625-632.

Djamali, A. (1984). Keadan Perikanan di Kabupaten Berau, Kalimantan Timur. *In* : S. Soctomo & A. Dajamali (eds.). *Penelitian Ekosixtem Mangrove Kalimantan Timur,* MABLIPI (Unpublished report).

Dixit, R.D. & J.N. Vohra (1984). *A Dictionary of Pteridophytes of India*, BSI, Howrah.

Dixon, R.G. (1959). *Working Plan for the Matang Mangrove Forest Reserve.* Perock For. Dept. Publi., Malaysia.

Doscas, A.E.C. (1927). The preparation of cigarette wrapper from nipah palm. *Malay Agr.* J. **15** : 420-432.

Downton, W.J.S. (1982). Growth and osmotic relations of the mangrove *Avicennia marina*, as influenced by salinity, *Aust. J. Plant Physiol.* **9** : 519-529.

Doyne, H.C. (1937). A note on the acidity of mangrove swamp soils. *Trop. Agric. Trin.* **14** : 236-237.

Drakenstein, H.A. Rheedle tot. (1678-1793). *Hortus Indicus Malabaricus*, 12 Vols. (*Aegiceras*, Vol. **6** Pl. 45 : *Rhizophora* and *Bruguiera*, Vol. **6**, Pls. 31, 32, 34, 35. : *Sonneratia* Vol. **3** . Pl. (40) van Someren R. van Dyk, Amsterdam.

Driessen, P.M. & M. Soepraptohardjo (1974). *Soils for agricultural Expansion in Indonesia*, Soil Research Institute, Bogor.

Duke, N.C. (1984). A mangrove hybrid, *Sonneratia* × *gulngai* (Sonneratiaceae) from north-eastern Australia. *Ausorobaileya.* **2** : 103-105.

Duke, N.C. (1990). Phenological trends with latitude in the mangrove tree, *Avicennia marina. Ecol.* **78** (1) : 113-133.

Duke, N.C. & J.S. Bunt (1979). The genus *Rhizophora* (Rhizophoraceae) in north-easter Australia, *Aust. J. Bot.* **27** : 657-678.

Duke, N.C., J.S. Bunt & W.T. Williams (1981). Mangrove litter fall in northeastern Australia. I. Annual totals by component in selected species. *Aust. J. Botany*, **29** : 547-553.

Duke, N.C., J.S. Bunt & W.T. Williams (1984). Observations on the floral and vegetative phenologies of north-eastern Australian mangroves, *Aust. J. Bot.* **32** : 87-99.

Durant, C. C.L. (1941). The growth of mangrove species in Malaya. *Malay Forester*, **10** : 3-15.

Dutrient, E., F. Martin & A. Debry (1990). Growth and mortality of *Sonneratia caseolaris* planted on an experimentally oil-polluted soil. *Mar. Pollut. Bull.* **21** (2) : 62-68.

Dutta, S.K. & A.R. Bal (1987). Mechanisms of salt tolerance in some plant species. *J. Indian Soc. Coastal agric. Res.* **5** (1) : 145-150.

Dwivedi, S.N. (1973). Some Biological Problmes of the West Coast of India. *Mahasagar–Bull.* NIO **6** (2) : 112-119.

Dwivedi, S.N., A.H. Parulekhar, S.C. Goswami & A.G. Untawale (1974). Ecology of Mangrove swamps of the Mandovi Estaury, Goa, India, *Proc. Intern. Symp. Biol. Mgt. Mangr. Hawaii*, **1** : 115-125.

Dwivedi, S.N. & D.V. Reddi (1975). Reclamation of Mangrove Swamps for Aquaculture. *Symp. Multiuse of Coastal Zone* **10** : 7.

Dwivedi, S.N. & K.G. Padmakar (1980). Ecology of Mangrove Swamp near Juhu Beach, Bombay with Reference to Sewage Pollution. *Abstract. 2nd International Symp. Biol. Mgt. of Mangrove and Tropical Shallow water Communities*, Papua New Guinea.

Dwivedi, S.N. & K.G. Padmakumar (1983). Ecology of a mangrove swamp near Juhu Beach, Bombay with special reference to sewage pollution. In : H.J. Teas (ed.) Biology and Ecology of Mangroves. The Hague : Dr. W. Junk Aubl.

Dwivedi, S.N., A.H. Parulekar & S.C. Goswami (1974). Ecology of mangrove swamps of the Mandovi Estuary, Goa, India. *Proc. Inter. Symp. Mang. Mangrove*, Hawaii, **1** : 115-194.

Egler, F.E. (1948). The dispersal and establishment of red mangrove, *Rhizophora* in Florida Caribb. For. **9** (4) 299-319.

Egler, F.E. (1952). South-East saline Evarglades vegetation Florida and its management. *Vegetatio*, **3** : 213-265.

Eldredge, L.G. (Comp.) (1987). Bibliography of marine ecosystems : Pacific Island. Rome, Italy, FAO : 72.

Ellison, J.C. & D.R. Stoddart (1991). Mangrove ecosystem collapse during predicted sea level rise. Holocene analogues and implications. *J. Coastal Research,* 7 (1) : 151-165.

El Wakeel, S.K. & J.P. Riley (1957). Determination of organic carbon in marine muds. *Jour. Cons. Perm. Int. Explor. Mr.* **22** : 180-183.

Encendencia, E.M. (1981). Mangrove species distribution. *Canopy,* 7 (8) : 10-11.

Eng, Chua-Thia, J.N. Paw & F.Y. Guarin (1989). The environmental impart of aquaculture and the effects of pollution of coastal aquaculture development in South East Asia. *Mar. Pollut. Bull.* **20** (7) : 335-343.

English, R.E., R.H Bradbury & R.E. Reichelt (1988). Management of large marine ecosystems. A multinational appraoch. Proceedings of the South International Coral Reef. Symp. Townville, Australia, 8th-12th Aug. 1988, Vol. 2, pp. 369-374.

Enos, P. (1989). Islands in the bay—a key habitat of Florida Bay. *Mar. Sci.* **44** (1) : 365-386.

Essig, F.B. (1993). Pollination in some New Guinea palms. *Princepes.* **17** : 75-83.

Eusebio, Mario A., Florentino O. Tesoro & Dexter M. Cabahug (Jr.) (1986). Environmental Impact of timber harvesting on Mangrove ecosystem in the Philippines. Tech. Reports on Mangroves UNDP/UNESCO, pp. 337-354.

Excell, A.W. (1954). Combretaceae. *In* : C.G.G.J. Van Steenis, *Flora Malesiana* 4 : 586.

Faber, F.C. Von (1923). Zur Physiologie der mangroven. Ber, d, bot. Gsesell. *41* : 227-234.

Faegri, K. & L. Van der Piji (1971). The Principles of Pollination Ecology Pergamon Press, New York. 291 p.

Fahn, A. & C. Shimony (1977). Development of glandular and non-glandular leaf hairs of *Avicennia marina* (Forsskal) Vierch. *Bot. J. Linn. Soc.* **74** : 37-46.

Fang, F.W. (1984). *Nypa* palm-A neglected mangrove resource. pp. 663-691. In : Proceeding Asian *Symp. Mangr. Environtes. Management,* University of Malaya, Kualalumpur, Malaysia.

Farrington, H.A. (1906). Third working plan for the Sundarbans Division, Bengal Secretarial Press, Calcutta.

Fatuchri, M. & A.B. Sutomo (1979). Perikanan tiramdi sekitar hutan mangrove perairan Gagra Menyan, pp. 165-175. *In* : S. Soemodihardjo *et al.* (eds.). Prosiding Seminar Ekosistem Hutan Mangrove. LIPI.

Fell, J.W. & I.M. Master (1973). Fungi associated with the degradation of mangrove (*Rhizophora mangle* L.) leaves in south Florida. *In* : Estuarine microbial ecology (eds. L.H. Stevensons & R.R. Colwell), University of South Carolina Press, pp. 455-465.

Fell, J.W., R.C. Cafalu, I.M. Master & A.S. Tallma (1975). Microbial activities in the mangrove (*Rhizophora mangale*) leaf detrital system. *In* : *Proceedings of the International Symposium on biology and management of Mangroves.* (eds. G.E. Walsh, S.C. Snedaker & H.J. Teas). Vol. 2, pp. 661-679.

Fell, J.W. & I.L. Hunter (1979). Fungi associated with the decomposition of the Black Rush, *Jumeus roemarianus* in South Florida. *Mycologia.* **71** : 322-342.

Fell, J.W. & I.M. Master (1980). The association and potential role of fungi in mangrove detrital systems. *Bot. Marina.* **23** : 257-263.

Fell, J.W. & S.Y. Newell (1980). Role of fungi in carbon flow and nitrogen immobilization in coastal marine plant litter systems. *In* : The fungal Community : Its Organization and role in the ecosystem. (eds. D.T. Wieklow & G.C. Caroll). Mareel Dekker, pp. 665-678.

Fell, J.W., I.M. Master & S.Y. Newell (1980). Laboratory model of the potential role of fungi (*Phytophthra* spp.) in the decomposition of the red mangrove (*Rhizophora mangale*) leaf litter. *In* : Marine benthic dynamics (eds. K.R. Tenore & B.C. Coull), University of South Caroline Press, pp. 359-372.

Fell, J.W., I.M. Master & R.G. Wiegert (1984). Litter decomposition and nutrient. **In** : The mangrove ecosystem : Research methods (eds. S.C. Snedaker & J.G. Snedaker), UNESCO, pp. 239-251.

Feller, C.,J. Trichet, J.C. Fontes & C. Marius (1989). Upon the role of vegetation in sulfur storage in mangrove soils (Senegal). Preliminary results. *Soil. Biol. Biochem.* **21** (7) : 947-952.

Fernanda, E.S. & J.V. Pandu (1980). Mangrove trees of the Philippines sylvatrop. *Philpp. For Res. J.* 5 (1) : 35-53.

Fernando, W.P.N. & M. Fonseka (1984). A preliminary analysis of small mesh gill net fishery in the inshore waters of Sri Lanka. *Nat. Aqua. Res. Agency*, Sri Lanka.

Field, C.G (1980). Ions in Mangroves. *Proc. Symp. Biol. Managem.* Papua New Guinea (under print by W. Junk Publisher, The Netherlands).

Field, C.D. (1984). Movements of ions in tropical mangroves, pp.514-520. In : *Proceeding Asian Symp. Mangr. Environ. Res. Mangaement*, University Malaya, Kualalumpur, Malaysia.

Field, C.D. (1984). Use of Mangroves for scientific research. *Proceedings of the Workshop on Productivity of the Mangrove Ecosystem, Management Implications Ong.*, Gong, W.K. eds., *Univrsity Sains, Malaysia, Penang, Malaysia* : 108-113.

Field, C.D. (1985). Salt Regulation in Mangroves. *In : The Mangroves : Proc. Nat. Symp. Biol. Util. Cons. Mangroves.* Nov. 1985 : 16-24.

Fiji Mangrove Management Committee (1987). Country Report : Fiji. Mangroves Asia and the Pacific : Status and Management. In : R.M. Umali *et al.* (eds.). Technical Report of the UNDP/UNESCO Research and Training Pilot Programme on Mang. Ecosystems in Asia and the Pacific (RAS/79/002).

Findlay, R. H., J. W. Fell, N.K. Coleman & J.R. Vestal Biochemical indicators of the role of fungi and *thraustochytrids* in mangrove detrital system. *The Biology* - OP. *Marine fungi, Moss, ST* ed. : 91-103.

Finlayson, M.& M. Moser (1991). Wetlands ; International Water-fowl and Wetlands Research Bureau (IWRB) : Facts on File ; Oxford,New York, pp.1-224.

Flores - Verdugo, F.G., J.W. Day (Jr.) & R. Briseno - Duenas (1987). Structure, litter fall, decomposition and detritus dynamics of mangroves in a mekilan coastal lagoon with an ephemeral inle. Mar. & Col. Prob., SER, *35* (1-2) : 83-90.

Flores - Verdugo, F., F. Gonzalez - Farias, O. Ramirez - Flores, A. Amezcua - Linares, A. Yanez - Arancibia, M. Alvarez - Rubio & J.W. Day (Jr.) (1990). Mangrove ecology, aquatic primary productivity and Fish Community dynamics in the Teacapan Agua Brava lagoon estuarine system Estuaries, 1990. *Vol.13*, No.2, pp. 219-230.

Flowers, T.J., P.F. Troke & A.R. Yeo (1977). The mechanism of salt tolerence in halophytes. *Ann. Rev. Plant Physiol.* *28* : 89-121.

Floyed, A.G. (1977) Ecology of the tidal forests in the Kikori-Romilly Sound Area. Gulf of Papua. Ecology Report No.4. Division of Botany, Office of Forests, Port Moresby, Papua New Guinea.

Flueer, T. (1993). A mangrove inventory in eastern Sri Lanka : A Case Study of man's pressure on natural resources. Paper presented at 4th Asian Conf. on Remote Sensing, November, 1983, Colombo, Sri Lanka.

Follasco, M,.P. & F.R. Uyenco (1984). Soil Fungi in relation to some ecological factors in a mangrove forest at Pagbilao, Quezon Provinee, *Natur. Appll. Sci. Bull.* University Philpp. *36* (2) : 77-102.

Foo, H.T. & J.T.S. Wong (1980). Mangrove swamp and fisheries in Sabah. *Trop. Ecol. Dev.* 2 : 1157-1161.

Food & Agriculture Organisation (1982). Management and Utilization of Mangroves in Asia and the Pacific Environmnt. Paper No.3, Rome.

Forestal (1960). Forest inventory 1958-1959. Sundarbans forest. Oregon : Forestal International Incorporated, Canada.

Forestal Forestry & Engineering International Ltd. 1(1960). Inventory of Sundarbans Forests. Vancourer, Canada.

Fortes,M. D. (1982). Productivity Studies on Mangrove sea grass and algae at Calatagan, Batangs (Philippines), pp.165-175. *In.*: Soemodihardjo *et al.* (ed.). Proc.

Fortes, M.D., R.C. Reyes, P.L. Gomez & G.H. Tiopes (1982). Litter production and decomposition nutrient transport and phenology in a mangrove forest at Pagbilao, Quezon, paper presented in the *National Mangr. Res. Coord. Symp. Workshop* held from 8-10 Dec. 1982 in the Covalatiolia Island Resort., Keuwit, Cavile, Philippines.

Fortes, M.D. (1988). Mangrove and sea grass beds of East Asia. Habitats under stress. *Ambio.* **17**(3) : 207-213.

Fosberg, F.R. (1939). Phytogeography of Micronesian mangroves. *In* : Proceedings of the International *Symp. on Biol. and Management of Mangroves,* (eds) G.E. Walsh, S.C. Snedaker & H.J. Teas, Gainesoillee : Institute of Food and Agricultural Sc., University of Florida.

Fosberg, F.R. (1949). Atoll Vegetation and Salinity. *Pacific Sci. 3* : 89-92.

Fosberg, F.R. (1971). Mangroves V. Tidal Waves, *Biol. Cons.*, **4** : 38-39.

Fosberg, F.R. (1975). Phytogeography of Micronesian Mangroves. *In : Proc. Inst. Symp. Biology and Mangement of Mangrove,* Vol.1. (eds.) G.E. Walsh, S.C. Snedaker & H.J. Teas, Univ. Florida, Gainesville. pp.23-42.

Fosberg, F.R. (1976). Coral Island Vegetation. *In* : O.A. Jones & R. Endeam (Editors). Biology and Geology of Coral reefs. Academic Press, N.Y. Vol.III. Biol. **2** : 256-279.

Fourmannoir, P. (1953). Notes sur la faune de la mangrove dans la region de Majunga. *Nat. malg.* **5** : 87-92.

Foxworthy, F.W. (1910). Distribution and Utilization of mangrove swamps in Malay. *Ann. Jard. Bot. Buitenz. Series.* 2, Suppl. 3, Part *I* : 319-344.

Foxworthy, F.W. & D.M. Mattews (1916). Mangrove and Nipa swamps of British North Borneo, *For. Bull. Br. North Borneo* I. 1-67.

Frankland, J.C., J.D. Ovington & O. Maerae (1963). Spatial and seasonal variation in soil, litter and ground vegetation in some Lake District woodlands. *Journal of Ecology,* **51**: 97-112.

Fredricks, L.J. & H.C. Lampe (1984). Socio-economic aspects of mangrove systerms in Asia. pp.16-26. ***In*** : Proceedings Asian Symp. Mangr. Environment Res. Management, University of Malaya, Kualalumpur, Malaysia.

Freyne, D.F. & G.S. Bell (1982). Forest and Land Resources of Manus and Los Negros Islands. *Dept. Lands Res. Bull.* 31 : 1-50.

Frusher, S.D. (1983). The ecology of juvenile penaeid prawns, mangrove crab *(Scylla serrata)* and the giant fresh-water prawn *(Macrobrachium rosenbergii)* in the Purari Delta. **In** : Petr. (ed.). The Purari. The Hague : Junk.

Fuller, C.C., J. A. Davis, D.J. Cain, P.J. Lamothe,T.L. Fries, G. Fernandez, J.A. Vargas & M.M. Murillo (1990). Distribution and transport of sediment-bound metal contaminants in the Riorande de Tarcoles, Costa Rica (Central America). *Water Res.* 1990. *Vol.24,* No. 7, pp.805-812.

Gabriel, B.C. (1977). Principles of coastal zone management. **In** : Proc. NEPC Identification and Planning Workshop in Environmental Management. Cavelandia Island Report, Kawit, Cavite, Philippines.

Galloway, R.W. (1982). Distribution and physiographic patterns of Australian mangroves, pp. 31-54. **In** : B.F. Clough (ed.) *Mangrove Ecosystems in Australia* : Structure, Function and Management. Canberra : *Austr. Nat. Univ. Press.*

Gamble, J.S. (1915-1936). Flora of the Presidency of Madras, Botanical Survey of India, Calcutta (Reprinted Edition : 1967) *1* : 409-492.

Gamble, J.S. (1915-1936). Flora of the Presidency of Madras, Botanical Survey of India, Calcutta (Reprinted Edition : 1967) II : 493-1389.

Gamble, J.S. (1967). Flora Presidency of Madras (2nd Repr. ed.) B.S.I., Calcutta, 3 Vols.

Ganapati, P.N. & M.V.L. Rao (1959). Incidence of marine borers in the mangroves of the Godavari estuary. *Curr. Sci. 28* : 232.

Ganapati, P.N. (1969). Prawn fishery of the Godavarine estuarine system sea Food Export J. 1. (9) : 11.

Ganguly, D.K. (1965). Sundarbans. *Indian Forester,* **91** : 661-663.

Ganguli, S.N. & S.M. Sircar (1974). Gibberellins from Mangrove Plants. Phytochemistry (Oxf.) 13 (c) : (Engl) : 1911-1914.

Gangwar, B., A.D. Mongia & A.K. Bandyopadhyay (1991). Cropping Systems and Management Practices for Humid Tropical Conditions of Bay Islands - Status and Future Challenges. *J. Indian Soc. Coastal agric. Rses.* 9(1 & 2) : 241-249.

Garg, K.L. (1989). Fungal succession on pneumatophore of mangrove plants. *Program. of The first International Marine Biotechnology Conference, IMBC* : 55.

Garlach, S.A. (1976). Mangrove Swamps. pp.323-332. **In** : B. Grzimak (ed.). Encyclopedia of Ecology. Van Nestrand/Reinhold Co. (English Edition).

Garnock - Jones, P.J. (1978). Plant communities on Lakeba and Southern Vanua Balavu. Lau Group, *Fiji Roy. Soc. N.Z.Bull.* **17** : 95-117.

Gearing, J., P. Gearing, M. Rodelli, N. Marshall & A. Sasekumar (1984). Initial findings from stable carbon isotope ratios in West Coast mangrove areas of Peninsular Malaysia. pp.488-495. **In** : *Proceedings Asian Symp. Mangr. Environ. Res. Management*, University Malaya, Kualalumpur.

Gedney, R.H., J.M. Kapetsuy & W.W. Kuhnhold (1982). Training on Assessment of Coastral Aquaculture Potential in Malaysia. South China sea fish *Dev. Coord. Prog.* s.e.s./GEN/82/35.

Gel, E.P. (1962). The management of India's wild life Sanctuaries and National Parks. Part.IV. *J. Bombay Nat. Hist. Soc.* **59** (2) : 453-485.

George, M.J. (1980). Status of coastal aquaculture in India. Present Status of Coastal Aquaculture in countries *boarding Indian Ocean*, 28-53. 1980.

Geeraldh, F.X. (1983). Biological discntrtion of Dominican coastal ecosystems, an aerial survey. *Proc. Assoc. ISI. Mar. Lab.* Caribb. (1985) : 17:15.

Geriech, S.A. (1973). Grzimek's Encyclopedia of Ecology. (Eds.) B. Grzimek, J. Illies, W. Klausewitz. Van Nostrand, Reinhold Company, New York.

Gessner, F. (1967). Unterschungen an der Mangrove in Ost-Venezuela. International Revue der gesamten. *Hydrobiologie, 52* : 769-781.

Ghate, N.S. (1939). Forests of the Godavari District. *Ind. Forester, LXV* : 760-764.

Ghosh, Apurba & K.R. Naskar (1981). Compost as a derivative of Integrated Farming. *Proc. of the Summer Institute*, Central Inland Fisheries Research Institute, Barrackpore, June : 23-29.

Ghosh, Apurba, S.K. Saha, R.K. Banerjee, A.B. Mukherjee & K.R. Naskar (1985). Rice-cum-Fish Farming System - Package of Practices For Increasing Production. Aquaculture Extension Manual (New Series) No.4. Dec. 1985, Central Inland Fisheries Research Institute (I.C.A.R.), Barrackpore - 743 101, W.B. : 1-14.

Ghosh, A., P.K. Chakraborti, K.R. Naskar, G.N. Chattopadhyay, D. Nath & M.L. Bhowmik (1987). Eecology and faunal association if intertidal mangrove habitats in the Hooghly-Matlah estuarine system. *Proc., Symposium on the Impact of Current Land Use Pattern & Water Resources Development on Riverine Fisheries* : 46.

Ghosh, Apurba, G.N. Chattopadhyay, P.K. Chakraborty & K.R. Naskar (1986). Studies on Rice-cum-Brackishwater Aquaculture in Reclaimed Mangrove Areas of the Sundarbans. *Proc. National Seminar on Biology, Utilization and Conservation on Mangroves*, Shivaji Univ., Kolhapur, pp. 439-444.

Ghosh, A., K.M. Das, G.N. Chattopadhyay, P.K. Chakraborty, Amitabh Ghosh, K.R. Naskar, R.K. Das & S.K. Mandal (1987). Best recycline through fish culture in estuarine wetlands—a productive method of land use. *Abst.* p. 78. *CICFRI Symposium* April 25-27, Barrackpore, W.B.

Ghosh, Apurba, G.N. Chattopadhyay, Amitabha Ghosh, R.K. Das, P.K. Chakraborty & K.R. Naskar (1988). Ecology of sewage fed saline wetlands in relation to fish production. *Abst.* p.9. *National Seminar on Managemnet of Fisheries in Inland Open Water System of India*, CICFRI/IFSI, Barrackpore, India.

Ghosh, A.K. (1940). Submerged Forests in Calcutta, *Sci. & Cult., 6* : 669-670.

Ghosh, A.K. (1964). Study of Calcutta Peat and Associated Sediment, *Ind. J. Power & River Valley Dev.* **13** : 14-21.

Ghosh, A.K., K.R. Naskar, T.K. Baguli & D.C. Ghosh (1995). Biodiversity of the Mangrove Habitat Fauna of the Sundarbans in West Bengal. *Abs.68. National Symposium on Perspective in Biodiversity*, The Zoological Society, Calcutta, Indian Institute of Chemical Biology.

Ghosh, D. (1967). Mangrove : A potential Tannin Material, *Tanner, Bombay, 21* : 379-383.

Ghosh, D.C., A.K. Sarkar & K.R. Naskar (1993). Physico-Chemical characteristics of the soil in the mangals of the Sundarbans, India. *Abs.142. Asia Pacific Symp. on Mangrove Ecosystem,* Research Centre and Dept. of Biology. The Hong Kong Univ. of Science and Technology, Hong Kong, 1-3 Sept. 1993.

Ghosh, Dolan Champa, A.K. Ghosh, G.N. Chattopadhyay & K.R. Naskar (1994). Nature and properties of some Mangrove soil of Sundarbans, India :*Abstract.* p.5. *National Seminar on Conservation & Sustainable Development of Coastal Resources,* Berhampur : 14-17, Dec.

Ghosh, D.C. K.R. Naskar & Debashis Saha (1994). Physico-Chemical Characteristics of Mangrove soils of the Sundarbans and their effect on the Ecology. *Abs.102. West Bengal State Science Congress.* West Bengal State Council of Science and Technology, Calcutta, 28th Feb. - 1st March, 1994.

Ghosh, Dolon Champa, R. Mandal, A.K. Ghosh & K.R. Naskar (1995). Studies on the utilities of Saline Resistant wild Rice variety, *Perteresia coasclatea* (Roxb.) Takeoka from the Intertidal Newly silted up River flats of the Indian Sundarbans. *Abs.28. National Seminar on Recent Advance in Plant Sciences,* National Bot. Society, Calcutta.

Ghosh, S., T.K. Jana, B.N. Singh & A. Choudhuri (1985). Comparative study of CO_2 system in virgin and reclaimed mangrove waters of Sundarbans during freshet. *Mahasagar.20* (3) : 155-161.

Ghosh, S.K. & K.R. Naskar (1989). Pollution and Drainage Problems in Calcutta Metropolis. *Monograph, 9* : 449-450. *Nat. Atl. & Themt. Map. Org.*, Calcutta.

Ghosh, S.K., K.R. Naskar & S.C. Santra (1990). Herbaceous flora of the tidal estuaries of the Sundarbans and their role in soil conservation. *Abstract.* pp.16-17. *National Symposium on Conservation & Management of Living Resources,* Unit of Ecology & Natural Resources, University of Agricultural Sc., Bangalore-560 065, January 10-12.

Ghosh, S.S. & B.S. Neogi (1958). Occurrence of *Carapa* sp. in the so-called Peat Bed near Calcutta, *Curr. Sci.,* 27 : 357-360.

Giao, H. V., D. Hiep & N.H. Dong (1986). A study on the use of computer processing of handsat MSS data for coastal zone mapping. *Proceedings of the Regional Seminar on the Application of Remote Sensing Techniques to Coastal Zone Management and Environmental Monitoring.* Dhaka, Bangladesh, 18-26 : 163-168.

Gidoon, P.W., P.K.S. Menon, S.R.V. Rao & K.V. Bose (1957). On the marine fauna of Gulf of Kutch. A Preliminary Survey, *J. Bombay Nat. Hist. Soc. 54* (3) : 690-706.

Giglioli, M.E.C. & I. Thornton (1965a). The mangrove swamps of Keneba, lower Gambia river basin-I. Descriptive notes on the climate, the mangrove swamps and the physical composition of their soils. *J. appl. Ecol.* 2 : 81-103.

Giglioli, M.E.C. & I. Thornton (1965b). The mangrove swamps of Kaneba, lower Gambia river basin-2, Sulphur and pH in the Profiles of swamp soils. *J. appl. Ecol.* 2 : 257-269.

Giglioli, M. E. C. & D.F. King (1966). The mangrove swamps of Keneba, lower Gambia river basins. 3. Seasonal variations in the chloride and water content of swamps soil, with observations on the water levels and chloride concentration of free soil water under a barren mud flat during the dry season. *J. appl. Ecol.*, 3 : 1-19.

Gill, A.M. & P.B. Tomlinson (1969). Studies on the growth of red mangrove (*Rhizophora mangle* L.). I., Habit and General Morphology. *Biotropica. 1* (1) : 1-9.

Gill, A.M. (1971). Endogenous control of growth ring development in *Avicennia. For. Sci. 17* : 462-465.

Gill, A.M. & P.B. Tomlinson (1971). Studies on the growth of red mangroves (*Rhizophora mangale* L.) and Phenology of the Shoot. *Biotropica.* 3 : 109-124.

Gill, A.M. & P.B. Tomlinson (1971). Studies on the growth of red mangrove (*Rhizophora mangle* L.) . 2. Growth and differentiation of aerial roots. *Biotropica.* 3 (1) : 63-77.

Gill,A.M. (1975). Australia's mangrove enclaves: A coastal resource. *Proc. Ecol. Soc. of Australia. 8* : 129-146.

Gill,A.M. & P.B. Tomlinson (1977). Studies on the growth of red mangrove (*Rhizophora mangle* L.). 4. The adult root system. *Biotropica. 9* (3) : 145-155.

Gill, K.S. & S.K. Dutt (1979). Tolerance of rice varieties at the germination stage to salt level in the Tidal waters of the Sundarban Region. Indian *J. agric Sci. 49* (5): 374-377.

Gill, K.S. & S.K. Dutt (1982). Physiological basis of salt tolerance : Effect of salinity on germination, gain yield, quality and chemical composition in barley. *Sci. & Cult.* **48** (9) : 323-325.

Gill, K.S. & S.K. Dutt (1982). Effect of saliity on stomatal number, size and opening in Barley genotypes. *Biol. Plantarum,* **24** (4) : 266-269.

Gledhill, D. (1963). The ecology of the Aberdeen creek mangrove swamp. *J. Ecol.* **51** : 639-703.

Godshalk, G.L. & R.G. Wetzel (1978). Decomposition in the littoral zones of lakes. **In** : *Freshwater wetlands : Ecological Processes and Management potential.* Academic Press, pp.131-143.

Gokhale, A.V. (1962). Control of Mangroves, Poona. *Agric. Col. Mag.* **53** : 30.

Golley, F. B., H.T. Odum & R.F. Wilson (1962). The structure and *reproduction* of a Puerto Rican red mangrove forest in May. *Ecology.* **43** : 9-19.

Gomes, H.R. & S. Mavinkurvi (1982). Studies on Mangrove Swamps of Goa. II. Microorganisms degrading phenolic compounds. *Mahasagar Bull. Nat. Inst. Oceanogr.* **15** (2) : 111-115.

Gomez - Pompa, A.&C. Vazquez - yanes (1974). Studies on the secondary succession of tropical low lands : the life cycle of secondary species, *Proc. First Int. Cong. Ecology,* pp. 336-342.

Gomez, E. D. (1980). The present state of Mangrove Ecosystems in south coast Asia and the Impact of Pollution. South China Sea Fish Div. Coord. Prog., Manila, Philippines.

Gomez, P.L. (1983). Transport of Organic Matter in Disturbed and undisturbed Mangrove Ecosystems at Paybilao, Quezon, M.S. Thesis, University of the Philippines, Quezon City.

Gong, W.K. (1984). Mangrove productivity. *Proc. Workshop Mangr. Prod.* October, 1983. University *Saim,* Malaysia.

Gong, W.K., J.E. Ong, C.H. Wong & G. Dhanarajan (1984). Productivity of mangrove trees and its significance in a managed mangrove ecosystem in Malayasia. *In* : E. Soepadmo, A.N. Rao & D.J. Macintosh (eds.). *Proc. Asian Symp. Mangr. Environ. Res. Manage.,* Univ. Malaya, Kualalumpur, Malayasia, pp. 216-225.

Gong, W.K., J.E. Ong & C.H. Ong (1985). The different uses of mangroves and their possible impacts on mangroves and adjacent coastal fisheries. pp.21-26. *In* : T. Saeki, A. Hino, T. Hirose, M. Sakomoto & K. Ruddle (eds.). *Proc. MAB/COMAR Regional Seminar Man's Impact on Coastal and Estuarine Ecosystems.* November 1984, Tokyo, MAB Coordinating Committee of Japan.

Gonzales, F.R. (1977). Mangrove and estuarine area development in the Philippines, pp. 72-82. In : *Proc. Nat. Symp. Workshop Mangro. Res. Div. Paranague, Rizal,* Philippines.

Goodall, J.A. & J.A. Stoddant (1989). Techniques for the electrophoresis of mangrove tissue. *Aqual. Bot.* **35** (2): 197-207.

Goon, W.K., I.E. Ong & Y.K. Leong (1984). The mangrove resources of the east coast of Peninsular, Malaysia; Botanical aspects. *Coastal Resources of East Coast Peninsular, Malaysia —An assessment in relation to potential oil spills Chua.* T.E. Charles J.K. (eds.) Palau Pinang Malaysia Peraerbit Univ., Sains, Malaysia : 91-109.

Gopalakrishnan, V., P. Ray & B.B. Ghosh (1970). Problem of estuarine pollution with special reference to Hooghly estuary. *Proc. Seminar on Pollution and Human Environment,* BARC, Trombay, August 26-27 (1970), pp.313-324.

Gopalakrishnan, V. (1973). The pollution problem in Indian estuaries. *Seafd. Export. J.,* **5** (1) : 83-86.

Gopalakrishnan, V., P. Ray & B.B. Ghosh (1973). Present status of pollution in the Hooghly estuary with special reference to the adverse effects observed on the fishery resources. *Proc. Symp. On Environmental Pollution,* Central Public Health Engineering Research Institute, pp.1-8.

Gopinath, C.P. & M.S. Rajagopalan (1983). Mangrove resources. *Bull. Cent. Mar. Fish. Res. Ins.* **34** : 44-46.

Gordon, D.M. (1993). A preliminary study of the mangroves of the Dampier Archipelago, Western Australia, Dept. of Conservation and Environment. *Env.Note* No. 141.

Gordon - Ish - Shalom, N.& Z. Dubins Ky. (1987). Ultra-structure and function of the salt gland in *Avicennia marina.* Isr. *Bot.* **36** (4) : 205.

Gosz, J.R., G.E. Likens & F.H. Bormann (1973). Nutrient release from decomposing leaf and branch litter in the Hubbard Brook Forest, New Hampshire, *Ecol. Monogr.* **43** : 173-191.

Goulter, P.F.E. & W.G. Allawan (1979). Litterfall and decomposition in a mangrove stand, *Avicennia marina* (Forsk.). Vierh. in middle Harbour, Sydney, *Aust. J. Mar. Freshwater Res.* **30** : 541-546.

Graham, R.M. (1931). Notes on the mangrove swamps of Kenya. Jl. E. Afr. Uganda.*Nat. Hist. Soc.* **1929** (36): 157-164.

Graham, Shirley A. (1964). The Genera of Rhizophoraceae and Combretaceae in the South Eastern United States. *Journal Arnold Arboretum* **XLV** (3) : 285-301.

Graham, M., J. Grimshaw, E. Hegerl, J. McNaity & R. Timmim (1975).Cairns - Wetlands - a preliminary report. *Operculum.* **4** : 117-148.

Greenway, H. & R. Munns (1980. Mechanism of salt tolerance in non-halophytes. *Ann. Rev. Plant Physiol.* **31** : 149-190.

Gregory, D.P. (1958). Rhizophoraceae. *Ann. Mo. Bot. Gdn.* **45** : 136-140.

Griffith, W. (1833). Notes on a Tour in South Bengal, W.B. Forest Dept., Calcutta, India.

Griffith, W. (1836). On the family Rhizophoraceae, *Trans.Linn. Soc.*, Calcutta, **8** : 15.

Griffith, W. (1851). On the Development of the Ovules in *Avicennia. Trans. Linn. Soc.* **20** : 1-6.

Griffith, W. (1883). On the family of Rhizophoraceae, *Trans. Med. Phys. Soc.*, Calcutta : 8.

Grindrod, J. (1988). The palynology of Holocene mangrove and salt marsh sediments, particularly in Northern Australia, *Quaternary-Palynology of Tropical Areas.* Wand, J.V. & Liu, Kam Biu (eds.). 1988. Vol.55, No.1-3, pp.229-245.

Guha Bakshi, D.N. & S. Sen (1969). A botanical exploration of Sagar Island in 24-Parganas. *Bull. Bot. Soc. Bengal,* **23** : 31-34.

Guha Bakshi, D.N. & K. R. Naskar (1985). Eco-floristic pattern of the Sundarbans mangrove forests in India. *Nat.Symp on Biology, Utilization and Conservation of Mangroves,* Kolhapur, 18-20 Nov., 1985. *Abst.No.4,* pp.6.

Guha Bakshi, D.N. & K.R. Naskar (1986). Vegetation Studies on the Bengal Basin and its Impact on Rural Economy. *ABS. P.6. All India Symposium on 'Role of Biology in Rural Development'*, Salem Inst. of Experimental Biology, 3rd -5th Oct., 1986.

Guha Bakshi, D.N. & K.R. Naskar (1987). Ecological balance in the Estuaries - A case study of wetlands of Sundarbans. *Proc. Fourth Regional Seminar on Water Resource Development-Environmental Issues and Solution,* W.P.C.S., New Delhi : 30 p.

Guha Bakshi, D.N. & K.R. Naskar (1987). Prospectives for conservation of the tidal mangrove forests in the Sundarbans Estuary of West Bengal. pp.1-11. Ibid.

Guha Bakshi, D.N. & K.R. Naskar (1987). Present needs to the Silvicultural practices in the tidal mangrove forests of Sundarbans in West Bengal, India. *Proc. Nat. Seminar Enviro. Prot.* Viswa Bharati, Santinekatan : 56-62.

Guha Bakshi, D.N. & K.R. Naskar (1989). Biology of the mangrove flora in the tidal environment of the Sundarbans delta in W.B. *Abst.48.* Rahara V.C. College, Rahara, 24-Parganas, W.B., Feb. 6-8th, 1989.

Guha Bakshi, D.N. & K.R. Naskar (1990). Ecosystem in Sundarbans, present Position and Future Strategies. *Abst.* p.4.*Nat. Seminar on Watershed Management & Ecological Dev.*, Vritti Vikas, 11th-12th April, 1990.

Guha Bakshi, D.N. & K.R. Naskar (1990). Needs of local awareness and legislature for the conservation of the mangrove with special reference to Sundarbans. *Abst. pp.45-46. National Seminar on Mangrove Awareness in India,* World wide fund *for Nature*, India, Bombay, Feb.21-23.

Guha Bakshi, D.N. & K.R. Naskar (1990). Estuarine wetland resource management of Sundarbans. *Abstract 3. National Seminar on Watershed Management and Ecological Development*, Vritti Vikash, 11-12th April, 1990.

Guha Bakshi, D.N. & K.R. Naskar (1991). Succession of the different species of *Sonneratia* L.f. in the Sundarban mangrove ecosystems. *J. Indian Soc. Coastal agric. Res.* **9** (1/2) : 451–454.

Guha Bakshi, D.N. & K.R. Naskar (1991). Mangrove Afforestation in the Sundarbans. *In : Plant Science and Man* : Problems and Prospects (Ed. Islam, AKMN, *et al.*). *Proceedings of the International Botanical Conference,* 10-12 January, 1991, Dhaka, pp. 75-83. Bangladesh Botanical Society.

Guha, M.K. (1991). Coastal Climate and Agriculture. *J. Indian Soc. Coastal agric. Res. 9* (1 & 2) : 317-321.

Guillamin, A. (1914). Revision des Rhizophoraceae, *Nat. Syst.* 3 : 55.

Gulati, K.L., K.K. Nagpaul & S.S. Bukhari (1979). Uranuan boon, nitrogen, phosphorous and potassium in leaves of mangrove. *Mahasagar Bull.* Natn. Inst. Oceanography, *12* (3) : 183-192.

Guppy, H.B. (1906). *Observation of a Naturalist in the Pacific between 1896 and 1899.* Vol. 2. Plant dispersal. Macmillan, London and New York, 627 p.

Gupta, A.C. (1957). The Sundarbans, its problems, its possibilities. *Indian Forester. 83* : 481-487.

Gupta, M.C. & S. Naik (1987). Wetland classification of the part of the Gulf of Kuchchh using Landsat M.S.S. digital data. *In* : Remote Sensing Application in Coastal Studies. IR.S-UP/SAC/ M.C.E./SN/03/87. pp. 13-25.

Gwyther, D. (1983). *The importance of the Purari River Delta to the prawn trawl fishery of the Gulf of Papua. In* : T. Petr (ed.) The Purari : The Hague : Junk.

Haberlandt, G. (1928). *Physiological plant anatomy.* London : Macmillan.

Hackett, H.E. (1977). Marine algae known from Maldive Island. *Atoll. Res. Bull.* No. 210.

Hadi, S.,A.J. Hanson, Purba M. Koesoebiono & S. Rahardjo (1977). Tidal patterns and resources use in Musi Bany Ussian Coastal Zone of Sundarban, *Mav. Res.*, Indonesia.

Hager, L.C. (1959). Anatomical study of *Aviçennia marina* (Forsk.) Vierh. (formerly called *A. officinalis* Linn.) MS. Botany Department, University of the Witwatersrand, Johannesburg.

Haines, A. K. (1983). Fish fauna and Ecology. *In* : T. Petr (ed.). The Hague : Dr. W. Junk Publ. As cited in Papua New Guinea : Mangrove of Asia and the Pacific Status and Usage : Papura New Guinea. Country Report.

Haines, H. H. (1921). The Botany of Bihar and Orissa. BSI, Vol.I–III.

Haines, H. H (1961). Botany of Bihar and Orissa (Rep. ed.) 3 Vols., BSI, Calcutta.

Halbrook. N.M. & F.E. Putz (1982). Vegetative sea ward expansion of *Sonneratia alba* trees in a Malaysian Mangrove forest. *Malay For. 45* : 278-281.

Halder, Sangeet & K.R. Naskar (1989). The Metropolis city of Calcutta in the Ganga Basin, Its past, present and future. *Monograph. 9* , pp. 454-456. *Nat. Atl. & Themt. Map. Org.*, Calcutta.

Halder, Sangeet & K.R. Naskar (1990). Pros and cons of water resource management in the lower Ganga delta with special reference to integrated farming. *Abstract.* p. 2. *National Seminar on Watershed Management and Ecological Development*, Vritti Vikash, 11-12th April. Institute of Chemical Engineers, Jadavpore University.

Halder, Sangeet & K.R. Naskar (1990). Problems and prospects of the mangrove reclaimed coastal lands in the Sundarbans of West Bengal. *Abstract.* p. 15. *National Seminar on Mangrove Awareness in India–World Wide Fund for Nature*–India, Maharastra and Goa States Committee, Bombay, Feb. 21-23.

Halder, S. & K. Naskar (1991). Mangroves–The Unique But Endangered Ecosystems in the Tropical Regions. *J. Indian Soc. Coastal agric. Res. 9* (1 & 2) : 447–450.

Hales, S.C. (1981). Nipa for alcohol production. *Canopy.* 7 (3) : 15.

Hamilton, L.S. & S.C. Snedaker (eds.) (1984). Handbook for Mangrove Area Management. IUCN-UNESCO-UNEP, EWC, Hawaii, Paris, France.

Hansa, C. (1984). Mangrove detritus. *Proceeding of the Workshop on Productivity of the Mangrove Ecosystem Management Implication.* J.E. Ong & W.E. Gong (eds.).University Sains, Malaysia, Penang, Malaysia, 48-59.

Hanson, A.J. & Koesoebiono (1977). Settling Coastal Swamp lands in Sumatra : A case study for Integrated Resource Management, PPSPL/Research No. 004. Bogor Agri. Univ.

Hanson, A.J. (1981). Transmigration and marginal land development. pp. 2219-2356. *In* : Gory E. Hansen (eds.), Agricultural and Rural Development in Indonesia, Colorador : Westview Press, Bonlder.

Hardjosuwamo, S., T. Notohadiprawio & Soerjowinolo (1982). The Cilacap Mangrove Ecosystem. pp. 57-65. *In* : Soe Giarto & K. Soegiarto (eds.). *Proc. Workshop on Coastal Resources Mangrove,* Cilacap Region, LIPI-UNU.

Harger, J.R.E. (1982). Major Problems in the functional analysis of mangrove in South East Asia. *In* : A.Y Kootirman & Sostrontom. Ecosystem Prod. S.E. Asia BIOTROP Spec. Publ. No. 17.

Haron, Hj, Abu Hassan (1981). A working plan for the Matang Mangroves, Perak, 1980-1989. *State For. Dept. Publ.*

Harrer, F. (1939). Die Mangroven des Rugiji deltas in Deutsh Ostafrika. Kolonialforstl. Mitt 2 : 160.

Harrigan, P., P. Zieman & S.A. Macko (1989). Petrographic/Botanical composition and significance of peat deposits of Florida Bay. *Mar. Sci.* ***44*** (1) : 515-516.

Hart, M.G.R. (1959). Sulphur oxidation in tidal mangrove soils of Sierre Leone. Pl. Soil II. 215-236.

Hart, M.G.R. (1962). Observations of the source of acid in empoldered mangrove soils. I. Formation of elemental sulphur. *Pl. Soil,* ***17*** : 107-135.

Hartnol, R.G. (1983). Substratum. *In* : R. Earll & D.G. Erwin (eds.) Subtidal Ecology : The Ecology of the Shallow Sub-tidal Benthos. Oxford : Clarendon Press.

Harun, Alrasyid (1984). Penelitian permvdoan alam hutan mangrove antara sungai Gany dar Sengai Mandoh Raiu, pp. 314-320. *In* : Prosiding Seminar Ecosystem Hutton Mangrove, LIPI.

Hassal, D.C. (1984). Variation in Fiji mangrove vegetation and its response to development, pp. 142-152. *In : Proceeding Asian Symp. Manger. Environ. Res. Management,* University of Malaya, Kualalumpur, Malaysia.

Heald, E. J. (1971). The production of organic detritus in South Florida Estuary. *Sea Grant Technical Bull.* Miami University, ***6*** : 1-10.

Harvey, H. W. (1960). The Chemistry and Fertility of Seawater, Cambridge University Press, London.

Heard, Richard W. (1982). Guide to common Tidal Marsh Invertebrates of the North Eastern Gulf of Mexico. Illustrations by–Linda B. Lutz.

Hegerl, E.J. (1984). Developing a conservation strategy for the mangrove ecosystems of Asia and Oceania pp. 43-56. *In : Proceeding Asian Symp. Mangr. Environment Res. Management,* University of Malaya, Kualalumpur, Malaysia.

Heining, R.L. (1883). Blind root suckers of the Sundarbans. *J. Asiat. Soc. Bengal.* ***62*** (2) : 158-161.

Henkel, P.A. (1963). On the ecology of mangrove vegetation, Mittul. Forestsozia. Arbutsgem. N.F. ***10*** : 201-206.

Hesse, P.R. (1961a), Decomposition of organic matter in a mangrove swamp soil. Pl. Soil., ***14*** : 249-263.

Hesse, P.R. (1961b). Some differences between the soil of *Rhizophora* and *Avicennia* mangrove swamps in Sierra Leone. Pl. Soil, ***14*** : 335-346.

Hesse, P.R. (1962). Phosphorous fixation in mangrove swamp muds, Nature. Lond. ***193*** : 295-296.

Hill, B.J. (1966). A contribution to the ecology of the Umlalazi estuary. Zool. Afr. 2 : 1-24.

Ho, S.Y. (1980). A note on Chinese mangrove, pp. 1103-1106. *In* : J.I. Furtado (ed.). Tropical Ecology and Development, Univ. Malaya, Kualalumpur.

Hodkiss, I.J. & H.C. Leung (1986). Cellulose associated with mangrove leaf decomposition. *Bot. Mar.* ***29*** (6) : 467-469.

Hoffman, J.S., D. Keya & J.G. Titus (1983). *Projecting Future Sea Level Rise*–Methodology, estimates to the year 2100 and research needs, U.S. Environmental Protection Agency, Washington, USA.

Hoffman, J.S., D. Keys & J.G. Titus (1983). Projecting future sea level rise. Methodology, estimates to the year 2100 and research needs. FPA 230-09-007. Environmental Protection Agency, Washington DC, USA.

Hong, W.C. & J.E. Ong (1984). Electron microscopy studies on the salt Glands of *Acanthus ilificolius* . pp. 172-182. *In* : Proceeding *Asian Symp.* Mangr. Environ. Res. Mangement, University of Malaya, Kualalumpur, Malaysia.

Hooker, J.D. (1872-1897). Flora British India, Vols. I-VII. BSI, India.

Hora, S. L. & K. K. Nair (1944). Suggestion for the Development of Saltwater Bheries or Bhasabada Fisheries in the Sundarbans, Fish. Dev. Pamphlet, Govt. of West Bengal, Calcutta.

Hora, S. L. & T. V. R. Pillay (1962). Handbook of Fish Culture in Indo-Pacific Fisheries Region, Fish. Biol. Tech. Paper No. 14, FAO.

Hosokawa, T. (1957). Outline of the mangrove and strand forests of the Micronesian Islands, *Mem. Fac. Sci.* Kyushu Univ. Ser. E.2. : 101-118.

Hosokawa, T., H. Tagawa & V.J. Chapman (1977). Mangals of Micronesia, Taiwan, Japan, the Philippines and Oceania. *In* : Chapman (ed.). Ecosystem of the World : Wet Coastal Ecosystems. Vol. II, New York, Elsevier.

Howe, M.A. (1911). A little known mangrove of Panama. J.N.Y. *Bot. Gard.* **12** : 61-72.

Hunter, W.W. (1893). Statistical Account of Bengal. Vol. I, Calcutta.

Hutchings, P.A. & P. Saenger (1987). The ecology of mangroves. *Brisbane, Ld.*, Australia Univesity of Queensland Press.

Hyde, K.D. (1988). Intertidal fungi from the mangrove fern, *Acrostichum speciosum*, including *Massarina acrostichi* Sp. nov. *Mycol. Res.* **93** (4) : 435-438.

Hyde, K.D. (1989). Intertidal mangrove fungi from North Sumatra. *Can. Bot.* **67** (10) : 3078-3082.

Hyde, K.D. (1990). A study of the vertical zonation of intertidal fungi on *Rhizophora apiculata* at Kampong Kapok Mangrove, Brunei. *Aquat. Bot.* **36** (3) : 255-262.

Hyde, K.D. & E.B.G. Jones (1988). Marine mangove fungi. P.S.Z.N.I. : *Mar. Ecol.* **9** (1) : 15-33.

Ho, S.Y. (1980). A note on Chinese mangrove. pp. 1103-1106. *In* : J.I. Furtado (ed.) *Tropical Ecology and Development*, Univ. Malaya, Kualalumpur.

Imam, S.A. (1982a). The Sundarbans and its future, *Proceeding Bangladesh Second National Conference on Forestry.* 1924 p.

Imam, S.A. (1982b). Mangrove afforestation project in the coastal belt and offshore islands. *Proc. Bangladesh Second National Forestry Conference*, pp. 99-102.

Imam, S.A. (1984). Sedimentation, soil conductivity, plant life and their interaction. Paper presented at the UNDP/UNESCO Training Seminar on Geology, Sedimentology, Erosion and Accretion in Mangrove Areas. Dhaka. 20 p.

Isarankura, K. (1976). Status report on faunistic aspects of mangrove forest in Thailand. *Paper presented in the First Seminar on Mangrove Ecosystems.* Held from 10-15 January, 1976 at Phuket Marine Biological Centre, Thailand.

Islam, A.K. M. Nurul (1973). The Algal Flora of Sundarbans Mangrove Forest. *Bangladesh J. Botany.* **2** (2) : 11-36.

Islam, A.K.M. Nurul (1976). Contribution to the study of the marine algae of Bangladesh, *Bibliotheca Psycologica,* **19** : 1-253.

Islam, A.K.M.N. (1982). Physicochemical properties of soils of Sundarbans mangrove forests. Proceeding of the Second National Forestry Conference. Dhaka, Bangladesh. pp. 50-52.

Islam, M.A. (1992). Some Relevant Information about Sundarbans. Government of the People's Republic of Bangladesh. Sundarbans Forest Division. 19 p.

Jadhav, S.P. (1957). A working scheme for Cher (Mangrove) forests of the Haldar Distrinct (Sourashtra).

Jagtap, T.G. (1985a). Studies on littoral flora of Andaman islands. *Marine Plants.* Papers presented at the *All India Symposium on Marine Plants ; their biology, chemistry and utilization,* Donapaula, Goa. Oct. 30-Nov 1, V. Krishnamurthy & A.G. Untawale (eds.). 1985, pp. 73-78.

Jagtap, T.G. (1985b). Studies on Littoral Flora of Andaman Islands. *In* : V. Krishnamurthy & A.G. Untawale (eds.), *Marine Plants*, pp. 43-50.

Jagtap, T.G. (1985c). Studies on the associate flora in the mangrove environment of Goa, India. The Mangroves : *Proc. Nat. Symp. Biol. Util. Cons. Mangroves.* Nov. 1985 : 180-187. **In** : L. J. Bhosale, Shivajee University, Kolhapur.

Jagtap. T.G. (1985d). Structure and composition of the Mangrove forest along the Goa Coast. The Mangroves : *Proc. Nat. Symp. Biol. Util. Cons. Mangroves.* Nov. 1985 : 193-195. **In** : L. J. Bhosale, Shivajee University, Kolhapur.

Jagtap, T.G. (1985d). Ecological Studies in relation to the mangrove environment along the Goa Coast, India. Ph.D. Thesis. Shivaji University, Kolhapur, pp. 1-212.

Jagtap, T.G. (1992). Marine flora of Nicobar Group of Islands in Andaman Sea. *Ind. J. Mar. Sci.* **21** : 56-58.

Jagtap, T.G. & A.G. Untawale (1980). Effect of Petroleum products on mangrove seedlings. Mahasagar Bull. of NIO. *13* (2) : 165-173.

Jagtap, G., Vishwas S. Chavan & Arvind G. Untawale (1993). Mangrove Ecosystems of India : A Need For Protection (Synopsis). AMBIO (A Journal of the Human Environment). Vol. XXII, Number 4, June 1993. pp. 252-254.

Jagtap, T.G. & S.N. Inamdar (1991). Mapping of Seagrass Meadows from the Lakshadweep Islands (India) Using Aerial Photographs. *J. Ind. Soc. Remote Sensing*, **19** (2) : 77-81.

Jain, S.K. (1960). The vegetation and succession of plant communities in Kutch, ***Gujarat Bull.*** *Bot. Surv. Ind.* **2** : 187-292.

Jana, M.M. & D. Chatterjee (1974). Some aspects of fishing in Sundarbans, its problem and future Development. *Geo. Rev. India.* **36** : 76-82.

Jansonnius, H.H. (1950). The vessles in the wood of Javan mangrove trees. *Blumea.* **6** : 465-469.

Jara, R.S. (1984a). Aquaculture and mangroves in the Philippines *Proceeding of the Workshop*, On productivity of the mangrove ecosystem : management implications. J.E. Ong & W.E. Gong (eds.). University Sain, Malaysia, Penang–Malaysia, pp. 97-107.

Jara, R.S. (1984b). Socio-economic aspects, management and conservation of Mangroves, pp. 193-220. *In* : Second Introductory Training Course on Mangrove Ecosystem, held from 1-25 November 1984 at Goa, India.

Jara, R.S. (1987). Utilization and management of mangrove resources. IPFC. Papers contributed to the Workshop on Strategies For the Mangement of Fisheries and Aquaculture on Mangrove Ecosystems. Bangkok, Thailand, 23-25 June 1986. And Country Status Reports on Inland Fisheries, Presented Bangkok, June 1986. (370) : 105-124.

Jara, R.S. & P.G. Tan (1986). Mangrove informaion management. Mangroves of Asia and the Pacific : Status and management. Technical Report of the UNDP-UNESCO Research and Training Pilot Programme Management Cent., Quezon City, Philippines : pp. 513-529.

Jayabalan, N., G.S. Thangaraj & K. Ramamoorthi (1985). Fin fish seed resources of Vellar Estuary. *Marine Biological Assoc. India*, Cochin, India, **6** : 742-749.

Jayewardene, R.P. (1984). Mangrove of Asia and the Pacific :Present Status and Multiple Practices : Sri Lanka Country Report. Paper presented for Second Regional Task Force Meeting. Bogor, Indonesia.

Jayewardene, R.P. (1986). Country Reports on Mangroves–Sri Lanka. Tech. Reports on Mangroves UNDP/UNESCO (RAS/79/002), pp. 219-230.

Jayasekara, A.M. (1987). Management strategies for fisheries and aquaculture in mangrove ecosystem in Sri Lanka. IPFC. (370) : 1-9.

Jennings, D.H. (1968). Halophytes, succulence and sodium—a unified theory. *New Physiologist*, **67** : 899-911.

Jimenez, JA. (1984). A hypothesis to explain the reduced range of the mangrove *Pelliciera rhizophorae* Tr. & Pl. *Biotropica.* **16** : 304-308.

Jimenez, J.A (1987). A clarification on the existence of *Rhizophora* sp. along the Pacific Coast of Central America. *Brenesia.* **28** : 25-32.

Jimenez, J.A. (1990). The structure and function of dry weather mangroves on the Pacific Coast of Central America, with emphasis on *Avicennia bicolor* forests. *Estuaries.* 1990. Vol. 13, No. 2, pp. 182-192.

Jimenez, J.A. & Soto-S.R. (1985). Regional patterns in the structure and floristic composition of pacific coast mangroves from Costa Rica. *Rev. Biol. Trop.* **33** (1) : 25-37.

Jhingran, A.G. (1989). Impact of environmental perturbations on the fisheries ecology of river Ganga, CICFRI, Barrackpore, India, pp. 1-26.

Jhingran, V.G. (1982). Fish & Fisheries of India (2nd edn.), Hindusthan Publishing Company, New Delhi, India.

Jhingran, A.G. & Apurba Ghosh (1987). Aquafarming in Coastal India. *J. Indian Soc. Coastal agric. Res.* **5** (1) : 195-203.

Jhingran, A.G. & P.K. Chakraborty (1990). An Approach to Coastal Zone Management, I.F.S.I., Barrackpore, pp. 1-54.

Johnstone, I.M. (1981). Consumption of leaves by herbivores in mixed mangrove stands. *Biotropica.* **13** : 252-259.

Johnstone, I.M. (1983). Mangrove succession and climax. Chap. 12 in Biology and Ecology of Mangroves, (ed.) H.J. Teas, Tasks for Vegetation Science 8. The Hague : Junk.

Johnstone, I.M. & D.G. Frodin (1982). Mangroves of the Papua sub region, pp. 513-528. In : J.L. Gressit (ed.). *Monographiae Biologicae.* Vol. 42 ed. The Hague : Junk.

Jones, W.T. (1971). The field identification and distribution of mangroves in eastern Australia. *Queensh. Nat.* **20** : 35-51.

Jonker, F.P (1959). The genus *Rhizophora* in Suriname. *Acta. Bot. Neerl.* **8** : 58-60.

Joshi, A.J. (1982). Ecophysiological aspects of some tropical salt marsh halophytes *In* : Contributions to the ecology of halophytes. (Eds.) David, N. Sen & Kishan S. Rajpurohit, Dr. W. Junk Publishers. The Hague, pp. 189-195.

Joshi, A.C. (1933). A suggested explanation of the prevalence of vivipary on the seashore. *Journal of Ecology,* **21** : 209-212.

Joshi, A.J., E.R. Iyengar & D.C. Bhattacharya (1978). Stomatal structure of helophytes in relation to salinity. *J. Indian Soc. Coastal agric. Res.* **1** (2) : 90.

Joshi, A.J. & A. Kumar Sagar (1985). Chemicai Characteristics of Mangrove Soils on Gujarat Coast. The Mangroves. *Proc. Nat. Symp. Biol. Util. Cons. Mangroves,* Nov. 1985 : 351-355. **In** : L. J. Bhosale (ed.).

Joshi, G.V. (1973). Soil plant Relationship in the plants of the saline soils of the Deccan. In : Symposium on Deccan Trap. Country. *Proc. Ind. Acad. Sci.,* **40** : 35-38.

Jothy, A.A. (1984). Capture fisheries and the mangrove ecosystem. *Proc. Workshop Mangr. Prod.* October, 1983, Univ. Sains, Malaysia.

Joshi, G.V. (1968). Soil plant Relationship in the Plants of the Saline Soils of the Deccan, Symposium on *'Deccan Trop. Country'.* (*National Acad. Sci.,* New Delhi, India. pp. 33-38.

Joshi, G.V., M. Pimplaskar & L.J. Bhosale (1972). Physiological studies in germination of mangroves. *Bot. Mar.* **15** : 91-105.

Joshi, G.V., L. Bhosale, B.B. Jamale & B.A. Karadge (1974). Photosynthetic carbon metabolism in Mangroves. *Intern. Symp. Biol. & Mgt. Mangrove,* Honolulu, Hawaii. **2** : 579-594.

Joshi, G.V., B.B. Jamale & L. Bhosale (1975). Ion regulations in mangroves. *In* : G.E. Walsh, S.C. Snedakar & N.J. Teas (eds.). *Proc. Intnl. Symp. Biol. & Mangmt.* of *Mangroves,* Inst. Food and Agril. Sci. Univ., Florida, Vol. 2 : pp. 595–607.

Joshi, G.V. & L.J. Bhosale (1982). Estuarine ecosystem of India. In contributions to the ecology of halophytes. (Eds.) David N. Sen & Kishan, S. Rajpurahit, Dr. W. Junk Publishers, The Hague, pp. 21-23.

Joshi, H.V. & V. Krishnamurthy (1972). The species of Enteromorpha from India. *Bot. J. Linn. Soc.,* **65** : 119-128.

Jr. Cabahug, Dexter M., Ambi M. Fidel, Nisperos O. Snsan & Truzan C. Nemesio Jr. (1986). Impact of community based mangrove forestation to mangrove dependent families and to nearby coastal areas in Central Visayas : A case example. Tech. Reports on Mangroves, UNDP/ UNESCO, pp. 441-466.

Juncosa, A.M. (1982a). Embryo and seedling development in the Rhizophoraceae. Ph.D. Thesis. Duke University, Durham, North Carolina.

Juncosa, A.M. (1982b). Developmental morphology of the embryo and seedlings of *Rhizophora mangle* L. (Rhizophoraceae) *Am. J. Bot.* **71** : 170-179.

Juncosa, A.M. (1984). Embryogenesis and seedling developmental morphology of the seedling in *Bruguiera exaristata* Ding Hou (Rhizophoraceae). *Am. J. Bot.* **71** : 180–191.

Kanakaratne, M.D., W.R.T. Perera & B.U.S. Fernando (1983). An attempt at determining the mangrove coverage around Puttalam lagoon, Dutch Bay and Portugal Bay in Sri Lanka through Remote Sensing. November, 1983. Colombo, Sri Lanka.

Kanapathy, K. (1971). Studies on the reclamation of acid swamp soil. *Malay Agric. J. 48* : 33-36.

Kannan, L. & K. Murugesan (1985). Studies of the style of *Lumintzera racemosa* Willd from the Pitchavaram Mangroves, South India. The Mangroves; *Proc. Nat. Symp. Biol. Util. Cons. Mangroves.* Nov. 1965 : 196-203. **In** : L. J. Bhosale (ed.), Shivajee University, Kolhapur.

Kapauan, A. (1977). Effect of pollution of mangrove ecosystem. pp. 27-32. *In : Proc. Nat. Symp. Workshop Mangr. Res. Dev. Paranaque. Rizal,* Philippines.

Kapetsky, J.M. (1986). Conversion of mangrove for pond aquacultures : Some short term and long term remedies. IPFC. Papers contributed to the Management of Fisheries and Aquaculture on Mangrove Ecosystem, Bangkok, Thailand, 23-25 June 1986. R.H. Mepham & T. Peto (eds.). FAO, Rome, Italy, 1987. (370), Sup. : 129–141.

Karim, A., Z. Hossain & K.J. White (1984). Study of the growth of mangrove plants in relation to edaphic factors in coastal afforestation plantation in Chittagong. pp. 195-199. *In* : Proceeding *Asian Symp Mangr. Environ. Rcs. Management,* University of Malay, Kualalumpur, Malaysia.

Karim, A. (1988). Environmental factors and the distribution of mangroves in Sundarbans with special reference to *Heritiera fomes* Buch.-Ham. Ph.D. Thesis (unpub.), University of Calcutta. 222 p.

Karkar, M.B. & L.J. Bhosale (1985). Studies on inorganic constituents of leaf, propagule, soil and flooding water of a mangrove. *K. candel.* The Mangroves : *Proc. Nat. Symp. Biol. Util. Cons. Mangroves.* Nov. 1985 : 147-150. (L. J. Bhosale (ed.).

Karmakar, S.M. (1985). Mangroves–A review. The Mangroves : *Proc. Nat. Symp Biol. Cons. Mangroves.* Nov. 1985 : 60-79. (**In** : L. J. Bhosale, (ed), Shivajee University, Kolhapur).

Karstedt, P. & N. Parameswaran (1976). Anatomy and systematics of *Rhizophora. Botanische Jahrbucher fur Systematic, Pflanzenges Chiohte und Pflanzengeographic.* **97** : 317-338.

Kartawináta, K. & E. Walugo (1977). A preliminary study of the mangrove forest and Palan Rambut, Jakarta Bay. *Mar. Res. Indonesia. 18* : 119-129.

Kartawinata, K., S. Adisoemarto, S. Soemodihardjo & I.G.M. Tantra (1979). Status Pengetahuan Hutan Bakau di Indonesia, pp. 21-39. **In** : S. Soemodihardjo, A. Nontji & A. Djamali (eds.). Prosiding Seminar Ekosistem Hutan Mangrove. LIPI.

Kathiresan, K. & L. Kannan (1985). Photosynthetic Productivity in three species of *Rhizophora.* The Mangroves : *Proc. Nat. Symp. Biol. Util. Cons. Mangroves.* 262-265.

Kathiresan, K. & T. Thangam Subramonia (1987). Biotoxicity of *Exoecaria agallocha* L. lates on marine organisms. *Curr. Sci. 56* (7) : 314-315.

Kaushik, N.K. & H.B.N. Hagnes (1971). The fate of dead leaves that fall into steams. *Arch. Hydrobiol 68* : 465-515.

Keay, R.W.J. (1953). Rhizophora in West Africa. *Kew. Bull.* 1953 : 121-127.

Kenneally, K.F., P.G. Wilson & J. Semeniuk (1978). A character for distinguishing vegetative material of the mangrove genera *Bruguiera* and *Rhizophora* (Rhizophoraceae) *Nuytsia,* **4** : 178-180.

Kerrest, R. (1978). Floristic Composition, Structure, Ecology and Dynamics of Mangrove Communities in Pichavaram (South India). International Congress in Ecology, Israel.

Kerrest, R. (1980). A simple instrument for recording tidal and water level fluctuations in Mangrove Swamps. *Mahasagar Bull.* NIO, 13 (1) : 53-60.

Khan, I.H. (1987). Conservation of endangered marine species in Andamans *In* : *Proc. Symp. Mangt.* of *Coastal Ecosystems* and *Oceanic Resources* of *Andamans.* Andaman Sci. Assoc., Port Blair, pp. 66-70.

Khan, M.A.R. (1986). Wildlife in Bangladesh Mangrove Ecosystem. *J. Bombay Nat. Hist. Soc.* **83** : 31-48.

Khan, M.S. (1945). On the Concept of vivipary in plants. *Proc. 32nd Ind. Sci. Congr.* **3** : 85.

Khan, M.S. (1978). Flora of Bangladesh. No. 7. Family–Rhizophoraceae. Bangladesh National Herbarium.

Khan, S.A. (1961). Regeneration of *Avicennia officinalis* in the coastal forests of West Pakistan. Pakistan, *J. For.* 2 (1) : 43-45.

Khan, S.A. (1964). Mangrove forests, their past and present management in Hyderabad region. *Pakistan J. Sci.* **16** : 329-332.

Khan, S.A. (1965). Mangrove forests, their past and present management in Hyderabad region. *Sci. Indus.* 3 (1) : 12-16.

Khan, S.A. (1966). Working Plan of the coastal zone afforestation Division from 1963-64 to 1983-84. Government of West Pakistan, Agriculture Dept., Lahore.

Khoo, H.W. (1980). Fertility of an estuary with particular attention to detritus production, utilization and decompositions of bottom sediments. *J. Sing Nat. Acad. Sci.* **9** : 4–10.

Kiener, A. (1973). Les mangroves du globe : aspects ecologiques biocenotiques et physiologiques particuliers, mise en valeur. *Bull. Mus, Nat. Hist. Nation* **20** : 317-331.

King, R.J. (1982). Mangroves and salt marsh plants. *In* : *Marine Botany*, Australian Perspective (eds. Clayton & King), Logman Cheshire, Singapore, 309-328.

King, R.J. (1990). Macroalgae associated with the mangrove vegetation of Papua New Guinea. *Bot. Mar.* **33** (1) : 55-62.

Kingston, B. (1981). Mangrove afforestation for Java. *FAO and Dir. Gen. For Convern.* Indonesia Occasional Paper No. 6.

Kjerfve, B. (1990). Manual for Investigation of hydrobiological process in Mangrove Ecosystems. 1 April, 1990. Supported by the UNESCO/UNDP Regional Project : Research and its application to the management of the Mangrove of Asia and the Pacific (RAS/86/120).

Klankamsom, B. & T. Charuppat (1982). Study on Changes of Mangrove Area in Thailand by Using LANDSAT Imageries. Royal Forest Department, Thailand.

Kloss, C.B. (1902). The Andaman and Nicobars (edn. 1971), p. 373, Vivek Publishing House, Delhi.

Koesoebiono, W.L. Collier & P.R. Burbridge (1982). Resource use and management in the coastal zone. pp. 115-133. *In* : C. Soysa, S.C. Lin & W.L. Collier (eds.). Man Land and Sea : Cosatal Resource Use and Management in Asia and the Pacific. Agri. Dev. Council, Bangkok.

Kogo, M. & C. Miyamoto (1983). A Report on Mangrove Research at Ras at Khafji, Saudi Arabia. 1981-82. Interime Report I. Al Gurm Research Centre.

Kogo, M. (1985). Report. Mangrove Research and Recommendation of Afforestation in Pakistan. Al Gurm Rsearch Centre, Tokyo.

Kogo, M., D. Kamimura & T. Miyagi (1986). Research for rehabilitation/reforestation of mangroves in Touk Islands. Mangroves of Asia and The Pacific : Status of Mangrove, Technical Report of the UNDP/UNESCO Research And Training Pilot Programme On Mangrove Eco-systems In Asia and the Pacific Ras. 79002. Natural Resources Management Cent., Quezon City, Philippines : 419–439.

Kogo, M.Y. & M. Hirata (1979). A Report of Mangrove Research in the Arabian Gulf. Japan Cooperation Center for the Middle East, Tokyo.

Kogo, M.Y., T. Mirgagi & Sato (1982). A report of the 3rd Research on Mangroves in the Middle East Kuwait, Saudi Arabia. U.A.E., Pakistan, India. Japan Cooperation Centre for the Middle East, Tokyo.

Komiyana, A., K. Ogino, S. Aksornkoae & S. Sabhasri (1987). Root biomass of a mangrove forest in Southern Thailand 1. Estimation of the trench method and the Zonal Str. of root to biomass. *J. Trop. Ecol.* 3 (2) : 97-108.

Kongasangchai, J. (1973). Some Soil Properties and Vegetation of Mangrove Forests at Different Tidal Zones at Phang-nga Province, M.S. Thesis, Kasetsartsart University, Bangkok, Thailand.

Kostermans, A.J.G.J. (1959). Monograph of the genus *Heritiera* Aiton (Sterant.). *Reinwardtia.* **4** : 465-583.

Kotmire, S.Y. & L.J. Bhosale (1979). Some aspects of chemical composition of mangrove, leaves and sediments. *Mahasagar*. Bull. NIO. **12** (13) : 149–154.

Kotmire, S.Y. & J. Bhosale (1980). Chemical composition of leaves of *Avicennia officinalis* L. and *A. marina* var. *acutissima* Stapp. & Muldenke. *Ind. J. Marine Sci.*, **9** (4) : 302.

Kotmire, S.Y. & L.J. Bhosale (1985). A study of Mangrove vegetation along Seogad estuary. *Marine Plants. Papers presented at the All India Symposium on Marine Plants, their Biology, Chemistry and Utilization*, Dona Paula, Goa. Oct. 30, Nov. 1, V. Krishnamurthy & A.G. Untawale (eds.) : 231-244.

Kotmire, S.Y. & L.J. Bhosale (1985). Leaf Senescence Behaviour of *Thespesia populnea* under saline and moisture condition. *The Mangroves : Proc. Nat. Symp. Biol. Util. Cons. Mangroves*, 266-271.

Krishnamurthy, K. (1975). Socio-economic aspects of Indian mangroves. *In* : G. E. Walsh, S. C. Snedaker & H. J. Teas (eds.). *Proc. Intnl. Symp. Biol. & Managt. of Mangroves*. Inst. of Food and Agril. Sci. Univ., Florida, Vol. 2 : pp. 729-731.

Krishnamurthy, K. (1983). The conservation of mangrove lands and water to other uses. *In* : Proc. UNDP / UNESCO Workshop on human population, mangrove resources, human stresses and human health, Bogor (Indonesia), pp. 1-12.

Krishnamurthy, K., V. Sundararaj & R. Santhanam (1974).A system approach in the study of Indian Mangroves. I. Environmental Factors. *Intern. Symp. Biol. & Mgt. Mangroves*, Hawaii.

Krishnamurthy, K., V. Sundararaj & R. Santhanam (1975). Aspects of Indian Mangrove forests. *In : Proc. Intnl. Symp. on Biol. & Mangt. of Mangroves*. G.E. Walsh, S.C. Snedaker & H.J. Teas (eds.). Inst. of Food and Agril. Sci., Univ. Florida, Gainesville, Florida, Vol. I. pp. 88-95.

Krishnamurthy, K., L. Kannan, M.J. Jeyaseelan Prince, R. Palaniappan & Ali Sultan (1981a). A floristic study of halophytes with special reference to the mangroves. *Bull. Bot. Surv.*, India, 23.

Krishnamurthy, K. & M.J.P. Jeyaseelan (1984). Human impact on the Pichavaram mangrove ecosystem : A case study from Southern India. pp. 624-632. *In : Proceeding Asian Symp. Mangr. Res. Management*. University Malaya, Kualalumpur, Malaysia.

Krishnamurthy, K. (1985). The changing landscape of the Indian Mangroves. *The Mangroves : Proc. Nat. Symp. Biol. Util. Cons. Mangroves*. Nov. 1985 : 119-126.

Krishnamurthy, K.L. & M.J.P. Jeyseelan (1986). Prospects of aquaculture in a mangrove ecosystem. *Proceedings of the Symp. on Coastal Aquaculture*, held at Cochin from Jan. 12 to 18, pp. 1059-1067.

Kugathasan, K.S. (1968). The distribution of mangrove vegetation in the lake. *Proc. Ann. Sess.* CAAS, Part-I, Sec. D.

Kugathasan, K.S., K. Rathnasabapathipillai & V. Ponnampalam (1972). Flora of the Thondaimannar Lagoon. *Proc. Ann. Sess.* CAAS, Sec. D.

Kulkarni, D.H. (1957). Extension of mangrove forests in Saurashtra and Kutch. *Proc. Mangrove Symp.*, Calcutta. pp. 30-32.

Kulkarni, D.H. (1959). Extension of mangrove species into the coastal saline banks. *Proc. Mangr. Symp.*, Calcutta : 32-35.

Kulkarni, D.H. & C.F. Junagad (1959). Utilization of mangrove Forests in Saurashtra and Kutch, *Proc. Mangrove Symp.*, Calcutta.

Kulkarni, S. & L.J. Bhosale (1985). Studies on excretion of salt in *Avicennia marina* and *Acanthus ilicifolius* under soil culture. *Marine Plants, Papers presented at All India Symposium on Marine Plants, their Biology, Chemistry and Utilization*, Dona Poula, Goa. Oct 30, Nov. 1, 1983, V. Krishnamurthy & A.G. Untawale (eds.) (1985). 295-298.

Kumar, Virendra & B. Subramanian (1987). Chromosome Atlas of Flowering Plants of the Indian Sub-Continent. Vol. I : Dicotyledons & Vol. II. Monocotyledons & Bibliography. BSI, Ministry of Environment & Forest, Govt. of India.

Kurlapkar, D.D. & L.J. Bhosale (1985). Preliminary report on ecology of *Rhizophora apiculata*. *The Mangroves : Proc. Nat. Symp. Biol. Util. Cons. Mangroves*. Nov. 1985 : 205-209.

Kurz, S. (1877). Forest Flora of Burma. 2 : 242.

Lacerda, L.D. de (1981). Mangrove wood pulp, an alternative food source for the tree crab *Aratus pisonii*. *Biotropica*. **13** : 317.

Lacerda,L. D. de & D.V. May (1982). Evolution of a new community type during the degradation of a mangrove ecosystem. *Bitropica*. **14** : 238-239.

Lakshmanan, K.K. & R. Jaylakshmi (1983). Floral biology in *Ceriops* (Rhizophoraceae). *Marine Plants, Papers presented at the All India Symposium on Marine Plants, their Biology, Chemistry and Utilization*, Dona Paula, Goa, Oct. 30-Nov. 1 ; V. Krishnamurthy & A.G. Untawale (eds.). 335-344.

Lakshmanan, K.K., K.M. Divakar & R. Jayalakshmi (1983). Floral variation in *Rhizophora apiculata* Bl. 10th Indian Cong. (Abstract). p. 157.

Lakshmanan, K.K. (1985). Mangrove as the source of Tannin—Past, Present and Prospects. *The Mangroves : Proc. Nat. Symp. Biol. Util. Cons. Mangroves*. Nov. 1985 : 105-111.

Lal, P.N., K. Swamy & P. Singh (1984). "Mangrove ecosystem" fisheries associated with mangroves and their management. Mangrove fishes in Wairiki Creek and their implications on the management of resources in Fiji, pp. 93-103. *In* : Productivity and Process in Island Marine Ecosystem. UNESCO *Rep. Mar. Sci.*, Paris.

Larue, C.D. & T.J. Muzik (1951). Does the mangrove really plant its seedlings ? *Science* **114** : 661-662.

Larue, C.D. & T.J. Mazuk (1954). Growth regeneration and precocious rooting in *Rhizophora mangle. Pap. Mich. Acad. Sci. Arts. Lett.* (Part-I), *Botany and Forestry*. **39** : 9-29.

Law, W.M. (1984). Productivity of the mangrove ecosystem : Management implications for agriculture. *Proc. Workshop Mangr. Prod.* Held from 4-6 October, 1983. Univ. Sains, Malaysia.

Lawrence, D.B. (1949). Self-erecting habit of seedling red mangroves (*Rhizophora mangle* L. *Am. J. Bot.* **36** (5) : 426-427.

Lawrence, S.H. & S.C. Snedakar (1974). The ecology of mangroves. *Ann. Rev. Eco. Syst.* **5** : 39-64.

Lawrence, S.H. & S.C. Snedaker (1984). Handbook of Mangrove Area Management. (Ed.) S.H. Lawrence & Samuel C. Snedaker, UNESCO.

Lear, Richard & Tom Turner (1977). *Mangrove of Australia*. University of Queensland Press. pp. 11-84.

Leach, G.J. & Burgin (1985). Litter production and seasonability of Mangroves in Papua New Guinea. *Aquat. Bot.* **23** : 215-224.

Lee, Kyu–Ho, M.A. Moran, R. Benner & R.E. Hodson (1990). Influence of soluble components of red mangrove (*Rhizophora mangle*) leaves–More–.*Bull. Mar. Sci.* 1990. Vol. 46, No. 2, pp. 374-386.

Lee, S.Y, (1989). Litter production and turnover of the mangrove *Kandelia candel* (L.) Druce in a Hong Kong tidal Shrimp pond *Estuar. Coast. Shelf. Sci.* **29** (1) : 75-87.

Lee, S.Y. (1990). Litter production and turnover of the mangrove *Kandelia candel* (L.) Druce in a Hong Kong tidal Shrimp pond. *3rd International Symposium on Marine Biogeography and Evolution in the Pacific–26 June to 3 July 1988, Vol. 47*, No. 1. p. 256.

Lee, S.Y., J.L. Furtado, A.R. Radzuan & A.T. Rambo (1978). Coastal Zone Resource Use and Management in Malaysia. University of Malaya (Mimeographed Paper).

Leh, C.M.U. & A. Sasekumar (1984). Feeding ecology of prawns in shallow waters adjoining mangrove shores. pp. 331-353. *In* : E. Soepadmo, A.N. Rao & D.J. Macintosh (eds.). *Proc. Asian Symp. Mangr. Environ. Res.* Manage. Univ., Malaya, Kualalumpur, Malaysia.

Leija–Tristan, A. & D.P. Sanchez-Vargas (1988). Biology and Ecology of the mud Shrimp Vpogebia dawsoni from the Requeson mangrove swamp, Conupcion Bay and Mulege River estuary, Baja California Sur, Mexico. *Rev. Biol. Trop. Vol. 36*, pp. 107-114.

Leoris, P. & G. Thanikaimoni (1972). Palynology in Pondicherry. *J. Palynology.* **8** : 56-162.

Lersten, N.R. & J.D. Curtis (1974). Colleter anatomy in red mangrove, *Rhizophora mangle* (Rhizophoraceae). *Can. J. Bot.* **52** : 2277-2280.

Lever, R.J.A.W. (1952). A Malayan leaf tying caterpillar from mangroves periodically submerged in sea water. *Entomologist's non Mag.* **88** : 95.

Levring, T., H.A. Hoppe & O.J. Schmid (1969). *Marine Alagae. Bot. Mar.* Handbook. *I*, 411.

Lewis, J. (1956). Rhizophoraceae. Flora trop. E. Afr. October 1956. 20 p.

Lewis, J.R. (1964). "The Ecology of Rocky Shores" English Universities Press, London.

Lewis, O.A.M. & G. Naidoo (1970). Tidal influence on the apparent transpirational rhythms of the white mangrove. South African Journal of Science, *66* : 268-270.

Lewis, R.R. (1983). Impact of oil spills on mangrove forests. Chap. 19 in Biology and Ecology of Mangroves, (Ed.) H.J. Teas. Tasks for Vegetation Science 8. The Jague, Junk.

Lewmanamont, K. (1983). Algal flora in mangrove community in Thailand. Paper presented in the Ist UNDP/UNESCO Regional Training Course on Introduction to Mangrove Ecosystem, held from 2-30 March, 1983 at National Research Council, Bangkok, Thailand.

Lezine, A.M. (1988). New pollen data from the Sahel, Senegal. *Quatunary Palynology of Tropical Areas.* J.V. Ward & Kam Biu Liu (eds.) 1988. *Vol.* 55, No. 1-3, pp. 141-154.

Li, F.X. & S.H. Gao (1983). The Malcofauna of *Hong Kong* and Southern *China. Proceedings of the Second International Workshop on the Malacofauna of Hong Kong and Southern Chines,* Hong Kong. 6-24 April. B. Morten & D. Dudgeon (eds.) : 449-455.

Liau, Yulou & Peng Lin (1987). Accumulation and biological cycle of Chloride in *Bruguiera sexangula* community. *J. Xiamen Univ. Nat. Sci. Xiamen Dexue Xuebao.* **26** (3) : 359-365.

Liebau, O. (1914). Bietrage Zur-Anatomic und morphologic der mangrove Pflanzen, insbesondere ihres Wurzelsystems. *Beit. Biol. Pflanz.* **12** : 181-213.

Liew, T.C. (1975). Mangrove exploitation and regeneration in Sabah, *Malayan Forester. 38* : 260-270.

Liew, T.C. (1980). Mangrove forests of Sabah. *Proc. Workshop Mangr. Estuar.* Veg. Malaysia. Univ. Pertanian Malaysia.

Lin, P. & X.M. Wei (1984). Ecological Notes on the Mangroves of Fujian, China. *In* : H.J. Teas (ed.) 1984. *Physiology and Management of Mangroves,* The Hague. Dr. W. Junk Publ.

Lin, P. & R.H. Chen (1986). Studies on the mangrove ecosystem of the Jiulongslang River estuary in China 3. Accumulation and biological cycle of Calcium and magnesium in *Kandelia candel* community. *Acta. Oceanol. Sin., Hajyang Xuebao.* 5 (3) : 447-455.

Lin, Peng, Su Lin & Qungyang Lin (1987). Studies on the mangrove ecosystem of Jivlong River estuary in China. 2. Accumulation and biological cycle of Potassium and Sodium element, *i.e., Kandelia candel* community. *Acta. Ecol. Sin.* Sheugtai Xuebeuo. 7 (2) : 102-110.

Lin, P., C.Y. Lu (1988). Study on biomass and productivity of a *Bruguiera sexangula* mangrove on Hainan Island, China. *3rd International Symposium on Marine Biogeography And Evolution In The Pacific,* 26 June–3 July 1988, *Vol.* 47, No. 1, p. 257.

Lindeman, J.C. (1953). Vegetation of the Coastal region of Suriname. Thesis Utrecht.

Ling, S.W. (1977). Aquaculture in South-east Asian : A *Historical Overview.* Washington Sea Grant Publ. Seattle : Univ. Washington Press.

Linnaeus, C. (1953). ***Species Plantarum.*** Stockholm : Holmiae.

Little, M.C., P.J Ready & S.J. Grove (1988). The fish community of an East African mangrove creek. *J. Fish. Biol.* 32 (5) : 729-747.

Loetschert, W. & F. Liemann (1967). Die Salzspeicherung in Keimling von *Rhizophora mangle* L. wahrend der. Entwicklung auf der mutterp flanze, *Planta* 72 : 142-156.

Lohwongwatans, B. (1979). The Industrial Pollution Control Programme of the Ministry of Industry for the Inner Gulf. *Proc. Seminar Protection Mar. Environ. Related Ecosystems 1979* : 153-172.

Lot Helgueras, A., C. Vazquez Yanes & F. Menendez (1975). Physiognomy and floristic changes near the northern limit of mangroves in the Gulf. Coast of Mexico. *In : Proceedings of the International Symposium on Biol. and Management of Mangroves,* (eds.) G.E. Walsh, S.C. Snedekar & H.J. Teas, pp. 52-61. Gainesville : Inst. of Food and Agril. Sci. University of Florida.

Lourensz, R.A. (1977). Tropical cyclones in the Australian region, July 1969 to June 1975. Canberra : *Austr. Govt. Pub. Office.*

Lu, Changyi, Fengzhong Zheng & Peng Lin (1988). Study on litter fall production of *Kandelia candel* mangrove community in estuary. *J. Xiamen Univ. Nat. Sci. Xiamen Dacile Xifao,* 27 (4) : 459-463.

Lu, Changyi & Peng Lin (1988). Litter leaf decomposition of two species of mangrove. *J. Xiamen. Univ. Nat. Sci.* 27 (6) : 679-683.

Lu, Changyi & P. Lin (1988, 1990). Study on litter fall and decomposition of *Bruguiera sexangula* mangrove of the Hainan Island of China. *3rd International Symposium On Marine Biogeography And Evolution In The Pacific*, 26 June–3 July 1988, *Vol. 47*, No. I, p. 257.

Ludlow-Wiechers, B. & J.L. Alvardo (1983). Palynological catalog of the Flora of Veracruz. No. 11. Rhizophoraceae family. *Biotica. 8* (1) : 7-14.

Lugo, A.E. (1981). The inland mangroves of Inagua. *J. Nat. Hist. 15* : 845-852.

Lugo, A.E. & S.C. Snedaker (1973). *The Role of Mangrove Ecosystems : Properties of Mangrove Forest in South Florida.* Resource Management Systems Programme, Univ. of Florida (Gainesville), U.S.A. Report No. D1-DEEP, 74-34, 62.

Lugo, A.E. & S.C. Snedaker (1974). The ecology of mangroves. *Ann. Rev. Ecol. Syst. 5* : 39-64.

Lugo, A.E. & S.C. Snedaker (1975). Properties of a mangrove forest in South Florida. *Proceeding of the International Symposium on Biology and Management of Mangroves* (eds. G.E. Walsh, S.C. Snedakar & H.J. Teas), *Vol. 1* : 176-212.

Lugo, A.E., G. Evink, M.M. Brinson, A. Broce & S.C. Snedaker (1975). Diurnal rates of photosynthesis, respiration and transpiration in mangrove forests of South Florida. Ecological Systems, New York, pp. 335-350.

Lugo, A.E. & C.P. Zucca (1977). The impact of low temperature stress on mangrove structure and growth. *Trop. Ecol. 18* : 149-161.

Maceda, E. (1986). A Key Note Address. *Workshop on the Conservation of Mangrove Areas to Aquaculture.* Held from 26-28 April, 1986 at Iloilo City, Philippines, UNDP/UNESCO Regional Mangrove Project for Asia and the Pacific. ,

Macintosh, Donald J. (1982). Fisheries and Aquaculture significance of mangrove swamps with special reference to the Indo West Pacific Region. Recent Advances in Aquaculture. (ed.) by James F. Muir & Ronald J. Roberts, Croom Helm London : 4-85.

Macintosh, Donald. J. (1984). Ecology and productivity of Malaysian mangrove crab populations. pp. 354-377. *In* : E. Soepadmo, A.N. Rao & D.J. Macintosh (eds.). *Proc. Asian Symp. Mangr. Environ. Res. Manage.*, Univ. Malaya, Kualalumpur, Malaysia.

Macintosh, S.B. (1987). The potential of *Merope augulata* (lvilldl). Swingle, a wild relative of the Citrus found in Malaysian mangroves. *Proceedings of the 10th Annual Seminar,* 28 March 1987, "Towards Conserving Malaysia's Marine heritage" : pp. 63-68.

MacNae, W. (1963). Mangrove swamps in South Africa. *J. Ecol. 51* : 1-15.

MacNae, W. (1966). Mangrove in Eastern and Southern Australia. *Aust. J. Bot. 15* : 67-104.

MacNae, W. (1967). Zonation within mangroves associated with estuaries in North Queensland. *In : "Estuaries"* (G.H. Lauff ed.), pp. 432-441. A.A.A.S., Washington.

MacNae, W. (1968). A General account of the fauna and flora of mangrove swamps and prospectives in the Indo-West Pacific Region. *Adv. Mar. Biol., 6* : 73-270.

MacNae, W. (1974). Mangrove forest and fisheries. FAO/UNDP Indian Ocean programme (Indian Ocean Fishery Commission) IOCF/DEV/74/34, Rome.

MacNae, W. & M. Kalk (1958). "A natural History of Inhaca Island, Mozambique," Witwatersrand University Press, Johannesburg.

MacNae, W. & M. Kalk (1962a). The ecology of the Mangrove swamps of Inhaca Island, Mozambique, *J. Ecol. 50* : 93-128.

MacNae, W. & M. Kalk (1962b). The fauna and flora of the sand flats at Inhaca Island, Mozambique. *J. anim. Ecol. 31* : 93-128.

MacNae, W. & F.R. Fosberg (1981). Sonneratiaceae. *In* : A revised handbook to the flora of Ceylon. *Vol. 3*, (ed.) M.D. Dassanayake, New Delhi : American Publishing.

MaCro, H.F. (1935). Systematic anatomy of the woods of the Rhizophoraceae. *Tropical Woods* (Yale University), *44* : 1-26.

Made Sandy, I (1976). *Atlas Indonesia,* Buku Ketiga. Yayasan Dwipura, Denpasar.

Mahabale, T.S. (1987). *In* : K.K. Chaudhari (ed.). Botany and Flora of Maharashtra. *Maharashtra State Gazetteer.*

Mahabale, T. S. & J.V. Despande (1959). The genus *Sonneratia* and its fosseil allies, *Paleobotanist. 6* : 51-64.

Mahapatra, A.K. & M.K. Mahapatra (1990). An Eco-floristic survey of some islands in Sundarbans of West Bengal. *Env. & Ecol. 8 (1)* : 171-174.

Mahapatra, A.K. & R.K. Chakraborty (1977). A survey of the Aquatic Flora of the Sagar Islands and their Economic aspects with reference to Integrated Rural Development, *J. Sci. Club. 31* (2) : 66-83.

Mahmood, N. (1990). Present State of the mangrove of Bangladesh. Report of the task team on the implication of climatic changes and the impact of rise in sea level in the South Karachi, Pakistan, 18 p.

Mahmud, M.D.B.H. (1984). Management and utilization of mangrove forests in the Asia-Pacific regions. *Proceeding of the Workshop on Productivity of the Mangrove Ecosystem : Management Implications*, J.E. Ong & W.K. Gong (eds.), University Sains, Malaysia Penang Malaysia : pp. 1-7.

Maity, H., G. Bandyopadhyaya & S.C. Santra (1987). Algal Flora of Saline Habitats of Sundarbans West Bengal and its Possible Role in Reclamation of Soil. *J. Indian Soc. Coastal agric. Res.* 5 (1) : 325-331.

Malaviya, C.V. (1963). On the distribution, structure and Ontogeny of stone cells in *Avicennia officinalis* L. *Proceedings of the Indian Academy of Science. 58* : 45-50.

Mall, L.P., S.K. Billore & D. Amritphale (1982). Certain ecological observations on mangroves of the Andaman Islands. *Trop. Ecol.* 23 : 225-233.

Mall, L.P., V.P. Singh, A. Garge & S.M. Pathak (1985). Mangrove frorest of Andamans and some aspects of its ecology. *The Mangroves : Proc. Nat. Symp. Biol. Util. Cons. Mangroves.* Nov. 1985 : 25-38.

Mall, L.P., V.P. Singh, A. Garge, S.M. Pathak & M.S. Dandoria (1985). A new approach towards the mangrove forests flora of Andaman Islands. *Indian For.* III : 290-300.

Mall, L.P., V.P. Singh, A. Garge & S.M. Pathak (1987). Ecological studies of mangrove forest of Ritchies Archipelago in relation to substratum. *Trop. Ecol.* 28 : 182-192.

Malley, D.F. (1978). Degradation of Mangrove leaf litter by the tropical sesarmid Crab (*Chiromathes onychophorum*). *Mar. Biol.* **49** : 377–386.

Mallik, N. (1969). Microfloristic studies of Calcutta peat. *J. Sen Memorial Vol. Bot. Soc. Bengal, Calcutta.* pp. 153-160.

Mandal, R.N. & K.R. Naskar (1991). Impact of Aquatic Flora for sustained productivity in humid tropics. *In : Proc. Symp. Farming For Sustained Productivity in Humid Tropics*, 16-17 Dec. 1991. Port Blair. pp. 207-209.

Mandal, R.N & K.R. Naskar (1993). Studies on the morpho-anatomical feature of mangrove with special reference to the halophytes. *Abs. 154. Asia Pacific Symp. on Mangrove Ecosystem.* Research Centre & Dept. of Biology, The Hong Kong Univ. of Science & Technology, Hong Kong, 1-3 Sept., 1994.

Mandal, R.N., D.C. Ghosh & K.R. Naskar (1994). Structural Features of the Tissue systems of Mangroves and their high saline Resistancy. *Abs. 101. West Bengal State Science Congress,* West Bengal State Council of Science & Technology, Calcutta, 28th Feb.–1st March, 1994.

Mandal, R.N., K.R. Naskar, Sipra Naskar, Bina Saha & D.N. Guha Bakshi (1994). Studies on the morpho-anatomical structures of 3 sp. of *Avicennia* L. commonly growing in the tidal mangals of the Indian Sundarbans. *Abs. 35,* pp. 10, *8th Biennial Botanical Conference, Bangladesh,* Dhaka, 12-13 Dec.

Mandal, R. N., D. Ghosh, A. K. Sarakar, D. Saha & K. R. Naskar (1994). Nutritional status of seven true mangrove sp. of the family Rhizophoraceae of the Indian Sundarbans Mangals. *Abstract 32,* P. 9. *8th Biennial Botanical Conference, Bangladesh,* Dhaka, 12-13 Dec.

Mandal, R.N. & K.R. Naskar (1994). Morpho-anatomical study of different sp. of *Sonneratia, National Seminar on Conservation & Sustainable Development of Coastal Resources,* Berhampur : 14-17 Dec.

Mandal, R.N., D.C. Ghosh, A.K. Ghosh & K.R. Naskar (1995). New Distribution Record of *Scyphiphore hydrophyllacea* Gaertn. *Atalanta Correa* M. Roem. from the Intertidal Mangrove Forests of the Indian Sundarbans. *J. Natl. Bot. Soc.* **49** : 71-74.

Martasubroto, P. & N. Namamin (1977). Relationship between tidal forest (mangrove) and Shrimp production in Indonesia. *Mar. Res. Ind. 18* : 81-86.

Masters, J.M. (1840). *Calcutta Flora, Trans. Agri. Hort. Soc.*, India, 7 : 39-85.

Mathauda, G.S. (1959). *The Mangroves of India. Proc. Mangrove Symp.*, Calcutta, pp. 66-87.

Mathauda, G.S. (1958). Silviculture of Mangrove Forests in Synopsis on Mangrove Vegetation. *Sci. & Cult.* 23 : 333.

Matilal, S., B.B. Mukherjee & N. Chatterjee (1986). Studies on soil and vegetation of Mangrove forests of Sundarbans. *Indian J. Mar. Sci. 15* (3) : 181-184.

Matondkar, S.G.P., S. Mahtani & S. Mavinkurvi (1980a). The fungal flora of the mangrove swamps of Goa. *Mahasagar Bull. Nat. Inst. Oceanogr. 13* (3) : 5.

Matondkar, S.G.P., S. Mahtani & S. Mavinkurve (1980b). Seasonal variations in microflora from mangrove swamps. *Indian J. Marine Sci. 9* (2) : 119-120.

Mazumdar, S.P. (1932). Gosaba Cooperative Commonwealth. West Bengal Cooperative Press, Calcutta, 2nd ed. 23 p.

McAlpine, J.R., G. King & R. Falls (1983). Climate of Papua New Guinea. Canberra : *Austr. Nat. Univ. Press.*

McMillan, C. (1971). Environmental factors affecting seedling establishment of the back mangrove on the Central Texas Coast. *Ecology. 52* : 927-930.

McMillan, C. (1975). Adaptive differentiation to Chilling in mangrove populations. *In* : G.E. Walsh, S.C. Snedaker & H.J. Teas (eds.). *Proceedings of the International Symposium on Biology and Management of Mangroves.* Institute of Food and Agricultural Sciences, University of Florida, Gainesville Florida, *Vol. 1*, pp. 62-70.

McNeely, J.A. & B. Lekagul (1976). Mammals of the Mangroves. *Paper presented in the First National Seminar on Mangrove Ecosystems.* Held from 10-15 January, 1976 at Phuket Marine Biological Center, Thailand.

Megeri, S. & J.H. Davis (1981). (Eds.) Global Status of Mangrove Ecosystem, *Paper No. 3*, Gland, Switzerland.

Meher Homji, V.M. (1973). Zonation of the mangrove vegetation in the Vellar Estuary (Abstract). *Proc. 60th Sci. Congr. 3* : 338.

Menasveta, P. (1982). Water Resources and Pollution Problems. Chulalongkorn University, Bangkok, Thailand.

Mepahm, R.H.(1983). Mangrove floras of the Southern Continents. Part-I : The geographical origin of Indo-Pacific mangrove genera and the development and present status of the Australian mangroves. *S. Afr. J. Bot. 2* : 1-8.

Mepham, R.H. & J.S. Mepham (1984). The flora of Tidal forests - a rationalization of the use of the term 'mangrove' S. *African J. Bot*. 1985, *51* : 77-99.

Merrill, E.D. (1945). *Plant Life of the Pacific World*. Macmillan, New York, 295 p.

Metcalfe, C.R. (1931). The breathing roots of *Sonneratia* and *Bruguiera*, a review of recent work by Troll and Dragendorff. *Kew Bull. 11* : 465-467.

Metcalfe, C.R. & L. Chalk (1950). Anatomy of the dicotyledous. Clarendon, Oxford.

Mehe Homji, V.M. (1979). On the subtropical climate vegetation of the Indian Sub-Continent, Geology. Ecology, *8* (2) : 137-148.

Miller, O.B. (1926). Mangroves at Sordwana Bay, N. Zululand S. Afr. *J. Nat. Hist. 6* : 54-55.

Miller, P.C., J. Hom & D.K. Bole (1975). Water relations of three mangrove species in South Florida. *Oecologia Plantarum. 10* : 355-367.

Milliman, J.D., J.M. Broadus & F. Gable (1989). Environmental and economic implications of rising sea level and subsiding deltas. The Nile and Bengal examples,*Ambio. 18* (6) : 340-345.

Misra, S., A. Ghosh, A.K. Dutta, J. Dutta & A. Choudhury (1993) Microbial oxidation of pencyclic triterpenes acids of *Avicennia officinalis* leaves in the mangrove detrital system of Sundarbans. *Proc. All India Symp. of Marine Plants*, No.10, Goa.

Misra, S., V.K. Yadav, A. Choudhury, A. Ghosh & J. Dutta (1985). Microbial decomposition of *Avicennia officinalis* leaf litter in and Mangrove forest biome : Bio-Chemical Changes during decomposition. *Marine Plants, Papers Presented at the All India Symposium on Marine Plants, their biology, chemistry and uitllization*, Donapaula, Goa, Oct.30 - Nov.1, 1983, V. Krishnamurthy & A.G. Untawale (eds.). 1985. 257-262.

Misra, S., A. Choudhury, S. Chattopadhyay & A. Ghosh (1988). Lipid composition of *Porteresia coarctata* from two different mangrove habitats in India. *Phytochemistry*. **27** (2) : 361-364.

Mitra, R. L. & L. K. Banerjee (1979). The Genus *Brugeciera* Lamk. (Rhizophoraceae in India) *Bull. Bot. Surv. India*, **21** : 142-150.

Mitre, Vishnu & H.P. Gupta (1972). Pollen Analysis Study of Quaternary Deposit in the Bengal Basin, *Paleobotanist*, **19** : 297-306.

Moldenke, H.N. (1960). Materials towards a monograph of the genus *Avicennia* I & II. *Phytologia* **7** : 123-168, 179-232, 259-263.

Moldenke, H.N. (1963). Avicenniaceae, Symphoremaceae and Verbenaceae Studies in the flora of Thailand, 1963. *Danrk bot. Ark.* 23, 83-92.

Moldenke, H.N. (1967). Additional notes on the genus *Avicennia*.I and II. *Phytologia*. **14** : 301-320, 326-336.

Mongia, A.D. &.A.K. Bandyopadhyay (1991). Response of Acid Sulphate Soil in Mangroves to Different Sources of P-Application for Low Land Rice Production in Andamans. *J. Indian Soc. Coastal agric. Res.*, **9** (1 & 2) 463-469.

Monkolprasit, S. (1983). Fish in mangroves and adjacent areas. *Paper presented in the First UNDP/UNESCO Regional Training Course in Introduction to Mangrove Ecosystems.* Held from 2-30 March, 1983 at National Research Council, Bangkok, Thailand.

Montecillo, E. & R. Rosito (Unpublished Report). Corals in the Mangrove Area Inside Arangas Island, Brief Accounts on Ecology and Systematics. Aras-asan Timber Co., Philippines.

Montgomery, S.Y. (1995). *Spell of the Tiger. The Man Eaters of Sundarbans.* A Peter Davison Book, Houghton Mifflin Company, Boston, New York. pp.1-230.

Morgan, J.P. & W.G. McIntire (1959). Quaternary Geology of the Bengal Basin, East Pakistan and Burma, *Bull. Geol.Sic. Am.* **70** :319-342.

Motoh, H & N. Solis (1997). The fauna of decapod, p.153. *In : Proc. Intern. Workshop Mangr. Estuar. Area Dev.*, Manila, Philippines.

Mukherjee, A.K.(1959). Pakhirala Sajnakhali —An introduction to a Bird Sanctuary in the Sundarbans. *J. Bengal Nat. Hist. Soc.* **30** : 161-165.

Mukherjee, A.K. (1965). Habits of the Rhesus *Macacue mulata* (Zimmlimann) in the Sundarbans, 24-Parganas, West Bengal. *J. Bombay Nat. Hist. Soc.* **62** (1) : 145-146.

Mukherjee, A.K. (1969). Food habits of water birds of the Sundarbans, 24-Parganas District, West Bengal, India. *J. Bombay Nat. Hist. Soc.* **66** (2) : 345-360.

Mukherjee, A.K. (1970). Wild life in the Lower West Bengal. *Indian Museum Bull.* **5** (2) : 28-34.

Mukerjee, A.K. (1971). Experience of Naturalist in the Sundarbans. *Indian Tourist*, 2 (1) : 33-35.

Mukherjee, A.K. (1975). Mangrove Wealth of Indian Sundarbans, utilization and conservation. *J. Eco. & Tax. Botany*. **5** : 227-230.

Mukherjee, A.K. & K.K. Tiwari (1984). Mangrove ecosystem changes under induced stress. The case history of Sundarbans, West Bengal, India, pp.633-643. *In : Proceeding Asian Symp. Mangr. Environ. Res. & Management*, University Malaya, Kualalumpur, Malaysia.

Mukherjee, B.B. (1969) Wood anatomy of a few mangrove and other plants (Recent and fossil). *Trans. Bose Res. Inst.* 32 (1) : 59-63.

Mukherjee, B.B. (1969). Vegetational history of southern part of West Bengal with an introduction of the pollen flora of Sundarbans. Ph.D. Thesis, Calcutta University.

Mukherjee, B.B. (1972). Pollen analysis of a few quaternary deposits of lower Bengal Basin. *Proc. Sem. Palaeopalynol*, India, Calcutta. pp.177-192.

Mukherjee, B.B. (1972). Quaternary pollen analysis as a possible indicater of prehistoric agriculture in Deltaic part of West Bengal. *J. Palynology. 8* : 144-151.

Mukherjee, B.B. (1980). Mangrove forest types and their distribution in Sundarbans, India. *Proc. Asian Symp. on Mangrove Environment*, Kualalumpur (Abst.).

Mukherjee, B.B. (1984). Mangrove forest types and their distribution in Sundarbans India. pp.75-81. *In : Proceeding Asian Symp. Mangr. Environ. Res. Management*, University Malaya, Kualalumpur, Malaysia.

Mukherjee, B.B. & J. Mukherjee (1978). Mangrove of Sundarbans, India. *Phytomorphologia, 28* (2) : 177-192.

Mukherjee, D. (1987). A note on coastal Region of West Bengal. *J. Indian Soc. Coastal agric. Res.* **5** (1) : 25-30.

Mukherjee, J. & S. Chanda (1973). Biosynthesis of *Avicennia* L. in relation to taxonomy. *Geophytology.* **3** : 85-88.

Mukherjee, R.K.(1938). Changing face of Bengal, Calcutta, pp.177-199.

Mukerjee, S.K. (1959) Associates of mangrove plants in the Sundarbans. *Proc. Mangrove Symp.*, Calcutta (1957), p.136.

Mukerjee, S.K. (1960). The genus *Bruguiera* in the Sundarbans. *Bull. Bot. Surv. India, 2* (1-2) : 173-174.

Mukerjee, S.K. (1966). *Bruguiera hainesii* Rogers from the Sundarbans, a new record. *Bull. Bot. Sun.*, India. *8* (3/4) : 357.

Mukhopadhyay, K. (1983). Preservation of natural environment—A case study in Sandeshkhali, P.S. (West Bengal). *Proc. Int. Seminer on Envo. Maps and Atlas*, Calcutta, Dec. 1-3, 1983, pp.134-135.

Mulik, N.G. & L.J. Bhosale (1985). Autecological Studies on the Seeds of *Acanthus ilicifolius Linn. The Mangroves : Proc. Nat Symp. Biol Util. Cons. Mangroves*, Nov. 1985 : 210-213.

Mullan, D.P. (1931a). Observations on the water storing devices in the leaves of some Indian Halophytes. *J. Ind. Bot. Soc. 10* : 184-189.

Mullan, D.P. (1931b). On the occurrence of glandular hairs (salt glands) on the leaves of some Indian halophytes, *J. Indian Bot. Soc. 10*: 184-189.

Mullan, D.P. (1932-33). Observations on the biology and physiological anatomy of some Indian halophytes. *J. Indian Bot. Soc.* II. 103-118 ; *12* : 165-182 ; 235-256.

Muller, J. (1969). A palynological study of the genus *Sonneratia* (Sonneratiaceae). *Pollen et Spores, XI* : 233-298.

Muller, J. & S.Y. Hou-Liu (1966). Hybrids and Chromosomes in the Genus *Sonneratia* (Sonneratiaceae). *Blumea, 14* : 337-343.

Murphy, D.H. (1965). Dehoutte *Viuen marina n. gen. n.* sp. *(Collembola, Sminthuridae)* from the intertidal zone of Singapore, *Bull. Nat. Mus.Sing. 36* (6) : 31-34.

Murty, C.S. & P.K. Das (1972). Premonsoon tidal flow characteristics of Mandovi estuary. *Ind. J. Mar. Sci. 1* : 148-151.

Mustafa, M., Soegondo H. Nurkin, N. Sutika & H. Sannsi (1979). Penelitian Komunitas Lingkungan dan Regenerasi serta Pengembangan Hutan Mangrove di Sulawesi Selatan-Universitas Hasanuddin (Unpublished Report).

Nabhitabhata, J. (1982). Ecological Studies of birds in mangrove forests. Songkhla Lake. *Paper presented in the fourth National Seminar on Mangrove Ecosystems.* Held from 7-11 July, 1982 at Sunat Thani, Thailand.

Naiyanetr, P. (1983). Crabs in mangroves and adjacent areas. *Paper presented in the First UNDP/UNESCO Regional Training Course on Introduction to Mangrove Ecosystems.* Held from 2-30 March, 1983 at National Research Council, Bangkok, Thailand.

Naskar, K.R. (1981). *Floristic Studies of the District 24-Parganas in West Bengal (India) with Special Reference to the Mangrove Vegetation of the Sundarbans*. Ph.D. Thesis (unpub.), Calcutta University, *3 Vols.* 1800 p.

Narkar, K. R. (1983). Halophytes and their unique adaptations on the Sundarbans Mangrove Swamps. *J. Indian Soc. Coastal agric Res.* 1 (2) : 91-105.

Naskar, K.R. (1984). *Mangroves and their plantation techniques.* "Service Centre", Development Research Communication & Service Centre, 18-B, Gariahat Road (s), Calcutta-700031.

Naskar, K. R. (1985a). Population Pressure on the Mangrove Swamps of Sundarbans in India—Its Causes & Problems, Population Mapping. Monograph 4 : 159-161. NATMO Calcutta.

Naskar, K.R. (1985b). Identification of some wild medicinal plants from the tidal mangrove forest of Sundarbans. *Abstract p.46. International Symposium on Medicinal, Aromatic and Spice Plants.* Directorate of Chinchona and other Medicinal Plants, Mungpoo, Darjeeling, W.B. 23rd - 26th February, 1985.

Naskar, K.R. (1985c). A Short History and the Present Trends of Brackishwater Fish Culture in Paddy Fields of the Kulti-Minakhan Areas of the Sundarbans in West Bengal.*J. Indian Soc. Coastal agric. Res.* 3 (2) : 115-126.

Naskar, K.R. (1986a). Role of *Avicennia* L. plantation on Brackishwater Fisheries with Reference to their Taxonomy and Ecological Notes in the Tidal Mangrove forests of Sundarbans. *J. Ind. Soc. Coastal agric. Res.* 4 (2) : 137-142.

Naskar, K.R. (1986b). Trends of Aquatic weeds management through proper utilization. *Abstract P.54. National Seminar on Recent Trends in Plant Science Research,* Visva Bharati, Dept. of Botany, Santiniketan-731 235, 18th-19th Feb.

Naskar, K.R. (1986c). Utilization of Some Aquatic weed Plants in the Rural Aquaculture Techniques. *Abs.* pp. 5-6. *National Seminar on 'Perspectives in Hydrobiology'* School of Studies in Zoology, Vikram Univ., Ujjain-456 010 (M.P.). 8th-9th, 1986.

Naskar, K.R. (1986d). Mangrove Environment of Sundarbans and its impact to the rural life. *Abstract p.8. All India Symposium on Role of Biology in Rural Development.* Salem Institute of Expt. Biol. 3rd-5th Oct.

Naskar, K.R. (1986e). *Sundarban Me laban joliya Matsya saha, Dhan Krishi Ke Bishesh Sandharbha Me Sanghatita Krishi Ke liya laban jal Ke Upoyog,* short paper, pp. 71-72. National Seminar on land use pattern, National Atlas & Thematic Mapping Organisation, Calcutta, 19th-20th Dec.

Naskar, K.R. (1986f). Ecology of Sewage Fed Brackishwater Fisheries in the Sundarbans Tidal delta of W.B. *Abstract- 7475. National Symposium on fish and their environment,* Dept. of Zoology, Gurukhal Kangri Viswa vidyalays, Hardwar-249404 (India), 15th-18th Dec.

Naskar, K.R. (1986g). Economic importance of the Dominant Mangrove Family Rhizophoraceae from the Sundarbans Delta of 24-Parganas District, W.B.*J. Econ. Tax. Bot.* 8(2) : 359-372.

Naskar, K.R. (1986h). Role of aquatic Weeds for waste water treatments. *Abstract p.21. All India Seminar on water quality in and around urban Ecosystems and their management,* Govt. Post-Graduate College, Chhindwara (M.P.).27th-29th Dec.

Naskar, K.R. (1987a). Recent trends of aquatic weeds management through proper utilization. *Plant Science Research.*Department of Botany, Visva Bharati, Santiniketan, India. pp.188-193.

Naskar, K.R. (1987b). Fish seeds prospectives in the Sundarbans Tidal Estuary and Nature of its Exploitation. *Abs.P.3. All India Symp. on Aquatic Organisms,* Dept. of Zool., A.V.V.M., Sri Pushpam College, Pondi-613 503, Thanjaware, Dist. Tamil Nadu. 9th-11th Feb., 1987.

Naskar, K.R. (1987). Mass production of Duck weeds for carp culture, *Proc. USAID Training Programme on Freshwater Aquaculture,* CIFA, Bhubaneswar. pp.56-65.

Naskar, K.R. (1988). Mangrove Ecology of the Sundarbans Delta in West Bengal and its Role on Brackishwater Fisheries. *Jour. Aqua.* 1 (1) : 1-9.

Naskar, K.R. (1989). Studies of macro vegetation in Beels. Bulletin No. (63) (CICFRI). July, 1989. pp. 32-56.

Naskar, K.R. (1990a). Role of Afforestation programme in the natural mangrove forests of the Sundarbans delta in India. *Abst.*pp.24-25. *National Seminar on Mangrove Awareness in India,* World Wide Fund for Nature - India, Bombay, February, 21-23.

Naskar, K.R. (1990b). *Aquatic and some aquatic plants of lower Ganga delta.* Daya Publishing House, Delhi-110006, pp.1-402.

Naskar, K.R. (1990c). Judicious management of the Aquatic Vegetation in the large water-bodies. *Abs.*p.2. *Nat. Seminar on Watershed Management & Ecological Development*, Vritti Vikas, 11-12th April, 1990. Inst.of Chemical Engineers, Jadavpur Univ.

Naskar, K.R. (1991). Plant Succession of the mangrove Ecosystem of Sundarbans in West Bengal, India, *Abstract* p.15. *National Symposium on Conservation & Management of Living Resources.* Unit of Ecology and Natural Resource, University of Agricultural Science, Bangalore-560 065, January 10-12.

Naskar, K.R. (1993). Mangrove ecosystem of the Sundarbans in India - its regeneration and socio-economic potentialities. *Abs.179. Asia - Pacific Symp. on Mangrove Ecosystems*, Research Centre & Dept. of Biology, The Hong Kong University of Science & Technology, Hong Kong, 1-3 Sept., 1993.

Nasbar, K.R. (1993). *Plant Wealth of the Lower Ganga Delta - An Eco-Taxonomical Approach*, 2 vols., Daya Publishing House, Delhi-110006, pp.1-810.

Naskar, K. R. (1996). Status of the Coastal Zone Vegetation in West Bengal with Special Reference to the Sundarbans. Status Paper Dept. of Env., Govt. of West Bengal (In Press.)

Naskar, K. R. (1996). Status of the Mangroves in Indian Sundarbans—In the Perspectives of India and the World Mangals. *Status Paper* : William Roxbugh Memories, CWS, Calcutta 8-10 Nov. 1996.

Naskar, K.R., R.N. Mandal, D. Ghosh, A.K. Barui & D.N. Guha Bakshi (1994). Studies on seedling growth,development and survivality of the true mangrove sp. of Sundarbans in the non-saline condition. *Abstract-36*, pp.10. *8th Biennial Botanical Conference*, Bangladesh, Dhaka, 12-13 Dec.

Naskar, K.R., D.N. Guha Bakshi & G.G. Maity (1979). On the occurrence of *Moghania macrophylla* (Willd). Kuntzo from the lower Ganga plain. *Bull. Bot. Soc. Bengal.* **33**: 39-43.

Naskar, K.R. & D.N. Guha Bakshi (1981). Some of the important medicinal plants of Apocynaceae from West Bengal. *Bull. Bot. Soc. Bengal.* **35** : 7-14.

Naskar, K. R. & D. N. Guha Bakshi (1982). Sundarbans—The World famous mangrove forests of the District 24-Parganas in West Bengal (India). *J. Econ. Tax. Botany* 3 : 883-918.

Naskar, K. R. & D. N. Guha Bakshi (1983a). A Brief Review on Some Less Familiar Plants of the Sundarbans. *J. Econ. Tax. Botany* **4** (3) : 699-711.

Naskar, K. R. & D. N. Guha Bakshi (1983b). An Eco-Taxanomical studies of the Typical Halophytic Flora of the Sundarbans in the District 24. Parganas, West Bengal with Special Reference to their Socio-Economic Impact. *J. Econ. Tax. Botony* **4** (3) : 713-746.

Naskar, K.R. & N.M. Chakraborty (1984). Studies on the economic fauna from the Sundarbans Delta in West Bengal. *J. Indian Soc. Coastal agric. Res.* **2** (1) : 17-25.

Naskar, K.R. & D.N. Guha Bakshi (1984). Identification and extraction technique of crude drugs (Phyto origin from the wild sources, commonly utilized in Ethnopharmacology. *Proc. Nat. Sem. Appl. Bio-Anatomy*, Cal. University ; 437-447.

Naskar, K.R. & D.N. Guha Bakshi (1984). Changing pattern of cultural landscape in the tidal mangrove swamps Sundarbans. *Abstract* p.56. *National Seminar on Changing Pattern of Cultural Landscape in India.* Geographical Society of India, Dept. of Geography. University of Calcutta, Cal-19.

Naskar, K.R. & Apurba Ghosh (1985). Role of *Wolffia arrhiza* (L.) Horkel en Winner as a fish-food and its influence on Algal food chain. *Abstract P.14. All India Symposium on 'Biology of Algae'*, Dept. of Biology, Univ. of Poona,Pune-7, 8th-10th Feb., 1985.

Naskar, K.R. & D.N. Guha Bakshi (1985). Medicinal plants from the District 24-Parganas. *Abs.p.26. International Aromatic and Spice Plants Organization*, Mungpoo, Darjeeling, W.B., 23rd-26th Feb., 1985.

Naskar, K.R. & S.C. Santra (1985). A note *Enteromorpha tubulosa* in Brackish Mixed Sewage Fed Fisheries from the Sundarbans in W.B.*J. Indian Soc. Coastal agric. Res.* **5** (2) : 471-472.

Naskar, K.R., S.K. Saha & P.K. Saha (1985). Compost of water hyacinth for the fields of Agriculture and aquaculture. *Environ, & Ecology*, **3** (2) : 247-253.

Naskar, K.R. & D.N. Guha Bakshi (1985). *Aeluropus lagopoides*—A new distributional record from the tidal mangrove forest of the Sundarbans. *J. Econ. Tax. Botany.* **7** (1) : 200-201.

Naskar, K.R. & D.N. Guha Bakshi (1985). *Parthenium hysterophorus* L. an exotic Poisonous weed in India — *A review J. Econ. Tax. Botany.* 7 (3) : 741-748.

Naskar, K. R., G. N. Chattophadhyay & Apurba Ghosh (1985). A Notes on *Ruppia maritima* L., a submerged aquatic weed from the Brackish water Fisheries of of Sundarbans. *J. Indian Soc. Coastal agric Res.* 3 (1) : 69-70.

Naskar, K.R. & S.K.Saha (1986). Role of Aquatic weeds in sewage treatment. *Abstract p.7 All India Symposium on Role of Biology in Rural Development.* Salem Institute of Experimental Biology, Habib Street, Fort Salem-636 001,3-5th Oct.

Naskar, K.R. & D.N. Guha Bakshi (1986). *Bhumi Ypoyoga Ke Paribekshya Me Bangal Ke 24 Parganas Ke Biuinna Paddy Paristhitik Khetra. Monograph.* 5 : 73-75. National Atlas & Thematic. Mapping Organisation, Calcutta.

Naskar, K.R., P.K. Saha, Rita Mazumder & D.N. Guha Bakshi (1986). Compost value of water hyacinth *(Eichornia crassipes (Maxt) Sotms.)* in the field of Agriculture & Aquaculture. *Bull. Bot. Soc. Bengal.* **40** (1 & 2) : 11-17.

Naskar, K.R., S.K. Saha & N.M. Chakraborty (1986). Role of sewage effluent in the field of Agriculture and Aquculture. *Abs.p.7. All India Symposium on Role of Biology in Rural Development,* Salem Inst. of Expt. Biol.,Salem, 3rd-5th Oct., 1986.

Naskar, K.R. & S.C. Santra (1986). Studies on *Entermerpha tubulosa* in brackishwater mixed sewage fed fisheries from Sundarbans, No.B, *Sci. & Cult.* 52 : 210.

Naskar, K.R., A. Banerjee, N.,M. Chakraborty & Apurba Ghosh (1986). Yield of *Wolffia arrhiza* (L.) Horkel ex Wimmer from cement cisterns with different sewage concentration and its efficacy as carp feed. ***Aquaculture*** (Netherland) *51* : 211-216.

Naskar, K.R. & D.N. Guha Bakshi (1986). *Bhumi upoyog Ke paripreksha Me Paschim Bengal Ke 24 Parana Ke Bivinna Padap Paristhitik Khetra* (in Hindi). Short Paper. pp.73-75. *National Seminar on Land Use pattern,* National Atlas & Thematic Mapping Organisation, Cal. 19th-20th Dec., 1986.

Naskar, K.R. & S.K. Saha (1986). Utilization of waste water for integrated Farming. *Abs.28. 6th All India Seminar on Ichthyology,* Dept. of Life Science, Manipur Univ., Imphal-795 003, Manipur, India. 21st Dec. -24th Dec., 1986.

Naskar, K.R. & D.N. Guha Bakshi (1986) . On the verge of extinction of some Important Mangrove Species from the Sundarbans Delta in West Bengal. *J. Econ. Tax. Botany.* 8 (2) : 433-440.

Naskar, K.R. & D.N. Guha Bakshi (1986). Different phyto-ecological zones in the 24-Pargana dist. of West Bengal with Special reference to its land utilization patterns. *Abstract* pp. 25-26. *National Seminar on Problem & Management of Coastal Ecosystem,* Indian Soc. Coastal Agril. Research, Canning Town, West Bengal, 29th-31st Dec.

Naskar, K.R. & D.N. Guha Bakshi (1987a). Different Phyto-Ecological zones in the 24-Parganas District of West Bengal with special reference to its land utilization patterns. J. *Indian Soc. Coastal Agric. Res.* 5(1) : 183-187.

Naskar, K.R. & D.N. Guha Bakshi (1987b). **Mangrove Swamps of the Sundarbans - An Ecological** Perspective, Naya Prakash, Calcutta-700 006 : 1-264.

Naskar, K.R. & P.R. Sen (1987). Utilization of Duck weed for carp seed rearing. *Abst.*p.5. The first Indian Fisheries Forum, *Asian Fish Soc. Ind. Branch,* C/o. College of Fishery, Mangalore, 4-8th Dec.

Naskar, K.R. & N.M. Chakraborty (1987). Studies on the Benthick Biota from the brackishwater fisheries in the Sundarbans delta of West Bengal, India. *National Symposium on Aquatic Biology,* Dept. of Zoology, Andhra University, Waltair, Vishakhapattanam-530 003.

Naskar, K.R. & P.K. Saha (1987). Studies of duck-weed culture utilizing domestic organic wastes. *Bull. Bot. Soc. Bengal,* **41** : 35-34.

Naskar, K.R. & N.M. Chakraborty (1987). Studies of the family Lemnaceae from the lower Ganga Delta. *Bull. Bot.Soc. Bengal.* **41** : 5-12.

Naskar, K.R., Apurba Ghosh, S.K. Saha, Santunu Biswas (1987). Obnoxus water hyacinth (*Eichhornia crascipes solum*) for the human welfare - a review. *Abs.p.16. 2nd Nat. Seminar on Models For Rational Ecological Development* (in waste lands). School of Fundamental Research, Cal, Dec. 6-8, 1987.

Naskar, K.R. G.N. Chattopadhyay, P.K. Chakraborty & N.N Muzumder (1988). Different typs of tidal mangroves of the Sundarbans and their role on estuarine fisheries. *Abs.*pp. 24-25. Dec. 14-16, 1988.

Naskar, K.R. & S.K. Saha (1988). Utilization of Domestic wastes for integrated farming in the Tropics. *J. Aqua. Trop.* 3 (1) : 37-41.

Naskar, K.R. & Santunu Biswas (1988). Hydrocharilaceae from the lower Ganga Delta and its role in Aquaculture and Agriculture, *Environ. & Ecol.* 6 (3). 620-626.

Naskar, K.R. & N.N. Mazumder (1988). Studies on some fish poisonous plants in the open water system. Contribution to the management of fisheries in Inland of India. (Ed.) A.G. Jhingran, V.K. Unithaan & Amitabha Ghosh, FSI Publication, 6 p.

Naskar, K.R. & Apurba Ghosh (1988). Status of the tidal aquatic environment in mangrove forests of the Sundarbans in India. *Abst. International Symp. on Mangrove,* Seamio Bogor, Indonesia.

Naskar, K.R. & N.N. Mazumder (1988). Studies on some fish poisonous plants in the open system. (CICFRI), 14-16 Dec., 1988. **In** : Contribution to the Fisheries of Inland Open Water System in India, pp. 184-188.

Naskar, K.R., S.K. Saha & A.K. Datta (1988). Utilization of Domestic Wastes for Rice Production. *Environ. & Ecol.* 6 (1) :125-128.

Naskar, K. R. & D. N. Guha Bakshi (1989). Distribution of Mangroves in India and Abroad and their Roles in the Estuarine Ecosystems. *Monograph 9* : 22-23 (NATMO, Calcutta).

Naskar, K. R. & D. N. Guha Bakshi (1989b). Stages of Land Formation and Comparative Studies of the Different Micro-Ecological Zones of Sundarbans (West Bengal). *Fragm. Flor. Geobot.* **34** : (3-4) : 341-354.

Naskar, K.R. & A.K. Ghosh (1989). Status of the tidal and intertidal mangrove forests of the Sundarbans in India. *Ist International Seminar on Coastal Fnvironment in Asian Region - Potentialities and Problems* ; Dec. 8. Asian Marine Conservation Association, Cal. pp. 23-29.

Naskar, K.R. & N.N. Mazumder (1989). Interdependence between plants and animals in the freshwater environment in lower Ganga delta. p.34. Rahara V.C. College, 240-Pgs., W.B., Feb. 6-8th.

Naskar, K.R. & Apurba Ghosh (1989). Mangrove forest of the Sundarbans and its impact on estuarine fisheries. *All India Workshop on Coastal Zone Management of West Bengal Sea Explorers Institute,*cal-21. Proceeding CZM-SEI A 47-A 59.

Naskar, K.R. & D.N. Guha Bakshi (1989). Joint endeavour for protection of estuarine of ecosystem of Sundarbans. *Abstract No. 102. 6th National Botanical Conference,*University of Chittagong, Chittagong,Bangladesh, Jan. 18-19.

Naskar, K. R. & N. N. Mazumder (1989). Weed fish eradication review on some piscides of plant origin. *Abs.p.52. Nat. Symp. on Modern Trends in Environmental Biology and Seminar on Environment and Animal life,* Dept. of Zool., Ramkrishna Mission, Vivekananda Centenary College, Rahara, 24-Parganas, W.B. Feb. 6-8th, 1989.

Naskar, K. R. & D. N. Guha Bakshi (1993). The Biological Spectrum of Floral Elements of the 24-Parganas District in West Bengal. *Bangladesh J. Botany* 22 (1) : 15-20.

Naskar, K.R.,R.N Mondal & D.C. Ghosh (1994) . A critical study on the morpho-anatomy of the different true mangrove Sp. under the family Rhizophoraceae of Sundarbans in West Bengal. OP. 2.4.A.7.*National Seminar on Conservation & Sustainable Development of Coastal Resources,* Berhampore, 14-17 Dec.

Naskar, K.R., R.N. Mandal & D.C. Ghosh (1994). A critical study on the morpho-anatomy of diff. true mangrove species under the family Rhizophoraceae of Sundarbans in West Bengal. OP. 2.4.A.7. *National Seminar on Conservation & Sustainable Development of Coastal Resources,* Berhampore. 14-17 Dec.

Naskar, K. R., A.K. Ghosh, D.C. Ghosh, T.K. Baguli & A.K. Sarkar (1995). Sundarbans Mangrove Ecosystem with special reference to Floral diversity. *Abs.52. National Symp. on Perspective in Biodiversity,* The Zoological Society, Calcutta, Indian Institute of Chemical Biology, Calcutta.National Mangrove Committee (1982). Pangangan Island revisit with self help, school retorest mangrove Bakawan. 2 (2) : 3.

Natarajan, A. V. (1983). Possibilities of Brachiskwater Paddy-Cum-Fish Farming in Coastal Saline Soils, *J. Indian Soc. Coastal Agricl. Res. 1* (1) : 27-30.

Navalkar, B.S. (1940). Studies in the ecology of mangroves. 2. Physical factors of the Mangrove Soil. *J. Univ. Bombay,* N.S. *9* : 78-92.

Navalkar, B.S. (1951). Succession of the Mangrove Vegetation of Bombay and Salsette Islands. *J. Bombay Nat. Hist. Soc. 50* : 157-160.

Navalkar, B.S. & F.R. Bharucha (1949). Studies in the ecology of mangroves : Chemical factors of mangrove soils. *J. Univ. Bombay, 18:* 17-35.

Nayak, S.R. & B. Sahani (1984). Coastal geomorphology of the Gulf of Khambhat (S.S. Mehr & N.M. Vashi eds.). *In the Proceeding of Quaternary Episodes in India,* Dept. of Geology, M.S. University of Baroda, Baroda, pp. 87-96.

Nayak, S.R. & B. Sahani (1985). Coastal Morphology : a case study in the Gulf of Khambhat. *Inter. J. Remote Sensing, 6* (3 & 4) : 559-568.

Nayak, S.R., M.C. Gupta, H.B. Chauhan, M.G. Shaikh, Anjali Pandeya & R.S. Rao (1986). Application of Landsat data for coastal zone monitoring : a case study of a part of the West Coast of India. *In the Proc. of the Regional Seminar on the Application of Remote Sensing Techniques to Coastal Zone Management*, BSR & RSO & UNDP / ESCAP, Reg. Remote Sensi. Progr. Dhaka, Bangladesh, pp. 320-327.

Nayak, S.R., A. Bahuguna, M.G. Shaikh & M.C. Gupta (1989). Coastal environmental studies of the Kavaratti and the Agatti coral islands of the Lakshadweep Archipelago using remote sensing data. *In the Proc. Nat. Symp. on Engin. Application of Remote Sensing and Recent Adv.* ISRS, Indore, pp. 184-188.

Nayak, S.R., Anjali Pandeya & J.P. Aggarwal (1989). Monitoring of Ecological Changes in the Marine National Park using satellite data. Scientific Note. Space Applications Centre, Ahmedabad. IRS-UP/SAC/MCE/SN/24/89. 20 p.

Nayak, S.R., Anjali Pandeya, M.C. Gupta, C.R. Trivedi, K.N. Prasad & S.A. Kadri (1989). Application of Satellite data for monitoring degradation of tidal wetlands of the Gulf of Kachchh, Western India. *Acta. Astron. Autica. 20* : 171-178.

Newherry, D.M.C. & M.G. Hill (1981). Numerical classification of "Mixed Scrub" vegetation on Aldabra Atoll. *Atoll. Res. Bull.* No. 246.

Newton, C.M. (1953). Marine Algae. *John Murry Expedition Reports, 9* (5) : 395-420.

Nishihira, M. & M. Urasaki (1976). Production, settlement and mortality of seedlings of a mangrove, *Kandelia candel* (L.) Druce in Okinawa. *Abstracts on Symposia and contributed Papers, International Symposium on the Ecology and Management of Some Tropical Shallow Water Communities,* Jakarta, July, 1976. 15 p.

Nixon, S.W., B.N. Furnas, V. Lee, N. Marshall, J.E. Ong, C.H. Wong, W.K. Gong & A. Sasekumar (1984). The role of mangroves in the carbon and nutrient dynamics of Malaysian estuaries, pp. 534-544. *In* : E. Soepadmo, A.N. Rao & D.J. Macintosh (eds.). *Proc. Asian Symp. Mangr. Environ. Res. Manage.* 25-29 August, 1980. Univ. Malaya, Kualalumpur.

Noakes, D.S.P. (1951). Notes on the silviculture of the mangrove forests of Matany. Perak. *Malay. Forest. 14* : 183-196.

Noakes, D.S.P. (1955). Methods of increasing growth and obtaining natural regeneration of the mangrove type in Malaya. *Malay. For. 8* : 23-30.

Northcote, K.H., G.D. Hubble, R.F. Isbell, C.H. Thompson & E. Bettenay (1975). A description of Australian soils. C.S.I.R.O., Australia.

ODA (1985). A forest inventory of the Bangladesh, Bangladesh (Main Report), Land Resources Development Centre, UK Overseas Development Administration. 196 p.

Odum, E.P. (1971). *Fundamentals of Ecology.* Philadelphia : W.B. Saunders.

Odum, W.E. & E.J. Heald (1972). Trophic Analysis of an estuarine mangrove community. *Bull. Mar. Sci.* 22 : 671-738.

Odum, W.E. & E.J. Heald (1975). Mangrove forests and aquatic productivity. pp. 129-136. *In* : P.D. Hasler (ed.). Coupling of Land and Water Systems. Berlin, Springer–Verlang.

Oertel, G. (1988). A geology do distritto de Goa *Common Serv. Geol.* Portugal 40 Lisbon.

Oliver, R.C.D. (1979). Wildlife Conservation and Management in Bangladesh. FAO/Forest Research Institute, Chittagong, Bangladesh, UNDP/FAO.

Oliver, J. (1982). The geographic and environmental aspects of mangrove communities : Climate, pp. 19-30. *In* : B.F. Clough (ed.). *Mangrove Ecosystem of Australia. Austr. Inst. Mar. Sci.* with A.N.U. Press.

Ong, J.E. (1978). The Malaysian Mangrove Environment. *Paper presented at the Regional Seminar of Human Use of the Mangrove Env. and Management Implication*, Bangladesh.

Ong, J.E., W.K. Gong & C.H. Gong (1980). Ecological Survey of the Sunger Merbok Estuarine Mangrove Ecosystem. *Sch. Bio. Sci.*, Univ. Sains, Malaysia Publ.

Ong, J.E., W.K. Gong & C.W. Wong (1981). Ecological monitoring of the Sungei Merbook estuarine mangrove ecosystem. *Sch. Biol. Sci.*, Univ. Sains, Malaysia Publ.

Ong, J.E. (1982). Mangrove and aquaculture in Malaysia. *Ambio, 11* (5) : 252-257.

Ong, J.E., W.K. Gong, C.H. Wong & G. Dhanaraj (1984). Contribution to aquatic productivity in a managed mangrove ecosystem in Malaysia. pp. 209-215. *In* : E. Soepadmo *et al.* (eds.), *Proc. Asian Symp. on Mangr. Environ. Res. Manage.* 25-29 August, 1980. Univ. Malaya, Kualalumpur, Malaysia.

Onuf, C.P., J.M. Teal & I. Valiela (1977). Interactions of nutrients, plant growth and herbivory in a mangrove ecosystem. *Ecology. 58* : 514-526.

Pahm, M.A. & R.B. Aspiras (1979). Nitrogen transformation in mangrove swamps. BIOTROP Spec. Publ *10* : 139–152.

Paigmans, K. (1975). Explanatory notes to the vegetation map of Papua New Guinea. CSIRO Land Series No. 35, CSIRO, Melbourne.

Paijmans, K. (1976). *New Guinea Vegetation.* Canberra : *Austr. Nat. Univ.* Press.

Pal, U.C., K.R. Naskar & S.C. Santra (1988). A check list of algal flora on Sundarban delta of West Bengal, India. *Phykos. 27* : 48-53.

Pandeya, Anjali, S.R. Nayak, C.R. Trivedi, S.A. Kadri & K.N. Prasad (1987). Wetland mapping of a part 'of Gulf of Kachchh using landsat data. In Remote Sensing Applications in Coastal Studies. Scientific Note. Space Applications Centre, Ahmedabad, IRS-UP/SAC/MCE/SN/08/87. pp. 1-12.

Pannier, F. (1962). Estudio fisiologico sobre la viviparia de *Rhizophora mangle* L. *Acta. Cient. Venezolana. 13* (6) : 184-197.

Pannier, F. & R.F. Pannier (1975). Physiology of vivipary in *Rhizophora mangle* L. *Proc. Int. Symp. Biol. Mgt. Mangroves* 2 : 632-639.

Panshin, A.J. (1932). An anatomical study of the woods of the Philippines mangrove swamps. *Philipp. J. Sci. 48* : 143-208.

Parham, J.W. (1972). *Plants of the Fiji Islands.* Suva, Fiji, Government Printer.

Parkinson, C.E. (1923). Forest Flora of the Andaman Islands, Simla, India.

Parkinson, C.E. (1934). The Indian species of *Xylocarpus. Ind. For 60* : 136-138.

Parulekar, A.H., S.N. Dwivedi & V.K. Dhargalkar (1973). Ecology of clam beds in Mandovi, Cumbarjua Canal and Zuari Estuarine System of Goa. *Ind. J. Marine Sci.* 2 (2) : 122-126.

Parulekar, A.H., V.K. Dhargalkar & S.Y.S. Singbal (1980). *Indian J. Marine Sci. 9* : 189-200.

Parulekar, A.H. & X.N. Verlencar (1984). Status and prospects of fish farming in Goa. *Proc. Symp. Coastal Aquaculture. 3* : 732-737.

Patil, R.P. (1961) A note on the vegetation at Sazinakhali in the Sundarbans. *Indian For. 87* : 481-483

Patil, R.P. (1961). Cultivation of mangrove seedlings in pots at Allahabad, U.P. *Sci. & Cult. 30* : 43-44.

Pennington, J.D. & B.T. Styles (1975). A. generic monograph of the Meliaceae. *Blumea.* 22 (3) : 481.

Percival, M. (1974). Floral ecology of coastal scrub in South East Jamaica. *Biotropica, 6* : 104-129.

Percival, M. & J.S. Womersley (1975). Floristics and ecology of the mangrove vegetation of Papua New Guinea. *Botany Bulletin* No. 8. Department of Forests, Papua New Guinea. p. 94.

Parnetta, J.C. & S. Burgin (1983). The status and ecology of Crocodiles. *In* : T. Petr (ed.). *The Purari*. The Hague : Junk.

Philippines Bureau of Forest Development (1984). Forestry Statistics Year Book Quezon City, Philippines.

Pickup, G. & V.H. Chewings (1983). The hydrology of the Purari and its environmental implications. *In*: T. Petr (ed.). *The Purari*. The Hague. Junk.

Pillai, G. (1985). Mangrove of Fiji : Their uses and management. *Paper presented in the Res. Dev. Seminar Mangr. Ecosys.* Held in May 1985 at Australia Institute of Marine Science, Cape Ferguson.

Pillay, T. V. R. (1954). The Ecology of Brackishwater Bheris with Special Reference to the fish culture practices and Biotic Interaction. *Proc. Nat. Inst. Sci. India, 20* : 899-927.

Pinto, M.L. (1978). Some Ecological Principles of Selected Mangrove Islets in Negombo Lagoon, Sri Lanka. *M. Phil Thesis,* Univ. of Colombo, Sri Lanka.

Pinto, M.L. (1982). Distribution and zonatin of mangroves in the northern part of the Negombo Lagoon. *J. Nat. Sci. Coun. 10* (2) : 245-255.

Polprasert, C. (1979). Final Report : Heavy Metals, DDT and PCBs in the Upper Gulf of Thailand, Phase–I. National Environmental Board, Bangkok, Thailand.

Popp, M. (1984a). Chemical composition of Australian Mangroves. I. Inorganic ions and organic acids. *Z. Pflanzenphysiol. 113* : 395-409.

Popp, M. (1984b). Chemical Composition of Australian Mangroves. II. Low molecular weight carbohydrates. *Z. Pflanzenphysiol. 113* : 411-421.

Pool, D.J., S.C. Snedakar & A.E. Lugo (1977). Structure of forests in Florida, Puerto Rico. Mexico and Costa Rica. *Biotropica, 9* (3) : 195-212.

Poriatuk, S. & S. Ulijaszek (1981). Molluscs in the Subsistence Diet Some Purari Delta People. Purari Environment Studies. Papu New Guinea.

Prain, D. (1903a). Flora of Sundribuns. *Rec. Bot. Surv. India*, 2 : 231-370.

Prain, D. (1903b). Bengal Plants. *2 Vols.* (Rep. Ed. 1963), B.S.I., Calcutta.

Prain, D. (1903c). Notes on Sundaribans Plants, *Proc. Asiat. Soc. Bengal, 72* : 107-108.

Prian, D. (1905). The vegetation of the District of Hugly, Howrah and 24-Pergunnahs, *Rec. Bot. Surv. India*, 3 : 143-329.

Pramanik, P. & G. Sinha (1972). Development of Coastal areas in Western Sunderbans. *Proc. of the Seminar on Geomorphlogy, Geohydrology and Geotechniques of the Lower Ganga Basin*, C-7 to C-14.

Puri, G.S. (1960). *Indian Forest Ecology, 2 Vols.*, Oxford Book and Stationery Co., New Delhi, India.

Puri, G.S. & S.K. Jain (1959). The mangrove vegetation of western India. *Proc. Mangr. Symp. Calcutta, 1957* : 51-57.

Quimba, L.L. (1975). Key to the species of Mangrove Swamps. College of Forestry, University of the Philippines, College Laguna.

Qureshi, J.M. (1959). Botanical Silvicultural features of mangrove forest of Bombay State. *Proc. Mangr. Symp.* Calcutta. *1957* : 20-26.

Qureshi, Tahir (1990). Experimental Plantation for Rehabilitation on Mangrove Forest in Pakistan. Mangrove Ecosystems. Occasional Papers No. 4. UNESCO & UNDP. (Ed.) C.D. Field.

Rabinowitz, D. (1978), Dispersal properties of mangrove propagules. *Biotropica. 10* : 47-57.

Radok, R. (1976). *Australia's Coast.* Rigby, Adelaide.

Rains, D.W. & E. Epstein (1967). Preferential absorption of potassium by leaf tissue of the mangrove *Avicennia marina* : an aspect of halophytic competence in copying with salt. *Aust. J. Biol. Sci.* 20 : 847-857.

Raj, U. & J. Seeto (1982). Report of an Investigation of Sweni Bay and the Coastal Environment for an Environmental Impact Statement. *Inst. Mar. Resources,* Univ. of South Pacific, Suva, Fiji.

Raj. U., S. Vodonivalu, J. Seeto, H. Hirta & S. Iwakiri (1984). Flora and Fauna of mangroves in Rewa Delta, Fiji, pp. 149-155. *In* : Prompt Report of the First Scientific Survey of the South Pacific. Research Centre for the South Pacific, Kogoshima Univ., Japan.

Rajagopalan, V.R. (1952). Ecological Adaptation of the Mangroe Vegetative at Pichavaram, *J. Annamalai Univ.* **20** : 56-59. Annamalai Nagar, India.

Ramamurthy, V.D. (1984). The effect of agricultural pollutants of fish products in the mangrove swamps of south west coast of India. p. 618-623. *In* : E. Soepadmo, A.N. Rao & D.J. Macintosh (eds.). *Proc. Asian Symp. Mangr. Env. Res. Mange.*, Univ. Malaya, Kualalumpur, Malaysia.

Ramchandran, K.K., C.N. Mohan, G. Balasubramonian, Kurien Johson & Jessy Thomes (1986). The Mangrove Ecosystem in Kerala. Centre of Earth Sci. Studies, Trivandrum. 38 p.

Rao, A.N. (1971). Morphology and Morphogenesis of foilar sclereids in Aegiceras corniculation Isr. J. Bot. **20** : 124-132.

Rao, A.N. (1986). Mangrove Ecosystems of Asia and the Pacific. *In* : Technical Report of the UNDP/UNESCO Research and Training Pilot Programme on Mangrove Ecosystems in Asia and the Pacific (RAS/79/002). Nov. 1986. Chapter I. pp. 1-48.

Rao, A.N. & M. Sharma (1968). The terminal sclereids and tracheids of *Bruguiera gymnorrhiza* Blume and the caulina sclerieds of *Ceriops roxburgiana* Arn. *Proc. Nat. Inst. Sci. India,* **34** (B) : 6 : 267-275.

Rao, A.N. & L.C. Wan (1969). A new record of foliar sclerieds in *Scyphiphora hydrophyllacea. Curr. Sci.* **38** 594-595.

Rao, A.N. & H. Tan (1984). Leaf structure and its ecological significance in certain mangrove plants. pp. 183-194. *In* : *Proceeding Asian Symp. Mangr. Res. Management.* University Malaya, Kualalumpur, Malaysia.

Rao, G.C. (1980). On the Zoogeography of the intertidal meiofauna of the Andaman and Nicobar Islands, *Indian Ocean. Rec. Zool. Surv. India* **77** : 153-178.

Rao, G.C. (1986). Meiofauna of the mangrove sediments in South Andamans. *J. Andaman Sci. Assoc.* **2** (2) : 23-32.

Rao, G.C. (1988). Meiofauna of the intertidal algae on Great Nicobar Islands. Ibid. **4** (1) : 18-31.

Rao, G.C. & Ganapati (1968). Intertidal fauna in habiting the beach of sands of Waltair Cost. *Proc. Natn. Sci. India.* **34** (2) : 82-125.

Rao, G.C. & A. Misra (1983). Studies on the meiofauna of Sagar Island. *Proc. Indian Acad. Sci.* (Anim. Sci.), **92** (1) 73-85.

Rao, G.C. & A. Misra (In Press). Macrofauna and microfauna of Digha Coast, West Bengal. *Rec. Zool. Survey India.*

Rao, K.P. (1958). Oxygen consumption as a function of size and salinity in *Metapenaeus monoceros.* Fab. from Marine and Brackishwater environment. *J. Expt. Biol.* **35** (2) : 307-313.

Rao, K.V.G.K. & R.S. Pandy (1982). Problem of Sub-surface draiange in coastal saline soils of Sundarbans. *J. Agric. Engg.* **4** : 85-89.

Rao, P.R. (1957). Utilization of mangrove bark extracts in the Tanning Industry. *Indian Forester,* **83** : 565-568.

Rao, P.S.N., V.D. Cahuhan & K.S Rao (1982). Effect of Nacl, pH and Carbon on the growth of brown algae. *Dunaliella* sp. *Ind. J. Mar. Sci.*, II : 262-263.

Rao, R.S. (1959). Observations on the mangrove vegetation of the Godawari Estuary. *Proc. Mangrove Symp.* Calcutta. **1957** : 36-44.

Rao, R.S., K. Lakshminarayana & P. Venkanna (1985). Mangrove Forest of the Godavari and the Krishna Estuaries of Andhra Pradesh. The Urgent Need For Their Development and Conservation. *The Mangroves : Proc. Nat. Symp. Biol. Util. Cons. Mangroves.* Nov. 1985 : 338-340.

Rao, T.A. (1957). Comparative morphology and ontogeny of foliar sclereids in seed plants. *Proceedings of the National Institue of Science,* India, Series B (Biological Science). **34** : 267-275.

Rao, T.A. (1971). Distributional pattern of the Indian Maritime Strand Flora. *Bull. Bot. Surv. India,* 192-202.

Rao, T. Ananda (1985). The veilet syndrome of certain Mangroves. *The Mangroves : Proc. Nat. Symp. Biol. Util. Cons. Mangroves.* Nov. 1985 : 49-59.

Rao, T.A., K.R. Agarwal & A.K. Mukherjee (1963a). Biological account of the vegetation of Rameswaram Islands. *Bull. Bot. Surv. India.* **5** : 319-323.

Rao, T.A., K.R. Aggarwal & A.K. Mukherjee (1963b). Ecological studies on the soil and vegetation of Krusadi Group of Islands in Gulf of Mannar, *Bull. Bot. Surv. Ind.* **5** : 141-148.

Rao, T.A., K.R. Aggarwal & A.K. Mukherjee (1963c). Biological Account of the vegetation of Rameswaram Islands, *Bull. Bot. Surv. Ind.* **5** : 301-323.

Rao, T.A. & K.R. Aggarwal (1964). Ecological Studies of Saurashtra Coast and Neighbouring Islands, *Bull. Bot. Surv. India.* **6** : 173-183.

Rao, T.A., K.R. Aggarwal & A.K. Mukherjee (1966). Ecological Studies of Saurashtra Coast and Neighbouring Islands, *Bull. Bot. Surv. India.* **8** : 60-67.

Rao, T.A., L.K. Banerjee & A.K. Mukherjee (1967). Some Plants Record from West Bengal, *Sci. & Cult.*, **33** : 125.

Rao, T.A. & P.G. Shanware (1967). Ecological studies of Saurashtra Coast and Neighbouring Islands. *Bull. Bot. Surv.* India, **9** (1 & 4) : 240-248.

Rao, T.A. & A.K. Mukherjee (1971). Ecological aspects along shores of Burabalanga tidal estuary, Orissa State, *Proc. Ind. Acad. Sci.* **76** (B) : 201-206.

Rao, T.A., A.R.K. Sastry & P.G. Shanware (1972). Analysis of the basic patterns of an estuarine showing the vicinity of Coringar Bay, Andhra Pradesh. *Proc. India Acad. Sci.* **75** (B) : 40-50.

Rao, T.A. & A.K. Mukherjee (1972). Ecological aspects along the shores of Buralalanga tidal estuary. Balasore District. *Proc. Indian Acad. Sci.* (B) : **76** (5) : 201-206.

Rao, T.A. & A.R.K. Sastry (1972). An ecological approach towards classification of coastal vegetation in India. I. Strand Vegetation. *Indian Forester.* **98** : 594-607.

Rao, T.A., A.K. Mukherjee & L.K. Banerjee (1973). *Acrostichum aureum* L. truely a mangrove fern ? *Curr. Sci.* **42** (15) : 546-547.

Rao, T.A. & A.R.K. Sastry (1974). An ecological approach towards classification of coastal vegetation of India II. Estuarine border vegetation, *Indian For.*, **100** : 438-452.

Rao, T.A. & A.R.K. Sastry (1974). Classifictaion and distribution of Indian Mangroves. *Indian Forester*, **100** : 438-452.

Rao, T.A., A.R. Sastry & P. Basu (1975). A contribution to the coastal Flora and vegetation of Tamil Nadu. *Ind. For.* **101** (B) : 460-475.

Rao, T.A. & A.R.K. Sastry (1976). An outline of the Coastal Vegetation of India. *Bull Bot. Surv. India.* **16** : 101–115.

Rao, T.A., P.V. Suresh & A.N. Sherieff (1986). Multiple viviparity in few taxa of mangrove. *Curr. Sci.* **55** : 295-261.

Rao, T.A. & Swapna Chakraborti (1987). Distributional resume of coastal floristic elements in the Andaman and Nicobar Islands. *Current Sci.* **56** (20) : 1045-1051.

Rao, T.A., P.V. Suresh & A.N. Sherieff (1987). Little known attributes of some mangrove plants of the Karnataka Coast. *Curr. Sci.* **56** (14) : 725-727.

Rao, R.S., S. Nayak, A.N. Sherieff, P.V. Suresh & T.A. Rao (1989). Tidal wetland mapping of the Karnataka Coast using Landsat TM data. Scientific Note. Space Application Centre. IRS-UP/SAC/MCE/SN/21/89.

Rao, V.S. (1950). Afforestation of wastelands as practiced in West Bengal in combating the soil erosion problem and meeting the scarcity of fuel and fodder. *Indian For.* **76** (8) : 323-326.

Rao, V.S. (1953). Floral Anatomy of some bicarpellatal 1. Acanthaceae. *J. Univ. Bombay*, **21** : 7-34.

Rao, V.S. (1959). Mangrove forests and the problem of reclaiming saline banks. *Proc. Mangr. Symp.* Calcutta (1957). pp. 58-65.

Raunkaier, C. (1934). *The life-forms of plants and statistical plant geography*, Oxford. 632 p.

Rau, M. (1984). Traditional uses of mangroves in the Central Province. *In* : Workshop on Human Population, Mangrove Resource, Human Induced Stress, Human Health. Oct. 2-7, Bagor, Indonesia.

Razani, Ujang (1982). The role of mangrove forests in the management of coastal zone *Proc. Workshop Ecol. Basis* Rational Resource Utilization Humid Trop. S.E. Asia, Univ. Pertanian Malaysia.

Reinders–Gouwentak, C.A. (1953). Sonneratiaceae and other mangrove swamp families; anatomical structure and water relations. *Flora Malesiana* Ser. I. **4** (4) : 513-515.

Richmond, T. de A. & J.M. Ackerman (1975). Flora and Fauna of mangrove formations in Viti Levu and Vanua Levu, Fiji, pp. 147-152. *In* : G.E. Walsh, S.C. Snedaker & M.J. Teas (eds.). *Proc Intern. Symp. Biol. Manage. Mangr.*, Univ. Florida, USA.

Rintz, R.E. (1980). *Blumea*. **26** (1) : 72.

Rodelli, M. R., J.N. Gearing, P.J. Gearing, N. Marshall & A. Sasekumar (1984). Stable isotope ratio as a tracer of mangrove carbon in Malaysian ecosystem. *Oceanologia* (Berlin). **61** : 326-333.

Rodin, L.E., N.I. Barilevich & N.N. Rozor (1975). Productivity of the world's main ecosystem. *Nat. Acad. Sci.*, Washington.

Ronquillo, I.A. & E.G. Llana (1979). Fisheries resources of the mangrove swamp area. NSDB Tech. *J.* ***1979***: 17-23.

Roth, I. (1965). Histogenese der Lentizellen am Hypocotyl von *Rhizophora mangle* L. *Ost. Bot. Z.* **112** : 640-653.

Roxburgh, W. (1814). Hortus Bengalensis or a Catalogue of Plants Growing in the Honourable East India Company Botanical Garden at Calcutta, Serampore Mission Press, India, p. 105.

Roxburgh, W. (1832). *Flora Indica. 3 Vols.*

Rumphius, G.E. (1741-1755). *Herbarium Amboinense*. Amsterdam : Amstebaedami.

Sabar, F., M. Djajasamita & A. Budiman (1979). Susunan danpenyebaran moluska dan Krustasea pada beberapa hutan rawa payau : Suatu studi pendahuluan, pp. 120-125. *In* : A. Soemodihardjo *et al.* (eds.). *Prosiding Seminar Ekosistem Hutan Mangrove*, Jakarta.

Saegiarto, A. (1979). The Indonesian marine environment. Its problems and management. *Ber. Ilmu. Peng. Teknol.* **23** (1) : 9-21.

Saenger, P. (1979). Mangrove ecosystems in Australia. *Proceeding of the Australian National Mangrove Workshop.* Australian Institute of Marine Science, pp. 153-191.

Saenger, P. (1982). Morphological, Anatomical and Reproductive Adaptations of Australian Mangroves. pp. 153-191. *In* : B.F. Clough (ed.). *Mangrove Ecosystem Australia : Structure, Function and Management.* Australian Institute of Marine Science.

Saenger, P. (1986). Bangladesh Mangrove Afforestation Project. Report prepared for Govt. of Bangladesh, Shedden Pacific Pty. Limited, Melbourne, Australia, 63 p.

Saenger, P., M.M. Specht, R.L. Specht & V.J. Chapman (1977). Mangal and coastal salt marsh communities in Australasia. Chap. 15. *In : Ecosystems of the World*, Vol. I. Wet Coastal Ecosystems. Amsterdam : Elsevier Scientific.

Saenger, P., E.J Hegerl & J.D.S. Davie (1983). *Global Status of Mangrove Ecosystems.* IUCN Commission on Ecology. Paper No. 3.

Saha, G.N., S.C. Thakurta, G.C. Laha, A.C. Nandi, K.R. Naskar, N.M. Mazumder, S.C. Karmakar & T.K. Chakraborty (1986). Ecology and Fishery Management of the Brackishwater Bheries in West Bengal. *Bull No.* 46, pp. 1–23 (CIFRI) Barrackpore, W.B.

Sahah, J.J. & K.P. Sunderaj (1965). Stipular Sclereids in *Rhizophora mucronata*. L. *Curr. Sci.* **34** : 1255-1256.

Sahani, K.C. (1953). Botanical Exploration in the Great Nicobar, *Ind. For.*, **79** : 3-16.

Sahani, K.C. (1959). Mangrove Forests in the Andaman & Nicobar Islands. *Proc. Mangr. Symp.*, Calcutta, India, pp. 114-123.

Salvoza, F.M. (1936). *Rhizophora, Natur, Appl. Sci. Bull.* (Univ. Philipp.) **5** : 179-237.

Santapau, H. & A.N. Henry (1973). *A dictionary of the flowering Plants in India.* CSIR, New Delhi.

Santisuk, T. (1983). Taxonomy of the terrestrial trees and shrubs in the mangrove formation in Thailand. *Paper presented in the First UNDP/UNESCO Regional Training Course on Introduction to Mangrove Ecosystem*, held from 2-30 March, 1983 at National Research Council, Bangkok, Thailand.

Sanyal, P. & A.R. Bal (1986). Some observations on abnormal adaptations of mangrove in Indian Sundarbans. *J. Ind. Soc. Coastal agric. Res.* **4** (1) : 9-16.

Sanyal, P., L.K. Banerjee & M.K. Chowdhury (1984). Dancing Mangals of Indian Sundarbans. *J. Ind. Soc. Coastal agric. Res.* **2** (1) : 10-16.

Sarkar, A.K., D.C. Ghosh & K.R. Naskar (1993). Phenological observations of the different mangrove flora of Sundarbns in India with special reference to their salinity tolerance. *Abs. 160. Asia Pacific Symp. on Mangrove Ecosystm*, Research Centre & Dept. of Biology, The Hong Kong Univ. of Science & Technology, Hong Kong, 1-3 Sept., 1993.

Sarkar, A.K., R.N. Mandal, D.C. Ghosh, Amoy Kumar Barui & K.R. Naskar (1994). Studies on the mangrove from Bakkhali forest Range of the Western part of Sundarbans, West Bengal. OP. 2.4.A.6 ; *National Seminar on Conservation & Sustainable Development of Coastal Resources*, Berhampur, 14-17 Dec., 1994.

Sarkar, A.K., R.N. Mandal, D.C. Ghosh, D. Saha, K.R. Naskar & A.K. Ghosh (1994). Phenological Observtions of 52 mangroves and associated flora of Back Mangals in the intertidal mangrove forest in West Bengal. *Abst. 38*, pp. 10, *8th Biennial Botanical Conference, Bangladesh*, Dhaka : 12-13 Dec.

Sarkar, A.K. & K.R. Naskar (1994). Some Steps For Undertaking Mangrove Afforestations. *Abs. 99*. West Bengal State Science Congress, West Bengal State Council of Science & Technology, Dept. of Science & Technology and Non-conventional energy source, Govt. of West Bengal, Calcutta.

Sasekumar, A. (1970). *Aspects of the Ecology of a Malayan Mangrove Fauna*, M.S. Thesis. University Malaya, Malaysia (Unpub.).

Sasekumar, A. (1974). Distribution of macro-fauna on a Malayan Mangrove Shore, *J. Anim. Ecol.* **43** : 51-69.

Sasekumar, A. (1984). Mangrove Secondary Productivity. *Proc. Workshop Mangr. Prod.* Univ. Sains, Malaysia.

Sasekumar, A. & J.J. Loi (1983). Litter production in three mangrove forest zones in the Malay Peninsula. *Aqu. Bot.* **17** : 238-290.

Sasekumar, A., T.L. Ong & K.L. Thong (1984). Predation of mangrove fauna by marine fishes, pp. 378-384. *In* : E. Soepadmo, A.N. Rao & D.J. Macintosh (eds.). *Proc. Asian Symp. Mangr. Environ. Res. Manage.*, Univ. Malaya, Kualalumpur, Malaysia.

Sattar, M.A. (1977). Sundri mortality in the Sundarbans. *Proc. of the First Bangladesh National Conference on Forestry*. pp. 84-100. Dhaka, Bangladesh.

Schlish, W. (1875). Remarks on the Sundarbans, *Ind. For.* **1** : 6-11.

Scholander, P.F. (1968). How mangroves desalinate water. *Physiol. Plant*, **21** : 251-261.

Scholander, P.F., L. Van Dam & S.I. Scholander (1955). Gas exchange in the roots of mangroves. *Am. J. Bot.* **42** : 92-98.

Scholander, P.F., H.T. Hammel, E.A. Hemmingsen & W. Garey (1962). Salt balance in mangroves, *Pl. Physiol.* **37** : 722-729.

Scimper, A. F. W. (1891). *Die Indo-Malayische, Bot. mitt. Trop. Jena* **3** : 204.

Schuster, W. H. (1952). *Fish Culture in Salt Water ond in Java;* Indo-Pacific Fisheries Council, Bangkok.

Seidensticker, J. & A. Hai (1985). *The Sundarbans wildlife management plan.* WWF, Switzerland. 120 pp.

Semeniuk, V. (1983). Mangrove distribution in north western Australia in relationship to regional and local freshwater seepage. *Vegetation*, **53** : 11-31.

Sen, P.K. & Manju Banerjee (1984). On the occurrence of Reptilian Remnants in the Peat Bed of Barrackpore, 24-Parganas, West Bengal, *Paleobot. Soc. Ind.* **29** : 47-51.

Sen, R., A. Sau, K.R. Naskar & A. Bhattacharya (1980). Studies on Nodulation on some useful leguminous plants. *Bull. Bot. Surv. India*, **22** (1-4) : 166-172.

Shafi, M. (1982a). Adverse effects of Farakka on the forests of south-west region on Bangladesh (Sundarbans). *Proceedings of Bangladesh Second National Conference on Forestry*, Dhaka, pp. 30-57.

Shafi, M. (1982b). Artificial regeneration of Keora in Bangladesh. *Proceeding Bangladesh Second National Conference on Forestry*, pp. 136-139.

Shah, G.L. (1962). The vegetation along the Sea Shores in Salsette Islands, Bombay, *Bull. Bot. Surv. India*, **4** : 239-240.

Shah, J.J. & K.P. Sundaraj (1965). Stipular sclereids in *Rhizophora mucronata* L. *Curr. Sci.* **34** : 1255-1256.

Shield, L.M. (1950). Leaf xeromorphy as related to physiological and structural influences. *Botanical Review,* **16** : 399-447.

Siddiqi, N.A. (1985). *Detailed Work Plan for "Phenological Studies of major mangrove species"*. Mangrove Silviculture Division, Bangladesh Forest Research Institute, 10 p.

Siddiqui, N.A. (1989). Seedling regeneration in relation to time and degree of salinity in the Sundarbans Mangrove Forests. *Bano Biggyan Patrika.* **18** : 9–17.

Siddiqi, N.A. & M.R. Islam (1988). Studies on fruit size, seed production and viability of seeds of Keora (*Sonneratia apetala* Buch. -Ham). *Bano Biggyan Patrika.* **17** : 15-19.

Siddiqui, N.A., M.R. Islam & M.A.S. Khan (1989). Effect of salinity on germination success in Keora (*Sonneratia apetala* Buch.-Ham.) seeds. *Bano Biggyan Patrika.* **18** : 57-62.

Siddiqi, N.A., M. Shahidullah & M.A.H. Shahjalal (1991). Studies on the size viability and germination of seeds of Sundri (*Heritiera fomes* Buch.–Ham.). *Indian J. Forestry.* **14** (2) : 119-124.

Sidhu, S.S. (1960). Studies on Mangroves. *National Acad. Sci.*, Allahabad, India. pp. 111-112.

Sidhu, S.S. (1961). Vegetation of East Godavari. *Proc. 48th Indian Sci. Congr.* **3** : 358-359.

Sidhu, S.S. (1962). Chromosomal Studies on some mangrove species. *Indian Forester,* **88** : 585-592.

Sidhu, S.S. (1963a). Studies on Mangroves. *Proc. Indian Academy Sci.* **33** (8) : 129-136.

Sidhu, S.S. (1963b). Studies on the Mangroves of India. I. East Godavari Region. *Indian For.* **89** : 337-351.

Sidhu, S.S. (1968). Further studies on the Cytology of Mangrove Species in India. *Caryologia,* **21** : 353-357.

Sidhu, S.S. (1974). Structure of epidermis and stomatal apparatus of some mangrove species. *Paper presented at the International Symposium on Biology and Management of Mangroves,* held at Honolulu, Hawaii.

Sidhu, S.S. (1975a). Structure of epidermis and stomatal apparatus of some mangrove species. *In* : G.E. Walsh, S.C. Snedaker & H.J. Teas (eds.). *Proceedings of the International Symposium on Biology and Management of Mangrove.* Inst. of Food & Agri. Sci., Univ. of Florida, Gainesville, Florida, *Vol. II.* pp. 569-578.

Sidhu, S.S. (1975b). Culture and growth of some mangrove species. *In* : G.E. Walsh, S.C. Snedaker & H.J. Teas (eds.). *Proceedings of the International Symposium on Biology and Management of Mangroves.* Inst. of Food & Agricultural Sciences, Univ. of Florida, Gainesville, Florida, *Vol. I,* pp. 394-401.

Simpson, D.A. (1955). Fish out of Water. *Aquar. J.* **26** : 181-183.

Smith, A.C. (1981). *Flora Vitiensis Nova, Vol. 2.* Lawai, Kauai, Hawaii : Pacific Tropical Botanical Garden.

Snedakar, S.C. (1978). Mangroves : Their value and perpetuation. *Nat. & Res.* **14** : 6-13.

Snedakar, S.C. (1982). Mangrove Species Zonation : Why ? Chap. 1. *In : Contribution to the ecology of halophytes.* (Eds.) D.N. Sen & K.S. Rajpurohit. Tasks for Vegetation Science, *Vol. 2.* The Hague: Junk.

Snedakar, S.C. (1984). The mangrove of Asia and Oceania. Status and Research Planning, pp. 5-15. *In* : E. Soepadmo, A.N. Rao & D.J. Macintosh (eds.) *Proc. Asian Symp. Mangr. Environ. Res. Manage.* Univ. Nataya, Kualalumpur, Malaysia.

Soemodihardjo, S. (1978). Utilization and Management of Mangrove Resources in Indonesia. *Paper presented at the UNESCO Regional Seminar on Human Uses of the Mangrove Environment and Management Implications.* 4-8 December, 1978, Dacca, Bangladesh.

Soemodihardjo, S. (1980). Laporan Ekspedisi Rumphius III. *Oceanologi di Indonesia.* **13** : 1-60.

Soemodihardjo. S. (1983). The impact of Transmigration on the Mangrove Ecosystem. *Paper presented in the Workshop on Productivity of the Mangrove Ecosystem Management Implication.* From 4-6 October, 1983, Penang, Malaysia.

Soemodihardjo, S. (1984). Mangrove Information system in Indonesia : Country Report. *Paper presented in the Asian Regional Workshop on Mangrove Information.* From 30 April–2 May, 1984. Manila, Philippines.

Soemodihardjo, S. (1987). Country Reports : Indonesia. *In* : R.M. Umali, *et. al.* (eds.). *Mangrove of Asia and the Pacific : Status and Management*. Technical Report of the UNDP/UNESCO Research and Training Pilot Programme on Mangrove Ecosystems in Asia and the Pacific (RAS/79/002). pp. 89-129.

Soemodihardjo, S. & W. Kastoro (1977). Notes on the *Terebralia Palustris* (Gastropoda) from the Coral Island in the Jakarta Bay area. *Mar. Ind. 18* : 131-148.

SPARRSO (1985). Mangrove Afforestation Project. *Album of land accretion and plantation map and statement of areas*. Dhaka. 31 p.

Specht, R.L. (1981). Biogeography of halophytic angiosperms (Salt-marsh, mangroves and sea-grass). *In* : A. Keast (ed.). *Ecological Biogeography of Australia*. Junk, The Hague, pp. 575-590.

Spicer, R.A. & D. MacNewberry (1979). The terrestrial vegetation of an Indian Ocean Coral Island. Willingili. Addu Atoll, Maldive Islands. I. Transect Analysis of Vegetation. *Atoll. Res. Bull.* No. 231.

Stace, C.A. (1966). The use of epidermal characters in phylogenetic considerations. *New Physiologist, 65* : 304-318.

Steinke, T. D. (1975). Some factors affecting dispersal and establishment of propagules of *Avicennia marina* (Forsk.) Vierh. *In* : G.E. Walsh, S.C. Snedaker & H.J. Teas (eds.). *Proceedings of the International Symposium on Biology and Management of Mangroves*. Inst. of Food & Agri. Sci., Univ. of Florida, Gainesville, Florida. *Vol. II*. pp. 402-414.

Steinke, T.D. & C.J. Ward (1973). *Ceriops tagal* (Perr.) C.B. Robinson at Kosi Bay. *Journal of South African Botany. 39* : 245-247.

Stern, W.L. & G.K. Voigt (1959). Effect of salt concentration on growth of red mangrove in culture. *Bot. Gaz. 12* (1) : 36-39.

Stocker, G.C. (1976). Report on Cyclone Damage to Natural Vegetation in the Darwin Area After Cyclone Tracy, 25 December, 1974. Forestry & Timber Bureau, Canberra, *Leaflet No. 127*. A.G.P.S.

Studdart, D.R. & F.R. Fosberg (1972). South Indian Sand Cays. *Atoll. Research Bulletin*. No 161 : 25 p.

Studdart, D.R., G.W. Bryan & P.E. Gibbs (1973). Inland mangroves and water chemistry, Barbuda, West Indies. *J. Nat. Hist. 7* : 33-46.

Studholme, W.P. & W.R. Philipson (1966). A comparison of the cambium in two woods with included phloem. *Heimerliodendron brunonianum* and *Avicennia resinifera* N. Z. *J. Bot. 4* (4) : 355-365.

Sukardjo, S. (1979). Mangrove forest in Banyuasin, Musi River Estuary, South Sumatra, pp. 61-79. *In* : Ahmad Srivastava, Ismail Dhanarajan & Hamzah (eds.). *Proc. Symp. Mangr. Estuar. Veg. in S.E. Asia*, Selangor, Malaysia.

Sukardjo, S. (1981). *Mangrove di Indonesia*. Duta Rimba 49/VII Nov.

Sukardjo, S. (1982). Soil in the mangrove forest of the Cimanuk delta, West Java (Indonesia). pp. 191-201. *In* : A.Y. Kostermans & Sastroutomo (eds.). *Proc. Symp. Mangr. For. Ecosys. Prod. S.E. Asia*, Bogor, 20-22 April, 1982.

Sukardjo, S. & S. Akhmad (1982). The mangrove forests of Java Bali (Indonesia), pp. 113-126. *In* : A.Y. Kostermans & Sastroutomo (eds.). *Proc. Symp. Mangr. For. Eco. Syst. Prod. S.E. Asia*, Bogor, Indonesia.

Sunaryo, I. (1982). A floristic study of mangrove in Segara Anakan, pp. 132-145. *In* : C.E.F. Bird, A. Soegiarto & K. Soegiarto (eds.). *Proc. Workshop Coastal Resource Manage*, Cilacap Region. LIPI-UNU.

Sunaryo, I. & C. Anwar (1984). Komposisi vegetasi hutan mangrove Pulau Guntung (Muara Berau) dan Pulau Tibi (Muara Sesayap). Kalimantan Timum. *In* : A.B. Sutomo & A. Djamali (eds.). Penelitian Ekosistem Mangrove Kalimantan Timur (Unpublished Report).

Sussex, I. (1975). Growth and metabolism of the embryo and attached seedling of the viviparous mangrove *Rhizophora mangle*. *Am. J. Bot. 62* : 948-53.

Sutomo, A.B. (1984). Penelitian zooplankton di Perairan Cilacap 1981-1982, pp. 286-297. *In* : A. Soem-odihardjo (ed.). *Prosiding Seminar II Ekosistem Mangrove*, LIPI, Jakarta.

Suvapeepan, S., P. Sripayak & Vicheinvorakul (1979). Zooplankton in mangrove forest. *Paper presented in the Third National Seminar on Mangrove Ecosystems*. Held from 8-12 April 1979 at University of Prince of Songkhla Nakarin, Thailand.

Swain, T. (1965). *The tannins. In* : J. Bonner & J.E. Varner (eds.), *Plant Biochemistry*. Academic Press, New York, 1054 pp.

Swingle, W.T. (1943). *The botany of citrus and the wild relatives of its orange sub-family*. In *The Citrus Industry*, Vol. 1. (eds.) H.J. Webber & L.D. Batchelor. Berkeley : Los Angles. Division of Agricultural Sciences.

Tan, A.S. & H. Keng (1969). Comparative morphological studies of four *Avicennia* species in Singapore. *J. Sing. Nat. Acad. Sci. 1* (3) : 8-29.

Teas, H.J. (1983). *The Biology and Ecology of Mangroves*. Junk.

Teas, H.J. & F. Montgomery (1968). Ecology of red mangrove seedling establishment. *Bulletin, Association of South-eastern Biologists, 15* : 56-57.

Teas, H.J & J. Kelly (1975). Effects of herbicides on mangroves of South Vietnam and florida. *In* : G.E. Walsh, S.C. Snedaker & H.J. Teas (eds.). *Proceedings of the International Symposium on the Biology and Management of Mangroves*. Institute of Food and Agricultural Sciences, Univ. of Florida, Gainesville, Florida, *Vol. II*. pp. 719-728.

Tham, A.K. (1968). Prawn culture in Singapore. *Paper presented in the FAO World Scientific Conference in the Biology and Culture of Shrimps and Prawns at New Mexico*. Held from 12-14. June, 1967. *FAO Fish. Rep. 57* (2) : 85-93.

Tham, A.K. (1973). *The Sea*. pp. 140-149. *In* : S.H. Chung (ed.). *Animal Life and Nature in Singapore*. Sing Univ. Press.

Thanikaimoni. G. (1972). Bibilographic index to the pollen morphology of angiosperms. Part II. *Inst. Fr. Pondicherry Trav. Sect. Sci. Techn. 13* (3) : 26-30.

Thanikaimoni, G. (1980). Mangrove formations in the Indian Peninsula. Current Trends in Botanical Research. Silver Jubilee Celebrations of the Botany Department, Central College, Bangalore. 1979.

Thanikaimoni. G. (1987). *Mangrove Palynology*. UNDP/UNESCO Regional Project on Training and Research on Mangrove Ecosystems, RAS/79/002 and the French Institute, Pondicherry. pp. 1-100 with 43 Plates.

Thirumalai Raj. K. (1959). Mangrove Forests of Tanjore Division, *Proc. Mangr. Symposium*, Faridabad, India, pp. 18-19.

Thom, B.G. (1967). *Mangrove ecology and deltaic geomorphology*. Tobasco, Mexico, *J. Ecol. 55* : 301-343.

Thom, B.G. (1975). Mangrove ecology from a geomorphological view point. pp. 469-81. In *Proc. Int. Symp. Biol. Manag. Mangroves*, (eds.) G. Walsh, S. Snedaker & H. Teas. Gainesville, University of Florida Press.

Thom, B.G. (1982). Mangrove eoclogy–a geomorphological perspectives, pp. 3-17. *In* : B.F. Clough (ed.). Mangrove Ecosystem in Australia : Stucture, Function and Management. Canberra : *Austr. Nat. Univ. Press*.

Thom, B.G. (ed.) (1984). *Coastal Geomorphology in Australia*. Australia Acad. Press.

Thom, B.G. & L.D. Wright (1983). Geomorphology of the Purari Delta. *In* : T. Petr. (ed.). *The Purari*. The Hague : Junk.

Thom, B.G.,L.D. Wright & J.M. Coleman (1975). Mangrove Ecology and deltaic estuarine geomorphology : Cambridge Gulf. Ord river, Western Australia. *J.E.* 01.63 : pp. 203-232.

Thong, K.L. & A. Sasekumar (1984). The trophic relationships of the fish community of the Angsa Bank Selangor, Malaysia, pp. 385-399. *In* : E. Soepadmo, A.N. Rao & D.J. Macintosh (eds.). *Proc. Asian Symp. Mangr. Environ. Res. Manage.*, Univ. Malaya, Kualalumpur, Malaysia.

Thotharthri, K. (1960a). Botanical Exploration in Car Nicobar and Nancouri Islands, *Bull. Bot. Surv. India* 2 : 341-346.

Thotharthri, K. (1960b). Studies on the Flora of Andaman Islands, *Bull. Bot. Surv. India*. 2 : 357-373.

Thotharthri, K. (1962). Contribution to the Flora of the Andaman and Nicobar Islands, *Bull. Bot. Surv. India*, 4 : 281-286.

Thotharthri, K. (1981). Studies on the mangroves of Peninsular India Vs. the Andaman & Nicobar Islands. *Bull. Bot. Surv. India*. 23 : 151-154.

Tieghem, Ph. Van (1898). Avicenniaceae et Symphoremaceae. Place Leces deus, nouvelles jamicies dans la classification. *J. Bot. Baris.* **12** : 345-365.

Tomlinson, P. B. (1961). *Palmae. In. : Anatomy of the Monocotyledons, Vol.* 2. (ed.) C.R. Metcalf, Oxford : Clarendon Press.

Tomlinson, P.B. (1971). The shoot apex and its dichotomous branching in the *Nypa* palm. *Ann. Bot. Lond.* **35** : 865-879.

Tomlinson, P.B. (1978). *Rhizophora* in Australia—some classification of taxonomy and distribution. *J. Arnold. Arbor.* **59** : 156-169.

Tomlinson, P. B. (1980). *The biology of Trees native to tropical Florida.* Petersham, Mass, published privately.

Tomlinson, P.B. (1986). *The Botany of Mangroves.* Cambridge University Press, Cambridge, London, New York. pp. 1-414.

Tomlinson, P.M., J.S. Bunt, R.B. Primack & N.C. Duke (1978). *Lumnitzera rosea*–its Status and flora morphology. *Arn. Arb.* **59** : 342-351.

Trimen, H. (1893). *Handbook to the Flora of Ceylon.* Oxford.

Trochain, J. & L. Dulau (1942). Quelques particularites anatomiques of *Avicennia nitida* (Verbinaceae) de la mangrove ouest africaine. *Bull. Soc. Hist. Nat. Toulousi,* **77** : 271-281.

Troll, W. & D. Dragendroff (1931). Uber die Luftwurzeln von *Sonneratia* und ihre biologische Bedenlurig. *Planta.* **13** : 311-473.

Tuck, Chang Hung (1987). Country Report : Malaysia. *In* : R.M. Umali, *et al.* (eds.). *Mangroves of Asia and the Pacific : Status and Management.* Technical Report of the UNDP/UNESCO Research and Training Pilot Programme on Mangrove Ecosystem in Asia and the Pacific (RAS/79/002). pp. 131-150.

Tuck, C.H. & S. Nor (1987). Traditional uses of the Mangrove Ecosystem in Malaysia. Mangrove Ecosystem Occasional Papers No. 1. UNESCO & UNDP. (Ed.) C.D. Field.

Turner, R.E. (1977). Intertidal vegetation and commercial Yields of Penaeid shrimp. *Trans. Amer. Fish Soc.* **106** : 411-416.

Twyford, I. & A. Wright (1965). The soil resources of the Fiji Islands. London : Eyre Spottiswoode.

Uhl, N.W. (1972). Inflorescence and flower structure in *Nypa fruticans* (Palmae). *Am. J. Bot.* **59** : 729-743.

Umali, R.M. (1947). Management Techniques and Methodologies. *In* : *Mangroves of Asia and The Pacific : Status and Management.* Technical Report of the UNDP/UNESCO Research & Training Pilot Programme on Mangrove Ecosystems (RAS/79/002).

UNESCO/SCOR, Working Group 60 on Mangrove Ecology (1984). The Mangrove Ecosystem : Research methods (eds.) Samuel C. Snedaker & Jane G. Snedaker. ISBN 92-3-102181-8. Published in 1984 by the United Nationas Educational Scientific and Cultural Organisation 7 Place de Fontenoy, 75700 Paris. Monograph on Oceanographic Methodology-8.

UNESCO (1984). Handbook of Mangrove Area Management. (Ed.) Lawrence S. Hamilton & S.C. Snedakar.

UNESCO (1985). Zooplankton fixation and preservation–Monographs on Oceanographic methodology–4. United Nations Educational, Scientific and Cultural Organisation, 7 Place de Fontenoy, 75700 Paris.

UNDP/UNESCO (1986). Mangroves of Asia and the Pacific Status and Management. *Technical Report of the UNDP/UNESCO Research and Training Pilot Programme on Mangrove Ecosystems in Asia and the Pacific* (RAS/79/002).

Untawale, A.G. (1978). Country Report on India. *Proc. Reg. Seminar on Human Uses of the Mangrove Environment and Management Implication,* UNESCO, Dacca, Bangladesh.

Untawale, A.G. (1980). Present Status of the mangrove along the west coast of India. *In* : *Proc. Asian Symp. Mangrove Environment Res. Management.* University of Malaya, Kualalumpur, pp. 1-26.

Untawale, A.G. (1984a). Present status of mangroves along the coasts of India. India. *In* : E. Soepadmo, A.N. Rao & D.J Macintosh (eds.). *Proc. Asian Symp. Mangr. Environ. Res. Manage.* Univ. Malaya, Kualalumpur, Malaysia. pp. 57-74.

Untawale, A.G. (1985). Status of Mangrove Research in India. *The Mangroves : Proc. Nat. Symp. Biol. Util. Cons. Mangrove.* Nov. 1985 : 127-134.

Untawale, A.G. (1985). Mangrove in India : Present status and multiple use practices, UNDP/UNESCO Regional Mangrove Project, 67 p.

Untawale, A.G. (1986). India (mangrove ecosystem). Mangroves of Asia and The Pacific : Status and Management. *Technical Report of the UNDP/UNESCO Research and Training Pilot Prog. on Mang. Ecosyst. In Asia and the Pacific.* Ras 791002 : 51-87.

Untawale, A.G. (1986). How to Grow Mangroves, October 1986. pp. 1-9.

Untawale, A.G. (1986). Research and Training needs (mangrove vegetation). Mangrove of Asia and the Pacific : Status and Management. *Technical Report of the UNDP/UNESCO Research and Training Pilot Programme on Mangrove Ecosystem.* In Asia and the Pacific, Res. 79002. Natural Resources Management Cent., Quezon City, Philippines : 467-470.

Untawale, A.G. (1987). Some observations on mangrove afforestation along the central west coast of India. *In : Afforestation of Salt-affected Soils.* International Symp. Papers. *Vol. II. Plant Resources.* 133-150.

Untawale, A.G. (1987). Conservation of Indian Mangroves–A National Perspectives. *In* : S.Z. Qasim (ed.). Contributions in Marine Sciences. pp. 85-104.

Untawale, Arvind G. (1992). Rehabilitation of coastal wetlands of India. *In : Ecosystem Rehabilitation, Vol.* 2 : Ecosystem Analysis and Synthesis (Edited by) Mohan K. Wali, pp. 333-348. Academic Publishing, The Hague, The Netherlands.

Untawale, A.G., S.N. Dwivedi & S.Y.S. Singbal (1973). Ecology of Mangrove in Mandovi and Zuari estuaries and the inter-connecting Cumbarjua Canal of Goa. *Indian J. Mar. Sci.* 2 : 47-53.

Untawale, A.G. & A.H. Parulekar (1976). Some observation on the ecology of an estuarine mangrove of Goa. *Mahasagar. Bull. Nat. Inst. Oceanography, 9* (1 & 2) : 57-62.

Untawale, A.G., N.B. Bhosale, V.K. Dhargalkar, S.G.P. Matondkar & S. Bukhari (1977). Biochemical changes in Mangrove foliage during growth and decomposition. *Ind. J. Mar. Sci.* 6 : 104-106.

Untawale, A.G. & A.H Parulekar (1978). Status Report of the Pirotan Islands in the Gulf of Kutchh for Development of a conservation Plan. *Report submitted to World Wildlife Fund,* Bombay.

Untawale, A.G., N.B. Bhosle, V.K. Dhargalkar, S.G.P Matondkar & Sayuda S. Bukahri (1978). Seasonal variation in major metabolites of mangrove foliage. *Mahasagar, Vol. 11,* No. 1 & 2 : 105.

Untawale, A.G., S. Wafer & T.G. Jagtap (1980). Application of remote sensing technique to study the distribution of mangrove along the estuaries of Goa. *Proc. 1st. Intern. Wetland Conf.* India.

Untawale, A.G. *et al.* (1982). Application of remote sensing techniques to study the distribution of mangroves along the estuaries of Goa. *In : Wetlands : Ecol. & Management. Proc. 1st Inter. Wetlands Conference,* New Delhi, pp. 51-57.

Untawale, A.G., K.K. Dhargalkar & V.V Agadi (1983). List of marine algae from India. National Institute of Oceanography Report.

Untawale, A.G. & T.J. Jagtap (1985). Marine macrophytes of Minicoy (Lakshadwip) Coral atoll of the Arabian Sea. *Aqua. Bot.* 19 : 97-103.

Untawale, A.G. & S. Wafar (1986). Distribution of Mangroves along the Gulf of Kuchchh. *In : Marine Archaeology of Indian Ocean Countries,* pp. 135-138.

Untawale, A.G., & S. Wafar & T.G. Jagtap (1986). Status of mangroves along the countries bordering the Arabian Sea. *In* : B.N. Desai (ed.). Oceanography of the Indian Ocean. pp. 239-245.

Untawale, A.G., Sayeeda Wafar & C.R.K. Reddy (1987). Management of coastal vegetation for integrated development of Islands. *In : Workshop on S. & T. Impacts for Island Development* Papers, Indian International Centre, New Delhi, Nov. Indian Association for Advancement of Science, DOD, Delhi.

Untawale, A.G., Sayeeda Wafar, C.R.K. Reddy (1988). Management of Coastal Vegetation for Integrated Development. *In* : S.N. Dwivedi & P. Chaturvedi (eds.). *Island Development. Technology Options,* IAAS Publications, pp. 59-69.

Untawale, A.G. & T.G. Jagtap (1989). Observations on marine macrophytes of the Republic of Scychelles. *Bot. Mar.* 32 (2) 115-119.

Untawale, A.G. & T.G. Jagtap (1991). *Memoirs Geol. Sci. India.* **22** : 243.

Untawale, A.G. & S. Wafar (1991). Socio-economic Significance of Mangrove in India *Proc. Symp. Significance Mangrov.* Pune, March, 1990.

Untawale, A.G. & T.G. Jagtap (1992). Floristic Composition of the Deltaic Regions of India. *Memoirs Geological Society of India.* No. **22** : 243-263.

Uphof, J.C. Th. (1941). *Halophytes. Bot. Rev.* **7** : 1-58.

Vaivanijkul, P. (1976). A general survey of insects at Bangpoo, Samut Prakarn. *Paper presented in the 1st National Seminar on Mangrove Ecosystems.* Held from 10-15 January 1976 at Phuket Marine Biological Centre, Thailand.

Vannucci, M. (1984). The conversion of mangroves to other uses (The Cochin Backwaters). *Paper presented at the Workshop on Human Populations ; Mangrove Resources; Human Induced Stresses and Human Health.* Held from 2-7 October, 1984 at Bogor, Indonesia.

Vannucci, M. (1989). *The Mangrove and Us;* A synthesis of insight; Indian Association for the Advancement of Science, New Delhi, pp. 1-203.

Van Steenis, C.G.S. (1958). Ecology of Mangroves, pp. 421-444. *In : Flora Malesiana* I. Noordhoff-Kolff, Djakarta.

Van Steenis, C.G.G. (1962). The distribution of mangrove plant genera and its significance for palaeo-geography. *Verh. K. ned. Akad. Wet. Ser. C.* **65** : (2) : 164-169.

Van Steveninck, R.F.M. W.D. Armstrong, P.D. Peters & T.A. Hall (1976). Ultrastructural location of ions. III. Distribution of Chloride in mesophyll cells of mangrove (*Aegiceras corniculatum* Blanco). *Australian Journ. of Plant Physiology.* **3** : 367-376

Venkatesan, T. (1966). The Mangroves of Madras State. *Indian For.* **92** : 27-34.

Venkateshwarlu, V. (1944). The estuarial flora of the Godavary. *J. Bomby Nat. Hist. Soc.* **44** : 431-435.

Venkateshwarlu, J. & R.S.P. Rao (1964). The wood anatomy and taxonomic position of Sonneratiaceae *Current Science* . 33 : 6-9.

Voight, J.O. (1845). *Hortus Suberbonus Calcutensis.* Calcutta, India.

Vuthisathirapinyo, P. (1976). A comparative study of chemical and physical parameters in mangroves and open sea water at Phuket Island. *Paper presented in the First National Seminar on Mangrove Ecosystems* at Phuket Marine Biological Centre, Thailand.

Wafar, M.V.M. (1986). Corals and coral Reefs of India. *In the Proc. of the Indian Acad. Sci.* (Anim. Sci./Plant Sci.) Suppl., pp. 19-43.

Wafar, Sayeeda (1987). Ecology of Mangroves along the Estuaries of Goa. Ph.D. Thesis, submitted to Karnataka University, Dharwad, India. pp. 1-299.

Wafar, S. (1985). Observation of the foliage production in some mangrove sp. of Goa. *Marine Plants, Papers presented at the All India Symposium on Marine plants, their biology, chemistry and utilization,* Donapaula, Goa, Oct. 30-Nov. 1, 1983, V. Krishnamurthy & A.G. Untawale (eds.) 1985 : 245-252.

Wafar, S. & A.G. Untawale (1991). Role of Mangrove Swamps in Brackishwater Fish Culture. *In : Aquculture Productivity* (ed.) V.R.P Sinha & H.C. Srivastava, OXFORD & IBM Publishing Co. Pvt. Ltd. pp. 235-239.

Wagale, B.G. (1982). Geomorphology of the Goa Coast. *Proc. Indian Acad. Sci.* (Earth, Planet Sci.). **91** : 105-117.

Waheed Khan, M.A. (1959). Ecological Studies of the Mangrove forests in India. *In : Proc. Mangs. Symposium.* Faridabad, pp. 97-113. 1957.

Walsh, C., S.C. Snedaker & H.J. Teas (eds.) (1975). Proceeding of the International Symposium on Biology and Management of Mangroves. Held from 8-11 October, 1974, at EWC, Hawaii.

Walter, H. & M. Steiner (1936). Oekologie der ost-afrikanschen Mangroven : *Z. Bot.* **30** : 65-93.

Walter, H (1971). *Ecology of Tropical and Sub-Tropical Vegetation,* New York, Van Nostrand. 539 pp.

Walter, H. (1973). Vegetation of the Earth in Relation to Climate and the Ecophysiological conditions, New York : Springer-Verlag.

Walter, H. (1973). Die Vegetation der Erde, 3rd ed. 1. Fischer, Stuttgart, 743 p.

Walter, H. (1977). Climate, pp. 54-66. *In* : V.J. Chapman (ed.). *Ecosystem of the World : Wet Coastal Ecosystems, Vol. I.* New York : Elsevier.

Watanachai, S. (1979). Species and Abundance of Fish Eggs and Fish Larvae in Petchburi Mangrove Area. Report 4/1979, Fisheries Department, Bangkok, Thailand.

Watling, D. (1985). A Mangrove Management Plan for Fiji, Phase I. Suva, Fiji : Government Press.

Watling, D. (1986) A. Mangrove Management Plan for Fiji, Phase II. Suva, Fiji : Government Press.

Watling, J.G. (1928). Mangrove forests of the Malay peninsula. *Malayan For. Rec.* **6** : 1-275. Illustr.

Watts, J.C.D. (1969). Phosphate retention in acid sulphate pond muds from the Malacca Area. *Malay Agri. J.* **47** : 187-202.

Way, W.K. (1977). Economically Important Herpetofauna of Thai Mangrove. Association for Conservation of Wildlife of Thailand.

Wells, A.G. (1983). Distribution of mangrove species in Australia. Chap. 6. *In : Biology and Ecology of Mangroves,* (ed.) H. J. Teas for Vegetation Science 8. The Hague, Junk.

Wester, L.L. (1967). The distribution of the mangrove in South Australia. Unpublished Honour Thesis, University of Adelaide.

White, C.T. (1926). A variety of *Ceriops tagal* C.B. Rob. (= *C. candolleana* W. & A.). J. Bot. Lond. **64** : 220-221.

Wirjodarmodjo, H. & Z. Hamzah (1984). Beberapa pengalaman perum Perhutani dalam pengelolaan hutan mangrove, pp. 29-40. In : Ș. Soemodihardjo *et al.* (eds.). Prosiding Seminar II Ekosistem Mangrove. LIPI.

Womersley, J.S. (1983). An introduction to the nomenclature and taxonomy of the mangrove flora in Papua New Guinea and Adjacent Areas. pp. 87-90. *In* : H.J. Teas (ed.). Biology and Ecology of Mangroves. The Hague : Dr. W. Junk Publ.

Wright, D. (1957). Pollen morphology of Australian Mangroves. (Unpublished thesis, James Cook University of North Queensland).

Wyatt, Smith (1960). Field Key to the trees of mangrove forests in Malaya. *Malay. For.* **23** : 126-132.

Wylie, R.B. (1949). Differences in foliar organization among leaves from four locations in the Crown of an isolated tree (Acerplantanoides). *Proc. of the Iowa Academy of Science,* **56** : 189-198.

Yadav, J. S. P., A. K. Bandopadhyay, K. V. G. K. Rao, T. S. Sinha & C. R. Biswas (1979). Coastal Saline Soils of Indian, CSSRI, *Bull.* No. 5 pp. 6-34. ICAR Karnal, India.

Yadav, J. S. P., A. K. Bandopadhyay, K. V. G. K. Rao, T. S. Sinha, C. R. Biswas, B. K. Bandopadhyay & S. K. Dutta (1981). Mangement of Coastal Saline Soils of Sundarbans. *Bull.* 7. pp. 1-32, CSSRI, ICAR, Karnal, India.

Yadav, J. S. P., A. K. Bandopadhyay & B. K. Bandopadhyay (1983). Extent of Coastal Saline Soils of India. *J. Indian Soc. Coastal agric. Res.* **1** (1) : 1-6.

Yadar, V.K., S. Mishra & A. Choudhury (1983). Microbial decomposition of leaf litter of *Avicennia* in a mangrove forest. Biochemical changes during the decomposition. *Proc. All India Symp. on Mar. Plants their Biol. Chem. Util.,* 30 Oct.–1 Nov., 1983, India.

Zamski, E. (1979). The mode of secondary growth and the three dimensional structure of the phloem in *Avicennia Bot. Coat.* **140** : 67-76.

Zelitch, I. (1971). Photosynthesis, *Photorespiration and Plant Productivity.* Academic Press, New York, 347 p.

Zimmermann, M.H. (1983), Xylem structure and the ascent of sap. Berlin : Springer-Verlag.

INDEX

PLATES

Plate-81. *Rhizophora mucronata* Lamk.

a. Herbarium, **b.** Plant in natural habitat along with stilt roots and hanging viviparous germinated seedlings, **c.** Medium well grown tree with dense stilt roots, **d.** Flowering twigs in hand

Plate-82. *Rhizophora apiculata* Blume

a. Herbarium, **b.** Branch with flowers and fruit, **c.** One tree in nature - Henry's Island / Bakkhali, **d.** Close view of twigs with flower buds and viviparous hypocotyle

Plate-83. *Bruguiera gymnorrhiza* (L.) Lamk.

a. Herbarium, **b.** one small tree on river side at Sajnakhali, Sundarbans, **c.** Flowering twigs - dominant with red persistant calyx, **d.** one viviparous germinated seedling in experimental garden.

Plate-84. *Bruguiera sexangula* (Lour.) Poir

a. Herbarium, **b.** Flowering twig, **c & d.** Branches with flowers - persistant calyx

Plate-85. *Bruguiera cylindrica* (L.) Blume.

a. Herbarium, **b.** Flowering and fruiting branch, **c.** one twig with flowers and viviparous germinated hypocotyle, **d.** seedling in earthern pot

Plate-86. *Bruguiera parviflora* W. & A. ex *Griffithi & Rhizophora mucronata* Lamk.

Plate-87. *Ceriops decandra* (Griff). Ding Hou

a. Herbarium, **b.** branch with viviparous germinating fruits, **c.** Plant in natural habitat, **d.** twig with viviparous germinated hypocotyle, **e.** stages of seedlings development

Plate-88. *Ceriops tagal* (Perr.) Robinson

a. Herbarium, **b.** conical small trees - forest floor inundate with tidal water, **c.** Small tree in the natural habitat, **d.** fruiting branch with hanging viviparous germinated seedlings.

Plate-89. *Kandelia candel* (L.) Druce

K. Herbarium, **L.** Natural Habitats, **M.** Flowering twigs, **N.** Branch with viviparous germinated hypocotyles

Plate-90. *Avicennia officinalis* L.

M. Herbarium, **N.** Flowering twig, **O.** Fruiting twig, **P.** Dense stilt root from trunk base

Plate-91. *Avicennia alba* Blume

M. Herbarium, **N.** Branch of the tree, **O.** Fruiting branch, **P.** Flowering twigs, **Q.** Pneumatophores

Plate-92. *Avicennia marina* (Forsk.) Vierh.

P. Herbarium, **Q.** Still roots, **R.** Flowering twigs, **S.** Germinated seed (Cryptoviviparous), **T.** Plant with pneumatophores

Plate-93. *Sonneratia apetala* Buch.-Ham.

S. Herbarium, **T.** Tree in the natural habitat, **U.** Tree in the eroding river bank, **V.** Twig with fruiting bunch

Plate-94. *Sonneratia griffithii* Kurz

Plate-95. *Sonneratia griffithii* Kurz

A. Herbarium, **B.** Flowering and fruiting branch, **C. & D.** Flowering and fruiting twigs.

Plate-96. *Sonneratia caseolaris* (L.) Engler

Plate-97. *Sonneratia abla* Smith and *Sonneratia griffithii* Kurz

a., b. & c. Flowering branches - white flowers

Plate-98. *Lumnitzera racemosa* Willd.

Plate-99. *Nypa fruticans* (Thunb.) Wurmb.

Plate-100. *Phoenix paludosa* Roxb.

Plate-101. *Heritiera fomes* Buch.-Ham.

Plate-102. *Heritiera fomes* Buch.-Ham.

Plate-103. *Xylocarpus granatum* Koen.

Plate-104. *Xylocarpus mekongensis* Pierre

Plate-105. *Excoecaria agallocha* L. Male plant

Plate-106. *Excoecaria agallocha* L. Female plant

Plate-107. *Aegiceras corniculatum* (L.) Blanco

Plate-108. *Scyphiphora hydrophyllacea* Gaertn. f.

Plate-109. *Hydrophyllax maritima* L.

Plate-110. *Aegialitis rotundifolia* Roxb.

Plate-111. *Acrostichum aureum* L.

Plate-112. *Acanthus ilicifolius* L.

Plate-113. *Acanthus volubilis* Wall.

Plate-114. *Brownlowia lanceolata* Benth.

Plate-115. *Hibiscus tortuosus* Roxb.

Plate-116. *Thespesia populnea* (L.) Sol. ex Corr.

Plate-117. *Atalantia correa* M. Roem.

Plate-118. *Caesalpinia crista* L.

Plate-119. *Caesalpinia bonduc* (L.) Roxb.

Plate-120. *Dalbergia spinosa* Roxb.

Plate-121. *Derris trifoliata* Lour.

Plate-122. *Derris scandens* Benth.

Plate-123. *Derris indica* Bennett

Plate-124. *Sarcolobus globosus* Wall.

Plate-125. *Sarcolobus carinatus* Wall.

Plate-126. *Pentatropis capensis* Bull.

Plate-127. *Crinum defixum* Ker Gawler

Plate-128. *Suaeda maritima* Dumort.

Plate-129. *Suaeda nudiflora* Roxb.

Plate-130. *Heliotropium curassavicum* L.

Plate-131. *Ipomoea pes-caprae* (L.) Sweet

Plate-132. *Salicornia brachiata* Roxb. & *Sesuvium portulacastrum* L.

Plate-133. *Clerodendron inerme* (L.) Gaertn.

Plate-134. *Viscum orientale* Willd.

Plate-135. *Dendrophthoe falcata* (Linn. f.) Etting.

Plate-136. *Porteresia coarctata* (Roxb.) Takeoka

Plate-137. *Porteresia coarctata* (Roxb.) Takeoka

Plate-138. *Tamarix gallica* L.

Plate-139. *Solanum trilobatum* L.

Plate-140. *Dolichandrone spathaceae* (Linn. f.) Sch.

Plate-141. *Aglaia cucullata* Pellegrin and *Hoya parasitica* Wall.

a., b. & c. Seedling development of *Aglaia cucullata* (Roxb.) Pelle. **d.** *Hoya parasitiea* wall—Epiphyte on *Xylocarpus* sp.

Plate-142. *Opuntia dillenii* Haw.

Plate-143. *Ruppia maritima* L.

Plate-144. *Manilkara hexandra* (Roxb.) Dub.

Plate-145. *Cryptocoryne ciliata* (Roxb.) Fish. & *Crinum detixum* Kar-Gawl. in the inner estuary at Kulti.

Plate-146. *Rhizophora mucronata* Lamk.

Plate-147. *Rhizophora apiculata* Blume

Plate-148. *Bruguiera gymnorrhiza* (L.) Lamk.

Plate-149. *Bruguiera sexangula* (Lour.) Poir

Plate-150. *Ceriops tagal* (Perr.) Robin

Plate-151. *Avicennia officinalis* L.

Plate-152. *Sonneratia griffithii* Kurz.
Plate-153. *Sonneratia caseolaris* (L.) Engler
Plate-154. *Heritiera fomes* Buch.-Ham.
Plate-155. *Xylocarpus mekongensis* Pierre
Plate-156. *Aegiceras corniculatum* (L.) Blanco
Plate-157. *Hydrophyllax maritima* L.
Plate-158. *Aegialitis rotundifolia* Roxb.
Plate-159. *Acrostichum aureum* L.
Plate-160. *Brownlowia lanceolata* Benth.
Plate-161. *Viscum orientale* Willd.
Plate-162. Microspocic view of Periphytic Algal Flora.

Plate-81

a

b

c

d

Plate-82

a
b

c d

Plate-83

Plate-84

a

b

c

d

Plate-85

a

b

c

d

Plate-8

Plate-87

Plate-88

a

b

c

d

Plate-89

K

L

N

M

Plate-90

M

N

O

P

Plate-91

M

N

O

P

Q

Plate-92

P

Q

R

T

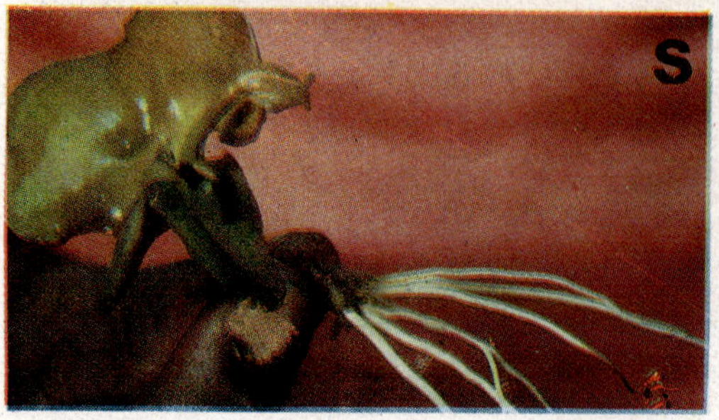
S

Plate-93

S

T

U

V

Plate-94

A

B

C

D

Plate-95

m

n

o

Plate-96

G

H

J

Plate-97

Plate-

a

b

c

d

Plate-99

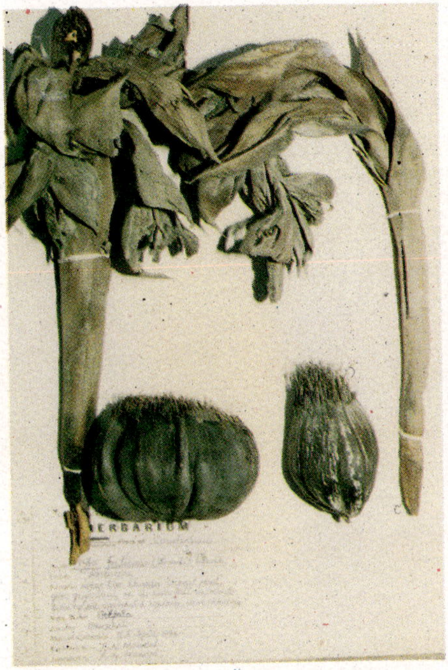

a

b

c

d

Plate-100

A

C

B

D

Plate-101

a

b

c

d

Plate-102

e

f

g

h

Plate-103

Plate-104

a

b

c

d

Plate-105

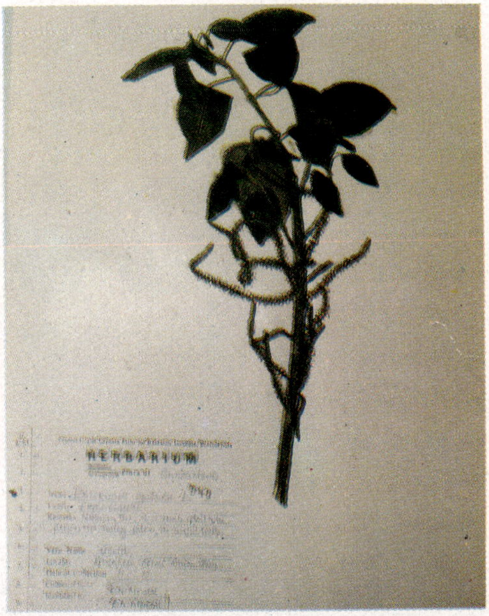

a

b

c

Plate-106

a

b

c

Plate-107

a

b

c

d

Plate-108

a

b

Plate-109

Central Inland Capture Fisheries Research Institute, Barrackpore
HERBARIUM
Flora of Sundarbans.
Name Hydrophylax maritima
Family Rubiaceae
Vern. Name
Locality
Date of Collection
Collected by
Identified by

C

D

Plate-110

A

B

D

C

Plate-111

A

D

C

B

Plate-112

A

B

C

Plate-113

a

b

c

Plate-11

HERBARIUM
Flora of
912
Vern. Name
Locality
Date of Collection

a

d

c

Plate-115

a

b

Plate-116

G

H

Plate-117

E

F

G

H

Plate-118

R

S

Plate-119

A

B

C

Plate-120

A

B

C

Plate-121

R

S

M

N

O

Plate-123

R

S

T

Plate-124

A

B

C

Plate-125

Plate-126

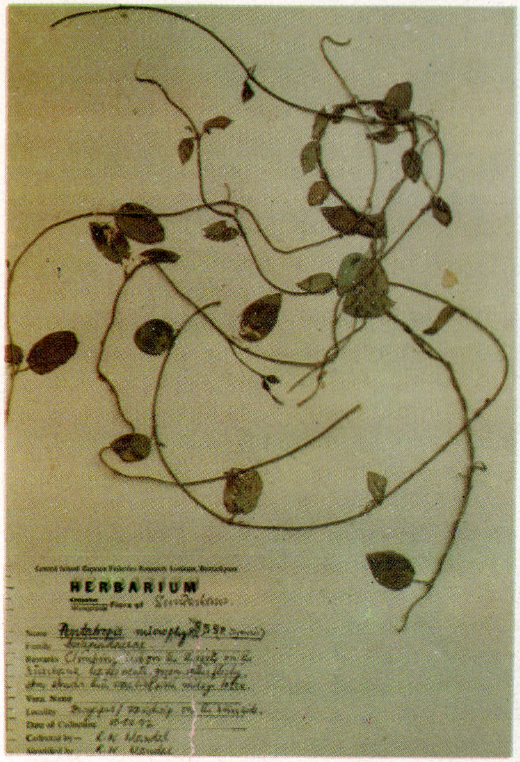

A

B

C

Plate-127

U

V

W

Plate-128

D

E

F

Plate - 129

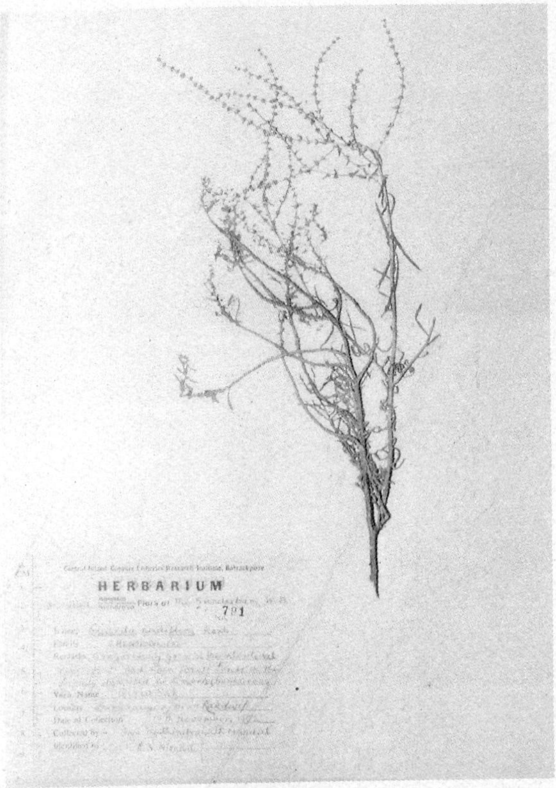

a

b

c

Plate - 130

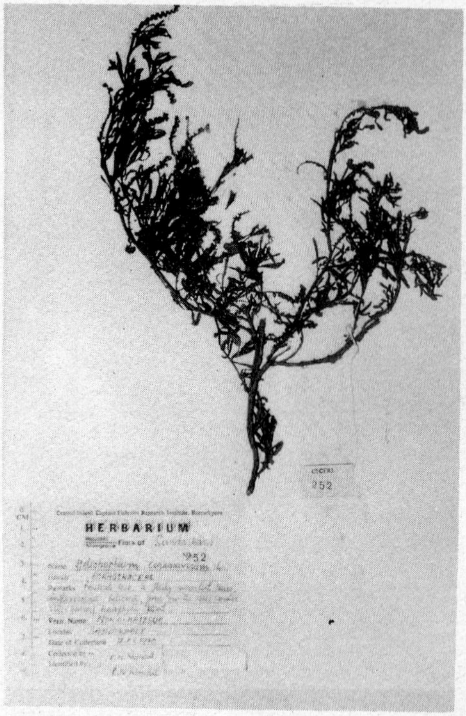

R

S

Plate - 131

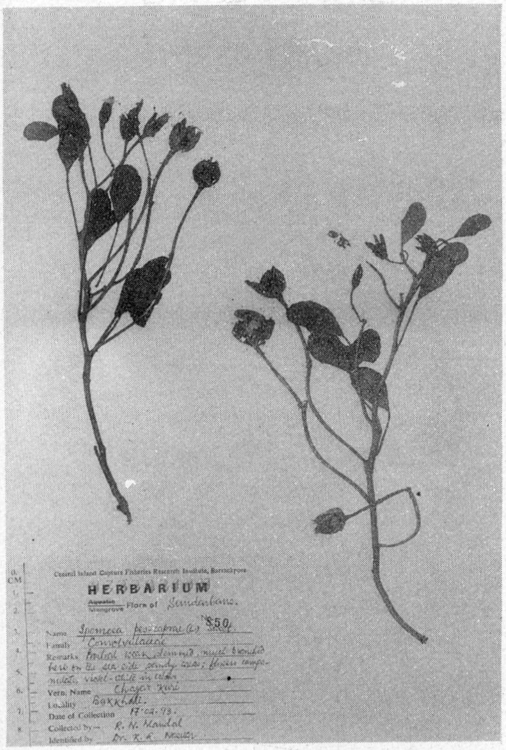

R

S

T

Plate - 132

P

Q

Plate - 133

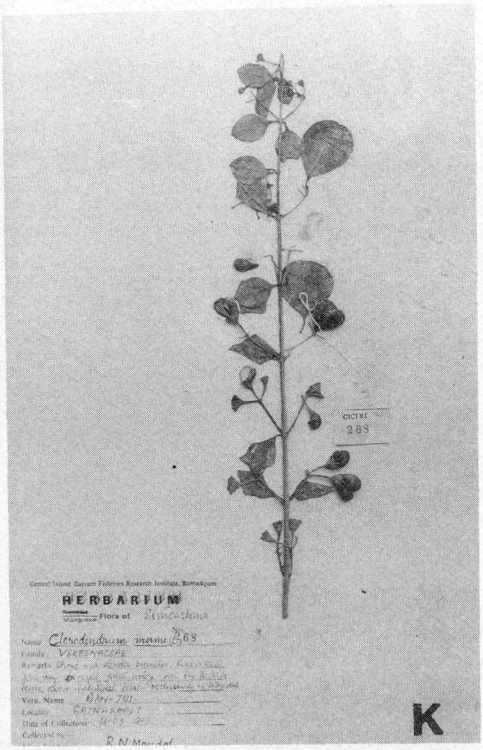

K

L

M

N

Plate-134

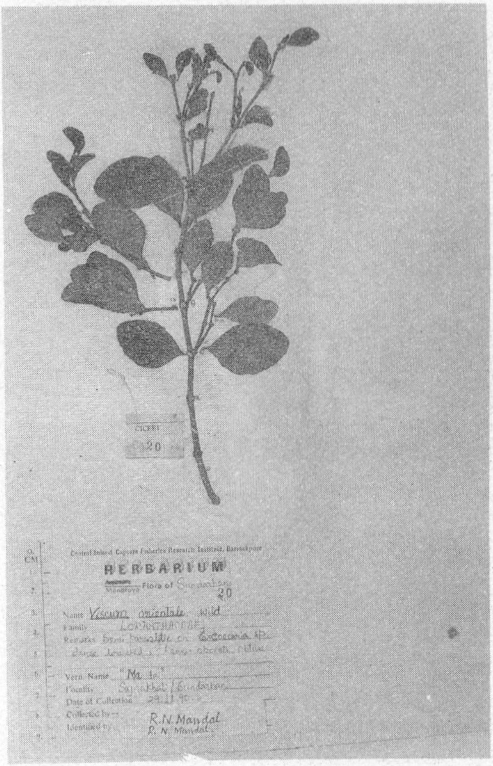

X

Y

Z

Plate - 139

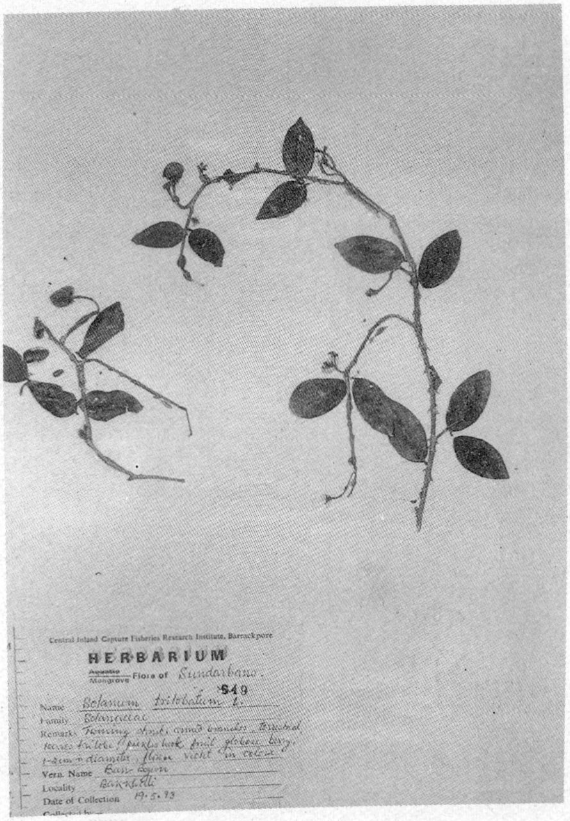
Central Inland Capture Fisheries Research Institute, Barrackpore
HERBARIUM
Mangrove Flora of Sundarbans.
549
Name Solanum trilobatum L.
Family Solanaceae
Vern. Name
Locality
Date of Collection 19.5.93

g

h

Plate - 140

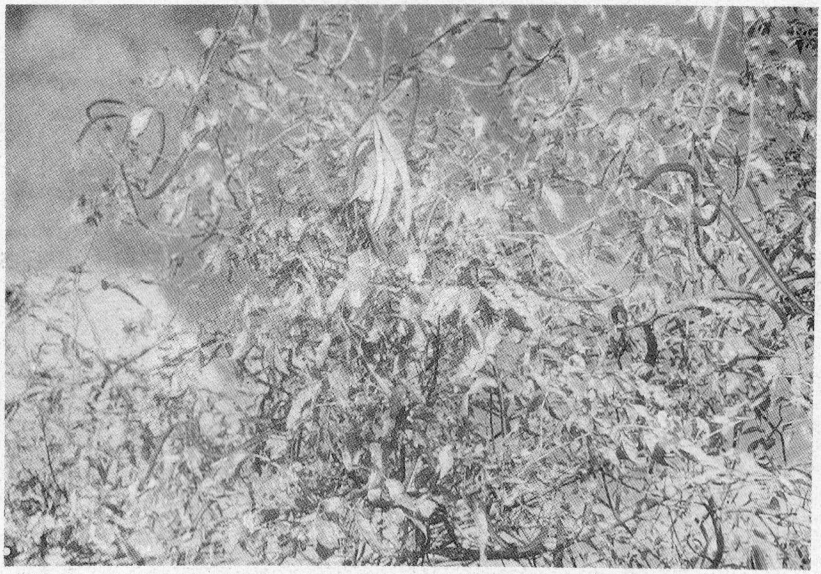

R

S

Plate - 141

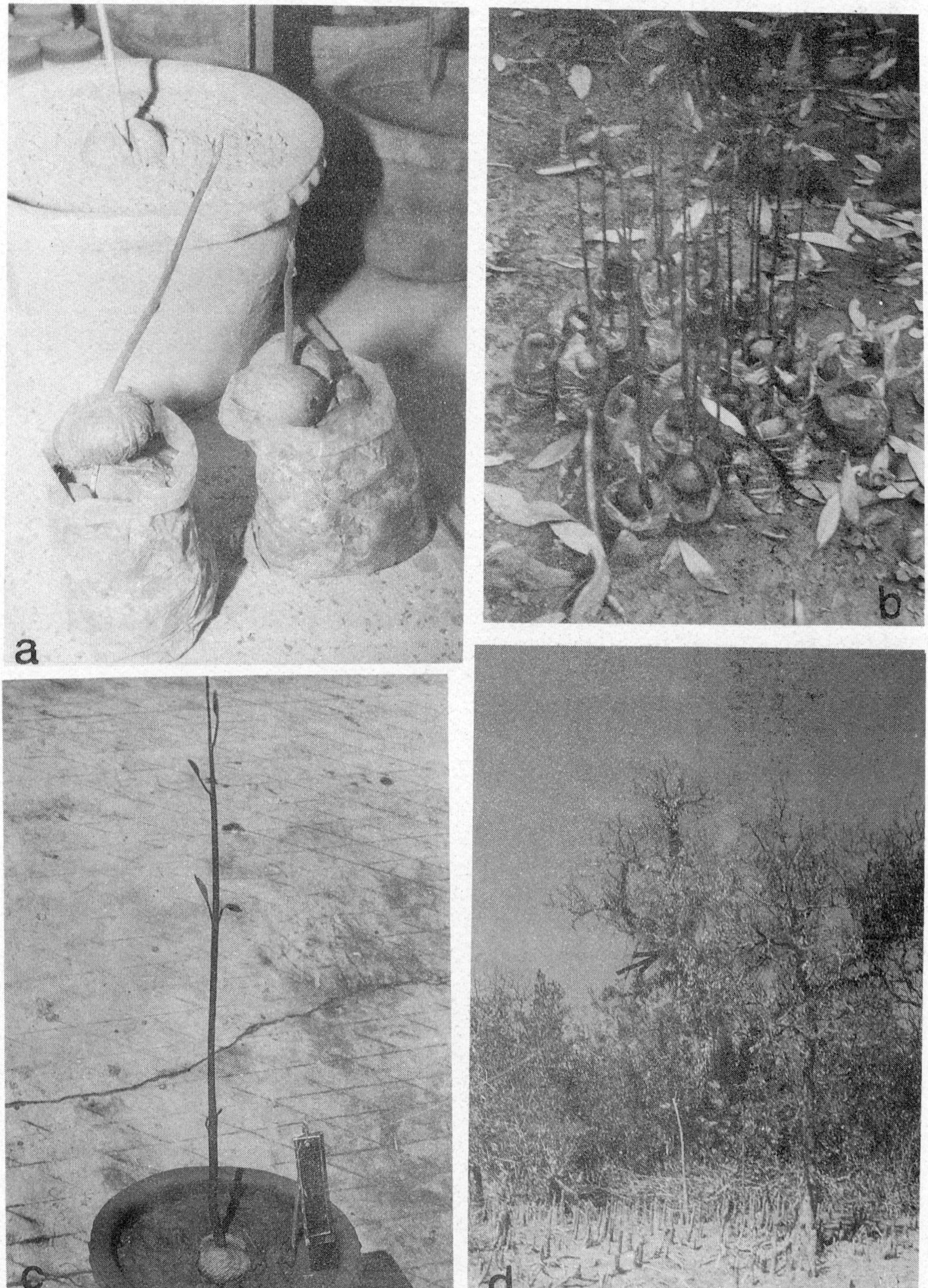

Plate - 142

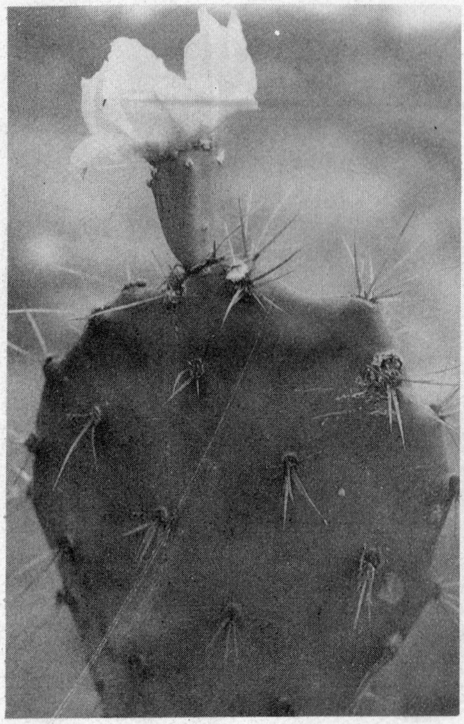

E

F

G

Plate - 143

Plate - 144

G

H

Plate - 145

D

E

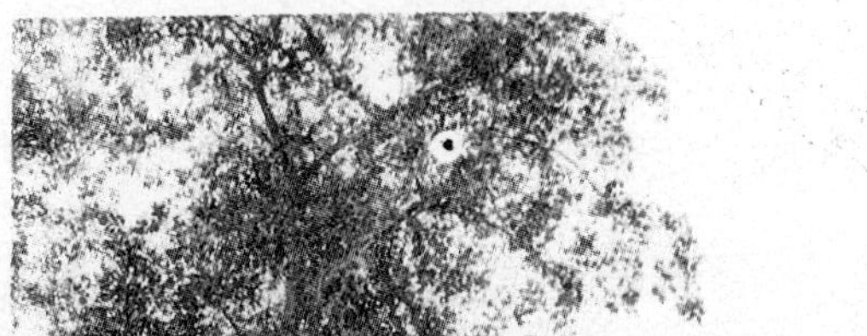

F

Plate - 146

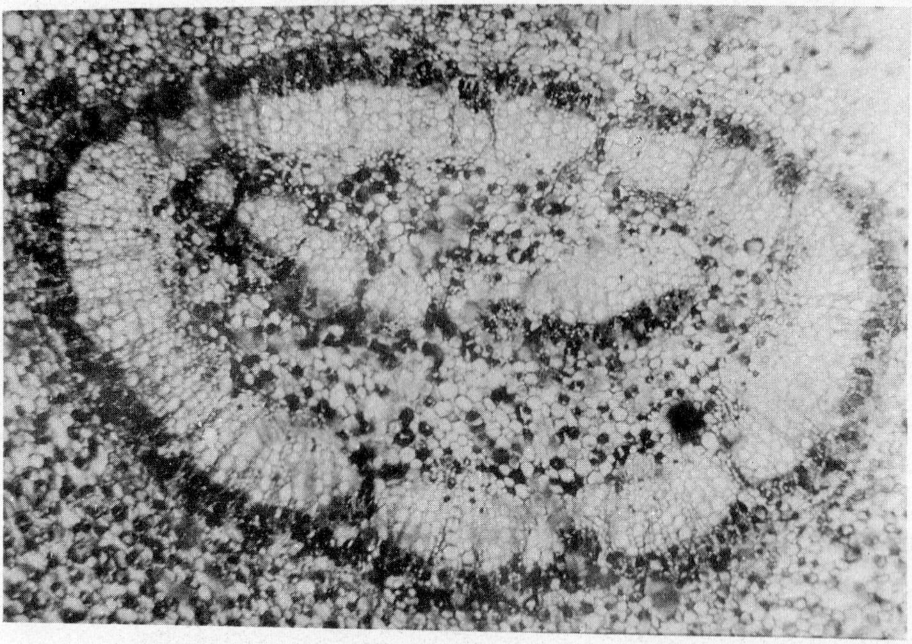

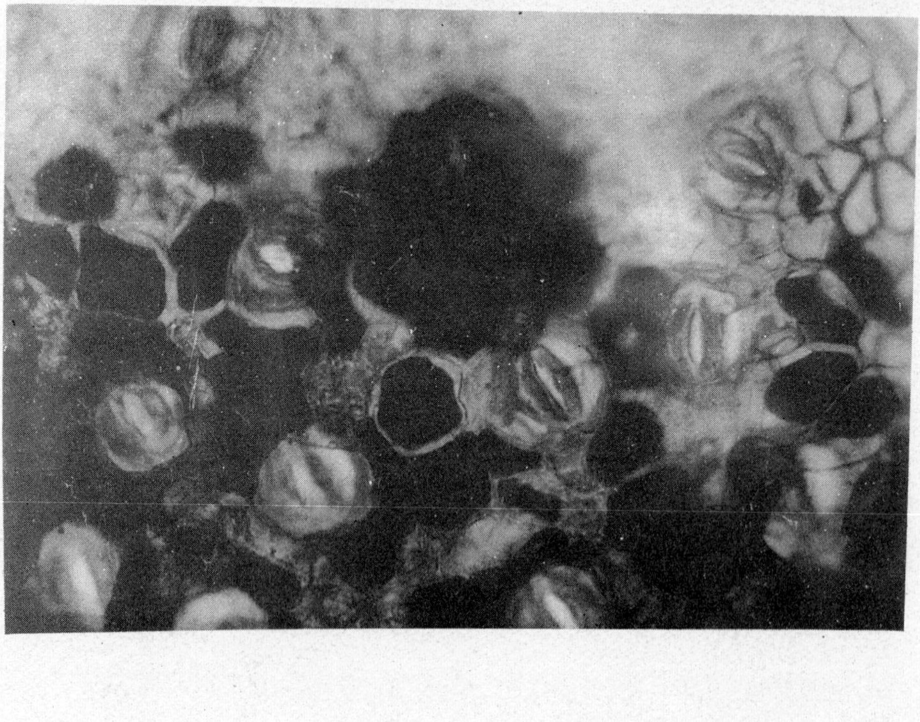

Plate - 147

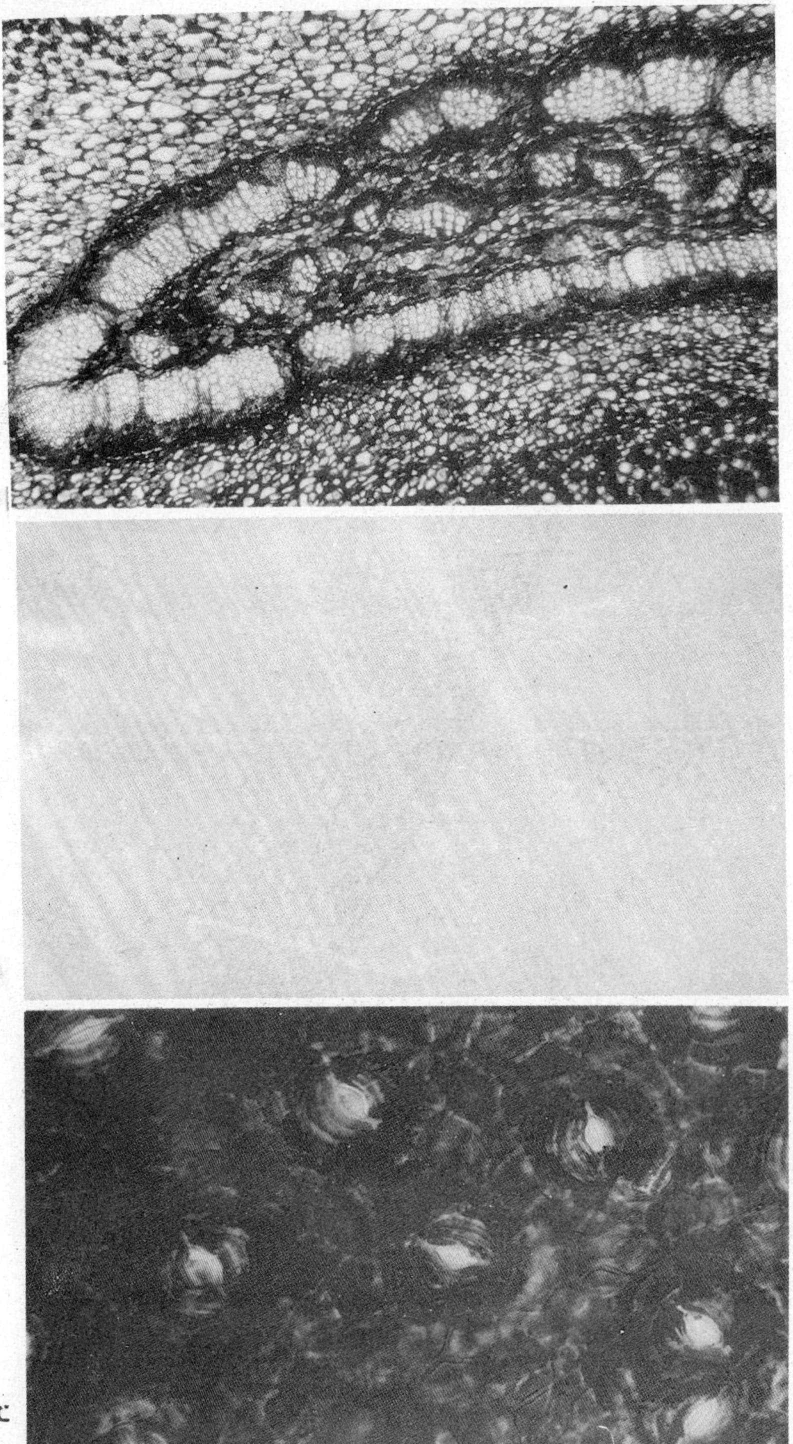

A

B

C

Plate - 148

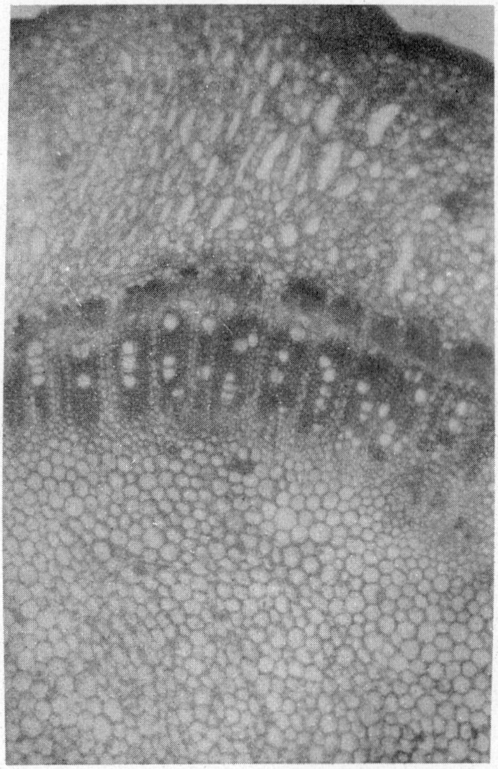

A

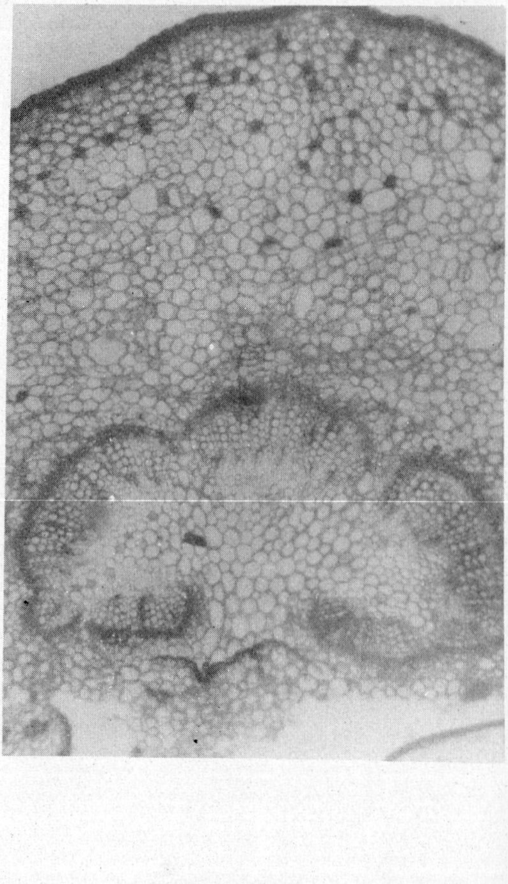

B

Plate - 149

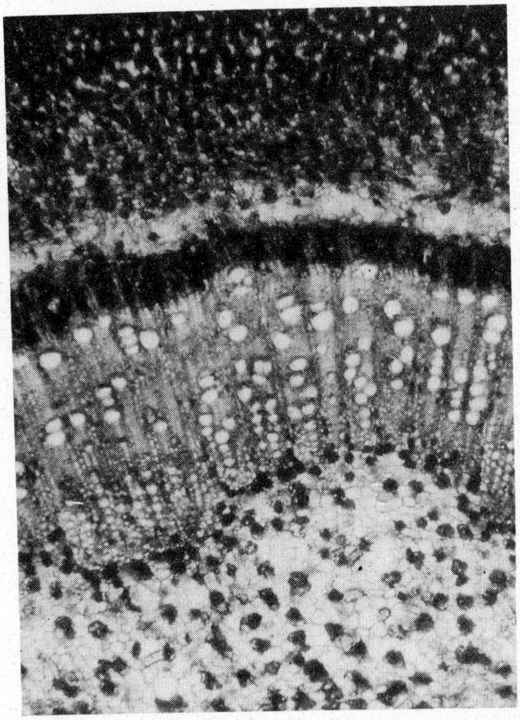

A

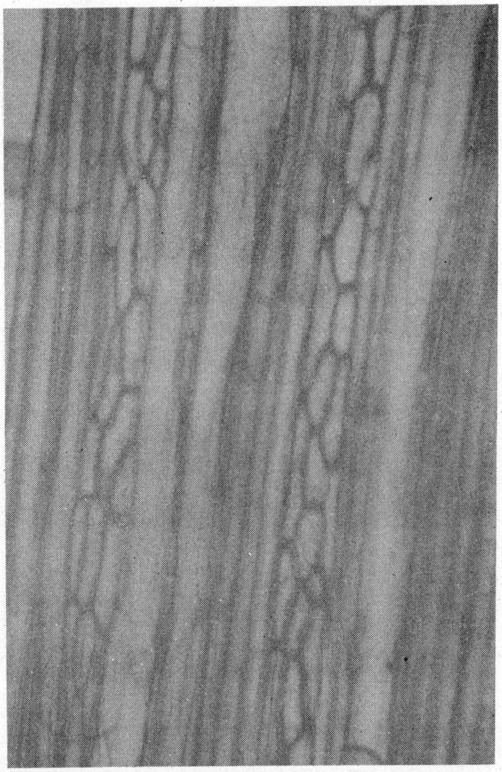

B

Plate - 150

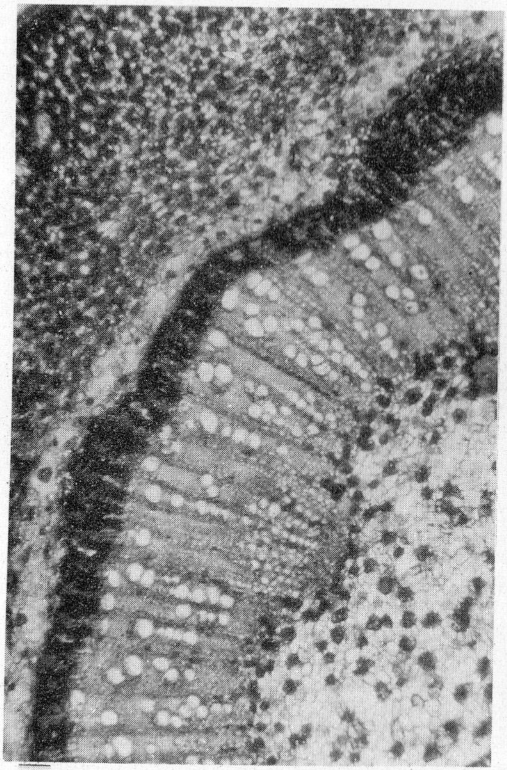

A

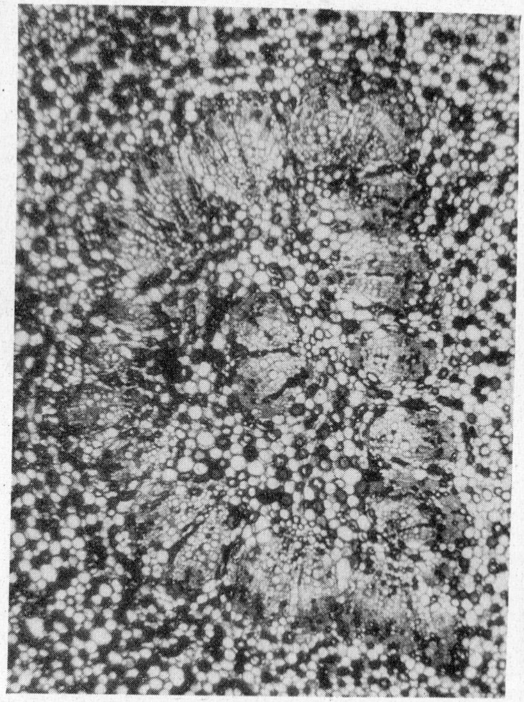

B

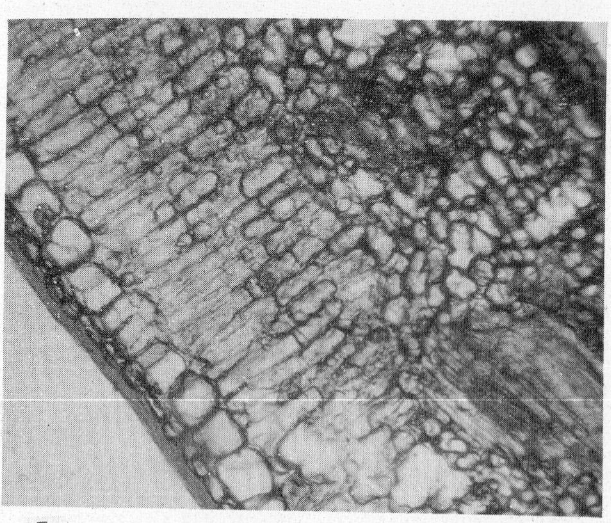

C

Plate - 151

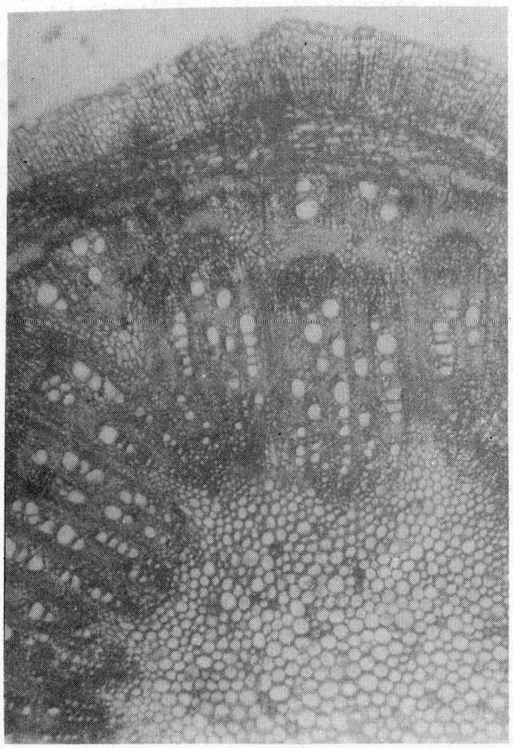

A

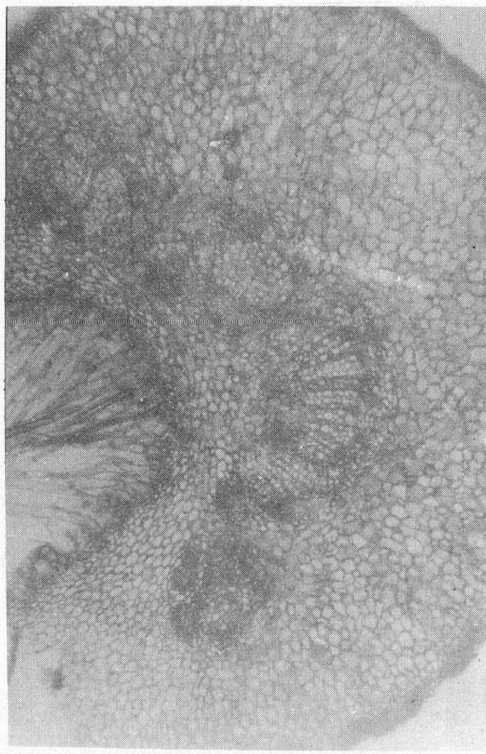

B

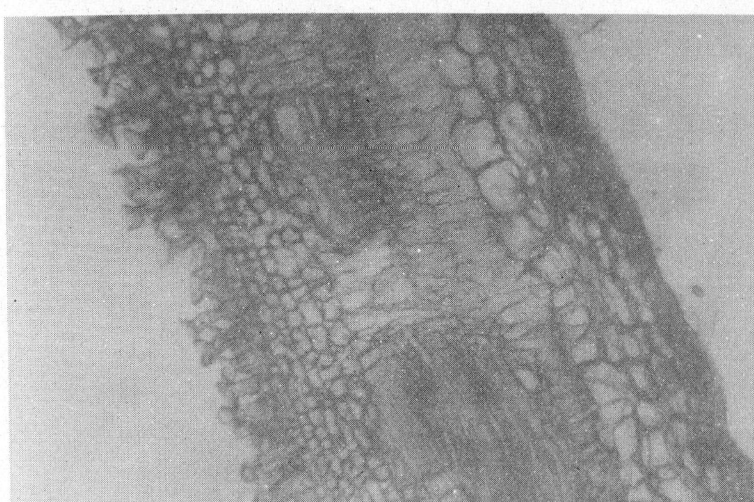

C

Plate - 152

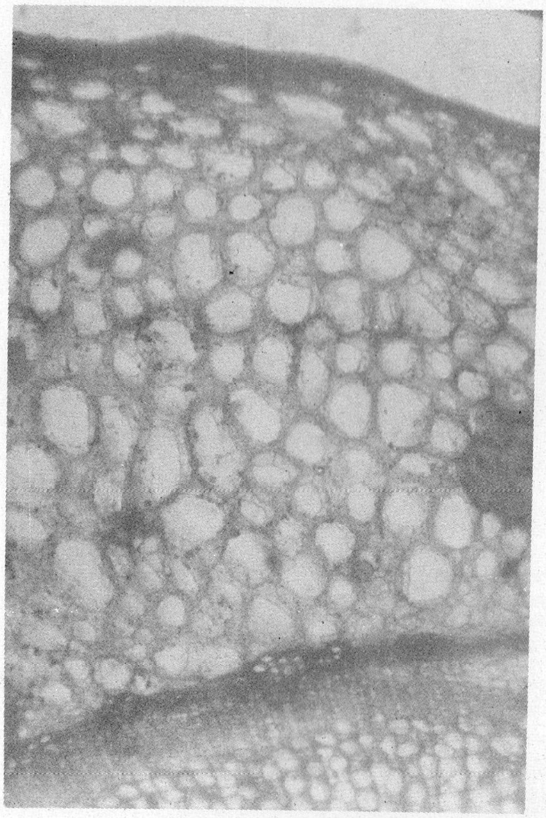

A

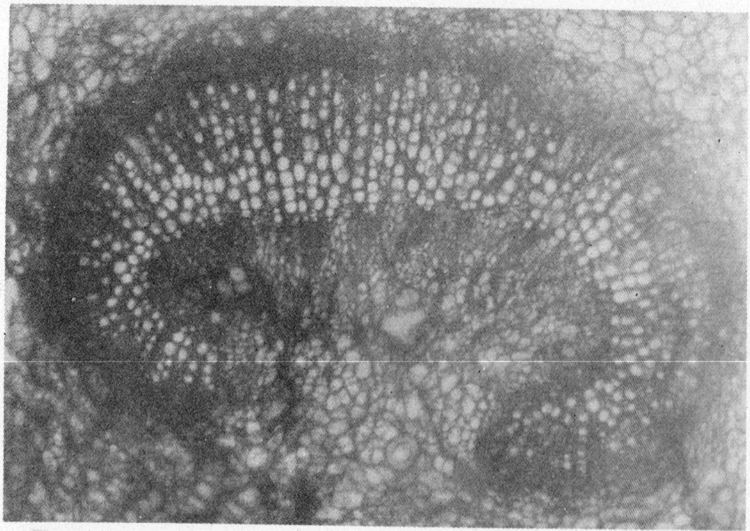

B

Plate - 153

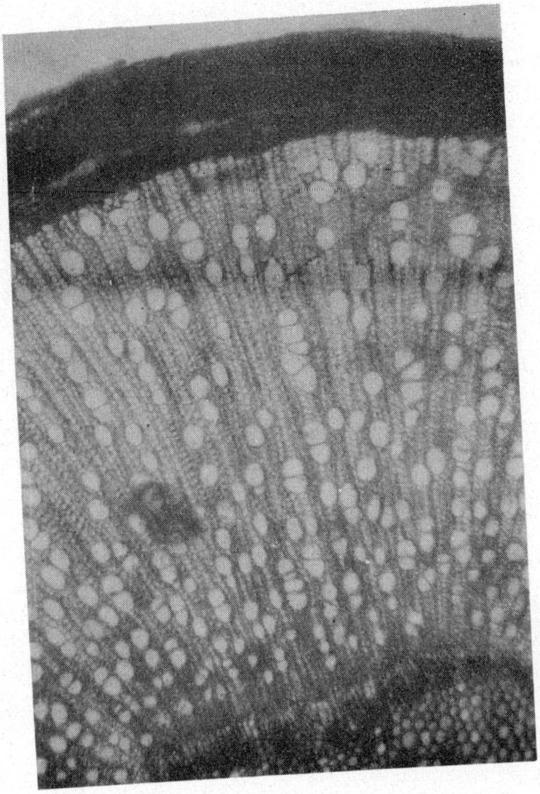

a

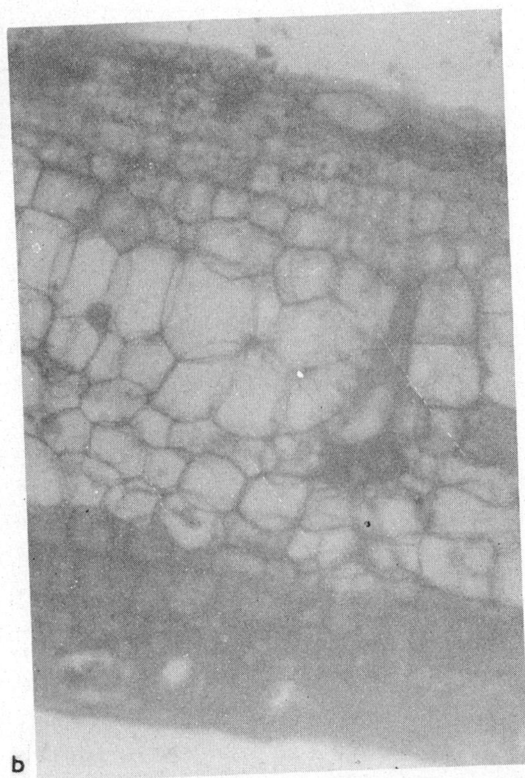

b

Plate - 154

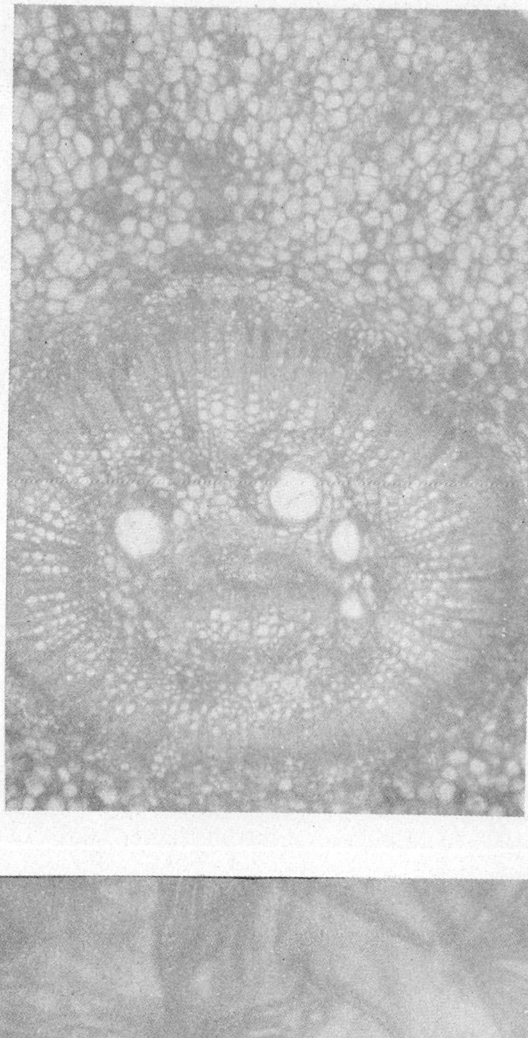

a

b

Plate - 155

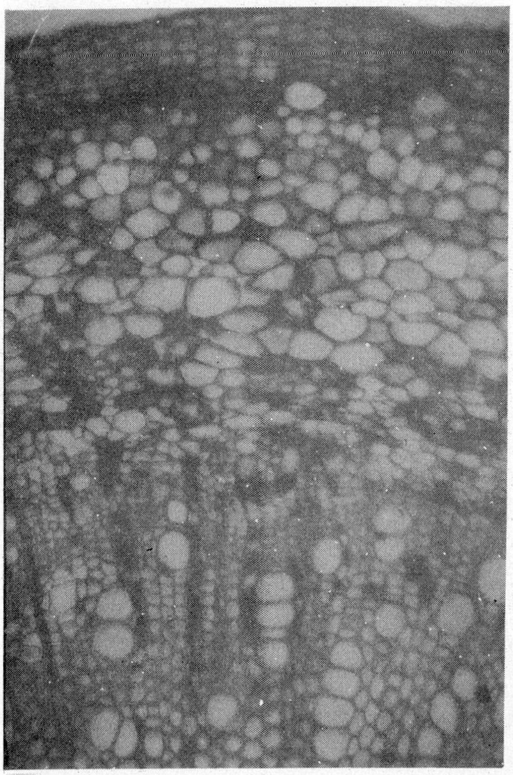

a

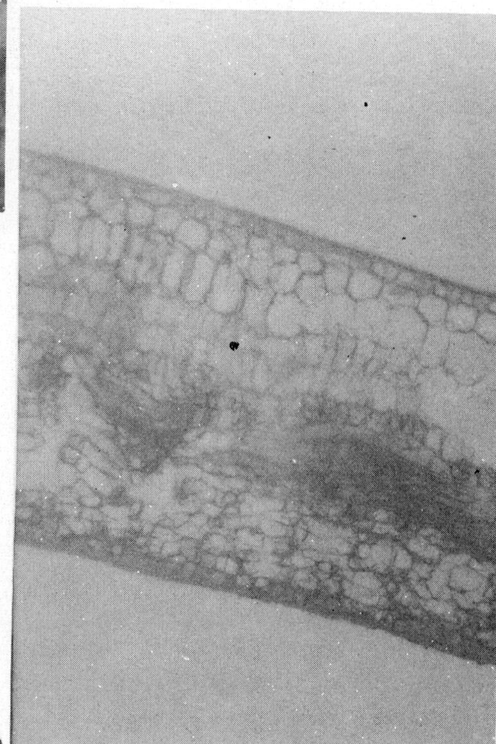

b

Plate - 156

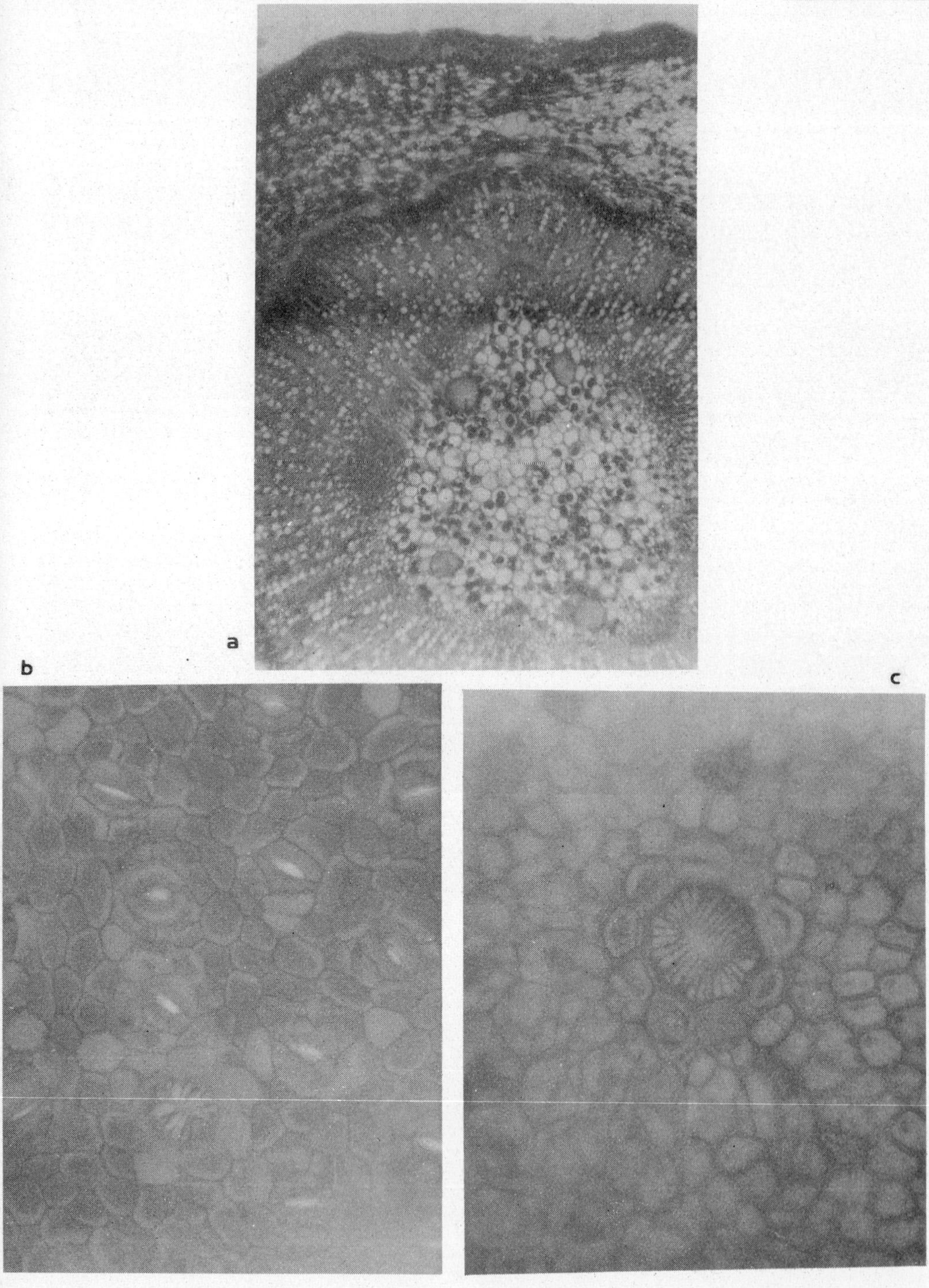

Plate - 157

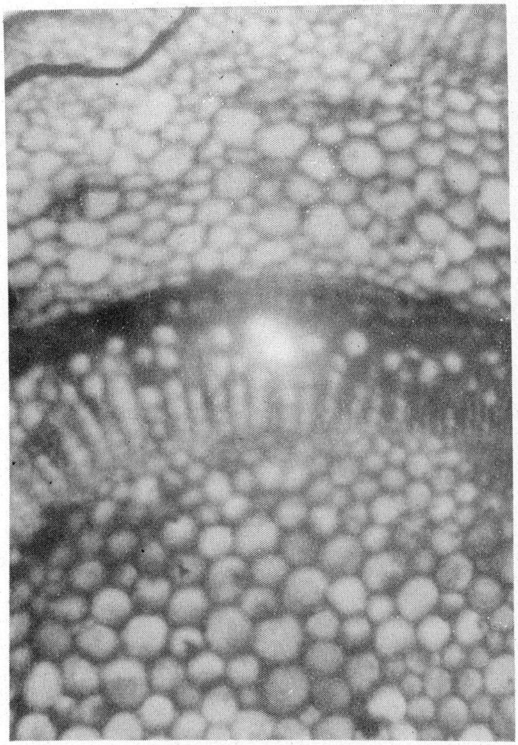

a

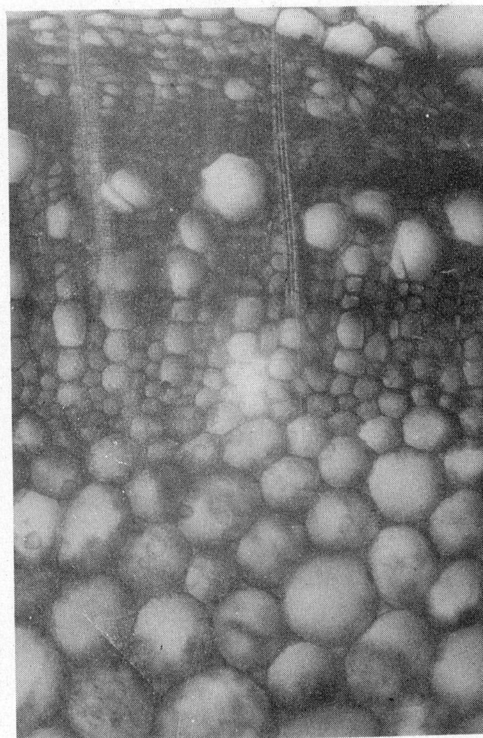

b

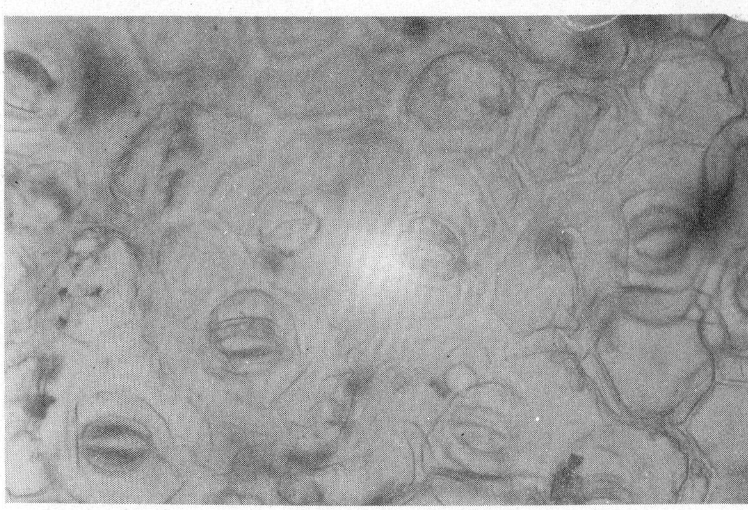

c

Plate - 158

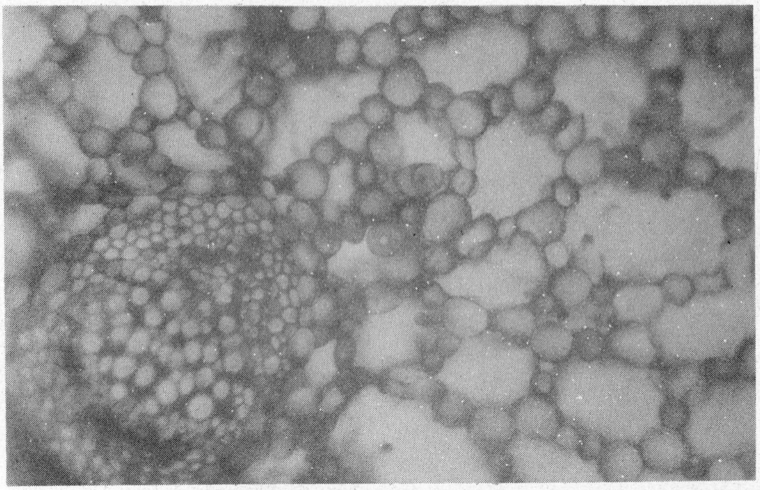

b

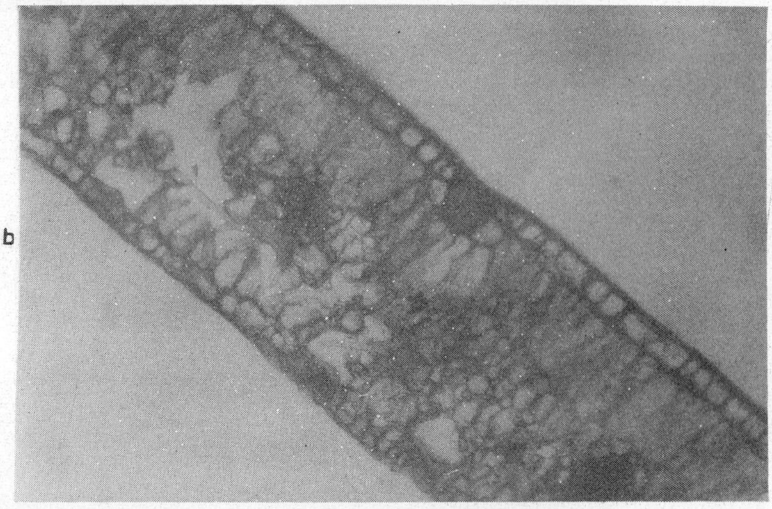

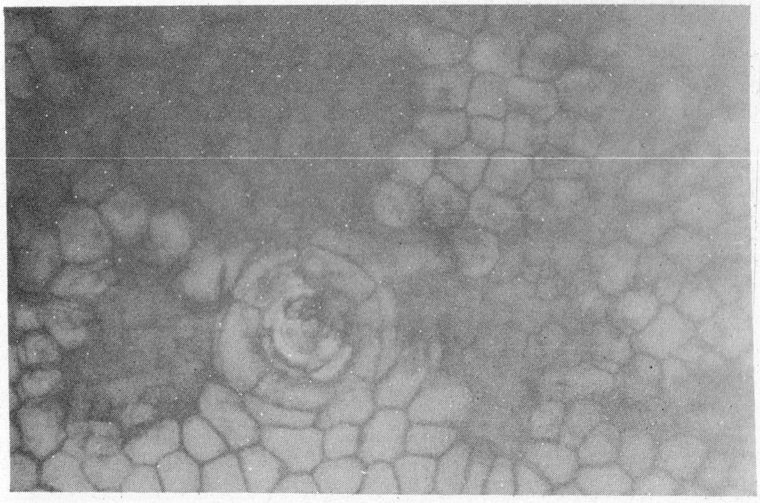

Plate - 159

a

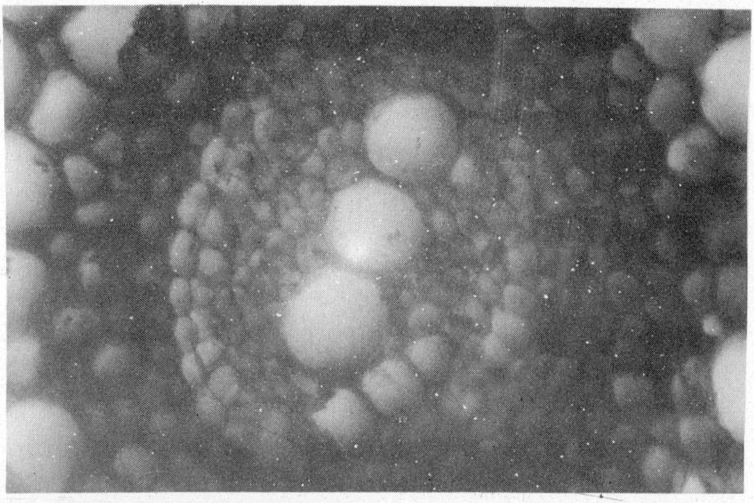

b

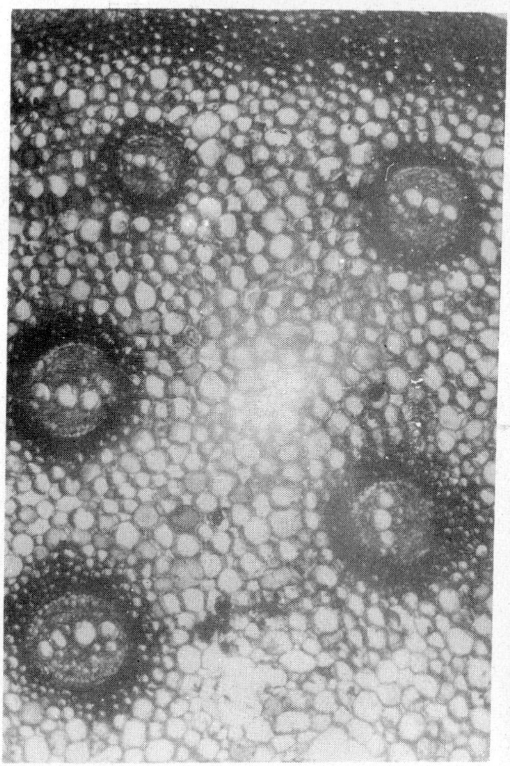

c

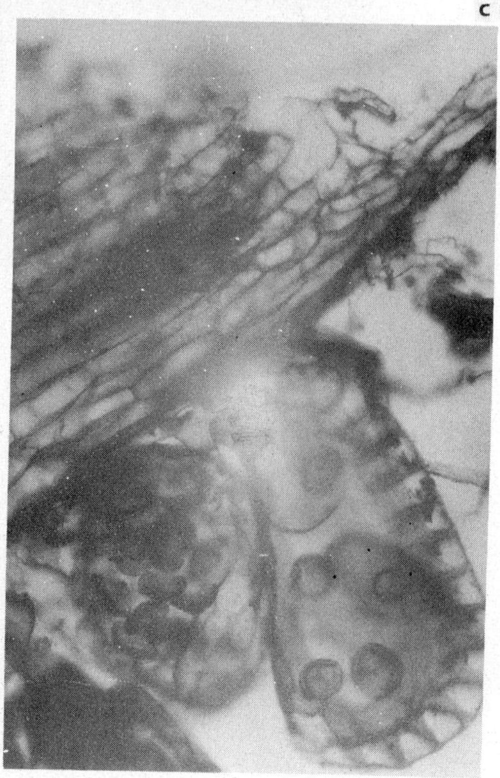

Plate - 160

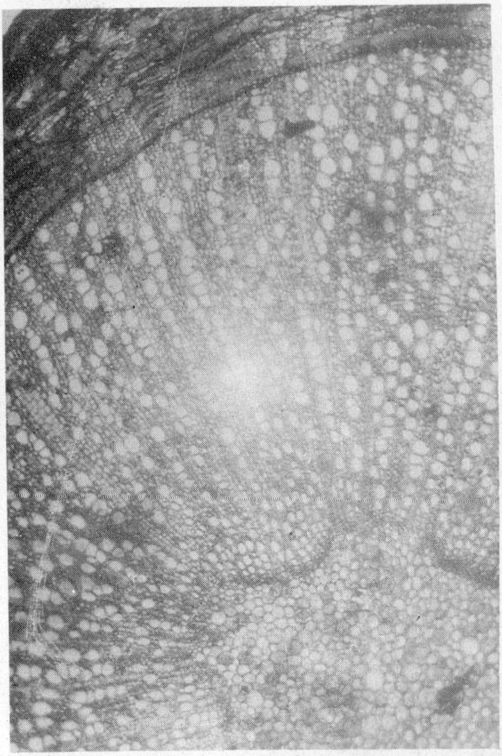

a

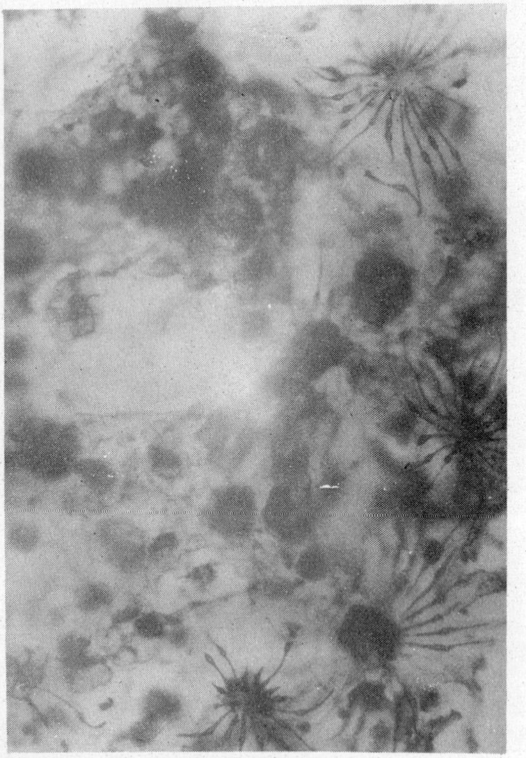

b

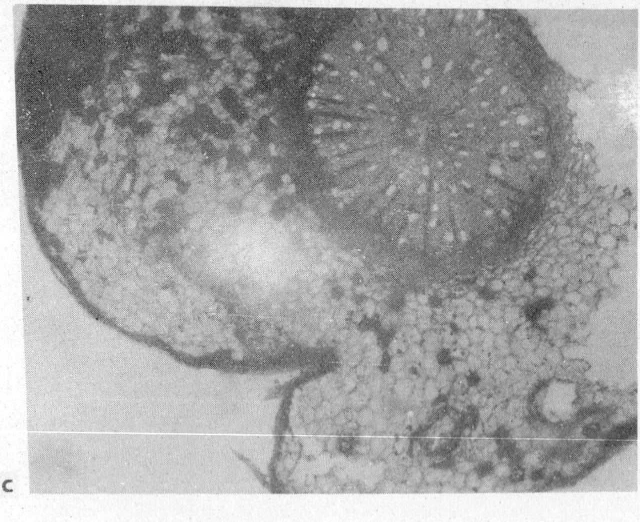

c

Plate - 162

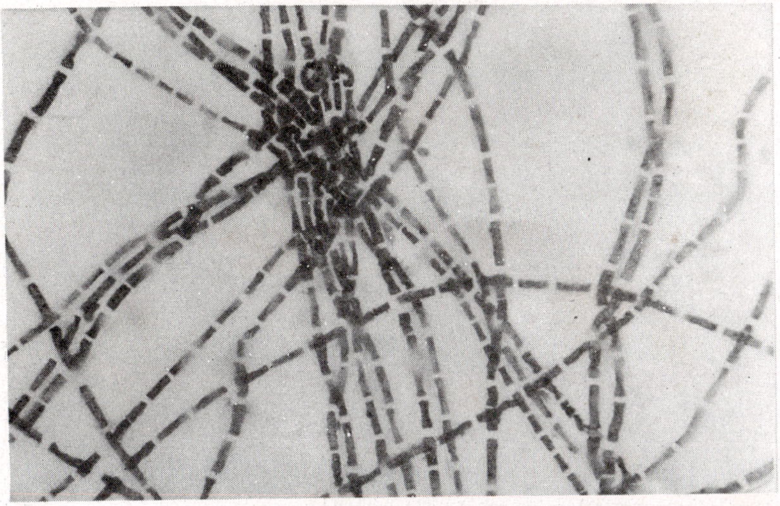

X

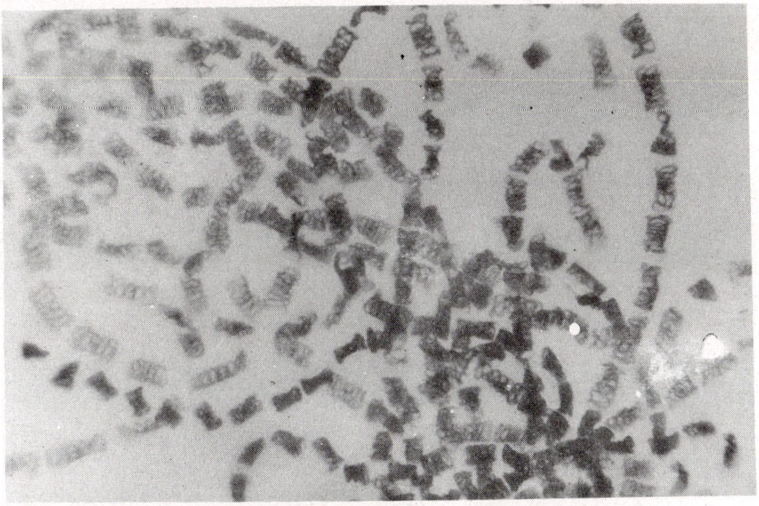

Y

Z

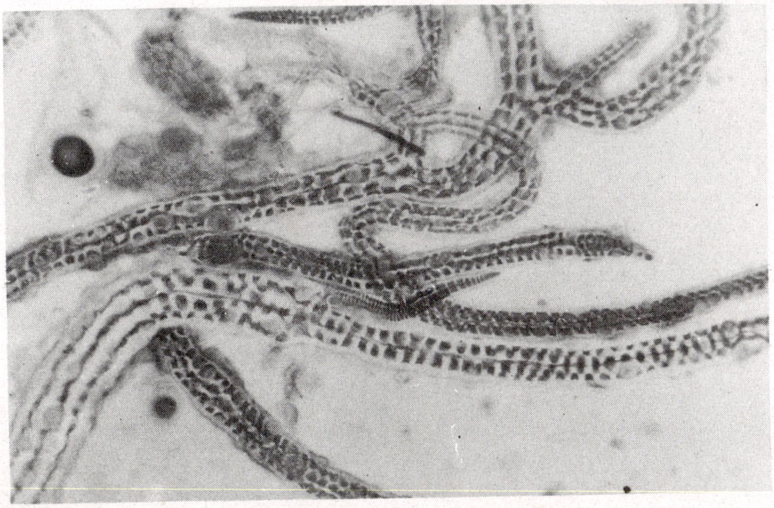

Plate - 161

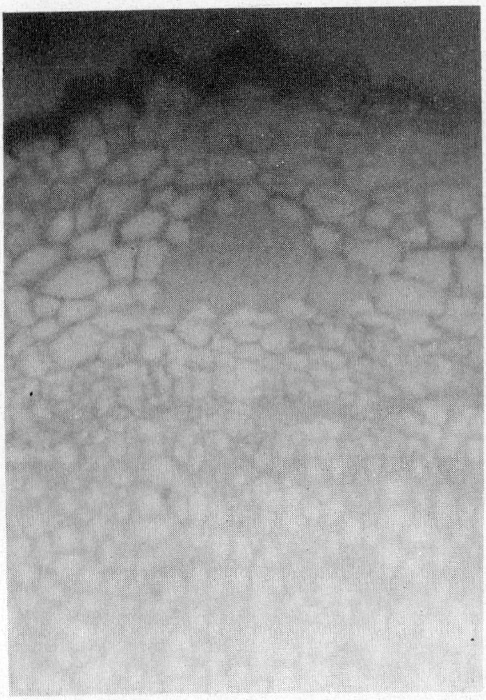

a

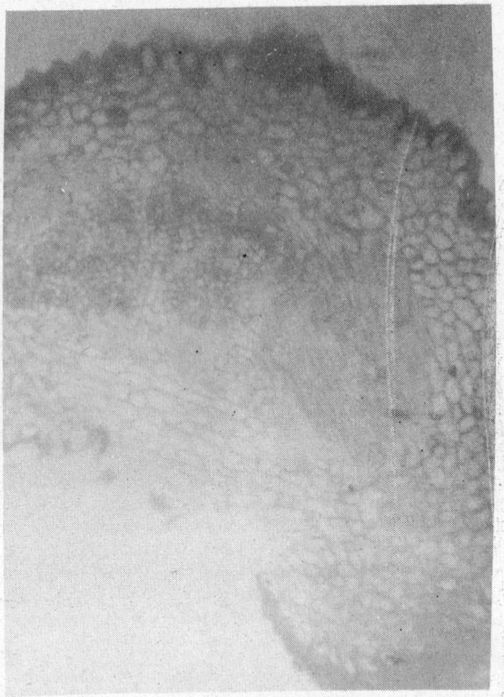

b

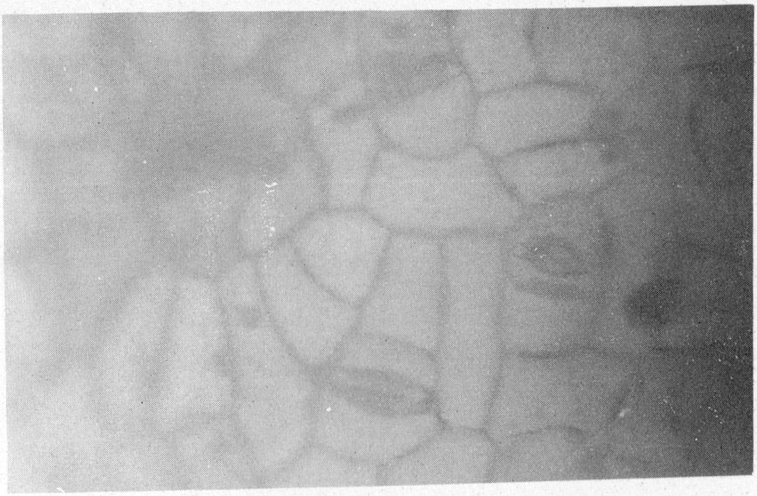

c